Graduate Texts in Mathematics 71

Springer
New York
Berlin
Heidelberg
Barcelona
Budapest
Hong Kong
London
Milan
Paris
Santa Clara
Singapore
Tokyo

Graduate Texts in Mathematics

1 TAKEUTI/ZARING. Introduction to Axiomatic Set Theory. 2nd ed.
2 OXTOBY. Measure and Category. 2nd ed.
3 SCHAEFER. Topological Vector Spaces.
4 HILTON/STAMMBACH. A Course in Homological Algebra. 2nd ed.
5 MAC LANE. Categories for the Working Mathematician. 2nd ed.
6 HUGHES/PIPER. Projective Planes.
7 SERRE. A Course in Arithmetic.
8 TAKEUTI/ZARING. Axiomatic Set Theory.
9 HUMPHREYS. Introduction to Lie Algebras and Representation Theory.
10 COHEN. A Course in Simple Homotopy Theory.
11 CONWAY. Functions of One Complex Variable I. 2nd ed.
12 BEALS. Advanced Mathematical Analysis.
13 ANDERSON/FULLER. Rings and Categories of Modules. 2nd ed.
14 GOLUBITSKY/GUILLEMIN. Stable Mappings and Their Singularities.
15 BERBERIAN. Lectures in Functional Analysis and Operator Theory.
16 WINTER. The Structure of Fields.
17 ROSENBLATT. Random Processes. 2nd ed.
18 HALMOS. Measure Theory.
19 HALMOS. A Hilbert Space Problem Book. 2nd ed.
20 HUSEMOLLER. Fibre Bundles. 3rd ed.
21 HUMPHREYS. Linear Algebraic Groups.
22 BARNES/MACK. An Algebraic Introduction to Mathematical Logic.
23 GREUB. Linear Algebra. 4th ed.
24 HOLMES. Geometric Functional Analysis and Its Applications.
25 HEWITT/STROMBERG. Real and Abstract Analysis.
26 MANES. Algebraic Theories.
27 KELLEY. General Topology.
28 ZARISKI/SAMUEL. Commutative Algebra. Vol.I.
29 ZARISKI/SAMUEL. Commutative Algebra. Vol.II.
30 JACOBSON. Lectures in Abstract Algebra I. Basic Concepts.
31 JACOBSON. Lectures in Abstract Algebra II. Linear Algebra.
32 JACOBSON. Lectures in Abstract Algebra III. Theory of Fields and Galois Theory.
33 HIRSCH. Differential Topology.
34 SPITZER. Principles of Random Walk. 2nd ed.
35 ALEXANDER/WERMER. Several Complex Variables and Banach Algebras. 3rd ed.
36 KELLEY/NAMIOKA et al. Linear Topological Spaces.
37 MONK. Mathematical Logic.
38 GRAUERT/FRITZSCHE. Several Complex Variables.
39 ARVESON. An Invitation to C^*-Algebras.
40 KEMENY/SNELL/KNAPP. Denumerable Markov Chains. 2nd ed.
41 APOSTOL. Modular Functions and Dirichlet Series in Number Theory. 2nd ed.
42 SERRE. Linear Representations of Finite Groups.
43 GILLMAN/JERISON. Rings of Continuous Functions.
44 KENDIG. Elementary Algebraic Geometry.
45 LOÈVE. Probability Theory I. 4th ed.
46 LOÈVE. Probability Theory II. 4th ed.
47 MOISE. Geometric Topology in Dimensions 2 and 3.
48 SACHS/WU. General Relativity for Mathematicians.
49 GRUENBERG/WEIR. Linear Geometry. 2nd ed.
50 EDWARDS. Fermat's Last Theorem.
51 KLINGENBERG. A Course in Differential Geometry.
52 HARTSHORNE. Algebraic Geometry.
53 MANIN. A Course in Mathematical Logic.
54 GRAVER/WATKINS. Combinatorics with Emphasis on the Theory of Graphs.
55 BROWN/PEARCY. Introduction to Operator Theory I: Elements of Functional Analysis.
56 MASSEY. Algebraic Topology: An Introduction.
57 CROWELL/FOX. Introduction to Knot Theory.
58 KOBLITZ. p-adic Numbers, p-adic Analysis, and Zeta-Functions. 2nd ed.
59 LANG. Cyclotomic Fields.
60 ARNOLD. Mathematical Methods in Classical Mechanics. 2nd ed.

continued after index

H. M. Farkas   I. Kra

Riemann Surfaces

Second Edition

With 27 Figures

Springer

Hershel M. Farkas
Department of Mathematics
Hebrew University
Jerusalem 91904
Israel

Irwin Kra
Department of Mathematics
State University of New York
Stony Brook, NY 11794-3651
USA

Mathematics Subject Classification (1991): 30F10, 32C10

Library of Congress Cataloging-in-Publication Data
Farkas, Hershel M.
Riemann surfaces / H.M. Farkas, I. Kra. — 2nd ed.
p. cm. — (Graduate texts in mathematics; 71)
Includes bibliographical references and index.
ISBN 0-387-97703-1 — ISBN 3-540-97703-1
1. Riemann surfaces. I. Kra, Irwin. II. Title. III. Series.
QA333.F37 1991
515′.223—dc20 91-30662

Printed on acid-free paper.

Production coordinated by Brian Howe and managed by Francine Sikorski; manufacturing supervised by Robert Paella.
Typeset by Asco Trade Typesetting Ltd., Hong Kong.

9 8 7 6 5 4 3

ISBN 0-387-97703-1 Springer-Verlag New York Berlin Heidelberg
ISBN 3-540-97703-1 Springer-Verlag Berlin Heidelberg New York SPIN 10677215

To
Eleanor
Sara

Preface to the Second Edition

It is gratifying to learn that there is new life in an old field that has been at the center of one's existence for over a quarter of a century. It is particularly pleasing that the subject of Riemann surfaces has attracted the attention of a new generation of mathematicians from (newly) adjacent fields (for example, those interested in hyperbolic manifolds and iterations of rational maps) and young physicists who have been convinced (certainly not by mathematicians) that compact Riemann surfaces may play an important role in their (string) universe. We hope that non-mathematicians as well as mathematicians (working in nearby areas to the central topic of this book) will also learn part of this subject for the sheer beauty and elegance of the material (work of Weierstrass, Jacobi, Riemann, Hilbert, Weyl) and as healthy exposure to the way (some) mathematicians write about mathematics.

We had intended a more comprehensive revision, including a fuller treatment of moduli problems and theta functions. Pressure of other commitments would have substantially delayed (by years) the appearance of the book we wanted to produce. We have chosen instead to make a few modest additions and to correct a number of errors. We are grateful to the readers who pointed out some of our mistakes in the first edition; the responsibility for the remaining mistakes carried over from the first edition and for any new ones introduced into the second edition remains with the authors.

June 1991
Jerusalem
and
Stony Brook

H.M. Farkas
and
I. Kra

Preface to the First Edition

The present volume is the culmination of ten years' work separately and jointly. The idea of writing this book began with a set of notes for a course given by one of the authors in 1970–1971 at the Hebrew University. The notes were refined several times and used as the basic content of courses given subsequently by each of the authors at the State University of New York at Stony Brook and the Hebrew University.

In this book we present the theory of Riemann surfaces and its many different facets. We begin from the most elementary aspects and try to bring the reader up to the frontier of present-day research. We treat both open and closed surfaces in this book, but our main emphasis is on the compact case. In fact, Chapters III, V, VI, and VII deal exclusively with compact surfaces. Chapters I and II are preparatory, and Chapter IV deals with uniformization.

All works on Riemann surfaces go back to the fundamental results of Riemann, Jacobi, Abel, Weierstrass, etc. Our book is no exception. In addition to our debt to these mathematicians of a previous era, the present work has been influenced by many contemporary mathematicians.

At the outset we record our indebtedness to our teachers Lipman Bers and Harry Ernest Rauch, who taught us a great deal of what we know about this subject, and who along with Lars V. Ahlfors are responsible for the modern rebirth of the theory of Riemann surfaces. Second, we record our gratitude to our colleagues whose theorems we have freely written down without attribution. In particular, some of the material in Chapter III is the work of Henrik H. Martens, and some of the material in Chapters V and VI ultimately goes back to Robert D. M. Accola and Joseph Lewittes.

We thank several colleagues who have read and criticized earlier versions of the manuscript and made many helpful suggestions: Bernard Maskit,

Henry Laufer, Uri Srebro, Albert Marden, and Frederick P. Gardiner. The errors in the final version are, however, due only to the authors. We also thank the secretaries who typed the various versions: Carole Alberghine and Estella Shivers.

August 1979 H.M. FARKAS I. KRA

Contents

Commonly Used Symbols

$\mathbb{Z}$	integers
$\mathbb{Q}$	rationals
$\mathbb{R}$	real numbers
$\mathbb{R}^n$	n-dimensional real Euclidean spaces
$\mathbb{C}^n$	n-dimensional complex Euclidean spaces
Re	real part
Im	imaginary part
$\|\cdot\|$	absolute value
C^∞	infinitely differentiable (function or differential)
$\mathscr{H}^q(M)$	linear space of holomorphic q-differentials on M
$\mathscr{K}(M)$	field of meromorphic functions on M
deg	degree of divisor or map
$L(D)$	linear space of the divisor D
$r(D)$	$\dim L(D) =$ dimension of D
$\Omega(D)$	space of meromorphic abelian differentials of the divisor D
$i(D)$	$\dim \Omega(D) =$ index of specialty of D
[ ]	greatest integer in
$c(D)$	Clifford index of D
$\mathrm{ord}_P f$	order of f at P
$\pi_1(M)$	fundamental group of M
$H_1(M)$	first (integral) homology group of M

$J(M)$	Jacobian variety of M
Π	period matrix of M
M_n	integral divisors of degree n on M
M_n^r	$\{D \in M_n; r(D^{-1}) \geq r + 1\}$
W_n	image of M_n in $J(M)$
W_n^r	image of M_n^r in $J(M)$
Z	canonical divisor
K	vector of Riemann constants (usually)
${}^t x$	transpose of the matrix x (vectors are usually written as columns; thus for $x \in \mathbb{R}^n$, ${}^t x$ is a row vector)

CHAPTER 0

An Overview

The theory of Riemann surfaces lies in the intersection of many important areas of mathematics. Aside from being an important field of study in its own right, it has long been a source of inspiration, intuition, and examples for many branches of mathematics. These include complex manifolds, Lie groups, algebraic number theory, harmonic analysis, abelian varieties, algebraic topology.

The development of the theory of Riemann surfaces consists of at least three parts: a topological part, an algebraic part, and an analytic part. In this chapter, we shall try to outline how Riemann surfaces appear quite naturally in different guises, list some of the most important problems to be treated in this book, and discuss the solutions.

As the title indicates, this chapter is a survey of results. Many of the statements are major theorems. We have indicated at the end of most paragraphs a reference to subsequent chapters where the theorem in question is proven or a fuller discussion of the given topic may be found. For some easily verifiable claims a (kind of) proof has been supplied. This chapter has been written for the reader who wishes to get an idea of the scope of the book before entering into details. It can be skipped, since it is independent of the formal development of the material. This chapter is intended primarily for the mathematician who knows other areas of mathematics and is interested in finding out what the theory of Riemann surfaces contains. The graduate student who is familiar only with first year courses in algebra, analysis (real and complex), and algebraic topology should probably skip most of this chapter and periodically return to it.

We, of course, begin with a definition: A *Riemann surface* is a complex 1-dimensional connected (analytic) manifold.

0.1. Topological Aspects, Uniformization, and Fuchsian Groups

Given a connected topological manifold M (which in our case is a Riemann surface), one can always construct a new manifold $\tilde{M}$ known as the universal covering manifold of M. The manifold $\tilde{M}$ has the following properties:

1. There is a surjective local homeomorphism $\pi: \tilde{M} \to M$.
2. The manifold $\tilde{M}$ is simply connected; that is, the fundamental group of $\tilde{M}$ is trivial ($\pi_1(\tilde{M}) = \{1\}$).
3. Every closed curve which is not homotopically trivial on M lifts to an open curve on $\tilde{M}$, and the curve on $\tilde{M}$ is uniquely determined by the curve on M and the point lying over its initial point.

In fact one can say a lot more. If M^* is any covering manifold of M, then $\pi_1(M^*)$ is isomorphic to a subgroup of $\pi_1(M)$. The covering manifolds of M are in bijective correspondence with conjugacy classes of subgroups of $\pi_1(M)$. In this setting, $\tilde{M}$ corresponds to the trivial subgroup of $\pi_1(M)$. Furthermore, in the case that the subgroup N of $\pi_1(M)$ is normal, there is a group $G \cong \pi_1(M)/N$ of fixed point free automorphisms of M^* such that $M^*/G \cong M$. Once again in the case of the universal covering manifold $\tilde{M}$, $G \cong \pi_1(M)$. (I.2.4; IV.5.6)

If we now make the assumption that M is a Riemann surface, then it is not hard to introduce a Riemann surface structure on any M^* in such a way that the map $\pi: M^* \to M$ becomes a holomorphic mapping between Riemann surfaces and G becomes a group of holomorphic self-mappings of M^* such that $M^*/G \cong M$. (IV.5.5–IV.5.7)

It is at this point that some analysis has to intervene. It is necessary to find all the simply connected Riemann surfaces. The result is both beautiful and elegant. There are exactly three conformally (= complex analytically) distinct simply connected Riemann surfaces. One of these is compact, it is conformally equivalent to the sphere $\mathbb{C} \cup \{\infty\}$. The non-compact simply connected Riemann surfaces are conformally equivalent to either the upper half plane U or the entire plane $\mathbb{C}$. (IV.4)

It thus follows from what we have said before that studying Riemann surfaces is essentially the same as studying fixed point free discontinuous groups of holomorphic self mappings of D, where D is either $\mathbb{C} \cup \{\infty\}$, $\mathbb{C}$, or U. (IV.5.5)

The simplest case occurs when $D = \mathbb{C} \cup \{\infty\}$. Since every non-trivial holomorphic self map of $\mathbb{C} \cup \{\infty\}$ has at least one fixed point, only the sphere covers the sphere. (IV.6.3)

The holomorphic fixed point free self maps of $\mathbb{C}$ are of the form $z \mapsto z + b$, with $b \in \mathbb{C}$. An analysis of the various possibilities shows that a discontinuous subgroup of this group is either trivial or cyclic on one (free) generator or a free abelian group with two generators. The first case corresponds to $M = \mathbb{C}$. The case of one generator corresponds to a cylinder which is conformally the same as a twice punctured sphere. Finally, the case of two generators $z \mapsto z + \omega_1$, $z \mapsto z + \omega_2$ with $\omega_2/\omega_1 = \tau$ and $\operatorname{Im} \tau > 0$ (without loss of generality) corresponds to a torus. We consider the case involving two generators. This is an extremely important example. It motivates a lot of future developments. The group G is to consist of mappings of the form

$$z \mapsto z + n + m\tau,$$

where $\tau \in \mathbb{C}$ is fixed with $\operatorname{Im} \tau > 0$, and m and n vary over the integers. (This involves no loss of generality, because conjugating G in the automorphism group of $\mathbb{C}$ does not change the complex structure.) If we consider the closed parallelogram $\mathcal{M}$ with vertices $0, 1, 1 + \tau$, and τ as shown in Figure 0.1, then we see that

1. no two points of the interior of $\mathcal{M}$ are identified under G,
2. every point of $\mathbb{C}$ is identified to at least one point of $\mathcal{M}$ ($\mathcal{M}$ is closed), and
3. each interior point on the line a (respectively, b) is identified with a unique point on the line a' (respectively, b').

From these considerations, it follows rather easily that $\mathbb{C}/G$ is $\mathcal{M}$ with the points on the boundary identified or just a torus. (IV.6.4)

These tori already exhibit a very important phenomenon. Every $\tau \in \mathbb{C}$, with $\operatorname{Im} \tau > 0$, determines a unique torus and every torus is constructed as above. Given two such points τ and τ', when do they determine the same torus? This is the simplest illustration of the general problem of *moduli* of Riemann surfaces. (IV.7.3; VII.4)

The most interesting Riemann surfaces have the upper half plane as universal covering space. The holomorphic self-mappings of U are $z \mapsto (az + b)/(cz + d)$ with $(a,b,c,d) \in \mathbb{R}$ and $\det\left[\begin{smallmatrix} a & b \\ c & d \end{smallmatrix}\right] > 0$. We can normalize so that $ad - bc = 1$. When we do this, the condition that the mapping be fixed point free is that $|a + d| \geq 2$. It turns out that for subgroups of the group of automorphisms of U, Aut U, the concepts of discontinuity and discreteness agree. Hence the Riemann surfaces with universal covering space U (and these are almost all the Riemann surfaces!) are precisely U/G for discrete fixed point free subgroups G of Aut U. In this case, it turns out that there exists a non-Euclidean (possibly with infinitely many sides and

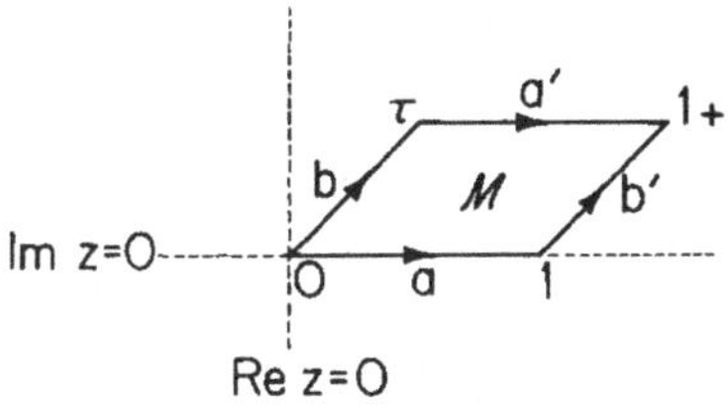

Figure 0.1

possibly open) polygon contained in U and that U/G is obtained by certain identifications on the boundary of the polygon. (IV.5 and IV.9)

We thus see that via the topological theory of covering spaces, the study of Riemann surfaces is essentially the same as the study of fixed point free discrete subgroups of Aut U, which is the canonical example of a Lie group, $SL(2,\mathbb{R})/\pm I$.

It turns out that the Riemann surfaces U/G are quite different from those with $\mathbb{C}$ as their holomorphic universal covering space. For example, a (topological) torus cannot have U as its holomorphic universal covering space. (III.6.3; III.6.4; IV.6)

Because we are mainly interested in analysis and because our objects of study have low dimensions, we shall also consider branched (= ramified) covering manifolds. The theory for this wider class of objects parallels the development outlined above. (IV.9)

In order to obtain a clearer picture of what is going on let us return to the situation mentioned previously where $\tilde{M} = \mathbb{C}$ and G is generated by $z \mapsto z + 1$, $z \mapsto z + \tau$, with $\tau \in U$. We see immediately that dz, since it is invariant under G, is a holomorphic differential on the torus $\mathbb{C}/G$. (Functions cannot be integrated on Riemann surfaces. The search for objects to integrate leads naturally to differential forms.) In fact, dz is the only holomorphic differential on the torus, up to multiplication by constants. Hence, given any point $z \in \mathbb{C}$ there is a point P in the torus and a path c from 0 to that point P such that z is obtained by integrating dz from 0 to P along c. Now this remark is trivial when the torus is viewed in the above way; however, let us now take a different point of view.

0.2. Algebraic Functions

Let us return to the torus constructed in the previous section. The meromorphic functions on this torus are the elliptic (doubly periodic) functions with periods 1, τ. The canonical example here is the Weierstrass $\wp$-function with periods 1, τ:

$$\wp(z) = \frac{1}{z^2} + \sum_{\substack{(n,m) \neq (0,0) \\ (n,m) \in \mathbb{Z}^2}} \left(\frac{1}{(z - n - m\tau)^2} - \frac{1}{(n + m\tau)^2} \right).$$

The $\wp$-function satisfies the differential equation

$$\wp'^2 = 4(\wp - e_1)(\wp - e_2)(\wp - e_3).$$

The points e_j can be identified as

$$e_1 = \wp\left(\frac{1}{2}\right), \qquad e_2 = \wp\left(\frac{\tau}{2}\right), \qquad e_3 = \wp\left(\frac{1 + \tau}{2}\right).$$

It is important to observe that $\wp'$ is again an elliptic function; hence a meromorphic function on the torus. If we now write $w = \wp'$, $z = \wp$, we obtain

$$w^2 = 4(z - e_1)(z - e_2)(z - e_3),$$

and we see that w is an algebraic function of z. The Riemann surface on which w is a single valued meromorphic function is the two-sheeted branched cover of the sphere branched over $z = e_j$, $j = 1, 2, 3$, and $z = \infty$. Now it is not difficult to show that on this surface dz/w is a holomorphic differential. Once again, given any point z in the plane there is a point P on the surface and a path c from ∞ to P such that z is the result of integrating the holomorphic differential dz/w from ∞ to P. That this is true follows at once by letting $z = \wp(\xi)$. So we are really once again back in the situation discussed at the end of the previous section. This has, however, led us to another way of constructing Riemann surfaces.

Consider an irreducible polynomial $P(z,w)$ and with it the set $S = \{(z,w) \in \mathbb{C}^2; P(z,w) = 0\}$. It is easy to show that most points of S are manifold points and that after modifying the singular points and adding some points at infinity, S is the Riemann surface on which w is an algebraic function of z; and S can be represented as an n-sheeted branched cover of $\mathbb{C} \cup \{\infty\}$, where n is the degree of P as a polynomial in w. The branch points of S alluded to above, and the points lying over infinity are the points which need to be added to make S compact. (IV.11.4–IV.11.11)

In the case of the torus discussed above, we started with a compact Riemann surface and found that the surface was the Riemann surface of an algebraic function. The same result holds for any compact Riemann surface. More precisely, given a compact Riemann surface (other than $\mathbb{C} \cup \{\infty\}$) there are functions w and z on the surface which satisfy an irreducible polynomial $P(z,w) = 0$. Hence every compact Riemann surface is the Riemann surface of an algebraic function. Another way of saying the preceding is as follows: We saw in the case of the torus that the field of elliptic functions completely determined the torus up to conformal equivalence. If M is any compact Riemann surface and $\mathscr{K}(M)$ is the field of meromorphic functions on M we can ask whether the field has a strictly algebraic characterization and whether the field determines M up to conformal equivalence. Now if

$$f: M \to N$$

is a conformal map between Riemann surfaces M and N, then

$$f^*: \mathscr{K}(N) \to \mathscr{K}(M)$$

defined by

$$f^*\varphi = \varphi \circ f, \qquad \varphi \in \mathscr{K}(N),$$

is an isomorphism of $\mathscr{K}(N)$ into $\mathscr{K}(M)$ which preserves constants. If M and N are conformally equivalent (that is, if the function f above, has an analytic inverse), then, of course, the fields $\mathscr{K}(M)$ and $\mathscr{K}(N)$ are isomorphic. If, conversely, $\Theta: \mathscr{K}(N) \to \mathscr{K}(M)$ is an isomorphism which preserves constants, then there is an f such that $\Theta\varphi = f^*\varphi$, and M can be recovered from $\mathscr{K}(M)$ in a purely algebraic manner. The above remarks hold as well in the case of non-compact surfaces. The compact case has the additional feature that the field of meromorphic functions can be characterized as an algebraic function field in one variable; that is, an algebraic extension of a transcendental extension of $\mathbb{C}$. (IV.11.10)

0.3. Abelian Varieties

Every torus is a compact abelian group. When we view the torus as $\mathbb{C}/G$ where G is the group generated by $z \mapsto z + 1$, $z \mapsto z + \tau$, addition of points is clearly well-defined modulo $m + n\tau$ with $m, n \in \mathbb{Z}$. What can we say about other compact surfaces? The only two compact surfaces we have actually seen are the sphere and the torus. The sphere is said to have genus zero and the torus genus one. In general a compact surface is said to have genus g, if its Euler characteristic is $2 - 2g$. Examples of compact Riemann surfaces of genus g are the surfaces of the algebraic functions

$$w^2 = \prod_{j=1}^{2g+2} (z - e_j), \qquad e_j \neq e_k \text{ for } j \neq k.$$

We will show that on the above surfaces of genus g, the g differentials $dz/w, \ldots, z^{g-1}\,dz/w$ are linearly independent holomorphic differentials. In fact, on any compact surface M of genus g, $\dim \mathscr{H}^1(M) = g$, where $\mathscr{H}^1(M)$ is the vector space of holomorphic differentials on M. Furthermore, the rank of the first homology group (with integral coefficients) on such a surface is $2g$. Let $a_1, \ldots, a_g, b_1, \ldots, b_g$ be a canonical homology basis on M. It is possible to choose a basis $\varphi_1, \ldots, \varphi_g$ of $\mathscr{H}^1(M)$ so that $\int_{a_j} \varphi_k = \delta_{jk}$ (= Kronecker delta).

In this case the matrix

$$\Pi = (\pi_{jk}), \qquad \pi_{jk} = \int_{b_j} \varphi_k$$

is symmetric with positive definite imaginary part. It then follows that $\mathbb{C}^g$ factored by the group of translations of $\mathbb{C}^g$ generated by the columns of the matrix (I,Π) is a complex g-torus and a compact abelian group. Hence we will see that each compact surface of genus g has associated with it a compact abelian group. (III.6)

In the case of $g = 1$, we saw that choosing a base point on the surface and integrating the holomorphic differentials from the base point to a variable point P on the surface gave an injective analytic map of the Riemann surface onto the torus. In the case of $g > 1$ we have an injective map into the torus by again choosing a base point on the surface and integrating the vector differential $\Phi = (\varphi_1, \ldots, \varphi_g)$ from a fixed base point to a variable point P. In this case the map cannot, of course, be surjective. If we want to obtain a surjective map, we must map unordered g-tuples of points into the torus by sending $(P_1, \ldots, P_g)$ into the sum of the images of the points P_k. This result is called the Jacobi inversion theorem. Two proofs of this theorem will be found in this book; one of them using the theory of Riemann's theta function. (III.6.6; VI.4.4)

A complex torus is called an abelian variety when the $g \times 2g$ matrix (A,B), whose columns are the generators for the lattice defining the torus, has associated with it a $2g \times 2g$ rational skew symmetric matrix P with the property that

$$(A,B)P\begin{pmatrix} {}^tA \\ {}^tB \end{pmatrix} = 0$$

and

$$i(A,B)P\begin{pmatrix} {}^t\bar{A} \\ {}^t\bar{B} \end{pmatrix}$$

is positive definite. In this case one can demonstrate the existence of multiplicative holomorphic functions. These functions then embed the torus as an algebraic variety in projective space. In our case the matrix P can always be chosen as the intersection matrix of the cycles in the canonical homology basis; that is, $\left[\begin{smallmatrix} 0 & I \\ -I & 0 \end{smallmatrix}\right]$.

0.4. More Analytic Aspects

The most important tools in studying (compact) Riemann surfaces are the meromorphic functions on them. All surfaces carry meromorphic functions. (II.5.3; IV.3.17)

What kind of singularities can a meromorphic function on a compact surface have? The answer is supplied by the Riemann–Roch theorem. (III.4.8–III.4.11; IV.10)

We finish this introductory chapter with one last remark. Let M be a compact Riemann surface. Assume that M is not the sphere nor a torus; that is, a surface of genus $g \geq 2$. For each point $P \in M$, we construct a sequence of positive integers

$$\nu_1 < \nu_2 < \cdots < \nu_k < \cdots,$$

as follows: ν_k appears in the list if and only if there exists a meromorphic function on M which is regular (holomorphic) on $M \setminus \{P\}$ and has a pole of order ν_k at P. Question: What do these sequences look like? Answer: For all but finitely many points the sequence is

$$g + 1,\ g + 2,\ g + 3, \ldots.$$

The finite number of exceptions are the *Weierstrass* points; they carry a lot of information about the surface M. One of the fascinating aspects of the study of Riemann surfaces is the ability to obtain such precise information on our objects. (III.5)

We shall see how to use the existence of these Weierstrass points in order to conclude that Aut M is always finite for $g \geq 2$. (V.1)

Another object of study which is extremely important is the Jacobian variety $J(M)$. It, together with the theory of Riemann's theta function, also is a source of much information concerning M. (III.6; III.8; III.11; VI; VII)

CHAPTER I

Riemann Surfaces

In this chapter we define and give the simplest examples of Riemann surfaces. We derive some basic properties of Riemann surfaces and of holomorphic maps between compact surfaces. We assume the reader is familiar with the elementary concepts in algebraic-topology and differential-geometry needed for the study of Riemann surfaces. To establish notation, these concepts are reviewed. The necessary surface topology is discussed. In later chapters we will show how the complex structure can help obtain many of the needed results about surface topology. The chapter ends with a development of various integration formulae.

I.1. Definitions and Examples

We begin with a formal definition of a Riemann surface and give the simplest examples: the complex plane $\mathbb{C}$, the extended complex plane or Riemann sphere $\hat{\mathbb{C}} = \mathbb{C} \cup \{\infty\}$, and finally any open connected subset of a Riemann surface. We define what is meant by a holomorphic mapping between Riemann surfaces and prove that if f is a holomorphic map from a Riemann surface M to a Riemann surface N, with M compact, then f is either constant or surjective. Further, in this case, f is a finite sheeted ramified covering map.

I.1.1. A *Riemann surface* is a one-complex-dimensional connected complex analytic manifold; that is, a two-real-dimensional connected manifold M with a maximal set of charts $\{U_\alpha, z_\alpha\}_{\alpha \in A}$ on M (that is, the $\{U_\alpha\}_{\alpha \in A}$ constitute an open cover of M and

$$z_\alpha : U_\alpha \to \mathbb{C} \tag{1.1.1}$$

is a homeomorphism onto an open subset of the complex plane $\mathbb{C}$) such that the *transition functions*

$$f_{\alpha\beta} = z_\alpha \circ z_\beta^{-1} : z_\beta(U_\alpha \cap U_\beta) \to z_\alpha(U_\alpha \cap U_\beta) \tag{1.1.2}$$

are holomorphic whenever $U_\alpha \cap U_\beta \neq \varnothing$. Any set of charts (not necessarily maximal) that cover M and satisfy condition (1.1.2) will be called a set of *analytic coordinate charts.*

The above definition makes sense since the set of holomorphic functions forms a pseudogroup under composition.

Classically, a compact Riemann surface is called *closed*; while a non-compact surface is called *open.*

I.1.2. Let M be a one-complex-dimensional connected manifold together with two sets of analytic coordinate charts $\mathfrak{A}_1 = \{U_\alpha, z_\alpha\}_{\alpha \in A}$, and $\mathfrak{A}_2 = \{V_\beta, w_\beta\}_{\beta \in B}$. We introduce a partial ordering on the set of analytic coordinate charts by defining $\mathfrak{A}_1 > \mathfrak{A}_2$ if for each $\alpha \in A$, there exists a $\beta \in B$ such that

$$U_\alpha \subset V_\beta \quad \text{and} \quad z_\alpha = w_\beta|_{U_\alpha}.$$

It now follows by Zorn's lemma that an arbitrary set of analytic coordinate charts can be extended to a maximal set of analytic coordinate charts. Thus to define a Riemann surface we need not specify a maximal set of analytic coordinate charts, merely a cover by any set of analytic coordinate charts.

Remark. If M is a Riemann surface and $\{U, z\}$ is a coordinate on M, then for every open set $V \subset U$ and every function f which is holomorphic and injective on $z(V)$, $\{V, f \circ (z|_V)\}$ is also a coordinate chart on M.

I.1.3. Examples. The simplest example of an open Riemann surface is the complex plane $\mathbb{C}$. The single coordinate chart ($\mathbb{C}$,id) defines the Riemann surface structure on $\mathbb{C}$.

Given any Riemann surface M, then a domain D (connected open subset) on M is also a Riemann surface. The coordinate charts on D are obtained by restricting the coordinate charts of M to D. Thus, every domain in $\mathbb{C}$ is again a Riemann surface.

The one point compatification, $\mathbb{C} \cup \{\infty\}$, of $\mathbb{C}$ (known as the *extended complex plane* or *Riemann sphere*) is the simplest example of a closed (= compact) Riemann surface. The charts we use are $\{U_j, z_j\}_{j=1,2}$ with

$$U_1 = \mathbb{C}$$
$$U_2 = (\mathbb{C}\backslash\{0\}) \cup \{\infty\}$$

and

$$z_1(z) = z, \qquad z \in U_1,$$
$$z_2(z) = 1/z, \qquad z \in U_2.$$

(Here and hereafter we continue to use the usual conventions involving meromorphic functions; for example, $1/\infty = 0$.) The two (non-trivial) transition functions involved are

$$f_{kj}:\mathbb{C}\backslash\{0\} \to \mathbb{C}\backslash\{0\}, \qquad k \neq j,\ k, j = 1, 2$$

with

$$f_{kj}(z) = 1/z.$$

I.1.4. Remark. Coordinate charts are also called *local parameters*, *local coordinates*, and *uniformizing variables*. From now on we shall use these four terms interchangeably. Furthermore, the local coordinate $\{U,z\}$ will often be identified with the mapping z (when its domain is clear or not material). We can always choose U to be simply connected and $f(U)$ a bounded domain in $\mathbb{C}$. In this case U will be called a *parametric disc*, *coordinate disc*, or *uniformizing disc*.

I.1.5. A continuous mapping

$$f:M \to N \tag{1.5.1}$$

between Riemann surfaces is called *holomorphic* or *analytic* if for every local coordinate $\{U,z\}$ on M and every local coordinate $\{V,\zeta\}$ on N with $U \cap f^{-1}(V) \neq \varnothing$, the mapping

$$\zeta \circ f \circ z^{-1}:z(U \cap f^{-1}(V)) \to \zeta(V)$$

is holomorphic (as a mapping from $\mathbb{C}$ to $\mathbb{C}$). The mapping f is called *conformal* if it is also one-to-one and onto. In this case (since holomorphic mappings are open or map onto a point)

$$f^{-1}:N \to M$$

is also conformal.

A holomorphic mapping into $\mathbb{C}$ is called a *holomorphic function*. A holomorphic mapping into $\mathbb{C} \cup \{\infty\}$, other than the mapping sending M to ∞, is called a *meromorphic function*. The ring ($\mathbb{C}$-algebra) of holomorphic functions on M will be denoted by $\mathscr{H}(M)$; the field ($\mathbb{C}$-algebra) of meromorphic functions on M, by $\mathscr{K}(M)$. The mapping f of (1.5.1) is called *constant* if $f(M)$ is a point.

Theorem. *Let M and N be Riemann surfaces with M compact. Let $f:M \to N$ be a holomorphic mapping. Then f is either constant or surjective. (In the latter case, N is also compact.) In particular, $\mathscr{H}(M) = \mathbb{C}$.*

PROOF. If f is not constant, then $f(M)$ is open (because f is an open mapping) and compact (because the continuous image of a compact set is compact). Thus $f(M)$ is a closed subset of N (since N is Hausdorff). Since M and N are connected, $f(M) = N$. □

Remark. Since holomorphicity is a local concept, all the usual local properties of holomorphic functions can be used. Thus, in addition to the openness property of holomorphic mappings used above, we know (for example) that holomorphic mappings satisfy the maximum modulus principle. (The principle can be used to give an alternate proof of the fact that there are no non-constant holomorphic functions on compact surfaces.)

I.1.6. Consider a non-constant holomorphic mapping between Riemann surfaces given by (1.5.1). Let $P \in M$. Choose local coordinates z on M vanishing at P and ζ on N vanishing at $f(P)$. In terms of these local coordinates, we can write

$$\zeta = f(z) = \sum_{k \geq n} a_k z^k, \qquad n > 0,\ a_n \neq 0.$$

Thus, we also have (since a non-vanishing holomorphic function on a disc has a logarithm) that

$$\zeta = z^n h(z)^n = (zh(z))^n,$$

where h is holomorphic and $h(0) \neq 0$. Note that $z \mapsto zh(z)$ is another local coordinate vanishing at P, and in terms of this new coordinate the mapping f is given by

$$\zeta = z^n. \tag{1.6.1}$$

We shall say that n (defined as above—this definition is clearly independent of the local coordinates used) is the *ramification number of f at P* or *that f takes on the value $f(P)$ n-times at P* or *f has multiplicity n at P*. The number $(n - 1)$ will be called the *branch number of f at P*, in symbols $b_f(P)$.

Proposition. *Let $f:M \to N$ be a non-constant holomorphic mapping between compact Riemann surfaces. There exists a positive integer m such that every $Q \in N$ is assumed precisely m times on M by f—counting multiplicities; that is, for all $Q \in N$,*

$$\sum_{P \in f^{-1}(Q)} (b_f(P) + 1) = m.$$

PROOF. For each integer $n \geq 1$, let

$$\Sigma_n = \left\{Q \in N;\ \sum_{P \in f^{-1}(Q)} (b_f(P) + 1) \geq n\right\}.$$

The "normal form" of the mapping f given by (1.6.1) shows that Σ_n is open in N. We show next that it is closed. Let $Q = \lim_{k \to \infty} Q_k$ with $Q_k \in \Sigma_n$. Since there are only finitely many points in N that are the images of ramification points in M, we may assume that $b_f(P) = 0$ for all $P \in f^{-1}(Q_k)$, each k. Thus $f^{-1}(Q_k)$ consists of $\geq n$ distinct points. Let $P_{k1}, \ldots, P_{kn}$ be n points in $f^{-1}(Q_k)$. Since M is compact, for each j, there is a subsequence of $\{P_{kj}\}$ that converges to a limit P_j. We may suppose that it is the entire sequence that converges. The points P_j need not, of course, be distinct. Clearly $f(P_j) = Q$, and since $f(P_{kj}) = Q_k$, it follows (even if the points P_j are not distinct) that $\sum_{P \in f^{-1}(Q)} (b_f(P) + 1) \geq n$. Thus each Σ_n is either all of N or empty. Let $Q_0 \in N$ be arbitrary and let $m = \sum_{P \in f^{-1}(Q)} (b_f(P) + 1)$. Then $0 < m < \infty$, and since $Q_0 \in \Sigma_m$, $\Sigma_m = N$. Since $Q_0 \notin \Sigma_{m+1}$, Σ_{m+1} must be empty. $\square$

Definition. The number m above, will be called the *degree* of f ($= \deg f$), and we will also say that f is an *m-sheeted cover of N by M* (or that f has m sheets).

Remarks

1. If f is a non-constant meromorphic function on M, then (the theorem asserts that) f has as many zeros as poles.
2. We have used the fact that (compact) Riemann surfaces are separable, in order to conclude that it suffices to work with sequences rather than nets. We will establish this in IV.5.
3. The above considerations have established the fact that *a single non-constant meromorphic function completely determines the complex structure of the Riemann surface.* For if $f \in \mathscr{K}(M)\backslash\mathbb{C}$, and $P \in M$, and $n - 1 = b_f(P)$, then a local coordinate vanishing at P is given by

$$(f - f(P))^{1/n} \quad \text{if } f(P) \neq \infty,$$

and

$$f^{-1/n} \quad \text{if } f(P) = \infty.$$

I.1.7. Since an analytic function (on the plane) is smooth (C^∞), every Riemann surface is a differentiable manifold. If $\{U,z\}$ is a local coordinate on the Riemann surface M, then $x = \operatorname{Re} z$, $y = \operatorname{Im} z$ ($z = x + iy$) yield smooth local coordinates on U. In I.3 we shall make use of the underlying C^∞-structure of M.

Remark. Every surface (orientable topological two-real-dimensional manifold with countable basis for the topology) admits a Riemann surface structure. We shall not prove (and not have any use for) this fact in this book.

I.2. Topology of Riemann Surfaces

Throughout this section M denotes an orientable two-real-dimensional manifold.

We review the basic notions of surface topology to recall for the reader the facts concerning the fundamental group of a manifold and the simplicial homology groups. This leads us naturally to the notion of covering manifold and finally to the normal forms of compact orientable surfaces. Covering manifolds lead us to the monodromy theorem, and the normal forms lead us to the Euler–Poincaré formula. As one application of these ideas, we establish the Riemann–Hurwitz relation.

I.2.1. We assume that the reader has been exposed to the general notions of surface topology, in particular to the fundamental group and the simplicial homology groups. We thus content ourselves with a brief review of these ideas. In this section the word *curve* on M will mean a continuous map c of the closed interval $I = [0,1]$ into M. The point $c(0)$ will be called the *initial point* of the curve, and $c(1)$ will be called the *terminal* or *end point* of the curve. Furthermore since we shall be primarily interested in compact Riemann surfaces, we shall (in general) assume that the manifold is compact, triangulable, and orientable. (All Riemann surfaces are triangulable and orientable.)

I.2.2. $\pi_1(M)$ = Fundamental Group of M. If P, Q are two points of M and c_1 and c_2 are two curves on M with initial point P and terminal point Q, we say that c_1 is *homotopic* to c_2 ($c_1 \sim c_2$) provided there is a continuous map $h: I \times I \to M$ with the properties $h(t,0) = c_1(t)$, $h(t,1) = c_2(t)$, $h(0,u) = P$ and $h(1,u) = Q$ (for all t, $u \in I$).

If P is now any point of M, we consider all closed curves on M which pass through P. This is the same as all curves on M with initial and terminal point P. We say that two such curves c_1, c_2 are *equivalent* whenever they are homotopic. The set of equivalence classes of closed curves through P forms a group in the obvious manner. The product of the equivalence class of the curve c_1 with the equivalence class of the curve c_2 is the equivalence class of the curve c_1 followed by c_2. The inverse of the equivalence class of the curve $t \mapsto c(t)$ is the curve $t \mapsto c(1 - t)$. The group of equivalence class so constructed is called the *fundamental group of M based at P*. It is easy to see that the fundamental group based at P and the fundamental group based at Q are almost canonically isomorphic as groups. The isomorphism between these two groups depends only on the homotopy class of the path from P to Q. The *fundamental group of M*, $\pi_1(M)$, is therefore defined to be the fundamental group of M based at P, for any $P \in M$. For most applications, the dependence of $\pi_1(M)$ on the base point P will be irrelevant.

Remark. It is easy to see that the fundamental group is a topological invariant.

I.2.3. Homology Groups. In a triangulation of a manifold we call the triangles *two-simplices*, the edges *one-simplices*, and the vertices *zero-simplices*. The orientation on the manifold induces an orientation on the triangles which in turn can be used to orient the edges bounding the triangle. An edge receives opposite orientation from the two triangles for which it is a common side. Further, the vertices $\{P_1, P_2, P_3, \ldots\}$ can be used to label the edges and triangles. Thus $\langle P_1, P_2 \rangle$ is the oriented edge from the vertex P_1 to P_2, and $\langle P_1, P_2, P_3 \rangle$ is the oriented triangle bounded by the oriented edges $\langle P_1, P_2 \rangle$, $\langle P_2, P_3 \rangle$, $\langle P_3, P_1 \rangle$. We identify the triangle $\langle P_1, P_2, P_3 \rangle$ with $-\langle P_3, P_2, P_1 \rangle$ and the edge $\langle P_1, P_2 \rangle$ with $-\langle P_2, P_1 \rangle$. An *n-chain* ($n = 0,1,2$) is a finite linear combination of n-simplices with integer coefficients. We define an operator δ from n-chains to $n - 1$ chains as follows: For $n = 0$, we define $\delta\langle P \rangle = 0$. For $n = 1$, we define $\delta\langle P_1, P_2 \rangle = \langle P_2 \rangle - \langle P_1 \rangle$. For $n = 2$, we define $\delta\langle P_1, P_2, P_3 \rangle = \langle P_2, P_3 \rangle - \langle P_1, P_3 \rangle + \langle P_1, P_2 \rangle$. The preceding defines δ on an n-simplex and we extend the definition to n-chains by linearity.

It is clear that the set of n-chains forms a group under addition and that δ is a group homomorphism of the group of n-chains to the group of $(n-1)$-chains. We denote the group of n-chains by C_n ($C_n = \{0\}$ for $n > 2$). Let Z_n denote the kernel of $\delta: C_n \to C_{n-1}$. Furthermore, let B_n denote the image of C_{n+1} in C_n under δ. Since $\delta^2 = 0$, it is clear that B_n is a subgroup of Z_n, and in fact since all groups in sight are abelian, a normal subgroup. It therefore follows that C_n/Z_n is isomorphic to B_{n-1}. The group we are interested in is, however, $H_n(M) = H_n = Z_n/B_n$ and we call this group the nth *simplicial homology group* (with integer coefficients). (By definition $H_n = \{0\}$ for $n > 2$.)

Let us now denote by $[z]$ the equivalence class in H_n of $z \in Z_n$. We shall say that $[z_j]$, $j = 1, \ldots, \beta_n$, are a *basis* for H_n provided each element of H_n can be written as an integral linear combination of the $[z_j]$ and provided the integral equation $\sum_{j=1}^{\beta_n} \alpha_i[z_j] = 0$ implies $\alpha_j = 0$. In this case we shall call the number β_n the nth-*Betti* number of the triangulated manifold.

It is very easy to describe the groups H_0 and H_2, and thus the numbers β_0 and β_2. In fact it is apparent that $\beta_0 = 1$, and that H_0 is isomorphic to the integers. As far as H_2 is concerned, a little thought shows that there are exactly two possibilities. If M is compact, then H_2 is isomorphic to the integers and $\beta_2 = 1$. If M is not compact, then H_2 is trivial and $\beta_2 = 0$.

The only non-trivial case to consider is H_1 and β_1. We have seen in the previous paragraph that H_0 and H_2 are independent of the triangulation. The same is true for H_1 although this is not at all apparent. One way to see this is to recall the fact that H_1 is isomorphic to the abelianized fundamental group. We shall not prove this résult here. Granting the result, however, and using the normal forms for compact surfaces to be described in I.2.5, it will be easy to compute $H_1(M)$ and hence β_1 for compact surfaces M.

I.2.4. Covering Manifolds. The manifold M^* is said to be a (ramified) *covering manifold* of the manifold M provided there is a continuous surjective map (called a (ramified) *covering map*) $f: M^* \to M$ with the following property: for each $P^* \in M^*$ there exist a local coordinate z^* on M^* vanishing at P^*, a local coordinate z on M vanishing at $f(P)$, and an integer $n > 0$ such that f is given by $z = z^{*n}$ in terms of these local coordinates. Here the integer n depends only on the point $P^* \in M^*$. If $n > 1$, P^* is called a *branch point of order* $n-1$ or a *ramification point of order* n. (Compare these definitions with those in I.1.6.) If $n = 1$, for all points $P^* \in M^*$ the cover is called *smooth* or *unramified*.

EXAMPLE. Proposition I.1.6 shows that every non-constant holomorphic mapping between compact Riemann surfaces is a finite-sheeted (ramified) covering map.

Continuing the general discussion, we call M^* an *unlimited* covering manifold of M provided that for every curve c on M and every point P^* with $f(P^*) = c(0)$, there exists a curve c^* on M^* with initial point P^* and $f(c^*) = c$. The curve c^* will be called a *lift* of the curve c.

There is a close connection between $\pi_1(M)$ and the smooth unlimited covering manifolds of M. If M^* is a smooth unlimited covering manifold of M, then $\pi_1(M^*)$ is isomorphic to a subgroup of $\pi_1(M)$. Conversely, every subgroup of $\pi_1(M)$ determines a smooth unlimited covering manifold M^* with $\pi_1(M^*)$ isomorphic to the given subgroup. (Conjugate subgroups determine homeomorphic covers.)

This is also a good place to recall the monodromy theorem which states: *Let M^* be a smooth unlimited covering manifold of M and c_1, c_2 two curves on M which are homotopic. Let c_1^*, c_2^* be lifts of c_1, c_2 with the same initial point. Then c_1^* is homotopic to c_2^*. In particular, the curves c_1^* and c_2^* must have the same end points.*

If M^* is a covering manifold of M with covering map f, then a homeomorphism h of M^* onto itself with the property that $f \circ h = f$ is called a *covering transformation* of M^*. The set of covering transformations forms a group, which is called *transitive* provided that whenever $f(P_1^*) = f(P_2^*)$ there is a covering transformation h which maps P_1^* onto P_2^*. For the smooth unlimited case, the group of covering transformations is transitive if and only if $\pi_1(M^*)$ is isomorphic to a normal subgroup of $\pi_1(M)$ and in this case the group of covering transformations is isomorphic to $\pi_1(M)/\pi_1(M^*)$.

The case where the cover M^* is determined by the trivial subgroup of $\pi_1(M)$ is of particular importance and is called the *universal, simply connected* or *homotopy* cover of M. It will be denoted by $\tilde{M}$. We note that the universal cover is indeed simply connected (that is, its fundamental group is trivial) and that the group G of covering (also called *deck*) transformations is isomorphic to the fundamental group of M. Since this isomorphism will be used extensively, we discuss it in some detail. Let $\rho: \tilde{M} \to M$ be a universal covering map. Let P be a point in M and $\tilde{P}$ a point of $\tilde{M}$ lying above P; that is, $\rho(\tilde{P}) = P$. Let $\pi_1(M,P)$ be the fundamental group of M based at P. Let c be a curve representing an element of $\pi_1(M,P)$. Lift the curve c to a curve $\tilde{c}_{\tilde{P}}$ on $\tilde{M}$ with initial point $\tilde{P}$. Since c is a closed path, the terminal point of $\tilde{c}_{\tilde{P}}$ projects to P and hence there is a deck transformation $T_c^{-1} \in G$ that takes $\tilde{P}$ to the terminal point of $\tilde{c}_{\tilde{P}}$. The map $T: c \mapsto T_c$ is the desired isomorphism. Thus by definition

$$T_c^{-1}(\tilde{P}) = \text{end point of } \tilde{c}_{\tilde{P}}.$$

Note that for any $A \in G$, the lift of a closed curve c through P to a curve with initial point $A(\tilde{P})$ is $\tilde{c}_{A(\tilde{P})} = A(\tilde{c}_{\tilde{P}})$, and therefore the terminal point of $\tilde{c}_{A(\tilde{P})}$ is just

$$A(\text{terminal point of } \tilde{c}_{\tilde{P}})$$

which is equal to $(A \circ T_c^{-1})(\tilde{P})$. It thus follows that $(\widetilde{c_1 c_2})_{\tilde{P}} = (\tilde{c}_1)_{\tilde{P}}(\tilde{c}_2)_{T_{c_1}^{-1}(\tilde{P})}$ and that its terminal point is $T_{c_1}^{-1}(\text{terminal point of } (\widetilde{c_2})_{\tilde{P}}) = (T_{c_1}^{-1} \circ T_{c_2}^{-1})(\tilde{P})$. We have therefore shown that

$$T_{c_1 c_2}^{-1} = T_{c_1}^{-1} \circ T_{c_2}^{-1} = (T_{c_2} \circ T_{c_1})^{-1}$$

and therefore that the map T is a group homomorphism. This map is easily seen to be both injective and surjective. We leave it to the reader to verify these facts and to generalize the discussion to a covering corresponding to an arbitrary normal subgroup of the fundamental group of M. The reason for the appearance of inverses in the definition of the map T should be clear to the reader: paths are composed from right to left and maps are composed (backwards according to some cultures) from left to right.

Let us now assume that M is a Riemann surface. Then the definition of covering manifold shows that M^* has a unique Riemann surface structure on it which makes f a holomorphic map. Furthermore, the group of covering transformations consists now of conformal self maps of M^*. In the converse direction things are not quite so simple. If M is a Riemann surface and G is a fixed point free group of conformal self maps of M, it is not necessarily the case that the orbit space M/G is even a manifold. However, if the group G operates discontinuously on M, then M/G is a manifold and can be made into a Riemann surface such that the natural map $f: M \to M/G$ is an analytic map of Riemann surfaces. More details about these ideas will be found in IV.5 and IV.9.

I.2.5. Normal Forms of Compact Orientable Surfaces. Any triangulation of a compact manifold is necessarily finite. Using such a triangulation we can proceed to simplify the topological model of the manifold. We can map successively each triangle in the triangulation onto a Euclidean triangle and by auxilliary topological mappings obtain at each stage k, a regular $(k + 2)$-gon, $k \geq 2$. This $(k + 2)$-gon has a certain orientation on its boundary which is induced by the orientation on the triangles of the triangulation. When we are finished with this process we have an $(n + 2)$-gon (n being the number of triangles in the triangulation). Since each side of this polygon is identified with precisely one other side, the polygon has an even number of sides. This polygon with the appropriate identifications gives us a topological model of the manifold M.

In order to obtain the *normal form* we proceed as follows: We start with an edge of the triangulation which corresponds to two sides of the polygon. The edge can be denoted by $\langle P,Q \rangle$, where both P and Q correspond to two vertices of the polygon. In traversing the boundary of the polygon we cross the edge $\langle P,Q \rangle$ once and the edge $\langle Q,P \rangle$ once. We will label one of these edges by c and the other by c^{-1}. In this way we can associate a letter with each side of the polygon, and call the word obtained by writing the letters in the order of traversing the boundary the *symbol* of the polygon. The remainder of the game is devoted to simplifying the symbol of the polygon.

If the sides a and a^{-1} follow one another in the polygon, and there is at least one other side then you can remove both sides from the symbol and the new symbol still is the symbol of a polygon which is a topological model for M.

The polygon's sides have been labeled and so have the vertices of the polygon. We now wish to transform the polygon into a polygon with all vertices identified. This is done by cutting up the polygon and pasting in a fairly straight forward fashion. To illustrate, suppose we have a vertex Q not identified with a vertex P, as in Figure I.1. Make a cut joining R to P and paste back along b to obtain Figure I.2. We note that the number of Q vertices has been decreased by one. Continuing, we end up after a finite number of steps with a triangulation with all the vertices identified.

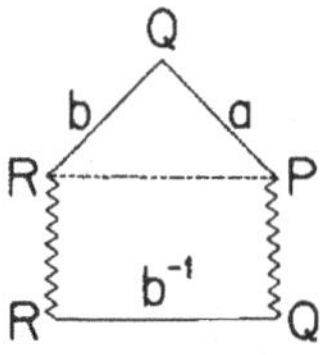

Figure I.1

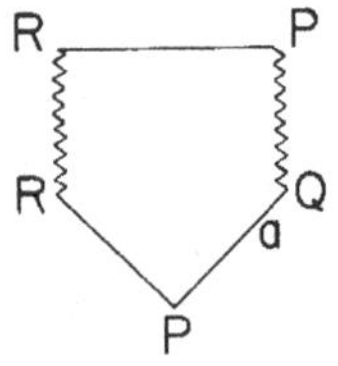

Figure I.2

The final simplification we need involves the notion of linked edges. We say that a pair of edges a and b are *linked* if they appear in the symbol of the polygon in the order $a \cdots b \cdots a^{-1} \cdots b^{-1} \cdots$. It is easy to see that each edge of the polygon is necessarily linked with some other edge (unless we are in the situation that there are only two sides in the polygon). We can then transform the polygon, by a cutting and pasting argument similar to the one used above so that the linked pair is brought together as $aba^{-1}b^{-1}$. We finally obtain the normal form of the surface. The normal form of a compact orientable surface is a polygon whose symbol is aa^{-1} or $a_1b_1a_1^{-1}b_1^{-1} \cdots a_gb_ga_g^{-1}b_g^{-1}$. In the former case we say that the *genus* of M is zero and in the latter case we say that the *genus* is g. It is clear that g is a complete topological invariant for compact orientable surfaces.

From the normal form we can, of course, reconstruct the original surface by a "pasting" process. Figures I.3 and I.4 explain the procedure.

In particular, a surface of genus g is topologically a sphere with g *handles.* A surface of genus 0 is topologically (also analytically—but this will not be

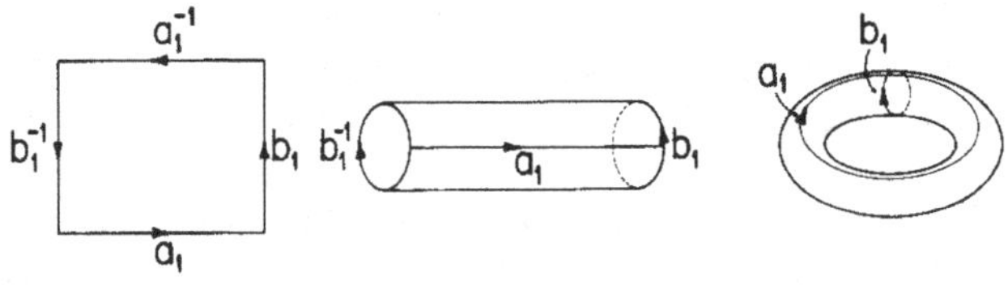

Figure I.3. Surface of genus 1.

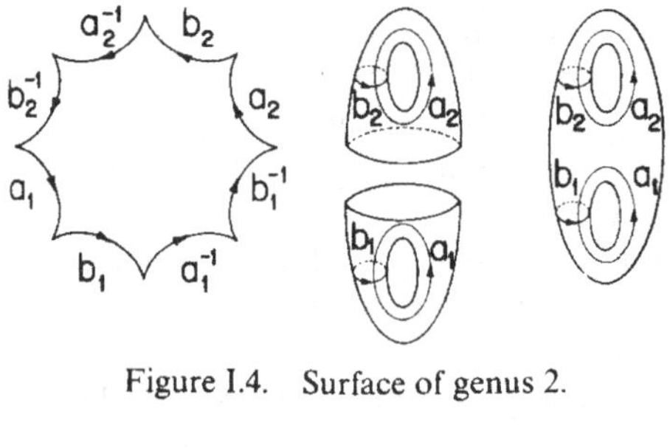

Figure I.4. Surface of genus 2.

seen until III.4 or IV.4) a *sphere*. A surface of genus 1 is topologically a *torus*. There are many complex tori. (In fact, as will be seen in IV.6, a one-complex-parameter family of tori.)

Using the common vertex of the normal form as a base point for the fundamental group, one shows that $\pi_1(M)$ is generated by the $2g$ closed loops $a_1, \ldots, a_g, b_1, \ldots, b_g$ subject to the single relation $a_1 b_1 a_1^{-1} b_1^{-1} \cdots a_g b_g a_g^{-1} b_g^{-1} = 1$. Hence $H_1(M)$ is the free abelian group on the generators $[a_j]$, $[b_j]$, $j = 1, \ldots, g$. In particular for a compact surface of genus g, $H_1(M) \cong \mathbb{Z}^{2g}$ and $\beta_1 = 2g$.

Remark. The "pasting" process is, of course, not uniquely determined by the symbol. For example, in the case of genus 1, after joining side a_1^{-1} to a_1, we may twist the resulting cylinder by 2π radians before identifying b_1^{-1} with b_1. The "twisted" surface is, of course, homeomorphic to the "untwisted" one. The homeomorphism is known as a *Dehn twist*.

Caution. The normal form for the polygon can lead to two different (but, of course, equivalent) presentations for $\pi_1(M)$. For the first form of the presentation (the one given above), we must properly interpret the symbol of the polygon. For the second presentation, it is helpful to use the isomorphism between the fundamental group $\pi_1(M,P)$, here P is the base point for the fundamental group, and the group G of deck transformations on the universal cover $\tilde{M}$.

To derive the defining relation listed above, let us give the counterclockwise orientation to the boundary of the polygon. Since all the vertices are identified on the surface, each side projects to a curve through the same point P on M and hence each of the sides a_j, b_j determines an element of $\pi_1(M,P)$. To see what elements they generate, we let α_j (β_j) be the (homotopy class of the) projection of the side a_j (b_j) to the surface M. We shift our point of view slightly. We can (and indeed will) consider the polygon as sitting in $\tilde{M}$, the universal covering space of M. Hence the sides of the polygon can be considered as lifts to $\tilde{M}$ of closed curves on M through P. Let z be the initial point of the side a_1. It is then obvious that a_1 is a lift of α_1 from z. In the language of I.2.4, which we will now use, $a_1 = \alpha_{1z}$. Now, b_1 is a lift of β_1 that starts at the end point of a_1 rather than its initial point. Thus $a_1 b_1$ is a lift of $\alpha_1 \beta_1$ that starts at z. Similarly, for the remaining elements. Thus $c = a_1 b_1 a_1^{-1} b_1^{-1} \cdots a_g b_g a_g^{-1} b_g^{-1}$ is a lift of $\gamma = \alpha_1 \beta_1 \alpha_1^{-1} \beta_1^{-1} \cdots \alpha_g \beta_g \alpha_g^{-1} \beta_g^{-1}$ that starts at z; since c is a closed curve on $\tilde{M}$, γ is homotopically trivial. This discussion allows us also to identify α_j with a_j and β_j with b_j.

To determine the second form of the presentation of the fundamental group suggested by the polygon, we observe that the sides a_j and a_j^{-1} of the polygon (sitting in $\tilde{M}$) project to the same curve on M. Hence there exists an element $A_j \in G$ with $A_j(a_j) = a_j^{-1}$. Similarly, there exists an element $B_j \in G$ with $B_j(b_j) = b_j^{-1}$. Let us order the vertices of the normal form of the polygon in the counterclockwise direction by $z = z_0, z_1, \ldots, z_{4g-1}, z_{4g} = z_0$.

It thus follows that for $j = 1, \ldots, g$ the elements A_j and B_j satisfy the following conditions:

$$A_j(z_{4j-4}) = z_{4j-1}, \qquad A_j(z_{4j-3}) = z_{4j-2}, \qquad B_j(z_{4j-3}) = z_{4j} \quad \text{and}$$
$$B_j(z_{4j-2}) = z_{4j-1}.$$

A simple chase of points shows that

$$\begin{aligned}(B_j \circ A_j^{-1} \circ B_j^{-1} \circ A_j)(z_{4j-4}) &= (B_j \circ A_j^{-1} \circ B_j^{-1})(z_{4j-1}) \\ &= (B_j \circ A_j^{-1})(z_{4j-2}) = B_j(z_{4j-3}) = z_{4j}.\end{aligned}$$

It follows that

$$B_g \circ A_g^{-1} \circ B_g^{-1} \circ A_g \cdots B_1 \circ A_1^{-1} \circ B_1^{-1} \circ A_1$$

fixes the point z_0 and must be the identity since the covering group acts fixed point freely.

To compare the two presentations (for the covering group, say), let $\mathscr{A}_j^{-1}$ ($\mathscr{B}_j^{-1}$) be the motion in G that identifies the initial point of $\tilde{\alpha}_{j_z}$ ($\tilde{\beta}_{j_z}$) with its terminal point. Since $\mathscr{A}_j = T_{\alpha_j}$ and $\mathscr{B}_j = T_{\beta_j}$, we have the following defining *new* relation for these *new* generators of the covering group:

$$\mathscr{B}_g^{-1} \circ \mathscr{A}_g^{-1} \circ \mathscr{B}_g \circ \mathscr{A}_g \cdots \mathscr{B}_1^{-1} \circ \mathscr{A}_1^{-1} \circ \mathscr{B}_1 \circ \mathscr{A}_1.$$

We want to express each of these new generators for G in terms of the *old* ones. It is useful to introduce the commutator C of two elements of G as

$$C = [B,A] = B^{-1} \circ A^{-1} \circ B \circ A$$

and to let

$$C_j = [B_j^{-1}, A_j], \qquad C_0 = I.$$

Observe that

$$a_j = (C_{j-1} \circ \cdots \circ C_1 \circ C_0)(\tilde{\alpha}_{j_z})$$

and

$$b_j = (A_j^{-1} \circ B_j^{-1} \circ A_j \circ C_{j-1} \circ \cdots \circ C_1 \circ C_0)(\tilde{\beta}_{j_z}).$$

Thus we see that

$$\text{end point of } \tilde{\alpha}_{j_z} = (C_0^{-1} \circ \cdots \circ C_{j-1}^{-1})(z_{4j-3})$$

and

$$\text{end point of } \tilde{\beta}_{j_z} = (C_0^{-1} \circ \cdots \circ C_{j-1}^{-1} \circ A_j^{-1} \circ B_j \circ A_j)(z_{4j-2});$$

from which we conclude that

$$\mathscr{A}_j^{-1} = C_0^{-1} \circ \cdots \circ C_{j-1}^{-1} \circ A_j^{-1} \circ B_j^{-1} \circ A_j \circ C_{j-1} \circ \cdots \circ C_1 \circ C_0$$

and

$$\mathscr{B}_j^{-1} = C_0^{-1} \circ \cdots \circ C_{j-1}^{-1} \circ A_j^{-1} \circ B_j \circ A_j \circ B_j^{-1} \circ A_j \circ C_{j-1} \circ \cdots \circ C_1 \circ C_0.$$

The relation for the new generators is hence given in terms of the old generators as

$$C_1^{-1} \circ \cdots \circ C_g^{-1} = I;$$

that is, the inverse of the old relation for the old generators.

I.2.6. Euler–Poincaré. The *Euler–Poincaré characteristic* χ of compact surfaces of genus g is given by $\chi = \alpha_0 - \alpha_1 + \alpha_2$, where α_k is the number of k simplices in the triangulation. A triangulation of the normal form gives $\chi = 2 - 2g$. The Euler–Poincaré characteristic is also given by $\beta_0 - \beta_1 + \beta_2$, where β_k is the kth Betti number. Thus the computations of the Betti numbers in I.2.4 and I.2.6 yield an alternate verification of the value of χ.

I.2.7. As an application of the topological invariance of the Euler–Poincaré characteristic, we establish a beautiful formula relating various topological indices connected with a holomorphic mapping between compact surfaces.

Consider a non-constant holomorphic mapping $f: M \to N$ between compact Riemann surfaces. Assume that M is a compact Riemann surface of genus g, N is a compact surface of genus γ. Assume that f is of degree n (that is, $f^{-1}(Q)$ has cardinality n for almost all $Q \in N$). We define the *total branching number* (recall definition preceding Proposition I.1.6) of f by

$$B = \sum_{P \in M} b_f(P).$$

Theorem (Riemann–Hurwitz Relation). *We have*

$$g = n(\gamma - 1) + 1 + B/2.$$

PROOF. Let $S = \{f(x);\ x \in M \text{ and } b_f(x) > 0\}$. Since S is a finite set, we can triangulate N so that every point of S is a vertex of the triangulation. Assume that this triangulation has F faces, E edges, and V vertices. Lift this triangulation to M via the mapping f. The induced triangulation of M has nF faces, nE edges, and $nV - B$ vertices. We now compute the Euler–Poincaré characteristic of each surface in two ways:

$$F - E + V = 2 - 2\gamma$$
$$nF - nE + nV - B = 2 - 2g.$$

From the above we obtain

$$1 - g = n(1 - \gamma) - B/2. \qquad \square$$

I.2.8. We record now (using the same notation as above) several immediate consequences.

Corollary 1. *The total branching number B is always even.*

Corollary 2. *Assume that f is unramified. Then*

a. $g = 0 \Rightarrow n = 1$ *and* $\gamma = 0$.
b. $g = 1 \Rightarrow \gamma = 1$ *(n arbitrary).*
c. $g > 1 \Rightarrow g = \gamma$ *for* $n = 1$.
$\Rightarrow g > \gamma > 1$ *for* $n > 1$ *(and n divides* $g - 1$).

Corollary 3

a. *If* $g = 0$, *then* $\gamma = 0$.
b. *If* $1 \leq g = \gamma$, *then either* $n = 1$ *and (thus)* $B = 0$ *or* $g = 1$ *and (thus)* $B = 0$.

I.3. Differential Forms

We assume that the reader is familiar with the theory of integration on a differentiable manifold. We briefly review the basic facts (to fix notation), and make use of the complex structure on the manifolds under consideration to simplify and augment many differential-geometric concepts. The necessity of introducing differential forms stems from the desire to have an object which we can integrate on the surface. The introduction of 1-forms allows us to consider line integrals on the surface, while the introduction of 2-forms allows us to consider surface integrals. Various operators on differential forms are introduced, and in terms of these operators we define and characterize different classes of differentials.

Remark. We shall use interchangeably the terms "form", "differential", and "differential form".

I.3.1. Let M be a Riemann surface. A *0-form* on M is a function on M. A *1-form* ω on M is an (ordered) assignment of two continuous functions f and g to each local coordinate z $(= x + iy)$ on M such that

$$f\,dx + g\,dy \tag{3.1.1}$$

is invariant under coordinate changes; that is, if $\tilde{z}$ is another local coordinate on M and the domain of $\tilde{z}$ intersects non-trivially the domain of z, and if ω assigns the functions $\tilde{f}$, $\tilde{g}$ to $\tilde{z}$, then (using matrix notation)

$$\begin{bmatrix} \tilde{f}(\tilde{z}) \\ \tilde{g}(\tilde{z}) \end{bmatrix} = \begin{bmatrix} \dfrac{\partial x}{\partial \tilde{x}} & \dfrac{\partial y}{\partial \tilde{x}} \\ \dfrac{\partial x}{\partial \tilde{y}} & \dfrac{\partial y}{\partial \tilde{y}} \end{bmatrix} \begin{bmatrix} f(z(\tilde{z})) \\ g(z(\tilde{z})) \end{bmatrix} \tag{3.1.2}$$

on the intersection of the domains of z and $\tilde{z}$. The 2×2 matrix appearing in (3.1.2) is, of course, the Jacobian matrix of the mapping $\tilde{z} \mapsto z$.

A 2-*form* Ω on M is an assignment of a continuous function f to each local coordinate z such that

$$f\, dx \wedge dy \tag{3.1.3}$$

is invariant under coordinate changes; that is, in terms of the local coordinate $\tilde{z}$ we have

$$\tilde{f}(\tilde{z}) = f(z(\tilde{z}))\frac{\partial(x,y)}{\partial(\tilde{x},\tilde{y})}, \tag{3.1.4}$$

where $\partial(x,y)/\partial(\tilde{x},\tilde{y})$ is the determinant of the Jacobian. Since we consider only holomorphic coordinate changes (3.1.4) has the simple form

$$\tilde{f}(\tilde{z}) = f(z(\tilde{z}))\left|\frac{dz}{d\tilde{z}}\right|^2. \tag{3.1.4a}$$

I.3.2. Many times it is more convenient to use complex notation for (differential) forms. Using the complex analytic coordinate z, a 1-form may be written as

$$u(z)\, dz + v(z)\, d\bar{z}, \tag{3.2.1}$$

where

$$\begin{aligned} dz &= dx + i\, dy, \\ d\bar{z} &= dx - i\, dy, \end{aligned} \tag{3.2.2}$$

and hence comparing with (3.1.1) we see that

$$\begin{aligned} f &= u + v, \\ g &= i(u - v). \end{aligned}$$

Similarly, a 2-form can be written as

$$g(z)\, dz \wedge d\bar{z}.$$

It follows from (3.2.2) that

$$dz \wedge d\bar{z} = -2i\, dx \wedge dy. \tag{3.2.3}$$

I.3.3. Remark. To derive (3.2.3), we have made use of the "exterior" multiplication of forms. This multiplication satisfies the conditions: $dx \wedge dx = 0 = dy \wedge dy$, $dx \wedge dy = -dy \wedge dx$. The product of a k-form and an l-form is a $k + l$ form provided $k + l \leq 2$ and is the zero form (still $k + l$) for $k + l > 2$.

In view of the last remark, we let $\bigwedge^k$ denote the vector space of k-forms. We see that $\bigwedge^k$ is a module over $\bigwedge^0$ and that $\bigwedge^k = \{0\}$ for $k \geq 3$. Further

$$\bigwedge = \bigwedge^0 \oplus \bigwedge^1 \oplus \bigwedge^2$$

is a graded anti-commutative algebra under the obvious multiplication of forms.

I.3.4. A 0-form can be "integrated" over 0-chains; that is, over a finite set of points. Thus, the "integral" of the function f over the 0-cycle

$$\sum n_\alpha P_\alpha, \qquad P_\alpha \in M,\ n_\alpha \in \mathbb{Z}$$

is

$$\sum n_\alpha f(P_\alpha).$$

A 1-form ω can be integrated over 1-chains (finite unions of paths). Thus, if the piece-wise differentiable path c is contained in a single coordinate disc $z = x + iy$, $c: I \to M$ (I = unit interval $[0,1]$), and if ω is given by (3.1.1), then

$$\int_c \omega = \int_0^1 \left\{ f(x(t),y(t)) \frac{dx}{dt} + g(x(t),y(t)) \frac{dy}{dt} \right\} dt.$$

By the transition formula for ω, (3.1.2), the above integral is independent of choice of z, and by compactness the definition can be extended to arbitrary piece-wise differentiable paths.

Similarly, a 2-form Ω can be integrated over 2-chains, D. Again, restricting to a single coordinate disc (and Ω given by (3.1.3)),

$$\iint_D \Omega = \iint_D f(x,y)\, dx \wedge dy.$$

The integral is well defined and extends in an obvious way to arbitrary 2-chains.

It is also sometimes necessary to integrate a 2-form Ω over a more general domain D. If D has compact closure, there is no difficulty involved in extending the definition of the integral. For still more general domains D, one must use partitions of unity.

Remark. We will see in IV.5 that evaluation of integrals over domains on an arbitrary surface M can always be reduced to considering integrals over plane domains.

I.3.5. For C^1-forms (that is, forms whose coefficients are C^1 functions), we introduce the differential operator d. Define

$$df = f_x\, dx + f_y\, dy$$

for C^1 functions f. For the C^1 1-form ω given by (3.1.1) we have (by definition)

$$\begin{aligned} d\omega &= d(f\, dx) + d(g\, dy) = df \wedge dx + dg \wedge dy \\ &= (f_x\, dx + f_y\, dy) \wedge dx + (g_x\, dx + g_y\, dy) \wedge dy \\ &= (g_x - f_y)\, dx \wedge dy. \end{aligned}$$

For a 2-form Ω we, of course, have (again by definition)

$$d\Omega = 0.$$

The most important fact concerning this operator is contained in Stokes' theorem. If ω is a C^1 k-form ($k = 0,1,2$) and D is a $(1 + k)$-chain, then

$$\int_{\partial D} \omega = \int_D d\omega.$$

(Of course, the only non-trivial case is $k = 1$.)

Note also that

$$d^2 = 0,$$

whenever d^2 is defined.

I.3.6. Up to now we have made use only of the underlying C^∞ structure of the Riemann surface M—except, of course, for notational simplifications provided by the complex structure. Using complex analytic coordinates we introduce two differential operators ∂ and $\bar{\partial}$ by setting for a C^1 function f,

$$\partial f = f_z\, dz \quad \text{and} \quad \bar{\partial} f = f_{\bar{z}}\, d\bar{z};$$

and setting for a C^1 1-form $\omega = u\, dz + v\, d\bar{z}$,

$$\partial\omega = \partial u \wedge dz + \partial v \wedge d\bar{z} = v_z\, dz \wedge d\bar{z},$$
$$\bar{\partial}\omega = \bar{\partial} u \wedge dz + \bar{\partial} v \wedge d\bar{z} = u_{\bar{z}}\, d\bar{z} \wedge dz = -u_{\bar{z}} dz \wedge d\bar{z},$$

where

$$f_z = \tfrac{1}{2}(f_x - if_y),$$
$$f_{\bar{z}} = \tfrac{1}{2}(f_x + if_y).$$

For 2-forms, the operators ∂ and $\bar{\partial}$ are defined as the zero operators.

Recall. The equation $f_{\bar{z}} = 0$ is equivalent to the Cauchy–Riemann equations for $Re\, f$, $Im\, f$; that is, $f_{\bar{z}} = 0$ if and only if f is holomorphic.

It is easy to check that the operators on forms we have defined satisfy

$$d = \partial + \bar{\partial}.$$

It is also easy to see that

$$\partial^2 = \partial\bar{\partial} + \bar{\partial}\partial = \bar{\partial}^2 = 0,$$

whenever these operators are defined.

I.3.7. In the previous paragraph the complex structure on M was still not used in any essential way. We shall now make essential use of it to define the operation of conjugation on smooth (C^1 or C^2—as is necessary) differential forms.

We introduce the *conjugation* operator * as follows: For a 1-form ω given by (3.1.1), we define

$$^*\omega = -g\, dx + f\, dy. \tag{3.7.1}$$

This is the most important case and the only one we shall need in the sequel. To define the operator * on functions and 2-forms, we choose a non-vanishing

2-form $\lambda(z)\,dx \wedge dy$ on the surface. (Existence of such a canonical 2-form will follows trivially from IV.8.) If f is a function, we set

$$*f = f(z)(\lambda(z)\,dx \wedge dy).$$

For a 2-form Ω, we set

$$*\Omega = \Omega/\lambda(z)\,dx \wedge dy.$$

It is clear that for each $k = 0, 1, 2,$

$$*: \textstyle\bigwedge^k \to \bigwedge^{2-k},$$

and $** = (-1)^k$. Further, if ω is given in complex notation by (3.2.1), then

$$*\omega = -iu(z)\,dz + iv(z)\,d\bar{z}. \tag{3.7.2}$$

The operation $*$ defined on 1-forms ω has the following geometric interpretation. If f is a C^1 function and $z(s) = x(s) + iy(s)$ is the equation of a curve parametrized by arc-length, then the differential df has the geometric interpretation of being $(\partial f/\partial \tau)\,ds$, where $\partial f/\partial \tau$ is the directional derivative of f in the direction of the tangent to the curve z. In this context, $*df$ has the geometric interpertation of being $(\partial f/\partial n)\,ds$, where $\partial f/\partial n$ is the directional derivative of f in the direction of the normal to the curve z. (Note that $ds = |dz|$ in this discussion.)

Remark. The reader should check that the above definitions (for example (3.7.1)) are all well defined in the sense that $*\omega$ is indeed a 1-form (that is, it transforms properly under change of local coordinates).

I.3.8. Our principal interest will be in 1-forms. Henceforth all differential forms are assumed to be 1-forms unless otherwise specified. A form ω is called *exact* if $\omega = df$ for some C^2 function f on M; ω is called *co-exact* if $*\omega$ is exact (if and only if $\omega = *df$ for some C^2 function f). We say that ω is *closed* provided it is C^1 and $d\omega = 0$; we say ω is *co-closed* provided $*\omega$ is closed.

Note that every exact (co-exact) differential is closed (co-closed). Whereas on a simply connected domain, closed (co-closed) differentials are exact (co-exact). Hence, closed (co-closed) differentials are locally exact (co-exact). If f is a C^2 function on M, we define the *Laplacian* of f, Δf in local coordinates by

$$\Delta f = (f_{xx} + f_{yy})\,dx \wedge dy.$$

The function f is called *harmonic* provided $\Delta f = 0$. A 1-form ω is *harmonic* provided it is locally given by df with f a harmonic function.

Remarks

1. It is easy to compute that for every C^2 function f,

$$-2i\,\bar{\partial}\partial f = \Delta f = d*df. \tag{3.8.1}$$

2. It must, of course, be verified that the Laplacian operator Δ is well defined. (Here, again, the fact that we are dealing with Riemann surfaces, and not just a differentiable surface, is crucial.)
3. The concept of harmonic function is, of course, a local one. Thus we know that locally every real-valued harmonic function is the real part of a holomorphic function. Further, real valued harmonic functions satisfy the maximum and minimum principle (that is, a non-constant real-valued harmonic function does not achieve a maximum nor a minimum at any interior point).

Proposition. *A differential ω is harmonic if and only if it is closed and co-closed.*

PROOF. A harmonic differential is closed (since $d^2 = 0$). It is co-closed by (3.8.1). Conversely, if ω is closed, then locally $\omega = df$ with f a C^2 function. Since ω is co-closed, $d(*df) = 0$. Thus, f is harmonic. □

At this point observe that we have a pairing between the homology group H_n and the group of closed n-forms of class C^1. This pairing is interesting for $n = 1$, and we describe it only in this case. If c is a 1-cycle and ω is a closed 1-form of class C^1, define $\langle c,\omega\rangle = \int_c \omega$. The homology group H_1 is defined by Z_1/B_1, where Z_1 is the kernel of δ and B_1 is the image of the 2-chains in the one chains. The operator d on functions of class C^2 and differentials of class C^1 gives rise also to subgroups of the group of closed 1-forms. In particular the exact differentials are precisely the image of C^2 functions (0-forms) in the group of 1-forms, and the closed forms themselves are the kernel of d operating on C^1 1-forms. Hence the quotient of closed 1-forms by exact 1-forms is a group and we have for compact surfaces a nonsingular pairing between H_1 and this quotient group. We shall soon see that this quotient group is isomorphic to the space of harmonic differentials. (See II.3.6.)

I.3.9. A 1-form ω is called *holomorphic* provided that locally $\omega = df$ with f holomorphic.

Proposition. a. *If u is a harmonic function on M, then ∂u is a holomorphic differential.*

b. *A differential $\omega = u\,dz + v\,d\bar{z}$ is holomorphic if and only if $v = 0$ and u is a holomorphic function (of the local coordinate).*

PROOF. If u is harmonic, then $\bar{\partial}\partial u = 0$. Since $\partial u = u_z\,dz$, we see that u_z is holomorphic ($u_{z\bar{z}} = 0$) and thus it suffices to prove only part (b), which is, of course, trivial (since we can integrate power series term by term). □

I.3.10. Assume $\omega = u\,dz + v\,d\bar{z}$ is a C^1 differential. Then (using $d = \partial + \bar{\partial}$) we see that

$$d\omega = (u_{\bar{z}} - v_z)\,d\bar{z} \wedge dz,$$

and

$$d^*\omega = -i(u_{\bar{z}} + v_z)\, d\bar{z} \wedge dz.$$

Thus, we see that ω is harmonic if and only if u and $\bar{v}$ are holomorphic. Hence, we see that if ω is harmonic, there are unique holomorphic differentials ω_1 and ω_2 such that

$$\omega = \omega_1 + \bar{\omega}_2.$$

I.3.11. Theorem. *A differential form ω is holomorphic if and only if $\omega = \alpha + i^*\alpha$ for some harmonic differential α.*

PROOF. Assume that α is harmonic. Then

$$\alpha = \omega_1 + \bar{\omega}_2$$

with ω_j $(j = 1,2)$ holomorphic. Thus

$$^*\alpha = -i\omega_1 + i\bar{\omega}_2.$$

Thus,

$$\alpha + i^*\alpha = 2\omega_1$$

is holomorphic. Conversely, if ω is holomorphic, then ω and $\bar{\omega}$ are harmonic and so is

$$\alpha = \frac{\omega - \bar{\omega}}{2}.$$

Further

$$^*\alpha = \frac{-i\omega - i\bar{\omega}}{2}.$$

Thus

$$\omega = \alpha + i^*\alpha. \qquad \square$$

Corollary. *A differential ω is holomorphic if and only if it is closed and $^*\omega = -i\omega$.*

PROOF. The forward implication has already been verified (every holomorphic differential is harmonic). For the reverse, note that if ω is given by (3.2.1) and $^*\omega = -i\omega$, then (3.7.2) implies that $\omega = u\, dz$. Since $d\omega = 0$, u is holomorphic. $\square$

I.4. Integration Formulae

In this section we gather several useful consequences of Stokes' theorem.

I.4.1. Theorem (Integration by Parts). *Let D be a relatively compact region on a Riemann surface M with piecewise differentiable boundary. Let f be a*

C^1 function and ω a C^1 1-form on a neighborhood of the closure of D. Then

$$\int_{\delta D} f\omega = \iint_D f\,d\omega - \iint_D \omega \wedge df. \tag{4.1.1}$$

PROOF. Apply Stokes' theorem to the 1-form $f\omega$ and observe that $d(f\omega) = f\,d\omega + df \wedge \omega$. □

Corollary 1. *If ω is a closed (in particular, holomorphic) 1-form, then (under the hypothesis of the theorem)*

$$\int_{\delta D} \omega = 0.$$

PROOF. Take f to be the constant function with value 1. □

Corollary 2. *Let f be C^1 function and ω a C^1 1-form on the Riemann surface M. If either f or ω has compact support, then*

$$\iint_M f\,d\omega - \iint_M \omega \wedge df = 0. \tag{4.1.2}$$

PROOF. If M is not compact, then take D to be compact and have nice boundary so that either f or ω vanishes on δD and use (4.1.1). If M is compact cover M by a finite number of disjoint triangles $\Delta_j, j = 1, \ldots, n$. Over each triangle (4.1.1) is valid. We obtain (4.1.2) by noting that

$$\sum_{j=1}^{n} \int_{\delta\Delta_j} f\omega = 0,$$

since each edge appears in exactly 2 triangles with opposite orientation. □

I.4.2. We fix a region D on M. By a *measurable 1-form* ω on D, we mean a 1-form (given in local coordinates by)

$$\omega = u\,dz + v\,d\bar{z},$$

where u and v are measurable functions of the local coordinates. As usual, we agree to identify two forms if they coincide almost everywhere (sets of Lebesgue measure zero are well defined on M!). We denote by $L^2(D)$ the complex Hilbert space of 1-forms ω with

$$\|\omega\|_D^2 = \iint_D \omega \wedge {}^*\bar{\omega} < \infty. \tag{4.2.1}$$

Note that in local coordinates

$$\begin{aligned}\omega \wedge {}^*\bar{\omega} &= i(u\bar{u} + v\bar{v})\,dz \wedge d\bar{z}\\ &= 2(|u|^2 + |v|^2)\,dx \wedge dy.\end{aligned}$$

We also define the inner product of $\omega_1, \omega_2 \in L^2(D)$ by

$$(\omega_1,\omega_2)_D = \iint_D \omega_1 \wedge {}^*\bar{\omega}_2. \tag{4.2.2}$$

Using obvious notational conventions, we see that

$$\omega_1 \wedge {}^*\bar{\omega}_2 = (u_1 dz + v_1 d\bar{z}) \wedge \overline{(-iu_2 dz + iv_2 d\bar{z})} = i(u_1\bar{u}_2 + v_1\bar{v}_2) dz \wedge d\bar{z},$$

and thus

$$\begin{aligned}(\omega_1,\omega_2)_D &= \iint_D \omega_1 \wedge {}^*\bar{\omega}_2 = i \iint (u_1\bar{u}_2 + v_1\bar{v}_2)\, dz \wedge d\bar{z} \\ &= \overline{i \iint (\bar{u}_1 u_2 + \bar{v}_1 v_2)\, dz \wedge d\bar{z}} = \overline{\iint_D \omega_2 \wedge {}^*\bar{\omega}_1} \\ &= \overline{(\omega_2,\omega_1)_D},\end{aligned}$$

as is required for a Hilbert space inner product. Further,

$$\begin{aligned}({}^*\omega_1,{}^*\omega_2)_D &= \iint_D {}^*\omega_1 \wedge -\bar{\omega}_2 \\ &= \overline{\iint_D \omega_2 \wedge {}^*\bar{\omega}_1} = \overline{(\omega_2,\omega_1)_D} = (\omega_1,\omega_2)_D.\end{aligned}$$

Remark. Whenever there can be no confusion, the domain D will be dropped from the symbols for the norm (4.2.1) and inner products (4.2.2) in $L^2(D)$.

I.4.3. Proposition. *Let D be a relatively compact region on M with piecewise differentiable boundary. Let φ be a C^1 function and α a C^1 differential on a neighborhood of the closure of D. Then*

$$(d\varphi,{}^*\alpha) = \iint_D \varphi\, d\bar{\alpha} - \int_{\delta D} \varphi\bar{\alpha}. \tag{4.3.1}$$

PROOF. By Stokes' theorem

$$\begin{aligned}\int_{\delta D} \varphi\bar{\alpha} &= \iint_D d(\varphi\bar{\alpha}) = \iint_D \varphi\, d\bar{\alpha} + \iint_D d\varphi \wedge \bar{\alpha} \\ &= \iint_D \varphi\, d\bar{\alpha} - (d\varphi,{}^*\alpha).\end{aligned}$$

□

I.4.4. Proposition. *If φ and ψ are C^2 functions on a neighborhood of the closure of D, then*

$$(d\varphi,d\psi) = -\iint_D \varphi\, \overline{\Delta\psi} + \int_{\delta D} \varphi\, {}^*\overline{d\psi}. \tag{4.4.1}$$

PROOF. By Stokes' theorem

$$\begin{aligned}\int_{\delta D} \varphi\, {}^*\overline{d\psi} &= \iint_D d(\varphi\, {}^*\overline{d\psi}) \\ &= \iint_D d\varphi \wedge {}^*\overline{d\psi} + \iint_D \varphi\, d\, {}^*\overline{d\psi} \\ &= (d\varphi,d\psi) + \iint_D \varphi\, \Delta\bar{\psi}.\end{aligned}$$

□

Corollary. *We have*

$$\iint_D (\varphi\, \Delta\psi - \psi\, \Delta\varphi) = \int_{\delta D} (\varphi\, {}^*d\psi - \psi\, {}^*d\varphi). \tag{4.4.2}$$

PROOF. Rewrite (4.4.1) with $\bar{\psi}$ replacing ψ. Rewrite the resulting expression by interchanging φ and ψ. Subtract one from the other, and use the fact that $(d\varphi, d\bar{\psi}) = (d\psi, d\bar{\varphi})$. □

I.4.5. Proposition. *We have*

$$(d\varphi, {}^*d\psi) = -\int_{\delta D} \varphi \, d\bar{\psi}.$$

PROOF. The notation is of Proposition I.4.4. Apply Proposition I.4.3 with $\alpha = d\psi$ (recall that $d^2 = 0$). □

I.4.6. We apply now the previous results (D is as defined in Proposition I.4.3) to analytic differentials.

Proposition. *If φ is a C^1 function and ω is closed (in particular, an analytic differential) on a neighborhood of the closure of D, then*

$$\int_{\delta D} \varphi\bar{\omega} = \iint_D d\varphi \wedge \bar{\omega}.$$

PROOF. Use (4.3.1) with $\alpha = \omega$ and observe that $\bar{\omega}$ is also closed. □

Corollary. *If f and g are C^2 (in particular, holomorphic functions) on a neighborhood of the closure D, then*

$$\int_{\delta D} f\,\overline{dg} = -\int_{\delta D} \bar{g}\,df.$$

PROOF. By the proposition

$$\int_{\delta D} f\,\overline{dg} = \iint_D df \wedge \overline{dg}.$$

Also,

$$\int_{\delta D} \bar{g}\,df = \iint_D \overline{dg} \wedge df.$$ □

I.4.7. Proposition. *If f is a holomorphic function and ω is a holomorphic differential on a neighborhood of the closure of D, then*

$$\iint_D df \wedge \bar{\omega} = 2\int_{\delta D} (\operatorname{Re} f)\bar{\omega} = 2i\int_{\delta D} (\operatorname{Im} f)\bar{\omega}.$$

PROOF. Observe that

$$\int_{\delta D} (\operatorname{Re} f - i \operatorname{Im} f)\bar{\omega} = \int_{\delta D} \bar{f}\bar{\omega} = 0$$

by Cauchy's theorem (or because $d(f\omega) = 0$ since $f\omega$ is a holomorphic form). Thus

$$2\int_{\delta D} \operatorname{Re} f\,\bar{\omega} = \int_{\delta D} (\operatorname{Re} f + i \operatorname{Im} f)\bar{\omega} = \iint_D df \wedge \bar{\omega}.$$

Thus, also

$$2\int_{\delta D} \operatorname{Re}(if)\bar{\omega} = i\iint_D df \wedge \bar{\omega}.$$ □

CHAPTER II
Existence Theorems

One way to study Riemann surfaces is through the meromorphic functions on them. Our first task is to show that every Riemann surface carries non-constant meromorphic functions. We do so by constructing certain harmonic differentials (with singularities). From the existence of harmonic differentials, it is trivial to construct meromorphic differentials. A ratio of two linearly independent meromorphic differentials produces a non-constant meromorphic function.

Our basic approach is through the Hilbert space $L^2(M)$ introduced in I.4.2. It is the key to the existence theorem for harmonic differentials with and without singularities. This method should be contrasted with the equally powerful method to be developed in Chapter IV.

II.1. Hilbert Space Theory—A Quick Review

We need only the first fundamental theorem about Hilbert space: the existence of projections onto arbitrary closed subspaces.

II.1.1. Let H be a (complex) Hilbert space with inner product $(\cdot,\cdot)$ and norm $\|\cdot\|$. If F is any subspace of H, then the *orthogonal complement* of F in H,

$$F^{\perp} = \{h \in H; (f,h) = 0 \text{ all } f \in F\}$$

is a closed subspace of H (hence a Hilbert space). About the only non-trivial result on Hilbert spaces that we will need is

II.1.2. Theorem. *Let F be a closed subspace of a Hilbert space H. Then every $h \in H$ can be written uniquely as*

$$h = f + g$$

with $f \in F$, $g \in F^{\perp}$. Furthermore, f is the unique element of F which minimizes

$$\|h - \varphi\|, \qquad \varphi \in F.$$

II.1.3. Writing $f = Ph$, we see that we also have the following equivalent form of the previous

Theorem. *Let $F \neq \{0\}$ be a closed subspace of a Hilbert space H. There exists a unique linear mapping*

$$P: H \to F$$

satisfying:

a. $\|P\| = 1$,
b. $P^2 = P$ (*P is a projection*), *and*
c. $\ker P = F^{\perp}$.

The mapping P is called the *orthogonal projection onto F*. The proofs of the above theorems may be found in any of the standard text books on Hilbert spaces.

II.2. Weyl's Lemma

We have introduced, in I.4.2, the Hilbert space $L^2(M)$ of square integrable (measurable) 1-forms on the Riemann surface M. In this section we lay the ground-work for characterizing the harmonic differentials in $L^2(M)$. The characterization is in terms of integrals. We show that a "weak solution" to Laplace's equation is already a harmonic function. The precise meaning of the above claim is the content of

II.2.1. Theorem (Weyl's Lemma). *Let φ be a measurable square integrable function on the unit disk D. The function φ is harmonic if and only if*

$$\iint_D \varphi \, \Delta\eta = 0 \tag{2.1.1}$$

for every C^∞ function η on D with compact support.

Remark. In the above theorem, the sentence "φ is harmonic" should, of course, be replaced by "φ is equal almost everywhere (a.e.) to a harmonic function". We will in similar contexts make similar identifications in the future.

Proof of Theorem. Assume that φ is harmonic. Let $D_r = \{z \in \mathbb{C};\ |z| < r\}$, and assume that η is supported in D_r. Use I(4.4.2) and conclude that

$$\iint_{D_r} (\varphi\, \Delta\eta - \eta\, \Delta\varphi) = \int_{\delta D_r} (\varphi * d\eta - \eta * d\varphi) = 0 \tag{2.1.2}$$

(because η and $*d\eta$ vanish on δD_r). Thus

$$\iint_{D_r} \varphi\, \Delta\eta = \iint_{D_r} \eta\, \Delta\varphi = 0 \tag{2.1.3}$$

(because $\Delta\varphi = 0$). But

$$\iint_{D} \varphi\, \Delta\eta = \iint_{D_r} \varphi\, \Delta\eta, \tag{2.1.4}$$

and thus the necessity of (2.1.1) is established.

To prove the converse, we assume first that φ is C^2. Choosing $0 < r < 1$ as before, we obtain (2.1.2) as a consequence of I(4.4.2), and hence (2.1.3) because of (2.1.1). As a consequence of (2.1.4) we may assume $r = 1$. Equation (2.1.3) shows that $\Delta\varphi = 0$ in D. To verify this claim it clearly suffices to consider only real-valued functions φ and η. Let $\psi(z) = 4\,\partial^2\varphi/\partial\bar{z}\,\partial z$. If $\psi(z_0) > 0$ for some $z_0 \in D$, we choose a neighborhood U of z_0 such that Cl U (= the closure of U) $\subset D$ and such that $\psi > 0$ in U. Select a C^∞ function η with $\eta(z_0) > 0$ and η supported in U. It is clear that for such η, $\iint_D \eta\, \Delta\varphi > 0$. Thus $\Delta\varphi = 0$ and φ is harmonic in D.

The heart of the matter is to drop the smoothness assumption on φ. Fix $0 < \varepsilon < \frac{1}{2}$. Construct a real-valued C^∞ (= smooth) function ρ on the positive real axis $[0,\infty)$ such that

$$0 \le \rho \le 1,$$

$$\rho(r) = 0 \quad \text{if } r > \varepsilon,$$

and

$$\rho(r) = 1 \quad \text{if } 0 \le r < \varepsilon/2.$$

(See Figure II.1)

Set for $r > 0$,

$$\omega(r) = -\frac{1}{2\pi}\,\rho(r)\log r.$$

(See Figure II.2)

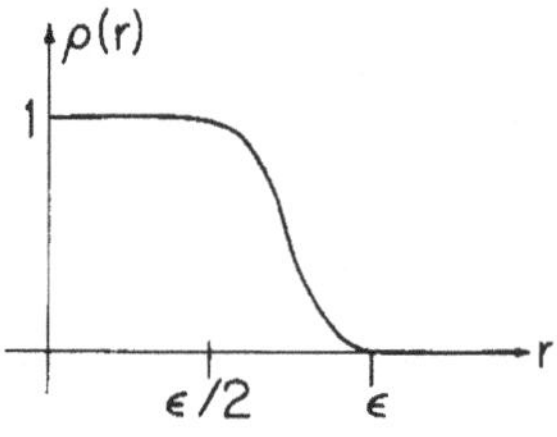

Figure II.1

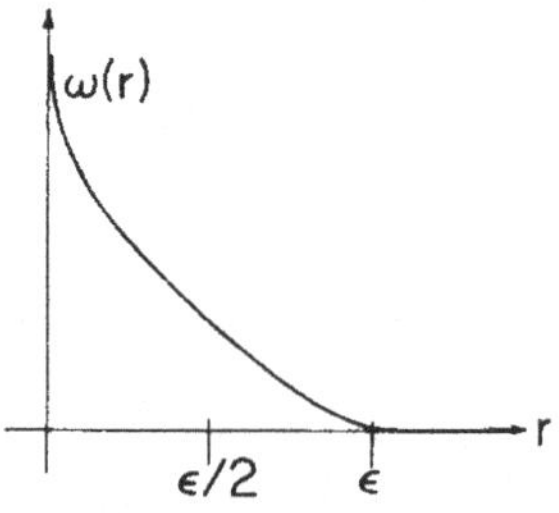

Figure II.2

Define on $\mathbb{C} \times \mathbb{C}$

$$\gamma(z,\zeta) = \begin{cases} 4\dfrac{\partial^2}{\partial \bar{z}\partial z}\omega(|z-\zeta|) & \text{if } z \neq \zeta, \\ 0 & \text{if } z = \zeta. \end{cases}$$

Since $\log|z-\zeta|$ is a harmonic function of z on $\mathbb{C}\backslash\{\zeta\}$, $\gamma(z,\zeta) = 0$ for $|z-\zeta| < \varepsilon/2$ (also for $|z-\zeta| > \varepsilon$).

Let μ be a C^∞ function with support in $D_{1-2\varepsilon}$. Consider

$$\psi(z) = \iint_D \omega(|\zeta - z|)\mu(\zeta)\frac{d\zeta \wedge d\bar{\zeta}}{-2i}, \qquad z \in \mathbb{C}. \tag{2.1.5}$$

It is clear that ψ is a continuous function on $\mathbb{C}$ with support in $D_{1-\varepsilon}$. Observe that we may integrate over $\mathbb{C}$ instead of over D, and that by a change of variable (after extending μ to be zero outside its support)

$$\psi(z) = \iint_{\mathbb{C}} \omega(|\zeta|)\mu(\zeta + z)\frac{d\zeta \wedge d\bar{\zeta}}{-2i}.$$

Thus

$$\begin{aligned}\frac{\partial\psi}{\partial\bar{z}} &= \iint_{\mathbb{C}} \omega(|\zeta|)\frac{\partial}{\partial\bar{z}}\mu(\zeta + z)\frac{d\zeta \wedge d\bar{\zeta}}{-2i} \\ &= \iint_{\mathbb{C}} \omega(|\zeta|)\frac{\partial}{\partial\bar{\zeta}}\mu(\zeta + z)\frac{d\zeta \wedge d\bar{\zeta}}{-2i} \\ &= \iint_{\mathbb{C}} \omega(|\zeta - z|)\frac{\partial}{\partial\bar{\zeta}}\mu(\zeta)\frac{d\zeta \wedge d\bar{\zeta}}{-2i}.\end{aligned}$$

Similarly,

$$\frac{\partial\psi}{\partial z} = \iint_{\mathbb{C}} \omega(|\zeta - z|)\frac{\partial}{\partial\zeta}\mu(\zeta)\frac{d\zeta \wedge d\bar{\zeta}}{-2i}.$$

Thus ψ is C^∞. We claim that

$$4\frac{\partial^2\psi}{\partial\bar{z}\,\partial z} = -\mu(z) + \iint_D \gamma(z,\zeta)\mu(\zeta)\frac{d\zeta \wedge d\bar{\zeta}}{-2i}. \tag{2.1.6}$$

Note that

$$\begin{aligned}\psi(z) &= -\frac{1}{2\pi}\iint_{|\zeta - z| < \varepsilon/2} \mu(\zeta)\log|\zeta - z|\frac{d\zeta \wedge d\bar{\zeta}}{-2i} \\ &\quad + \iint_{|\zeta - z| \geq \varepsilon/2} \omega(|\zeta - z|)\mu(\zeta)\frac{d\zeta \wedge d\bar{\zeta}}{-2i} \\ &= \alpha(z) + \beta(z).\end{aligned}$$

Note further that

$$4\frac{\partial^2\beta}{\partial\bar{z}\,\partial z} = \iint_{|\zeta - z| \geq \varepsilon/2} \gamma(z,\zeta)\mu(\zeta)\frac{d\zeta \wedge d\bar{\zeta}}{-2i}$$

$$= \iint_{\mathbb{C}} \gamma(z,\zeta)\mu(\zeta)\frac{d\zeta \wedge d\bar{\zeta}}{-2i}. \qquad (2.1.7)$$

We have already shown that $\Delta\alpha$ exists (ψ is C^∞). Thus we may compute formally. Now $\partial\alpha/\partial z$ can be computed formally without any difficulty (since differentiating under the integral sign leads to a Lebesgue integrable function). First

$$\frac{\partial}{\partial z}\log|\zeta - z| = \frac{1}{|\zeta - z|}\frac{\partial}{\partial z}|\zeta - z| = -\frac{1}{2}\frac{1}{\zeta - z},$$

and thus

$$\frac{\partial\alpha}{\partial z} = \frac{1}{4\pi}\iint_{|\zeta - z| < \varepsilon/2} \frac{\mu(\zeta)}{\zeta - z}\frac{d\zeta \wedge d\bar{\zeta}}{-2i}. \qquad (2.1.8)$$

That

$$4\frac{\partial^2\alpha}{\partial\bar{z}\,\partial z} = -\mu \qquad (2.1.9)$$

will follow from

II.2.2. Lemma (Cauchy's Integral Formula). *Let B be connected open subset of $\mathbb{C}$ bounded by finitely many C^1 Jordan curves. If $u \in C^1(\text{Cl } B)$, then for $z \in B$,*

$$2\pi i u(z) = \int_{\delta B} \frac{u(\zeta)}{\zeta - z}d\zeta + \iint_B \frac{\partial u/\partial\bar{\zeta}}{\zeta - z}d\zeta \wedge d\bar{\zeta}. \qquad (2.2.1)$$

PROOF. Let $\varepsilon > 0$ be chosen so that the closed ball of radius ε about z is contained in B. Let B_ε be the complement in B of this ball. Start with Stokes' theorem

$$\int_{\delta B_\varepsilon} \frac{u(\zeta)}{\zeta - z}d\zeta = \iint_{B_\varepsilon} \frac{\partial}{\partial\bar{\zeta}}\left(\frac{u(\zeta)}{\zeta - z}\right)d\bar{\zeta} \wedge d\zeta,$$

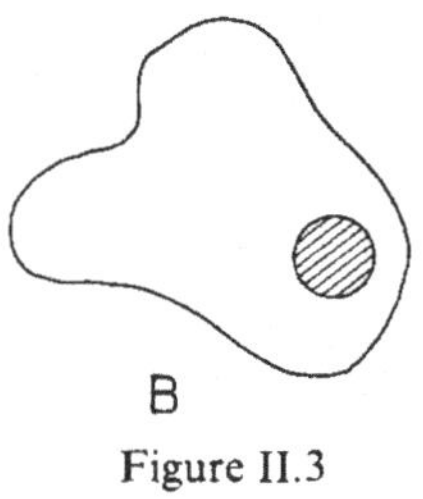

Figure II.3

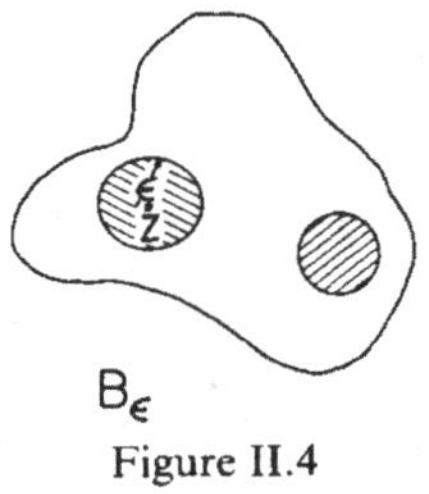

Figure II.4

and obtain

$$\iint_{B_\varepsilon} \frac{\partial u/\partial \bar{\zeta}}{\zeta - z}\, d\bar{\zeta} \wedge d\zeta = \int_{\delta B} \frac{u(\zeta)}{\zeta - z}\, d\zeta - i \int_0^{2\pi} u(z + \varepsilon e^{i\theta})\, d\theta.$$

Letting $\varepsilon \to 0$, we obtain (2.2.1). □

II.2.3. Proof of Weyl's Lemma (Conclusion). To verify formula (2.1.9), it suffices to assume that μ has support in $|\zeta - z| < \varepsilon/2$, and thus that the integral in (2.1.8) extends over $\mathbb{C}$. To check this last claim, define

$$v(\zeta) = \rho(2|\zeta - z|)\mu(\zeta), \qquad \zeta \in \mathbb{C}.$$

Note that

$$\iint_{|\zeta - z| < \varepsilon/2} \frac{\mu(\zeta)}{\zeta - z} \frac{d\zeta \wedge d\bar{\zeta}}{-2i} = \iint_{|\zeta - z| < \varepsilon/2} \frac{v(\zeta)}{\zeta - z} \frac{d\zeta \wedge d\bar{\zeta}}{-2i}$$
$$+ \iint_{|\zeta - z| < \varepsilon/2} \frac{\mu(\zeta) - v(\zeta)}{\zeta - z} \frac{d\zeta \wedge d\bar{\zeta}}{-2i}, \qquad (2.3.1)$$
$$v(\zeta) = \mu(\zeta) \quad \text{for } |\zeta - z| < \varepsilon/4,$$

and that

$$v(\zeta) = 0 \quad \text{for } |\zeta - z| \geq \varepsilon/2.$$

The third integral in (2.3.1) represents a holomorphic function of z, and thus the $\bar{z}$-derivative of the first and second integral in (2.3.1) must coincide. So now we assume that μ has support in $|\zeta - z| < \varepsilon/2$. As before (when we computed $\partial\psi/\partial\bar{z}$),

$$\frac{\partial \alpha}{\partial z} = \frac{1}{4\pi} \iint_{\mathbb{C}} \frac{\mu(\zeta + z)}{\zeta} \frac{d\zeta \wedge d\bar{\zeta}}{-2i},$$

and thus

$$\frac{\partial^2 \alpha}{\partial \bar{z}\, \partial z} = \frac{1}{4\pi} \iint_{\mathbb{C}} \frac{\mu_{\bar{z}}(\zeta + z)}{\zeta} \frac{d\zeta \wedge d\bar{\zeta}}{-2i} = \frac{1}{4\pi} \iint_{\mathbb{C}} \frac{\mu_{\bar{\zeta}}(\zeta + z)}{\zeta} \frac{d\zeta \wedge d\bar{\zeta}}{-2i}$$
$$= \frac{1}{4\pi} \iint_{\mathbb{C}} \frac{\partial \mu/\partial \bar{\zeta}}{\zeta - z} \frac{d\zeta \wedge d\bar{\zeta}}{-2i} = -\frac{1}{4}\mu(z)$$

by (2.2.1).

We now use condition (2.1.1) with $\eta = \psi$ (of the previous construction via (2.1.5)). Thus we obtain

$$0 = \iint_D \varphi\, \Delta\psi = -\iint_D \varphi(z)\mu(z) \frac{dz \wedge d\bar{z}}{-2i} + \iint_D \varphi(z)\left(4 \frac{\partial^2 \beta}{\partial \bar{z}\, \partial z}\right) \frac{dz \wedge d\bar{z}}{-2i},$$

where $(4\partial^2\beta/\partial\bar{z}\,\partial z)$ is given by (2.1.7). Hence for every C^∞ function μ with support in D, we have by Fubini's theorem

$$\iint_{D(z)} \varphi(z)\mu(z)\frac{dz\wedge d\bar{z}}{-2i} = \iint_{D(z)} \varphi(z) \iint_{D(\zeta)} \gamma(z,\zeta)\mu(\zeta)\frac{d\zeta\wedge d\bar{\zeta}}{-2i}\frac{dz\wedge d\bar{z}}{-2i}$$

$$= \iint_{D(\zeta)} \mu(\zeta) \iint_{D(z)} \varphi(z)\gamma(z,\zeta)\frac{dz\wedge d\bar{z}}{-2i}\frac{d\zeta\wedge d\bar{\zeta}}{-2i}. \quad (2.3.2)$$

(In the above $D(z)$ is just another symbol for D, and the (z) is supposed to remind the reader that we are integrating with respect to z.) It thus follows (because C^∞ functions of compact support are dense in the L^2 functions) that

$$\iint_{D(z)} \varphi(z)\gamma(z,\zeta)\frac{dz\wedge d\bar{z}}{-2i} = \varphi(\zeta), \text{ a.e. } \zeta\in D. \quad (2.3.3)$$

Clearly the left-hand side is C^∞ (in ζ), and thus the proof is complete.

Remark. We do not need to know that C^∞ functions of compact support are dense in $L^2(D)$. We show that we can do with slightly less. Let $\hat{\varphi}(\zeta)$ denote the left-hand side of (2.3.3). If (2.3.3) is not true then (for real φ) one of the sets

$$D_+ = \{z\in\mathbb{C};\ \hat{\varphi}(z) > \varphi(z)\}$$
$$D_- = \{z\in\mathbb{C};\ \hat{\varphi}(z) < \varphi(z)\}$$

has positive measure. Hence we may assume that one of these sets has interior. If we now choose μ to be non-negative and to have support in this set, then we obtain a contradiction to (2.3.2) as in the arguments that established sufficiency for a C^2 function φ. Thus μ could have been assumed to have support in an arbitrarily small set to begin with, simplifying slightly the reasoning at the beginning of this paragraph.

Exercise

Prove the following alternate form of Weyl's lemma: *Let φ be a measurable square integrable function on the unit disk D. The function φ is holomorphic if and only if*

$$\iint_D \varphi(z)\frac{\partial\eta}{\partial\bar{z}}dz\wedge d\bar{z} = 0 \quad (2.3.4)$$

for every C^∞ function η on D with compact support.

Hints.

(1) Let φ be holomorphic and η, smooth with compact support. Then

$$0 = \int_{\delta D}\varphi\eta\,dz = \iint_D \bar{\partial}(\varphi\eta\,dz). \quad (2.3.5)$$

This establishes necessity of (2.3.4). (The fact that φ is not defined on δD should not cause any trouble.)

(2) For sufficiency, first assume φ is C^1. Use that for every η with compact support (2.3.5) holds, and thus

$$-\iint_D \varphi_{\bar{z}}\eta\, dz \wedge d\bar{z} = \iint_D \varphi\eta_{\bar{z}}\, dz \wedge d\bar{z} = 0.$$

From this equation deduce the Cauchy–Riemann equations.

(3) Now take arbitrary φ. Note that for arbitrary η with compact support,

$$\iint_D \varphi \frac{\partial^2 \eta}{\partial \bar{z}\,\partial z}\, dz \wedge d\bar{z} = \iint_D \varphi \frac{\partial}{\partial \bar{z}}\left(\frac{\partial \eta}{\partial z}\right) dz \wedge d\bar{z}.$$

Thus by the form of Weyl's lemma at our disposal, φ is C^∞.

The above form of Weyl's lemma can, of course, also be proven directly, and our form recovered (with a few more technical complications) from this form.

II.2.4. Exercise

Let $f \in L^2([0,1])$. Show that f equals almost everywhere a constant if and only if

$$\int_0^1 f(x)g'(x)\, dx = 0$$

for all C^∞ functions g on (0,1) with compact support.

This is the one-dimensional analogue of Weyl's lemma.

II.3. The Hilbert Space of Square Integrable Forms

We decompose the space of square integrable 1-forms into closed subspaces. The basic tool is Weyl's lemma. The decomposition will prove to be most useful for compact surfaces. In general, we establish a sufficient condition for the existence of a non-zero square integrable harmonic 1-form. For compact surfaces, the condition is also necessary.

II.3.1. Throughout this section, M will denote a Riemann surface and $L^2(M)$, the Hilbert space of (measurable) square integrable 1-forms with inner product $(\cdot,\cdot)$ and norm $\|\cdot\|$ (as defined in I.4.2). Throughout this chapter the word smooth will denote C^∞.

Definition. We denote by E the $L^2(M)$ closure of

$$\{df; f \text{ is a smooth function on } M \text{ with compact support}\},$$

and by

$$E^* = \{\omega \in L^2(M); {}^*\omega \in E\}.$$

Thus, for every $\omega \in E(E^*)$, there exists a sequence of smooth functions f_n on M with compact support such that

$$\omega = \lim_n df_n \qquad \left(= \lim_n {}^*df_n\right).$$

Corresponding to these closed subspaces we have orthogonal decompositions of $L^2(M)$,

$$\begin{aligned} L^2(M) &= E \oplus E^{\perp} \\ &= E^* \oplus E^{*\perp}, \end{aligned} \tag{3.1.1}$$

where, as usual,

$$E^{\perp} = \{\omega \in L^2(M); (\omega, df) = 0, \text{ for all smooth functions } f \text{ on } M \text{ with compact support}\},$$

$$E^{*\perp} = \{\omega \in L^2(M); (\omega, {}^*df) = 0, \text{ all } f \text{ as above}\}.$$

II.3.2. Proposition. *Let $\alpha \in L^2(M)$ be of class C^1. Then $\alpha \in E^{*\perp}$ (respectively, $E^{\perp}$) if and only if α is closed (co-closed).*

Proof. Assume that α is closed. Let f be a smooth function on M with support inside D (with Cl D compact). Then

$$\begin{aligned} (\alpha, {}^*df) &= -\iint_D \alpha \wedge d\bar{f} = -\iint_D [d(\alpha\bar{f}) - d\alpha \wedge \bar{f}] \\ &= -\iint_D d(\alpha\bar{f}) = -\int_{\delta D} \alpha\bar{f} = 0. \end{aligned}$$

Thus $\alpha \in E^{*\perp}$. Conversely, we have starting from the second equality

$$\iint_M d\alpha \wedge \bar{f} = 0, \quad \text{all smooth } f \text{ on } M \text{ with compact support.}$$

This, of course, is sufficient to conclude $d\alpha = 0$. The argument for $\alpha \in E^{\perp}$ is similar. □

We let

$$H = E^{\perp} \cap (E^*)^{\perp},$$

and obtain the following orthogonal decomposition

$$L^2(M) = E \oplus E^* \oplus H. \tag{3.2.1}$$

Note that from Proposition II.3.2 we deduce that E and E^* are orthogonal subspaces. It then follows that the direct sum $E \oplus E^*$ is closed and thus also a Hilbert subspace of $L^2(M)$. By orthogonal decomposition (Theorem II.1.1), we therefore certainly have $L^2(M) = (E \oplus E^*) \oplus (E \oplus E^*)^{\perp}$, and all we need to establish (3.2.1) is to verify the easy identity $(E \oplus E^*)^{\perp} = E^{\perp} \cap (E^*)^{\perp}$. Comparing (3.2.1) with (3.1.1) we see that

$$\begin{aligned} E^{\perp} &= E^* \oplus H \\ E^{*\perp} &= E \oplus H. \end{aligned}$$

II.3.3. Let c be a simple closed curve on M. Cover c by a finite number of coordinate disks and obtain a region Ω containing c. We call this region Ω, a *strip* around c. By choosing Ω sufficiently small, we may assume that it is an annulus and that $\Omega \backslash c$ consists of two annuli Ω^-, Ω^+. We orient c

so that Ω^- is to the left of c. We put a smaller strip, Ω_0, (with corresponding one-sided strips Ω_0^-, Ω_0^+) around c in Ω. We construct a real-valued function f on M with the following properties (see Figure II.5):

$$f(P) = 1, \qquad P \in \Omega_0^-,$$
$$f(P) = 0, \qquad P \in M \backslash \Omega^-,$$

and

$$f \text{ is of class } C^\infty \text{ on } M \backslash c.$$

We now define a C^∞ differential

$$\eta_c = \begin{cases} df & \text{on } \Omega \backslash c, \\ 0 & \text{on } (M \backslash \Omega) \cup c. \end{cases}$$

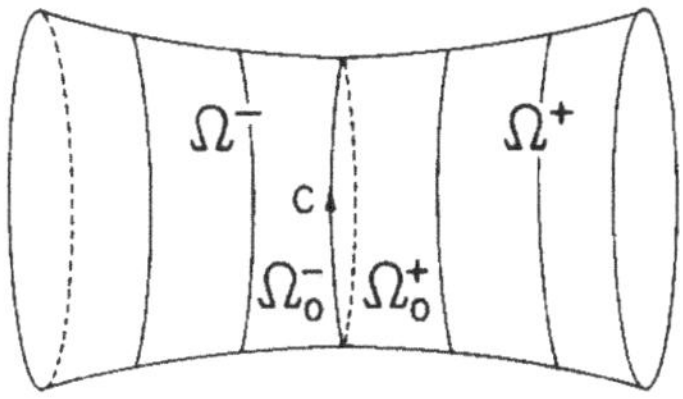

Figure II.5

It is clear that η_c is a closed, smooth, real differential form with compact support. It is, in general, not exact. We call it *the differential form associated with the closed curve c*.

Proposition. *Let $\alpha \in L^2(M)$ be closed and of class C^1. Then*

$$\int_c \alpha = (\alpha, {}^*\eta_c). \tag{3.3.1}$$

PROOF. We compute

$$\begin{aligned}(\alpha, {}^*\eta_c) &= -\iint_M \alpha \wedge \eta_c = -\iint_{\Omega^-} \alpha \wedge df = \iint_{\Omega^-} df \wedge \alpha \\ &= \iint_{\Omega^-} d(f\alpha) - \iint_{\Omega^-} f \wedge d\alpha = \iint_{\Omega^-} d(f\alpha) \\ &= \int_{\delta\Omega^-} f\alpha = \int_c \alpha. \end{aligned}$$

□

II.3.4. Proposition. *Let $\alpha \in L^2(M)$ be of class C^1. Then α is exact (respectively, co-exact) if and only if $(\alpha,\beta) = 0$ for all co-closed (closed) smooth differentials β of compact support.*

PROOF. If α is C^1 and exact, then $\alpha = df$ with f of class C^2. If β is co-closed, smooth, with support in D (with Cl D compact), then

$$\begin{aligned}(\alpha,\beta) &= \iint_D df \wedge {}^*\bar{\beta} = \iint_D [d(f{}^*\bar{\beta}) - fd{}^*\bar{\beta}] \\ &= \int_{\delta D} f{}^*\bar{\beta} = 0.\end{aligned}$$

To establish the converse, it suffices to show (because α is closed by Proposition II.3.2) that $\int_c \alpha = 0$ for all simple closed curves. But this follows from the hypothesis and (3.3.1). The assertion for the co-exact differentials follows from the part of the proposition already established. □

II.3.5. The most important result about $L^2(M)$ is contained in the following

Theorem. *The Hilbert space H consists of the harmonic differentials in* $L^2(M)$.

PROOF. If $\omega \in L^2(M)$ is harmonic, then ω is smooth, closed, and co-closed. Thus by Proposition II.3.2, $\omega \in E^\perp \cap (E^*)^\perp = H$.

For the converse, let $\omega \in H$. Choose a coordinate disk D on M with local coordinate $z = x + iy$. Choose a real-valued function η that is smooth and supported in D. Let $\varphi = \partial\eta/\partial x$ and $\psi = \partial\eta/\partial y$. Then φ and ψ are C^∞ functions on M with support in D and $\partial\varphi/\partial y = \partial\psi/\partial x$. Write ω as $p\,dx + q\,dy$ (with p and q measurable) on D. Since $\omega \in E^\perp \cap (E^*)^\perp$,

$$0 = (\omega, d\varphi) = \iint_D (p\varphi_x + q\varphi_y)\,dx \wedge dy, \tag{3.5.1}$$

$$0 = (\omega, {}^*d\psi) = \iint_D (-p\psi_y + q\psi_x)\,dx \wedge dy. \tag{3.5.2}$$

Thus

$$\begin{aligned} 0 = (\omega, d\varphi - {}^*d\psi) &= \iint p(\varphi_x + \psi_y)\,dx \wedge dy \\ &= \iint p\,\Delta\eta. \end{aligned} \tag{3.5.3}$$

By Weyl's lemma, p is harmonic and hence C^1. Applying this result to ${}^*\omega$ (which also belongs to $E^\perp \cap E^{*\perp}$) we see that q is C^1. Hence ω is of class C^1. Proposition II.3.2 now yields that ω is closed and co-closed (that is, harmonic). □

Remark. The space $L^2(M)$ can best be represented by the "three" dimensional "orthogonal" diagram given in Figure II.6.

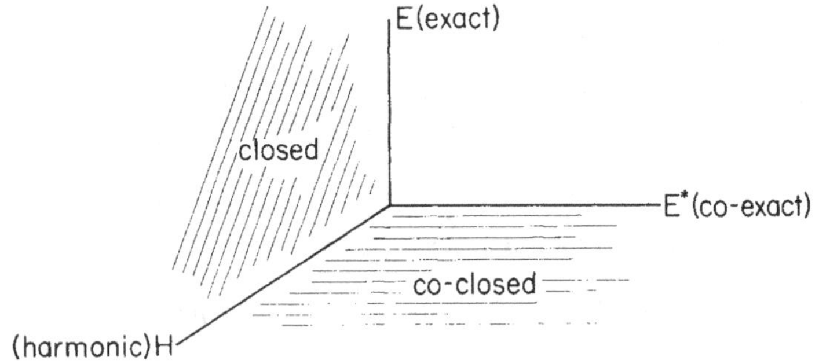

Figure II.6

Corollary

a. *The $L^2(M)$ closure of the square integrable closed (respectively, co-closed) differentials is $E \oplus H$ ($E^* \oplus H$).*
b. *The square integrable smooth differentials are dense in $L^2(M)$.*
b′. *The smooth differentials with compact support are dense in $L^2(M)$.*

The verification of (a) and (b) is at this stage trivial. For example, if ω is closed, then $\omega \in E^{*\perp} = E \oplus H$. Conversely, if $\omega \in E \oplus H$, then $\omega = \omega_1 + \omega_2$ with $\omega_1 \in E$ and $\omega_2 \in H$. Thus, ω_2 is closed (it is harmonic) and $\omega_1 = \lim_n df_n$ with f_n smooth of compact support. We have shown that ω is the limit of closed differentials.

We cannot at this point establish (b′). We need to know that

$$M = \bigcup_n D_n,$$

where

$$D_n \text{ is open,}$$
$$D_n \subset D_{n+1},$$

and

$$\text{Cl } D_n \text{ is compact}$$

(a fact that will follow from IV.5). Having such an "exhaustion" of M, we construct a sequence of smooth functions $\{f_n\}$ on M with

$$0 \le f_n \le 1,$$
$$f_n = 1 \quad \text{on } D_n,$$
$$(\text{support of } f_n) \subset D_{n+1}.$$

It clearly suffices to show that every $\alpha \in H$ can be approximated in $L^2(M)$ by smooth forms with compact support. Now $f_n\alpha \in L^2(M)$ is smooth and has compact support. By the Lebesgue dominated convergence theorem,

$$\lim_{n\to\infty} \|f_n\alpha - \alpha\| = 0.$$

Caution. *Not* every exact (co-exact) differential is in $E(E^*)$. For example: consider the unit disk D. Let f be a function holomorphic in a neighborhood of the closure of D. Then $df \in H$. Clearly, df is exact as well as harmonic.

For compact surfaces E does, of course, contain all the exact differentials. Thus, our picture is quite accurate in this case. (The above analysis shows that a compact surface carries no non-constant harmonic functions. The differential of such a function would have to be in both E and H, which would imply the function is constant. This fact can, of course, also be established as a consequence of the maximum and minimum principles for real valued harmonic functions.)

II.3.6. As a digression, we consider the situation of a compact Riemann surface M, and expand slightly our discussion of I.3.8. We define the *first* *de Rham cohomology group of M*, $H^1(M)$, as the smooth closed differentials modulo the smooth exact differentials. If α is a closed smooth differential, then its equivalence class in $H^1(M)$ is called *the cohomology class of* α. There is a pairing

$$H_1(M) \times H^1(M) \to \mathbb{C}$$

which maps the pair (c,α), where c is a closed piecewise differentiable path on M and α is a smooth 1-form, onto $\int_c \alpha$. It is quite clear that this integral depends only on the homology class of c and the cohomology class of α. It vanishes for a given cohomology class α over all curves c if and only if α is the zero class (represented by an exact differential). Thus the above pairing is non-singular. Further, every homomorphism of $H_1(M)$ into $\mathbb{C}$ is given by integration over curves against some $\alpha \in H^1(M)$. (See also III.2).

Note also that

$$H^1(M) \cong H, \tag{3.6.1}$$

and that the cohomology class of the 1-form η_c constructed in II.3.3 is uniquely determined by equation (3.3.1) and depends only on the homology class of c. The isomorphism of (3.6.1) is the surface version of the much more general Hodge theorem.

II.3.7. Theorem. *A sufficient condition for the existence of a non-zero harmonic differential on a Riemann surface M is the existence of a square integrable closed differential which is not exact. If M is compact, the condition is also necessary. Explicitly, for every closed* $\omega \in L^2(M)$, *there exists a* $\omega_2 \in H$ *such that* $\int_c \omega = \int_c \omega_2$ *for all closed curves c on M.*

Proof. Let $\omega \in L^2(M)$ be closed and not exact. Then $\omega \in (E^*)^{\perp} = E \oplus H$. Thus, $\omega = \omega_1 + \omega_2$ with $\omega_1 \in E$ and $\omega_2 \in H$. Note that ω_1 must be C^1. Since ω is not exact, there is a closed curve c with $\int_c \omega \neq 0$. Since $\int_c \omega_1 = 0$, we must have $\int_c \omega_2 = \int_c \omega \neq 0$. Hence $\omega_2 \neq 0$.

If M is compact, then there are no exact harmonic differentials so that any $0 \neq \omega \in H$ is closed and not exact. □

II.3.8. In view of Theorem II.3.7 we see that in order to construct harmonic differentials, we must construct closed differentials that are not exact.

Let c be a simple closed curve on the Riemann surface M that does not separate (that is, $M \backslash c$ is connected). We construct now a closed curve

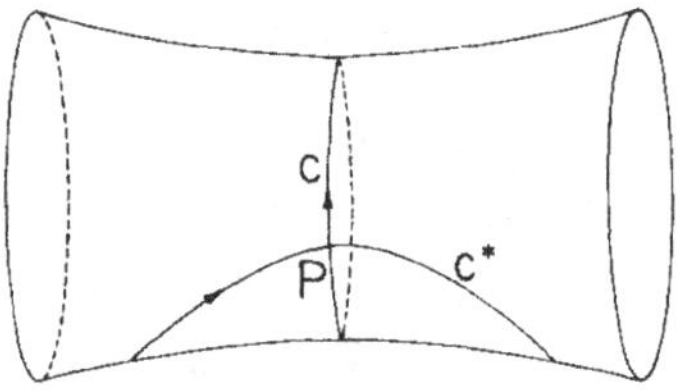

Figure II.7

c^* (a dual curve to c) by starting on the right(+) side of c and ending up on the other (= left (−)) side of c. (See Figure II.7).

The curve c^* will intersect c in exactly one point P. Using the construction and notation of II.3.3, we see that

$$\int_{c^*} \eta_C = \lim_{Q \to P^-} f(Q) - \lim_{Q \to P^+} f(Q) = 1$$

(here, of course, $\lim_{Q \to P^+}$ means you approach P through Ω^+). This remark together with Theorem II.3.7 establishes the following

Theorem. *If the Riemann surface M carries a closed curve that does not separate, then there exists a closed non-exact differential in $L^2(M)$ and therefore a non-zero harmonic differential.*

II.4. Harmonic Differentials

We have seen in the previous sections that on a compact surface there do not exist exact (non-zero) harmonic differentials. To construct "exact harmonic differentials" we must hence allow singularities. In this section we construct harmonic functions and differentials with a prescribed isolated singularity.

II.4.1. Let D be a parametric disc with local coordinate $z = x + iy$ on a Riemann surface M. It involves no loss of generality to assume that z maps D onto the open unit disc. We assume that $z = 0$ corresponds to the point $P_0 \in M$. We define a function h on M by choosing a real number a, $0 < a < 1$, an integer $n \geq 1$, and setting

$$h(P) = \begin{cases} z(P)^{-n} + \dfrac{\overline{z(P)}^n}{a^{2n}} & \text{if } P \in D \text{ and } |z(P)| \leq a, \\ 0 & \text{otherwise.} \end{cases}$$

(In the future we will identify the point $P \in D$ with its image $z(P) \in \mathbb{C}$ and hence write the above equation for h

$$h(z) = \begin{cases} z^{-n} + \dfrac{\bar{z}^n}{a^{2n}} & \text{for } |z| \leq a, \\ 0 & \text{for } |z| > a. \end{cases}$$

In this convention the points in $M \backslash D$ are denoted by $\{|z| \geq 1\}$.)

We define another function θ on M by setting

$$\theta(z) = h(z) \quad \text{for } |z| \geq a/2,$$

and requiring θ to be smooth in $\{|z| < a\}$.

We form the differential $d\theta \in L^2(M)$ (this terminology involves some abuse of notation since θ is C^∞ only on $M\backslash\{|z| = a\}$). It is smooth whenever θ is. Note that because of the discontinuity of θ, $d\theta$ does not have to belong to $E^{*\perp}$. Write

$$d\theta = \alpha + \omega, \tag{4.1.1}$$

with $\alpha \in E$ and $\omega \in E^\perp = E^* \oplus H$.

Let U be a parametric disk about the point $Q_0 \in M$. Let $\zeta = \xi + i\eta$ be a local coordinate vanishing at Q_0. We choose a real-valued smooth function ρ on M with support in U and set $\varphi = \partial\rho/\partial\xi$, $\psi = \partial\rho/\partial\eta$ (as in II.3.5). Write $\alpha = p\,d\xi + q\,d\eta$ in U. Since $\alpha \in E^{*\perp}$, obtain as in (3.5.2),

$$0 = (\alpha,{}^*d\psi) = \iint_U (-p\psi_\eta + q\psi_\xi)\,d\xi \wedge d\eta.$$

Also since $\omega \in E^\perp$, we obtain as in (3.5.1),

$$(d\theta,d\varphi) = (\alpha,d\varphi) = \iint_U (p\varphi_\xi + q\varphi_\eta)\,d\xi \wedge d\eta.$$

The difference of the last two expressions yields (as in (3.5.3))

$$(d\theta,d\varphi) = \iint_U p\,\Delta\rho. \tag{4.1.2}$$

Similarly, by interchanging the roles of φ and ψ and adding (rather than subtracting) the two resulting expressions, we obtain

$$(d\theta,d\psi) = \iint_U q\,\Delta\rho. \tag{4.1.3}$$

We know that $\alpha \in E^{*\perp}$. Thus α is annihilated by all co-exact differentials with compact support. Thus if α is C^1 on a given open subset of M, it must be a closed form on that subset by Proposition II.3.2. As an illustration we assume that Q_0 does not belong to $\{|z| \le a\} = \mathrm{Cl}\,D_a$, and we take U also to belong to the complement of $\mathrm{Cl}\,D_a$. In this case (since $d\theta$ vanishes on $\{|z| > a\}$)

$$(d\theta,d\psi) = 0 = (d\theta,d\varphi),$$

and we conclude from (4.1.2) and (4.1.3) and Weyl's lemma that p and q are harmonic (thus C^∞) in U. In particular, α is a smooth closed differential on $M\backslash\mathrm{Cl}\,D_a$. (The above was not necessary because of the stronger statement we are about to prove. We included it to help the reader.)

Our first step is to show that α is harmonic on $M\backslash\mathrm{Cl}\,D_{a/2}$. So we assume that $U \subset M\backslash\mathrm{Cl}\,D_{a/2}$. We compute (using (4.1.2))

$$\begin{aligned}\iint_U p\,\Delta\rho &= (d\theta,d\varphi)_M\\ &= (d\theta,d\varphi)_{D_{a/2}} + (d\theta,d\varphi)_{D_a\backslash D_{a/2}} + (d\theta,d\varphi)_{M\backslash D_a}.\end{aligned}$$

We have

$$\begin{aligned}(d\theta,d\varphi)_{D_{a/2}} &= 0, \quad \text{since } d\varphi = 0 \text{ on } D_{a/2},\\ (d\theta,d\varphi)_{M\backslash D_a} &= 0, \quad \text{since } d\theta = 0 \text{ on } M\backslash D_a.\end{aligned}$$

It remains to evaluate the middle integral. We will show that is it zero. If $U \subset M \backslash D_a$, the integral vanishes because $d\varphi = 0$ on D_a. It hence involves no loss of generality to assume that $U \subset D$ and that $z = \zeta$.

We recall the formula I(4.4.1)

$$(d\varphi, d\theta)_{D_a \backslash D_{a/2}} = \iint_{D_a \backslash D_{a/2}} d\varphi \wedge {}^*d\bar{\theta}$$

$$= -\iint_{D_a \backslash D_{a/2}} \varphi \, \Delta\bar{\theta} + \int_{\delta(D_a \backslash D_{a/2})} \varphi({}^*d\bar{\theta}).$$

Again, the double integral (on the right-hand side) vanishes because θ is harmonic on $D_a \backslash \mathrm{Cl}\, D_{a/2}$. Similarly, the line integral vanishes because $\varphi = 0$ on $\{|z| = a/2\}$, while by a simple calculation

$${}^*d\theta = \frac{\partial\theta}{\partial n}|dz| = \frac{\partial\theta}{\partial r}|dz|$$

(where $\partial\theta/\partial n$ is the normal derivative of θ and $\partial\theta/\partial r$ is the radial derivative of θ) vanishes on $\{|z| = a\}$. Thus we see that α is closed on $M \backslash \mathrm{Cl}\, D_{a/2}$.

We compute next for ρ with compact support inside $M \backslash \mathrm{Cl}\, D_{a/2}$:

$$(\alpha, d\rho) = (d\theta, d\rho) - (\omega, d\rho) = (d\theta, d\rho)$$

$$= (d\theta, d\rho)_{D_a \backslash D_{a/2}} = 0,$$

by the argument just used. Thus α is co-closed on $M \backslash \mathrm{Cl}\, D_{a/2}$ by Proposition II.3.2; that is, α is harmonic on $M \backslash \mathrm{Cl}\, D_{a/2}$.

Finally, what happens close to the singularity of h? Assume that the support of ρ is now contained inside D_a. We have

$$(d\theta, d\varphi) = \iint_{D_a} (\theta_x \rho_{xx} + \theta_y \rho_{xy})\, dx \wedge dy,$$

and

$$0 = (d\theta, {}^*d\psi) = -\iint_{D_a} (\theta_x \rho_{yy} - \theta_y \rho_{yx})\, dx \wedge dy.$$

The first equality on the last line follows from the fact that $d\theta$ is closed on D_a and hence $\in E^{*\perp}$ (with respect to D_a). As a consequence we obtain from the above two equations:

$$(d\theta, d\varphi) = \iint_{D_a} \theta_x(\rho_{xx} + \rho_{yy})\, dx \wedge dy = \iint_{D_a} \theta_x \, \Delta\rho.$$

It follows therefore from (4.1.2) that

$$0 = \iint_{D_a} (p - \theta_x)\, \Delta\rho.$$

Similarly, we can obtain

$$0 = \iint_{D_a} (q - \theta_y)\, \Delta\rho.$$

Thus, by Weyl's lemma, $p - \theta_x$ and $q - \theta_y$ are harmonic in D_a. In particular, p and q are smooth on D_a.

We have shown that $\alpha \in E = (E^* \oplus H)^\perp$ is smooth. Since $E^* \oplus H$ contains the co-closed differentials with compact support, α is exact by Proposition II.3.4. Hence there is a smooth function f on M with $df = \alpha$. Since α is harmonic on $M \backslash \text{Cl}\, D_{a/2}$, so is f.

By (4.1.1), $d(\theta - f) = \omega \in E^\perp$. Since $\theta - f$ is smooth on D_a, $d(\theta - f)$ is a co-closed differential on D_a. Since $d(\theta - f)$ is clearly closed on D_a, we conclude that $d(\theta - f)$ is a harmonic differential on D_a or that $\theta - f$ is harmonic in D_a. We now define a function u on M

$$u = f - \theta + h.$$

For $0 < |z| < a$, $f - \theta$ and h are harmonic. For $|z| > a/2$, f is harmonic and $h - \theta = 0$. Thus u is harmonic on $M \backslash \{P_0\}$.

We summarize our conclusions in the following

Theorem. *Let M be a Riemann surface with z a local coordinate vanishing at $P_0 \in M$. There exists on M a function u with the following properties:*

u is harmonic on $M \backslash \{P_0\}$, (4.1.4)

$u - z^{-n}$ is harmonic on every sufficiently small neighborhood N of P_0, (4.1.5)

$\iint_{M \backslash N} du \wedge {*\overline{du}} < \infty$, and (4.1.6)

$(du, df) = 0 = (du, {*df})$ for all smooth functions f on M that have compact support and vanish on a neighborhood of P_0. (4.1.7)

PROOF. Only (4.1.7) needs verification, and this follows because du is in H with respect to the surface $M \backslash \text{Cl}\, N$. □

II.4.2. Note that condition (4.1.5) is invariant under a limited class of coordinate changes. (Determine this class!) It can, of course, also be replaced by

$u - \operatorname{Re} z^{-n}$ *(resp., $u - \operatorname{Im} z^{-n}$) is harmonic in a neighborhood N of P_0.* (4.2.1)

In this case we may require u to be real valued.

Observe also that if M is compact, then u is unique up to an additive constant.

II.4.3. We note that most of our arguments to prove Theorem II.4.1 did not depend on the chioce of the particular form of the function h with singularity as long as

$$h \text{ is harmonic in } \{a/2 < |z| < a\},$$

and

$$*dh = 0 \text{ on } \{|z| = a\}.$$

Thus, another candidate for h is the function

$$h(z) = \log\left|\frac{z - z_1}{z - z_2}\,\frac{z - a^2/\bar{z}_1}{z - a^2/\bar{z}_2}\right|, \qquad |z| \le a,\ 0 < |z_1|, |z_2| < a/2.$$

Theorem. *Let M be a Riemann surface and P_1 and P_2 two points on M. Let z_j ($j = 1,2$) be a local coordinate vanishing at P_j. There exists on M a real-valued function u with the following properties:*

$$u \text{ is harmonic on } M\backslash\{P_1,P_2\}, \tag{4.3.1}$$

$$u - \log|z_1| \text{ is harmonic in a neighborhood of } P_1 \text{ and } u + \log|z_2| \text{ is harmonic in a neighborhood of } P_2, \tag{4.3.2}$$

$$\iint_{M\backslash N} du \wedge *\overline{du} < \infty, \text{ for every open set } N \text{ containing } P_1 \text{ and } P_2, \text{ and} \tag{4.3.3}$$

$$(du,df) = 0 = (du,*df) \text{ for all smooth functions } f \text{ on } M \text{ that have compact support and vanish on a neighborhood of } P_1 \text{ and } P_2. \tag{4.3.4}$$

Note that condition (4.3.2) is invariant under certain coordinate changes. See, for example, IV.3.6.

Proof of Theorem. The arguments preceding the statement of the theorem establish it for P_1 and P_2 sufficiently close. If P_1 and P_2 are arbitrary, we can join P_1 to P_2 by a chain $P_1 = Q_0, Q_1, \ldots, Q_n = P_2$ so that Q_j is close to $Q_{j-1}, j = 1, \ldots, n$. For each pair of points there is a function u_j with appropriate singularity at Q_{j-1} and Q_j. Let

$$u = u_1 + \cdots + u_n,$$

and note that u is regular at $Q_1, \ldots, Q_{n-1}$, and has appropriate (logarithmic) singularities at Q_0 and Q_n. □

Remark. For compact M, the function u is unique up to an additive constant.

II.4.4. We can also let

$$h(z) = \arg\left(\frac{z - z_1}{z - z_2}\frac{z - a^2/\bar{z}_2}{z - a^2/\bar{z}_1}\right), \qquad |z| \le a, 0 < |z_1|, |z_2| < a/2.$$

This case is slightly different from the previous one. The function h is not well defined on M. It is well defined on $M\backslash\{\text{slit joining } z_1 \text{ to } z_2\}$. Of course, the differential du obtained by this procedure is well defined. We leave it to the reader to formulate the analogue of Theorem II.4.3 in this case.

II.4.5. The decomposition (3.2.1) is closely related to the *Dirichlet Principle*, which we proceed to explain. Let D be a domain on a compact Riemann surface M_0 whose boundary δD consists of finitely many simple closed analytic arcs. Thus, topologically D is a compact Riemann surface of genus $g \ge 0$ from which $n > 0$ discs have been removed. Consider now two copies D and D' of D. We shall construct a compact Riemann surface $M = \text{Cl}\, D \cup D'$ known as *the double of D*. We use the usual local coordinates on D. A function z is a local coordinate at $P' \in D'$ if and only if $\bar{z}$ is a local coordinate at the corresponding point $P \in D$. We now identify each point $P \in \delta D$ with the corresponding point $P' \in \delta D'$. To construct local coordinates at points of δD, we map a neighborhood U of $P \in \delta D$ by a conformal mapping z into the closed upper half plane such that $U \cap \delta D$ goes into a segment of the real axis. By the reflection principle z is a local coordinate at $P \in M$. Note that M is a compact Riemann surface of genus $2g + n - 1$, and that there exists on M an anti-conformal involution j such that $j(D) = D'$ and $j(P) = P$ for all $P \in \delta D$. From now on we may forget about M_0 and think of D as a domain on its double M. (The above discussion is not strictly necessary for what follows. It was introduced for its own sake.) We consider now the following problem:

Fix a C^2 function φ_0 defined on a neighborhood of Cl D. *Among all functions φ, C^2 on a neighborhood of* Cl D, *find (if it exists) one u that has the same boundary values (on δD) as φ_0 and minimizes $\|d\varphi\|_D$ over this class.*

Assume that u is a harmonic function with the same boundary values as φ_0. We compute for arbitrary φ with the same boundary values;

$$\begin{aligned}\|d\varphi\|^2 &= (d(\varphi - u) + du, d(\varphi - u) + du)\\ &= \|d(\varphi - u)\|^2 + \|du\|^2 + 2\,\mathrm{Re}(d(\varphi - u), du).\end{aligned}$$

Now use Proposition I.4.4 to conclude that $(d(\varphi - u), du) = 0$ (since $\varphi - u = 0$ on δD and $\Delta u = 0$ on D). Thus

$$\|d\varphi\|^2 = \|d(\varphi - u)\|^2 + \|du\|^2 \geq \|du\|^2.$$

Thus our problem has a unique solution—if we can find a harmonic function with the same boundary values as φ_0. We shall in IV.3 solve this problem by Perron's method. Here we outline how the decomposition of $L^2(D)$ given by (3.2.1) can be used to solve our problem. Finding a function for which a given non-negative function on $L^2(D)$ achieves a minimum is known as the *Dirichlet Principle.*

Consider the function φ_0. Since $d\varphi_0$ is exact (and thus closed), $d\varphi_0 \in E \oplus H$. Now let ω be the orthogonal projection of $d\varphi_0$ onto H. We have already seen that ω is exact and harmonic. Thus $\omega = du$, for some harmonic function u on D. Now $d(u - \varphi_0) \in E$. Thus

$$(d(u - \varphi_0), \alpha) = 0, \quad \text{all } \alpha \in H.$$

Let $u_{P_1P_2}$ be the function produced by Theorem II.4.3. By letting α run over the set of differentials

$$\{du_{P_1P_2};\ P_1 \in D', P_2 \in D'\},$$

one can show that $u - \varphi_0$ is C^2 on a neighborhood of Cl D and that $u - \varphi_0 = 0$ on δD.

II.5. Meromorphic Functions and Differentials

Using the results of the previous section, we construct first meromorphic differentials on an arbitrary Riemann surface M and then (non-constant) meromorphic functions.

II.5.1. By a *meromorphic differential* on a Riemann surface we mean an assignment of a meromorphic function f to each local coordinate z such that

$$f(z)\,dz$$

is invariantly defined.

Theorem

a. *Let $P \in M$ and let z be a local coordinate on M vanishing at P. For every integer $n \geq 1$, there exists a meromorphic differential on M which is holomorphic on $M\backslash\{P\}$ and with singularity $1/z^{n+1}$ at P.*
b. *Given two distinct points P_1 and P_2 on M and local coordinates z_j vanishing at P_j, $j = 1, 2$, there exists a meromorphic differential ω, holomorphic on $M\backslash\{P_1,P_2\}$, with singularity $1/z_1$ at P_1 and singularity $-1/z_2$ at P_2.*

PROOF. Let $\alpha = du$ where u is the function whose existence is asserted by Theorem II.4.1 for part (a) and by Theorem II.4.3 for part (b). In the former case set

$$\omega = \frac{-1}{2n}(\alpha + i^*\alpha),$$

and in the latter

$$\omega = \alpha + i^*\alpha. \qquad \square$$

II.5.2. Let q be an integer. By a (*meromorphic*) *q-differential* ω on M we mean an assignment of a meromorphic function f to each local coordinate z on M so that

$$f(z)\,dz^q \tag{5.2.1}$$

is invariantly defined. For $q = 1$, these are just the meromorphic differentials previously considered, and they are called *abelian* differentials.

If ω is a q-differential on M, and z is a local coordinate vanishing at $P \in M$, and ω is given by (5.2.1) in terms of z, then we define the *order* of ω at P by

$$\operatorname{ord}_P \omega = \operatorname{ord}_0 f.$$

(If we write $f(z) = z^n g(z)$ with g holomorphic and non-zero at $z = 0$, then $\operatorname{ord}_0 f = n$.) It is an immediate consequence of the fact that local parameters are homeomorphisms that the order of a q-differential at a point is well defined.

Note that $\{P \in M;\ \operatorname{ord}_P \omega \neq 0\}$ is a discrete set on M; thus a finite set, if M is compact.

If ω is an abelian differential, then we define the *residue* of ω at P by

$$\operatorname{res}_P \omega = a_{-1},$$

where ω is given by (5.2.1) in terms of the local coordinate z that vanishes at P, and the Laurent series of f is

$$f(z) = \sum_{n=N}^{\infty} a_n z^n.$$

The residue is well-defined since

$$\operatorname{res}_P \omega = \frac{1}{2\pi i}\int_c \omega,$$

where c is a simple closed curve in M that bounds a disk D containing P such that c has winding number 1 about P and ω is holomorphic in $\operatorname{Cl} D\backslash\{P\}$.

II.5.3. Theorem. *Let $P_1, \ldots, P_k$ be $k > 1$ distinct points on a Riemann surface M. Let $c_1, \ldots, c_k$ be complex numbers with $\sum_{j=1}^k c_j = 0$. Then there exists a meromorphic abelian differential ω on M, holomorphic on $M\backslash\{P_1, \ldots, P_k\}$ with*

$$\operatorname{ord}_{P_j} \omega = -1, \qquad \operatorname{res}_{P_j} \omega = c_j.$$

PROOF. Let $P_0 \in M$, $P_0 \neq P_j$, $j = 1, \ldots, k$. Choose a local coordinate z_j vanishing at P_j, $j = 0, \ldots, k$. For $j = 1, \ldots, n$, let ω_j be an abelian differential with singularities $1/z_j$ at P_j and $-1/z_0$ at P_0 and no other singularities. Set

$$\omega = \sum_{j=1}^{k} c_j \omega_j. \qquad \square$$

Corollary. *Every Riemann surface M carries non-constant meromorphic functions.*

PROOF. Let P_1, P_2, P_3 be three distinct points on M. Let ω_1 be a differential holomorphic on $M\backslash\{P_1,P_2\}$ with

$$\operatorname{ord}_{P_1} \omega_1 = -1 = \operatorname{ord}_{P_2} \omega_1$$
$$\operatorname{res}_{P_1} \omega_1 = +1, \qquad \operatorname{res}_{P_2} \omega_1 = -1.$$

Let ω_2 be a differential holomorphic on $M\backslash\{P_2,P_3\}$ with

$$\operatorname{ord}_{P_2} \omega_2 = -1 = \operatorname{ord}_{P_3} \omega_2,$$
$$\operatorname{res}_{P_2} \omega_2 = 1, \qquad \operatorname{res}_{P_3} \omega_2 = -1.$$

Set $f = \omega_1/\omega_2$. Note that f has a pole at P_1 and a zero at P_3. $\square$

II.5.4. Proposition. *Let M be a compact Riemann surface and ω an abelian differential on M. Then*

$$\sum_{P \in M} \operatorname{res}_P \omega = 0. \tag{5.4.1}$$

PROOF. Triangulate M so that each singularity of ω is in the interior of one triangle. Let $\Delta_1, \Delta_2, \ldots, \Delta_k$ be an enumeration of the 2-simplices in the triangulation. Then

$$\sum_{P \in M} \operatorname{res}_P \omega = \frac{1}{2\pi i}\sum_{j=1}^{k}\int_{\delta\Delta_j} \omega, \tag{5.4.2}$$

where $\delta\Delta_j$ is the (positively oriented) boundary of Δ_j. Since each 1-simplex appears twice, with opposite signs, in the sum (5.4.2), we conclude that (5.4.1) holds. □

EXERCISE

Using only formal manipulations of power series show that the residue of a meromorphic abelian differential is well defined.

Remark. The above proposition shows that the sufficient condition in Theorem II.5.3 is also necessary if the surface M is compact.

EXERCISE

Give an alternate proof of Proposition I.1.6 in the special case that $N = \mathbb{C} \cup \{\infty\}$. First reduce to showing that it suffices to establish that f has as many poles as zeros. Then relate $\text{ord}_P f$ to $\text{res}_P(df/f)$.

II.5.5. Remark. The existence of meromorphic functions shows *at once* that every compact Riemann surface M is triangulable. Let

$$f: M \to \mathbb{C} \cup \{\infty\}$$

be a non-constant meromorphic function. Triangulate $\mathbb{C} \cup \{\infty\}$ with a triangulation $\Delta_1, \Delta_2, \ldots, \Delta_k$ such that the image under f of every ramified point (points $P \in M$ with $b_f(P) > 0$) is a vertex of the triangulation and such that f restricted to the interior of each component of $f^{-1}(\Delta_j)$ is injective. It is clear that this triangulation of $\mathbb{C} \cup \{\infty\}$ lifts to a triangulation of M.

The existence of meromorphic functions has many other important consequences. We will discuss these in IV.3 and IV.5. Also, in IV.3 we will establish the existence of meromorphic functions without relying on integration (for which triangulations are needed). Thus we will be able to derive the above topological facts from the complex structure on the Riemann surface.

CHAPTER III

Compact Riemann Surfaces

This is one of the two most important chapters of this book. In it, we prove (based on the existence theorems of the previous chapter) the three most important theorems concerning compact Riemann surfaces: the Riemann–Roch theorem, Abel's theorem, and the Jacobi inversion theorem. Many applications of these theorems are obtained; and the simplest compact Riemann surfaces, the hyperelliptic ones, are discussed in great detail.

III.1. Intersection Theory on Compact Surfaces

We have shown (in I.2) that a single non-negative integer (called the genus) yields a complete topological classification of compact Riemann surfaces. Every surface of genus 0 is topologically the sphere, while a surface of positive genus, g, can be obtained topologically by identifying in pairs appropriate sides of a $4g$-sided polygon. The reader should at this point review Figures I.3 and I.4 in I.2.5. Thus, with each surface of genus $g > 0$ we associate a $4g$-sided polygon with symbol $b_1 a_1 b_1^{-1} a_1^{-1} \cdots b_g a_g b_g^{-1} a_g^{-1}$ (as in I.2.5). The side of the polygon correspond to curves (homology classes) on the Riemann surface. These curves intersect as shown in the figures referred to above. Our aim is to make this vague statement precise. This involves the introduction of a cononical homology basis (basis for H_1) on M.

III.1.1. Let c be a simple closed curve on an arbitrary Riemann surface M. We have seen (II.3.3) that we may associate with c a (real) smooth closed differential η_c with compact support such that

$$\int_c \alpha = (\alpha, {}^*\eta_c) = -\iint_M \alpha \wedge \eta_c, \qquad (1.1.1)$$

for all closed differentials α. Since every cycle c on M is a finite sum of cycles corresponding to simple closed curves, we conclude that to each such c, we can associate a real closed differential η_c with compact support such that (1.1.1) holds.

Let a and b be two cycles on the Riemann surface M. We define the *intersection number of a and b* by

$$a \cdot b = \iint_M \eta_a \wedge \eta_b = (\eta_a, -{}^*\eta_b). \tag{1.1.2}$$

Proposition. *The intersection number is well defined and satisfies the following properties (here a, b, c are cycles on M):*

$$a \cdot b \text{ depends only on the homology classes of } a \text{ and } b, \tag{1.1.3}$$

$$a \cdot b = -b \cdot a, \tag{1.1.4}$$

$$(a + b) \cdot c = a \cdot c + b \cdot c, \tag{1.1.5}$$

and

$$a \cdot b \in \mathbb{Z}. \tag{1.1.6}$$

Furthermore, $a \cdot b$ "counts" the number of times a intersects b.

Proof. To show that (1.1.2) is well defined, choose η_a' and η_b' also satisfying (1.1.1). We must verify

$$\iint_M \eta_a' \wedge \eta_b' = \iint_M \eta_a \wedge \eta_b.$$

Note that while η_a and η_a' are only closed, their difference $\eta_a' - \eta_a = df$, where f is a C^∞ function which is constant on each connected component of the complement of a compact set on M. Thus it suffices to show

$$\iint_M df \wedge \eta_b = 0, \quad \text{all } f \text{ as above.}$$

Assume that M is compact. In this case, $\iint_M df \wedge \eta_b = -(df, {}^*\eta_b) = 0$, because $df \in E$ and ${}^*\eta_b \in E^\perp$ (by Proposition II.3.2). For the general case, assume that the support of η_b is contained in D, where D is a domain on M with smooth boundary and compact closure. Then

$$\begin{aligned}\iint_M df \wedge \eta_b &= \iint_M d(f\eta_b) = \iint_D d(f\eta_b)\\ &= \int_{\delta D} f\eta_b = 0\end{aligned}$$

because η_b vanishes on δD.

The verification of (1.1.3), (1.1.4), and (1.1.5) is straightforward. To check (1.1.6), it suffices to assume that a and b are simple closed curves. Therefore we have (as in II.3.3),

$$\begin{aligned}a \cdot b &= \iint_M \eta_a \wedge \eta_b = -\iint_M \eta_b \wedge \eta_a\\ &= (\eta_b, {}^*\eta_a) = \int_a \eta_b.\end{aligned}$$

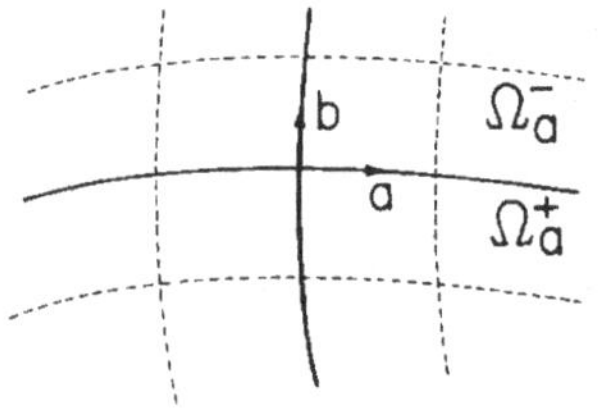

Figure III.1

But $\int_a \eta_b$ contributes ± 1 for each "intersection" of a with b. (This can be verified as in II.3.8 using Figure III.1.) □

III.1.2. We now consider a compact Riemann surface of genus $g \geq 0$, and represent this surface by its symbol

$$aa^{-1} \quad \text{(genus 0)},$$

$$\prod_{j=1}^{g} b_j a_j b_j^{-1} a_j^{-1} \quad \text{(genus } g \geq 1).$$

The sides of the polygon corresponding to the symbol give a basis for the homology, $H_1 = H_1(M)$, of M. Assume now that $g \geq 1$. It is easy to check that this basis has the following intersection properties

$$a_j \cdot b_k = \delta_{jk} = \begin{cases} 0 & j \neq k \\ 1 & j = k, \end{cases}$$

$$a_j \cdot a_k = 0 = b_j \cdot b_k.$$

We can represent all this information in an intersection matrix J. This J is a $2g \times 2g$ matrix of integers. If we label

$$\aleph_j = a_j, j = 1, \ldots, g \quad \text{and} \quad \aleph_j = b_{j-g}, j = g+1, \ldots, 2g, \qquad (1.2.1)$$

then the (j,k)-entry of J is the intersection number $\aleph_j \cdot \aleph_k$. Thus J is of the form

$$\begin{bmatrix} 0 & I \\ -I & 0 \end{bmatrix},$$

where 0 is the $g \times g$ zero matrix and I is the $g \times g$ identity matrix.

Any basis $\{\aleph_1, \ldots, \aleph_{2g}\}$ of H_1 with intersection matrix J will be called a *canonical homology basis* for M. Given a canonical homology basis we can use (1.2.1) to define the "a" and "b" curves. Note that we *do not* claim that these curves come from a polygon in normal form.

III.2 Harmonic and Analytic Differentials on Compact Surfaces

We compute the dimensions of the spaces of holomorphic and harmonic differentials on a compact Riemann surface. Certain period matrices are introduced and we determine some of their basic properties. The key tool is Theorem II.3.5.

III.2.1. Theorem. *On a compact Riemann surface M of genus g, the vector space H of harmonic differentials has dimension $2g$.*

PROOF. The theorem is easy to prove if $g = 0$. For in this case, let α be a harmonic differential on M. Fix $P_0 \in M$ and define

$$u(P) = \int_{P_0}^{P} \alpha, \qquad P \in M.$$

The function u is well defined (since M is simply connected). By the maximum principle for harmonic functions, u is constant. Thus $\alpha = 0$. (The maximum principle implies, of course, that there are no non-constant harmonic functions on any compact Riemann surface.)

Assume $g > 0$. Let $\{\aleph_1, \ldots, \aleph_{2g}\}$ be a canonical homology basis on M. Construct a map

$$\Phi: H \to \mathbb{C}^{2g} \quad \text{or} \quad H \to \mathbb{R}^{2g},$$

depending on whether we are interested in complex-valued harmonic differentials or real-valued harmonic differentials, by sending $\alpha \in H$ into the $2g$-tuple

$$\left(\int_{\aleph_1} \alpha, \ldots, \int_{\aleph_{2g}} \alpha\right).$$

If $\dim H > 2g$, then Φ has a non-trivial kernel; that is, there exists an $\alpha \in H$, all of whose periods are zero. Such an α must be the differential of a harmonic function. Since there are no non-constant harmonic functions on a compact surface, $\dim H \leq 2g$. The above argument also establishes the injectivity of Φ. It remains to verify surjectivity. As we saw in the previous chapter, (Theorem II.3.7) it suffices to find a closed differential with period 1 over a cycle $\aleph_j$ and period 0 over cycles $\aleph_k$, $k \neq j$.

Let

$$\alpha_j = \eta_{b_j}, \qquad j = 1, \ldots, g,$$
$$\alpha_j = -\eta_{a_{j-g}}, \qquad j = g + 1, \ldots, 2g.$$

Then we see (by(1.1.1)) that for $k = 1, \ldots, g$,

$$\int_{a_k} \alpha_j = -\iint_M \alpha_j \wedge \eta_{a_k}$$
$$= \iint_M \eta_{a_k} \wedge \alpha_j = \begin{cases} a_k \cdot b_j = \delta_{kj}, & j = 1, \ldots, g, \\ -(a_k \cdot a_{j-g}) = 0, & j = g + 1, \ldots, 2g, \end{cases}$$

and

$$\int_{b_k} \alpha_j = -\iint_M \alpha_j \wedge \eta_{b_k} = \iint_M \eta_{b_k} \wedge \alpha_j$$
$$= \begin{cases} b_k \cdot b_j = 0, & j = 1, \ldots, g, \\ -(b_k \cdot a_{j-g}) = a_{j-g} \cdot b_k = \delta_{j-g,k}, & j = g + 1, \ldots, 2g. \end{cases}$$

We summarize the above information in

$$\int_{a_k} \alpha_j = \begin{cases} 1, & k = j, \\ 0, & \text{otherwise}, \end{cases} \qquad \int_{b_k} \alpha_j = \begin{cases} 1, & k = j - g, \\ 0, & \text{otherwise}. \end{cases}$$

In other words,

$$\int_{\aleph_k} \alpha_j = \delta_{jk}, \qquad j, k = 1, \ldots, 2g. \tag{2.1.1}$$ □

Thus we have proven the following.

Corollary. *Given a canonical homology basis $\{\aleph_1, \ldots, \aleph_{2g}\}$ for $H_1(M)$ there is a unique dual basis $\{\alpha_1, \ldots, \alpha_{2g}\}$ of H; that is, a basis satisfying* (2.1.1). *Furthermore, each α_j is real.*

Thus given any set of $2g$ complex numbers $c_1, \ldots, c_{2g}$,

$$\alpha = \sum_{j=1}^{2g} c_j \alpha_j$$

is the unique harmonic differential whose $\aleph_k$ period (that is $\int_{\aleph_k} \alpha$) is c_k, for $k = 1, \ldots, 2g$.

III.2.2. We have seen that the $2g \times 2g$ matrix with (k,j)-entry $\int_{\aleph_j} \alpha_k$ is the identity. We note that

$$\int_{\aleph_j} \alpha_k = -\iint_M \alpha_k \wedge \eta_{\aleph_j} = \begin{cases} \iint_M \alpha_k \wedge \alpha_{j+g}, & j = 1, \ldots, g, \\ -\iint_M \alpha_k \wedge \alpha_{j-g}, & j = g+1, \ldots, 2g. \end{cases}$$

From this it is evident that the matrix whose (k,j)-entry is $\iint_M \alpha_k \wedge \alpha_j$ is of the form $\left[\begin{smallmatrix} 0 & I \\ -I & 0 \end{smallmatrix}\right] = J$. We conclude that

$$(\alpha_k, -{}^*\alpha_j) = \aleph_k \cdot \aleph_j.$$

We will investigate a companion matrix Γ whose (k,j)-entry is $(\alpha_k, \alpha_j) = \iint_M \alpha_k \wedge {}^*\alpha_j$. We note immediately that Γ is a real matrix. Further from I.4,

$$(\alpha_j, \alpha_k) = ({}^*\alpha_j, {}^*\alpha_k) = \iint_M {}^*\alpha_j \wedge {}^{**}\alpha_k = \iint_M \alpha_k \wedge {}^*\alpha_j = (\alpha_k, \alpha_j).$$

Thus, Γ is symmetric. Before continuing our investigation of the matrix Γ, we establish the following

III.2.3. Proposition. *If θ and $\tilde{\theta}$ are two closed differentials on M (compact of genus g), then*

$$\iint_M \theta \wedge \tilde{\theta} = \sum_{j=1}^{g} \left[\int_{a_j} \theta \int_{b_j} \tilde{\theta} - \int_{b_j} \theta \int_{a_j} \tilde{\theta} \right]. \tag{2.3.1}$$

PROOF. The right-hand side of (2.3.1) is obviously unchanged if we replace θ by $\theta + df$ with f a C^2-function. The left-hand side is also unchanged since

$$\iint_M (\theta + df) \wedge \tilde{\theta} = (\theta + df, -*\bar{\tilde{\theta}}) = (\theta, -*\bar{\tilde{\theta}}) = \iint_M \theta \wedge \tilde{\theta},$$

by Proposition II.3.2. Replacing thus θ and $\tilde{\theta}$ by harmonic differentials with the same periods, we may assume

$$\theta = \sum_{j=1}^{2g} \mu_j \alpha_j, \qquad \tilde{\theta} = \sum_{j=1}^{2g} \tilde{\mu}_j \alpha_j, \qquad \mu_j, \tilde{\mu}_j \in \mathbb{C}, \tag{2.3.2}$$

where $\{\alpha_j\}$ is the basis dual to the canonical homology basis $\{a_1, \ldots, a_g, b_1, \ldots, b_g\} = \{\aleph_1, \ldots, \aleph_{2g}\}$. Thus,

$$\iint_M \theta \wedge \tilde{\theta} = \sum_{k,j=1}^{2g} \mu_j \tilde{\mu}_k \iint_M \alpha_j \wedge \alpha_k = \sum_{k,j=1}^{2g} \mu_j \tilde{\mu}_k (\aleph_j \cdot \aleph_k)$$

$$= \sum_{j=1}^{g} \mu_j \tilde{\mu}_{g+j} (\aleph_j \cdot \aleph_{j+g}) + \sum_{j=g+1}^{2g} \mu_j \tilde{\mu}_{j-g} (\aleph_j \cdot \aleph_{j-g}). \tag{2.3.3}$$

Now it follows immediately from (2.3.2) that

$$\mu_j = \int_{\aleph_j} \theta \quad \text{and} \quad \tilde{\mu}_j = \int_{\aleph_j} \tilde{\theta}. \tag{2.3.4}$$

We now substitute (2.3.4) into (2.3.3) and obtain, using the fact that for $j = 1, \ldots, g$, $(\aleph_j \cdot \aleph_{g+j}) = 1$, and for $j = g + 1, \ldots, 2g$, $(\aleph_j \cdot \aleph_{j-g}) = -1$,

$$\iint_M \theta \wedge \tilde{\theta} = \sum_{j=1}^{g} \int_{\aleph_j} \theta \int_{\aleph_{g+j}} \tilde{\theta} - \sum_{j=g+1}^{2g} \int_{\aleph_j} \theta \int_{\aleph_{j-g}} \tilde{\theta}$$

$$= \sum_{j=1}^{g} \int_{a_j} \theta \int_{b_j} \tilde{\theta} - \int_{b_j} \theta \int_{a_j} \tilde{\theta}.$$

□

Corollary. *If θ is a harmonic* 1*-form on M, then*

$$\|\theta\|^2 = \sum_{j=1}^{g} \left[\int_{a_j} \theta \int_{b_j} *\bar{\theta} - \int_{b_j} \theta \int_{a_j} *\bar{\theta} \right]. \tag{2.3.5}$$

PROOF. Since θ is harmonic, $*\bar{\theta}$ is also closed. Thus, we compute

$$\|\theta\|^2 = \iint_M \theta \wedge *\bar{\theta}$$

by (2.3.1) and obtain (2.3.5). □

Remark. We may view the Riemann surface M as a polygon $\mathcal{M}$ whose symbol is $\prod_{j=1}^{g} a_j b_j a_j^{-1} b_j^{-1}$. Since $\mathcal{M}$ is simply connected $\theta = df$ on $\mathcal{M}$. Thus,

$$\iint_M \theta \wedge \tilde{\theta} = \iint_{\mathcal{M}} df \wedge \tilde{\theta} = \iint_{\mathcal{M}} d(f\tilde{\theta}) = \int_{\delta\mathcal{M}} f\tilde{\theta}$$

$$= \sum_{j=1}^{g} \left[\int_{a_j} f\tilde{\theta} + \int_{b_j} f\tilde{\theta} + \int_{a_j^{-1}} f\tilde{\theta} + \int_{b_j^{-1}} f\tilde{\theta} \right].$$

Let $z_0 \in \mathcal{M}$ be arbitrary, then for $z \in \mathcal{M}$,

$$f(z) = \int_{z_0}^{z} \theta.$$

Letting z and z' denote two equivalent points on the sides a_j and a_j^{-1} of $\delta\mathcal{M}$ we see that

$$\int_{a_j} f\tilde{\theta} + \int_{a_j^{-1}} f\tilde{\theta} = \int_{a_j} \left[\int_{z_0}^{z} \theta - \int_{z_0}^{z'} \theta\right]\tilde{\theta}$$

$$= \int_{a_j} -\left[\int_{b_j} \theta\right]\tilde{\theta} = -\int_{b_j} \theta \int_{a_j} \tilde{\theta}.$$

The remaining terms can be treated similarly to obtain an alternate proof of Formula (2.3.1).

III.2.4. We return to the matrix Γ and note that its (k,j)-entry γ_{kj} is given by

$$\gamma_{kj} = \iint_M \alpha_k \wedge {}^*\alpha_j = \sum_{l=1}^{g} \left[\int_{a_l} \alpha_k \int_{b_l} {}^*\alpha_j - \int_{b_l} \alpha_k \int_{a_l} {}^*\alpha_j\right]$$

$$= \begin{cases} \int_{b_k} {}^*\alpha_j = \int_{\aleph_{g+k}} {}^*\alpha_j, & k = 1, \dots, g, \\ -\int_{a_{k-g}} {}^*\alpha_j = -\int_{\aleph_{k-g}} {}^*\alpha_j, & k = g+1, \dots, 2g. \end{cases}$$

We show next that the real and symmetric matrix Γ is positive definite ($\Gamma > 0$). Let

$$\theta = \sum_{k=1}^{2g} \xi_k \alpha_k \quad \text{with } \xi_k \in \mathbb{C}, \ \sum_{k=1}^{2g} |\xi_k|^2 > 0.$$

Recall that the differentials α_k are real and that $\|\theta\| \neq 0$. Thus, by (2.3.5),

$$0 < \sum_{j=1}^{g} \left[\int_{a_j} \sum_{k=1}^{2g} \xi_k \alpha_k \int_{b_j} \sum_{l=1}^{2g} \bar{\xi}_l {}^*\alpha_l - \int_{b_j} \sum_{k=1}^{2g} \xi_k \alpha_k \int_{a_j} \sum_{l=1}^{2g} \bar{\xi}_l {}^*\alpha_l\right]$$

$$= \sum_{k,l=1}^{2g} \xi_k \bar{\xi}_l \sum_{j=1}^{g} \left[\int_{a_j} \alpha_k \int_{b_j} {}^*\alpha_l - \int_{b_j} \alpha_k \int_{a_j} {}^*\alpha_l\right]$$

$$= \sum_{k,l=1}^{2g} \xi_k \bar{\xi}_l \gamma_{kl}.$$

It is now convenient to write

$$\Gamma = \begin{bmatrix} A & B \\ C & D \end{bmatrix}$$

with A, B, C, D, $g \times g$ matrices. Note that we have established (since ${}^t\Gamma = \Gamma$ and $\Gamma > 0$) these are real with

$$B = {}^tC, \qquad A = {}^tA, \qquad D = {}^tD, \tag{2.4.1}$$

and

$$A > 0, \qquad D > 0. \tag{2.4.2}$$

III.2.5. Let us consider * as an operator on the space of complex-valued harmonic differentials. It is clearly $\mathbb{C}$-linear and $*^2 = -I$. We represent * by a $2g \times 2g$ real (since * preserves the space of real harmonic forms) matrix $\mathscr{G}$ with respect to the basis $\alpha_1, \ldots, \alpha_{2g}$:

$$\mathscr{G} = (\lambda_{kj}), \qquad k, j = 1, \ldots, 2g,$$

Thus

$$*\alpha_k = \sum_{j=1}^{2g} \lambda_{kj}\alpha_j, \qquad k = 1, \ldots, 2g.$$

If we represent by $\mathscr{A}$ the column vector of the basis elements $\alpha_1, \ldots, \alpha_{2g}$, then we can consider $\mathscr{G}$ as defined by the equation

$$*\mathscr{A} = \mathscr{G}\mathscr{A}.$$

Since $*^2 = -I$, $\mathscr{G}^2 = -I$. We wish to compute the matrix $\mathscr{G}$. We note that

$$\gamma_{lk} = (\alpha_l,\alpha_k) = (*\alpha_l,*\alpha_k) = \left(\sum_{j=1}^{2g} \lambda_{lj}\alpha_j, *\alpha_k\right)$$

$$= \sum_{j=1}^{2g} \lambda_{lj}(\alpha_j,*\alpha_k) = \sum_{j=1}^{2g} \lambda_{lj} \iint_M \alpha_k \wedge \alpha_j.$$

In III.2.2 we saw that, the matrix with (k,j)-entry $\iint_M \alpha_k \wedge \alpha_j$ is given by $J = \left[\begin{smallmatrix} 0 & I \\ -I & 0 \end{smallmatrix}\right]$. It therefore follows that the above equation can be written as

$$\Gamma = \mathscr{G}\,{}^tJ. \tag{2.5.1}$$

If we now write

$$\mathscr{G} = \begin{bmatrix} \lambda_1 & \lambda_2 \\ \lambda_3 & \lambda_4 \end{bmatrix},$$

we find as a consequence of (2.5.1) that

$$\Gamma = \begin{bmatrix} \lambda_2 & -\lambda_1 \\ \lambda_4 & -\lambda_3 \end{bmatrix}.$$

We therefore conclude, because of (2.4.1) and (2.4.2), that

$$\lambda_4 = -{}^t\lambda_1, \qquad \lambda_2 = {}^t\lambda_2, \qquad \lambda_3 = {}^t\lambda_3, \tag{2.5.2}$$

and

$$\lambda_2 > 0, \qquad -\lambda_3 > 0. \tag{2.5.3}$$

Since $\mathscr{G}^2 + I_{2g} = 0$, we see that $\mathscr{G}$ satisfies the additional equations:

$$\lambda_1^2 + \lambda_2\lambda_3 + I_g = 0, \qquad \lambda_1\lambda_2 = \lambda_2{}^t\lambda_1, \qquad \lambda_3\lambda_1 = {}^t\lambda_1\lambda_3. \tag{2.5.4}$$

III.2.6. Up to now we have essentially used only the space of real-valued harmonic forms. (A basis over $\mathbb{R}$ for the space of real-valued harmonic forms is also a basis over $\mathbb{C}$ for the space of complex-valued harmonic forms.) We construct the holomorphic differentials

$$\omega_j = \alpha_j + i*\alpha_j, \qquad j = 1, \ldots, 2g,$$

and a matrix whose (k,j)-entry is

$$\begin{aligned}\tfrac{1}{2}(\omega_k,\omega_j) &= \tfrac{1}{2}(\alpha_k,\alpha_j) + \tfrac{1}{2}(\alpha_k,i^*\alpha_j) + \tfrac{1}{2}(i^*\alpha_k,\alpha_j) + \tfrac{1}{2}(i^*\alpha_k,i^*\alpha_j)\\ &= (\alpha_k,\alpha_j) - i(\alpha_k,{}^*\alpha_j) = (\alpha_k,\alpha_j + i^*\alpha_j) = (\alpha_k,\omega_j).\end{aligned}$$

We see from the above that this is the matrix

$$\Gamma + iJ,$$

and since (recall I.3.11)

$$\begin{aligned}(\alpha_k,\omega_j) &= i\iint_M \alpha_k \wedge \bar{\omega}_j = i\sum_{l=1}^{g}\left[\int_{a_l}\alpha_k\int_{b_l}\bar{\omega}_j - \int_{b_l}\alpha_k\int_{a_l}\bar{\omega}_j\right]\\ &= \begin{cases} i\displaystyle\int_{b_k}\bar{\omega}_j, & k = 1, \ldots, g,\\ -i\displaystyle\int_{a_{k-g}}\bar{\omega}_j, & k = g+1, \ldots, 2g,\end{cases}\end{aligned}$$

this matrix can be viewed as a period matrix.

Observe also that

$$\begin{aligned}\tfrac{1}{2}(\omega_k,\omega_j) &= \tfrac{1}{2}\overline{(\omega_j,\omega_k)} = \overline{(\alpha_j,\omega_k)}\\ &= \begin{cases} -i\displaystyle\int_{b_j}\omega_k, & j = 1, \ldots, g,\\ i\displaystyle\int_{a_{j-g}}\omega_k, & j = g+1, \ldots, 2g.\end{cases}\end{aligned} \tag{2.6.1}$$

Before continuing our investigation of the matrix $\Gamma + iJ$, we establish

III.2.7. Proposition. *On a compact Riemann surface of genus g, the vector space $\mathscr{H} = \mathscr{H}^1(M)$ of holomorphic differentials has dimension g. Furthermore, $\{\omega_1, \ldots ,\omega_g\}$ forms a basis for $\mathscr{H}$.*

Proof. We show that we have a direct sum decomposition of the space H of complex-valued harmonic forms

$$H = \mathscr{H} \oplus \bar{\mathscr{H}}, \tag{2.7.1}$$

where $\bar{\mathscr{H}}$ represents the anti-holomorphic (= complex conjugates of holomorphic) differentials. It is obvious that $\mathscr{H} \cap \bar{\mathscr{H}} = \{0\}$. It remains to verify that the decomposition (2.7.1) is possible. If $\alpha \in H$, then $\alpha + i^*\alpha \in \mathscr{H}$, $\alpha - i^*\alpha \in \bar{\mathscr{H}}$ (since $\bar{\alpha} + i^*\bar{\alpha} \in \mathscr{H}$), and $\alpha = \frac{1}{2}(\alpha + i^*\alpha) + \frac{1}{2}(\alpha - i^*\alpha)$.

Since $\omega \mapsto \bar{\omega}$ is an $\mathbb{R}$-linear isomorphism of $\mathscr{H}$ onto $\bar{\mathscr{H}}$, it follows from (2.7.1) that

$$\dim_{\mathbb{C}} \mathscr{H} = \frac{1}{2}\dim_{\mathbb{R}} \mathscr{H} = \frac{1}{4}\dim_{\mathbb{R}} H = g.$$

To show that $\{\omega_1, \ldots ,\omega_g\}$ form a basis for $\mathscr{H}$ it suffices to show that this is a linearly independent set. Let

$$\begin{gathered}{}^tC = (c_1, \ldots ,c_g), \text{ with } c_j \in \mathbb{C},\ A = \operatorname{Re} C,\ B = \operatorname{Im} C,\\ {}^t\Omega = (\omega_1, \ldots ,\omega_g),\\ {}^t\mathfrak{A}_1 = (\alpha_1, \ldots ,\alpha_g), \qquad {}^t\mathfrak{A}_2 = (\alpha_{g+1}, \ldots ,\alpha_{2g}).\end{gathered}$$

Assume (recall that λ_1 and λ_2 have been defined in III.2.5)

$$\begin{aligned} 0 = {}^tC\Omega &= ({}^tA + i{}^tB)(\mathfrak{A}_1 + i{}^*\mathfrak{A}_1) = ({}^tA + i{}^tB)[\mathfrak{A}_1 + i(\lambda_1\mathfrak{A}_1 + \lambda_2\mathfrak{A}_2)] \\ &= {}^tA\mathfrak{A}_1 - {}^tB(\lambda_1\mathfrak{A}_1 + \lambda_2\mathfrak{A}_2) + i[{}^tB\mathfrak{A}_1 + {}^tA(\lambda_1\mathfrak{A}_1 + \lambda_2\mathfrak{A}_2)]. \end{aligned}$$

Thus, we obtain two equations:

$$({}^tA - {}^tB\lambda_1)\mathfrak{A}_1 = {}^tB\lambda_2\mathfrak{A}_2, \qquad ({}^tB + {}^tA\lambda_1)\mathfrak{A}_1 = -{}^tA\lambda_2\mathfrak{A}_2.$$

Since the differentials in $\mathfrak{A}_1$ are linearly independent from the differentials in $\mathfrak{A}_2$, we conclude that

$${}^tB\lambda_2 = 0 = {}^tA\lambda_2.$$

Since λ_2 is non-singular,

$${}^tB = 0 = {}^tA. \qquad \square$$

III.2.8. We return now to the matrix $\Gamma + iJ$. In particular, we restrict our attention to the first g rows of this matrix; that is, the g by $2g$ matrix

$$(\lambda_2, -\lambda_1 + iI).$$

We have shown by (2.6.1) that the (j,k)-entry of λ_2 is $-i\int_{b_k}\omega_j$; and the (j,k)-entry of $-\lambda_1 + iI$ is $i\int_{a_k}\omega_j$.

We conclude that if we consider $i\omega_1, \ldots, i\omega_g$ as a basis for the vector space $\mathscr{H}$, of holomorphic differentials on M, then the period matrix (whose (j,k)-entry is $\int_{\aleph_k} i\omega_j$) with respect to this basis is

$$(-\lambda_1 + iI, -\lambda_2).$$

Similarly, the holomorphic differentials $\omega_{g+1}, \ldots, \omega_{2g}$ are linearly independent (over $\mathbb{C}$), and the last g rows of the matrix $\Gamma + iJ$ is the $g \times 2g$ matrix

$$(-{}^t\lambda_1 - iI, -\lambda_3).$$

Thus, the period matrix of $i\omega_{g+1}, \ldots, i\omega_{2g}$ (whose (j,k)-entry is $\int_{\aleph_k} i\omega_{j+g}$) is of the form

$$(-\lambda_3, {}^t\lambda_1 + iI).$$

We now make another change of basis. Let

$$\Xi = {}^t(\zeta_1, \ldots, \zeta_g) = (-\lambda_3)^{-1}\,{}^t(i\omega_{g+1}, \ldots, i\omega_{2g}).$$

With respect to the basis Ξ of $\mathscr{H}$, we obtain the period matrix

$$(I, (-\lambda_3)^{-1}\,{}^t\lambda_1 + i(-\lambda_3)^{-1}) = (I, \Pi). \tag{2.8.1}$$

Proposition. *There exists a unique basis $\{\zeta_1, \ldots, \zeta_g\}$ for the space of holomorphic abelian differentials (= space $\mathscr{H}$) with the property $\int_{a_j}\zeta_k = \delta_{jk}$. Furthermore, for this basis, the matrix $\Pi = (\pi_{jk})$ with $\pi_{jk} = \int_{b_j}\zeta_k$ is symmetric with positive definite imaginary part.*

PROOF. We must only verify that Π is symmetric and $\operatorname{Im} \Pi > 0$. To show that Π is symmetric it suffices (because ${}^t\lambda_3 = \lambda_3$) to show that $(-\lambda_3)^{-1}\,{}^t\lambda_1$ is symmetric. But (recall (2.5.4))

$$ {}^t[(-\lambda_3)^{-1}\,{}^t\lambda_1] = \lambda_1(-{}^t\lambda_3)^{-1} = \lambda_1(-\lambda_3)^{-1} = (-\lambda_3)^{-1}\,{}^t\lambda_1. $$

Note that since $\operatorname{Im} \Pi = (-\lambda_3)^{-1}$, positive definiteness is not an issue. □

Remark. Note that $(-\lambda_3) = I$ if and only if for $k, j = g+1, \ldots, 2g$,

$$ \frac{1}{2}(\omega_k, \omega_j) = i \int_{a_{j-g}} \omega_k = \begin{cases} 1, & k = j \\ 0, & k \neq j \end{cases}, $$

if and only if $(\omega_{g+1}/\sqrt{2}, \ldots, \omega_{2g}/\sqrt{2})$ is an orthonormal basis for $\mathscr{H}$ viewed as a Hilbert subspace of $L^2(M)$. Can this happen? Yes, if and only if $\operatorname{Im} \Pi = I$ (see (2.8.1)).

III.3. Bilinear Relations

We start with a compact Riemann surface M of positive genus $g > 0$, and a canonical homology basis $\{a_1, \ldots, a_g, b_1, \ldots, b_g\}$ on M. Let $\{\zeta_1, \ldots, \zeta_g\}$ be the dual basis for holomorphic differentials (that is, $\int_{a_k} \zeta_j = \delta_{jk}$). Represent the Riemann surface M by a $4g$-sided polygon $\mathscr{M}$ with identification. Our starting point is Formula (2.3.1) for the "inner product" of closed differentials. Let θ and $\tilde{\theta}$ be closed differentials. Since $\mathscr{M}$ is simply connected, $\theta = df$ on $\mathscr{M}$ with f a smooth function on $\mathscr{M}$ (note that at equivalent points on $\delta\mathscr{M}$, f need not take on the same value). As we have seen in the remark in III.2.3, formula (2.3.1), may be viewed as a consequence of two identities:

$$ \iint_M \theta \wedge \tilde{\theta} = \int_{\delta\mathscr{M}} f\tilde{\theta}, \tag{3.0.1} $$

and

$$ \int_{\delta\mathscr{M}} f\tilde{\theta} = \sum_{l=1}^{g} \left[\int_{a_l} \theta \int_{b_l} \tilde{\theta} - \int_{b_l} \theta \int_{a_l} \tilde{\theta} \right]. \tag{3.0.2} $$

We shall see in this section that normalizing a set of meromorphic differentials (with or without singularities) forces certain identities between their periods. These are the "bilinear relations" of Riemann. They will turn out to be useful in the study of meromorphic functions on M.

III.3.1. Let us assume that θ and $\tilde{\theta}$ are holomorphic differentials, then

$$ 0 = \sum_{l=1}^{g} \left[\int_{a_l} \theta \int_{b_l} \tilde{\theta} - \int_{b_l} \theta \int_{a_l} \tilde{\theta} \right]. \tag{3.1.1} $$

Equation (3.1.1) follows from (3.0.2) by Cauchy's theorem, or from (2.3.1) since $\theta \wedge \tilde{\theta} = 0$ for holomorphic θ and $\tilde{\theta}$. We now let $\theta = \zeta_j$ and $\tilde{\theta} = \zeta_k$, and obtain

$$\int_{b_j} \zeta_k = \int_{b_k} \zeta_j. \tag{3.1.2}$$

Of course, (3.1.2) is just another way of saying that the matrix Π introduced in III.2.8 is symmetric.

III.3.2. Next we let $\theta = \zeta_j$ and $\tilde{\theta} = \bar{\zeta}_k$. We note that ($df = \zeta_j$ on $\mathscr{M}$) by Stokes' theorem (Formula (3.0.1))

$$\int_{\delta \mathscr{M}} f\bar{\zeta}_k = \iint_{\mathscr{M}} \zeta_j \wedge \bar{\zeta}_k.$$

We conclude that

$$\begin{aligned} -i(\zeta_j,\zeta_k) &= \int_{b_j} \zeta_k - \overline{\int_{b_k} \zeta_j} \\ &= -2i(\operatorname{Im} \pi_{jk}) \end{aligned}$$

(as usual π_{jk} is the (j,k)-entry of the matrix Π). In particular,

$$\operatorname{Im} \pi_{jj} > 0.$$

Applying the same argument to

$$\theta = \sum_{k=1}^{g} c_k \zeta_k, \qquad c_k \in \mathbb{C},$$

we conclude that

$$\operatorname{Im} \Pi > 0.$$

These facts have already been established in III.2.

III.3.3. Proposition. *Let θ be a holomorphic differential. Assume either*

a. *all the "a" periods of θ are zero (that is, $\int_{a_j} \theta = 0, j = 1, \ldots, g$), or*
b. *all the periods of θ are real.*
Then $\theta = 0$.

PROOF. We compute

$$\begin{aligned} \|\theta\|^2 &= \iint_M \theta \wedge {}^*\bar{\theta} = i \iint_M \theta \wedge \bar{\theta} \\ &= i \sum_{l=1}^{g} \left[\int_{a_l} \theta \overline{\int_{b_l} \theta} - \int_{b_l} \theta \overline{\int_{a_l} \theta} \right]. \end{aligned}$$

In either case (a) or (b) we conclude that $\|\theta\|^2 = 0$, and hence $\theta = 0$. □

Remark. The above observation (plus the fact that $\dim \mathscr{H} = g$, where $\mathscr{H} = \mathscr{H}^1(M)$) can be used to give an alternate proof of the theorem that there is a basis for $\mathscr{H}$ dual to a specific canonical homology basis. (The reader is invited to do so!)

III.3.4. We shall now adopt the following terminology: Recall that meromorphic one-forms are called *abelian* differentials. The abelian differentials which are holomorphic will be called of the *first kind*; while the meromorphic abelian differentials with zero residues will be called of the *second kind.* Finally, a general abelian differential (which may have residues) will be called of the *third kind.* Before proceeding let us record the following consequence of the previous proposition.

Corollary. *We can prescribe uniquely either*
a. *the "a" periods or*
b. *the real parts of all the periods*
of an abelian differential of the first kind.

Proof. Consider the maps $\mathscr{H} \to \mathbb{C}^g$ and $\mathscr{H} \to \mathbb{R}^{2g}$ defined by $\varphi \mapsto (\int_{a_1} \varphi, \ldots, \int_{a_g} \varphi)$ and $\varphi \mapsto (\operatorname{Im} \int_{a_1} \varphi, \ldots, \operatorname{Im} \int_{a_g} \varphi, \operatorname{Im} \int_{b_1} \varphi, \ldots, \operatorname{Im} \int_{b_g} \varphi)$, respectively. The map $\mathscr{H} \to \mathbb{C}^g$ is a linear transformation of $\mathscr{H}$ viewed as a g-dimensional vector space over $\mathbb{C}$, and the map $\mathscr{H} \to \mathbb{R}^{2g}$ is a linear transformation of $\mathscr{H}$ viewed as a $2g$-dimensional vector space over $\mathbb{R}$. The proposition tells us that these maps have trivial kernels and thus are isomorphisms (since the domains and targets have the same dimensions). □

Remark. Parts (a) of the proposition and its corollary have previously been established (Proposition III.2.8).

III.3.5. Let us consider abelian differentials of the third kind on M. Choose two points P and Q on M. It involves no loss of generality to assume that the canonical homology basis has representatives that do not contain the points P and Q. Let us consider a differential τ, regular (= holomorphic) on $M \backslash \{P, Q\}$, with

$$\begin{aligned} &\operatorname{ord}_P \tau = -1 = \operatorname{ord}_Q \tau, \\ &\operatorname{res}_P \tau = 1, \qquad \operatorname{res}_Q \tau = -1. \end{aligned} \tag{3.5.1}$$

Let c be a closed curve on M. It is, of course, no longer true that $\int_c \tau$ depends only on the homology class of the curve c. However (assuming P, Q are not on the curves in question) if c and c' are homologous, then there is an integer n such that

$$\int_c \tau - \int_{c'} \tau = 2\pi i n. \tag{3.5.2}$$

The easiest way to see (3.5.2) is as follows: Let θ be an arbitrary abelian differential of the third kind on M. Let $P_1, \ldots, P_k$ $(k \geq 1)$ be the singularities of θ. Assume that the canonical homology basis for M does not contain any of the points P_j. Let c_j, $j = 1, \ldots, k$, be a small circle about P_j. We may assume that the curves

$$a_1, \ldots, a_g, \qquad b_1, \ldots, b_g, \qquad c_1, \ldots, c_k$$

are mutually disjoint except that a_j and b_j must cross for $j = 1, \ldots, g$. It is easy to see that on $M' = M \backslash \{P_1, \ldots, P_k\}$, c_k is homotopic to

$$\prod_{j=1}^{g} b_j a_j b_j^{-1} a_j^{-1} \prod_{j=1}^{k-1} c_j,$$

and that the curves

$$a_1, \ldots, a_g, \qquad b_1, \ldots, b_g, \qquad c_1, \ldots, c_{k-1}$$

form a basis for $H_1(M')$. Hence, if c is a curve on M which is homologous to zero, then on M' it is homologous to a linear combination of $c_1, \ldots, c_{k-1}$. Thus there are integers n_j with c homologous to $\sum_{j=1}^{k-1} \eta_j c_j$. The differential θ is closed on M'. Hence

$$\int_c \theta = \sum_{j=1}^{k-1} n_j \int_{c_j} \theta = 2\pi i \sum_{j=1}^{k-1} n_j \operatorname{res}_{P_j} \theta. \tag{3.5.3}$$

Clearly (3.5.2) is a special case of (3.5.3).

In particular, for τ as before, $\int_{\aleph_j} \tau$ is defined only modulo $2\pi i\mathbb{Z}$ (hereafter, mod $2\pi i$).

III.3.6. To get around the above ambiguity, we consider M as represented by the polygon $\mathscr{M}$ with identifications. We choose two points P and Q in the interior of $\mathscr{M}$. Let τ be a differential of the third kind satisfying (3.5.1). By subtracting an abelian differential of the first kind, we normalize τ so that

$$\int_{a_j} \tau = 0, \qquad j = 1, \ldots, g, \tag{3.6.1}$$

or

$$\int_{\aleph_j} \tau \text{ is purely imaginary}, \qquad j = 1, \ldots, 2g. \tag{3.6.2}$$

We denote the (unique) differential τ with the first normalization by τ_{PQ} and the one with the second normalization by ω_{PQ}.

Warning. In (3.6.1) we think of a_j as a definite curve—not its homology class. If we want to think of it as a homology class, (3.6.1) must be replaced by

$$\int_{a_j} \tau = 0 \ (\mathrm{mod}\ 2\pi i), \qquad j = 1, \ldots, g, \tag{3.6.1a}$$

and τ_{PQ} is no longer unique.

Applying (3.0.2) with $\theta = \zeta_j$ and $\tilde{\theta} = \tau_{PQ}$, we obtain

$$\int_{\delta\mathscr{M}} f\tau_{PQ} = \int_{b_j} \tau_{PQ}.$$

The line integral around the boundary of $\mathscr{M}$ can be evaluated by the residue theorem, since

$$f(z) = \int_{z_0}^{z} \zeta_j$$

with the path of integration staying inside $\mathscr{M}$. Thus we obtain

$$\int_{\delta\mathscr{M}} f\tau_{PQ} = 2\pi i(f(P) - f(Q)),$$

or

$$2\pi i \int_Q^P \zeta_j = \int_{b_j} \tau_{PQ}, \tag{3.6.3}$$

as long as we integrate ζ_j from Q to P along a path lying in $\mathscr{M}$.

Similarly,

$$2\pi i \int_Q^P \zeta_j = \int_{b_j} \omega_{PQ} - \sum_{l=1}^{g} \pi_{jl} \int_{a_l} \omega_{PQ},$$

where π_{jl} is the (j,l)-entry of the period matrix Π.

III.3.7. We treat now the case where

$$\theta = \tau_{PQ}, \qquad \tilde{\theta} = \tau_{RS}$$

(P, Q, R, S are all interior points of $\mathcal{M}$). Here we cannot assert that $\theta = df$ on $\mathcal{M}$. To get around this little obstruction, we cut $\mathcal{M}$ by joining a point 0 on $\delta\mathcal{M}$ to P by one curve and to Q by another curve. We obtain this way a simply connected region $\mathcal{M}'$ (see Figure III.2).

In $\mathcal{M}'$, $\theta = df$ where

$$f(z) = \int_{z_0}^{z} \theta.$$

Now a simple calculation yields

$$\begin{aligned}\int_{\delta\mathcal{M}'} f\tilde{\theta} &= 2\pi i[\operatorname{res}_R f\tilde{\theta} + \operatorname{res}_S f\tilde{\theta}]\\ &= 2\pi i(f(R) - f(S))\\ &= 2\pi i \int_S^R \theta.\end{aligned}$$

The formula analogous to (3.0.2) is

$$\int_{\delta\mathcal{M}'} f\tilde{\theta} = \sum_{l=1}^{g} \left[\int_{a_l} \theta \int_{b_l} \tilde{\theta} - \int_{b_l} \theta \int_{a_l} \tilde{\theta}\right] + \int_c f\tilde{\theta},$$

where c is a curve from 0 to Q back to 0 ("on the other side") to P and back to 0. Now the value of f on the $+$ side of c differs from the value of f on the $-$ side by $2\pi i$ (by the residue theorem).

Thus

$$\int_c f\tilde{\theta} = 2\pi i\left[\int_0^P \tilde{\theta} - \int_0^Q \tilde{\theta}\right] = 2\pi i \int_Q^P \tilde{\theta}.$$

We summarize our result in

$$\int_S^R \tau_{PQ} = \int_Q^P \tau_{RS}$$

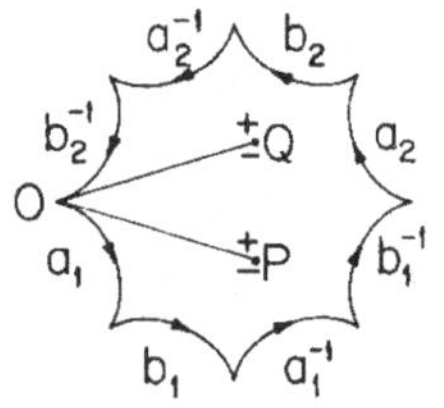

Figure III.2. Illustration for genus 2.

(where as before each path of integration is restricted to lie in $\mathcal{M}'' = \mathcal{M}' \backslash$ {lines joining R and S to 0}). For the differential ω_{PQ}, a similar formula can be derived; namely

$$\operatorname{Re} \int_S^R \omega_{PQ} = \operatorname{Re} \int_Q^P \omega_{RS}.$$

III.3.8. Let $P \in M$ and choose a local coordinate z vanishing at P. We have seen that there exists on $M \backslash \{P\}$ a holomorphic differential $\tilde{\theta}$ whose singularity at P is of the form

$$\tilde{\theta} = \frac{dz}{z^n}, \qquad n \geq 2.$$

Assume that $\tilde{\theta}$ is normalized so that it has zero periods over the cycles $a_1, \ldots, a_g$. Let

$$\theta = \zeta_j = \left(\sum_{l=0}^{\infty} \alpha_l^{(j)} z^l \right) dz \quad \text{at } P. \tag{3.8.1}$$

Then, as before,

$$\int_{b_j} \tilde{\theta} = \frac{2\pi i}{n-1} \alpha_{n-2}^{(j)}. \tag{3.8.2}$$

We will denote the differential $\tilde{\theta}$ considered above by the symbol

$$\tau_P^{(n)},$$

and note it depends on the choice of local coordinate vanishing at P. (For our applications, this ambiguity will not be significant.)

III.3.9. A few other possibilities remain. We will not have any use for other bilinear relations. The reader who is looking for further amusement may derive more such relations.

III.4. Divisors and the Riemann–Roch Theorem

We come now to one of the most important theorems on compact Riemann surfaces—the Riemann–Roch theorem, which allows us to compute the dimensions of certain vector spaces of meromorphic functions on a compact Riemann surface. The beauty and importance of this theorem will become apparent when we start deriving its many consequences in subsequent sections. As immediate corollaries, we obtain the fact that every surface of genus zero is conformally equivalent to the Riemann sphere and give a new (analytic) proof of the Riemann–Hurwitz relation.

Although many definitions will make sense on arbitrary Riemann surfaces, most of the results will apply only to the compact case. We fix for the duration of this section a compact Riemann surface M of genus $g \geq 0$. We let $\mathcal{K}(M)$ denote the field of meromorphic functions on M.

III.4.1. A *divisor* on M is a formal symbol

$$\mathfrak{A} = P_1^{\alpha_1} P_2^{\alpha_2} \cdots P_k^{\alpha_k}, \tag{4.1.1}$$

with $P_j \in M$, $\alpha_j \in \mathbb{Z}$. We can write the divisor $\mathfrak{A}$ as

$$\mathfrak{A} = \prod_{P \in M} P^{\alpha(P)}, \tag{4.1.1a}$$

with $\alpha(P) \in \mathbb{Z}$, $\alpha(P) \neq 0$ for only finitely many $P \in M$.

We let $\operatorname{Div}(M)$ denote the *group of divisors* on M; it is the free commutative group (written multiplicatively) on the points in M. Thus, if $\mathfrak{A}$ is given by (4.1.1a) and

$$\mathfrak{B} = \prod_{P \in M} P^{\beta(P)},$$

then

$$\mathfrak{A}\mathfrak{B} = \prod_{P \in M} P^{\alpha(P)+\beta(P)},$$

and

$$\mathfrak{A}^{-1} = \prod_{P \in M} P^{-\alpha(P)}.$$

The unit element of the group $\operatorname{Div}(M)$ will be denoted by 1.

For $\mathfrak{A} \in \operatorname{Div}(M)$ given by (4.1.1a), we define

$$\deg \mathfrak{A} = \sum_{P \in M} \alpha(P).$$

It is quite clear that deg establishes a homomorphism

$$\deg : \operatorname{Div}(M) \to \mathbb{Z}$$

from the multiplicative group of divisors onto the additive group of integers.

If $f \in \mathscr{K}(M) \backslash \{0\}$, then f determines a divisor $(f) \in \operatorname{Div}(M)$ by

$$(f) = \prod_{P \in M} P^{\operatorname{ord}_P f}. \tag{4.1.2}$$

It is clear that we have established a homomorphism

$$(\quad) : \mathscr{K}(M)^* \to \operatorname{Div}(M)$$

from the multiplicative group of the field $\mathscr{K}(M)$ into the subgroup of divisors of degree zero (see Proposition I.1.6). A divisor in the image of $(\quad)$ is called *principal*. The group of divisors modulo principal divisors is known as the *divisor class group*. It is quite clear that the homomorphism, deg, factors through to the divisor class group. The (normal) subgroup of principal divisors introduces an equivalence relation on $\operatorname{Div}(M)$. Two divisors $\mathfrak{A}$, $\mathfrak{B}$ are called *equivalent* ($\mathfrak{A} \sim \mathfrak{B}$) provided that $\mathfrak{A}/\mathfrak{B}$ is principal. If $f \in \mathscr{K}(M) \backslash \mathbb{C}$, then

$$f^{-1}(\infty) = \prod_{P \in M} P^{\max\{-\operatorname{ord}_P f, 0\}}$$

defines the *divisor of poles of* (or *polar divisor of*) f. Similarly,

$$f^{-1}(0) = \prod_{P \in M} P^{\max\{\operatorname{ord}_P f, 0\}}$$

defines the *divisor of zeros of* (or *zero divisor of*) f. Both divisors have the same degree as the function f, and since

$$(f) = \frac{f^{-1}(0)}{f^{-1}(\infty)},$$

they are equivalent. More generally, for $c \in \mathbb{C}$,

$$f^{-1}(c) = (f - c)^{-1}(0).$$

Thus the divisor class of $f^{-1}(c)$ is independent of $c \in \mathbb{C} \cup \{\infty\}$. (It will be clear from the context when $f^{-1}(c)$ stands for a divisor or just for the underlying point set.)

One more remark about equivalent divisors: Let f_1 and f_2 be meromorphic non-constant functions on M. Assume that $f_1 = A \circ f_2$ for some Möbius transformation A. Then $f_1^{-1}(\infty) = f_2^{-1}(A^{-1}(\infty))$. Thus $f_1^{-1}(\infty) \sim f_2^{-1}(\infty)$.

If $0 \neq \omega$ is a meromorphic q-differential, then we define the divisor of ω in anology to (4.1.2) by

$$(\omega) = \prod_{P \in M} P^{\operatorname{ord}_P \omega}.$$

A divisor of a meromorphic q-differential is called a *q-canonical divisor*, or simply a *canonical divisor* if $q = 1$. We note that if ω_1 and ω_2 are two non-(identically) zero meromorphic q-differentials, then $\omega_1/\omega_2 \in \mathscr{K}(M)^*$ and hence the divisor class of (ω_1) is the same as the divisor class of (ω_2). We will call it the *q-canonical class* (*canonical class*, if $q = 1$). Since abelian differentials exist, the q-canonical class is just the qth power of the canonical class.

III.4.2. The divisor $\mathfrak{A}$ of (4.1.1a) is *integral* (in symbols, $\mathfrak{A} \geq 1$) provided $\alpha(P) \geq 0$ for all P. If, in addition, $\mathfrak{A} \neq 1$, then $\mathfrak{A}$ is said to be *strictly integral* (in symbols, $\mathfrak{A} > 1$). This notion introduces a partial ordering on divisors; thus $\mathfrak{A} \geq \mathfrak{B}$ (or $\mathfrak{A} > \mathfrak{B}$) if and only if $\mathfrak{A}\mathfrak{B}^{-1} \geq 1$ (or $\mathfrak{A}\mathfrak{B}^{-1} > 1$).

A function $0 \neq f \in \mathscr{K}(M)$ (resp., a meromorphic q-differential $0 \neq \omega$) is said to be a *multiple* of a divisor $\mathfrak{A}$ provided $(f)\mathfrak{A}^{-1} \geq 1$ (resp., $(\omega)\mathfrak{A}^{-1} \geq 1$).

In order not to make exceptions of the zero function and differential, we will introduce the convention that $(0)\mathfrak{A}^{-1} \geq 1$, for all divisors $\mathfrak{A} \in \operatorname{Div}(M)$. Thus, f is a multiple of the divisor $\mathfrak{A}$ of (4.1.1a) provided $f = 0$ or

$$\operatorname{ord}_P f \geq \alpha(P), \quad \text{all } P \in M.$$

Hence such an f must be holomorphic at all points $P \in M$ with $\alpha(P) = 0$; f must have a zero of order $\geq \alpha(P)$ at all points P with $\alpha(P) > 0$; and f may have poles of order $\leq -\alpha(P)$ at all points P with $\alpha(P) < 0$.

III.4.3. For a divisor $\mathfrak{A}$ on M, we set

$$L(\mathfrak{A}) = \{f \in \mathscr{K}(M); (f) \geq \mathfrak{A}\}.$$

It is obvious that $L(\mathfrak{A})$ is a vector space. Its dimension will be denoted by $r(\mathfrak{A})$, and we will call it the *dimension of the divisor* $\mathfrak{A}$.

Proposition. *Let* $\mathfrak{A}, \mathfrak{B} \in \text{Div}(M)$. *Then*

$$\mathfrak{B} \geq \mathfrak{A} \Rightarrow L(\mathfrak{B}) \subset L(\mathfrak{A}).$$

PROOF. Write

$$\mathfrak{B} = \mathfrak{A}\mathfrak{J}$$

with $\mathfrak{J}$ integral. If $f \in L(\mathfrak{B})$, then $(f)\mathfrak{B}^{-1} \geq 1$. But

$$(f)\mathfrak{A}^{-1} = (f)\mathfrak{B}^{-1}\mathfrak{J} \geq \mathfrak{J} \geq 1.$$

Thus $f \in L(\mathfrak{A})$. □

III.4.4. Proposition. *We have* $L(1) = \mathbb{C}$, *and thus* $r(1) = 1$.

PROOF. If $f \in L(1)$, then $(f) \geq 1$; that is,

$$\text{ord}_P f \geq 0 \quad \text{all } P \in M.$$

Since f has no poles it is constant by Proposition I.1.6. □

III.4.5. Proposition. *If* $\mathfrak{A} \in \text{Div}(M)$ *with* $\deg \mathfrak{A} > 0$, *then* $r(\mathfrak{A}) = 0$.

PROOF. If $0 \neq f \in L(\mathfrak{A})$, then $\deg(f) \geq \deg \mathfrak{A} > 0$, contradicting Proposition I.1.6. □

III.4.6. For $\mathfrak{A} \in \text{Div}(M)$, we set

$$\Omega(\mathfrak{A}) = \{\omega; \omega \text{ is an abelian differential with } (\omega) \geq \mathfrak{A}\},$$

and

$$i(\mathfrak{A}) = \dim \Omega(\mathfrak{A}).$$

We call $i(\mathfrak{A})$, the *index of specialty* of the divisor $\mathfrak{A}$.

Theorem. *For* $\mathfrak{A} \in \text{Div}(M)$, $r(\mathfrak{A})$ *and* $i(\mathfrak{A})$ *depend only on the divisor class of* $\mathfrak{A}$. *Furthermore, if* $0 \neq \omega$ *is any abelian differential, then* $i(\mathfrak{A}) = r(\mathfrak{A}(\omega)^{-1})$.

PROOF. Let $\mathfrak{A}_1$ be equivalent to $\mathfrak{A}_2$ (that is, $\mathfrak{A}_1\mathfrak{A}_2^{-1}$ is a principal divisor or $\mathfrak{A}_1\mathfrak{A}_2^{-1} = (f)$ for some $0 \neq f \in \mathscr{K}(M)$). Then the mapping

$$L(\mathfrak{A}_2) \ni h \mapsto hf \in L(\mathfrak{A}_1)$$

establishes a $\mathbb{C}$-linear isomorphism, proving that

$$r(\mathfrak{A}_1) = r(\mathfrak{A}_2). \tag{4.6.1}$$

Next, the mapping

$$\Omega(\mathfrak{A}) \ni \zeta \mapsto \frac{\zeta}{\omega} \in L(\mathfrak{A}(\omega)^{-1})$$

also establishes a $\mathbb{C}$-linear isomorphism proving that

$$i(\mathfrak{A}) = r(\mathfrak{A}(\omega)^{-1}). \tag{4.6.2}$$

Finally, if $\mathfrak{A}_1$ is equivalent to $\mathfrak{A}_2$, then (4.6.1) and (4.6.2) yield (upon choosing a non-zero abelian differential ω)

$$i(\mathfrak{A}_1) = r(\mathfrak{A}_1(\omega)^{-1}) = r(\mathfrak{A}_2(\omega)^{-1}) = i(\mathfrak{A}_2). \qquad \square$$

III.4.7. It is quite easy to verify the following

Proposition. *We have $\Omega(1) = \mathscr{H}^1(M)$, the space of holomorphic abelian differentials (see Proposition* III.2.7*) and thus $i(1) = g$.*

III.4.8. We will first prove our main result in a special case.

Theorem (Riemann–Roch). *Let M be a compact Riemann surface of genus g and $\mathfrak{A}$ an integral divisor on M. Then*

$$r(\mathfrak{A}^{-1}) = \deg \mathfrak{A} - g + 1 + i(\mathfrak{A}). \tag{4.8.1}$$

Proof. Formula (4.8.1) holds for $\mathfrak{A} = 1$ by Proposition III.4.4 and Proposition III.4.7. Thus, it suffices to assume that $\mathfrak{A}$ is strictly integral.

$$\mathfrak{A} = P_1^{n_1} \cdots P_m^{n_m}, \qquad P_j \in M, n_j \in \mathbb{Z}, n_j > 0,$$

$$\deg \mathfrak{A} = \sum_{j=1}^{m} n_j > 0.$$

Note that $\mathbb{C} \subset L(\mathfrak{A}^{-1})$. Furthermore, if $f \in L(\mathfrak{A}^{-1})$, then f is regular on $M\backslash\{P_1, \ldots, P_m\}$ and f has at worst a pole of order n_j at P_j. Choosing a local coordinate z_j vanishing at P_j, we see that for such an f, the Laurent series expansion at P_j is of the form

$$\sum_{k=-n_j}^{\infty} c_{jk} z_j^k.$$

Consider the divisor

$$\mathfrak{A}' = P_1^{n_1+1} \cdots P_m^{n_m+1}.$$

Let $\{a_1, \ldots, a_g, b_1, \ldots, b_g\}$ be a canonical homology basis on M. We think of the elements of this basis as fixed curves in their homology classes and assume they avoid the divisor $\mathfrak{A}$. Let $\Omega_0(\mathfrak{A}'^{-1})$ be the space of abelian differentials of the second kind that are multiples of the divisor $\mathfrak{A}'^{-1}$ and have zero "a" periods. We can easily compute the dimension of $\Omega_0(\mathfrak{A}'^{-1})$. Recall the abelian differentials $\tau_P^{(n)}$ introduced in III.3.8. For $P = P_j$ and $2 \le n \le n_j + 1$, $\tau_P^{(n)} \in \Omega_0(\mathfrak{A}'^{-1})$. Thus

$$\dim \Omega_0(\mathfrak{A}'^{-1}) \ge \sum_{j=1}^{m} n_j = \deg \mathfrak{A}.$$

Conversely, if $\omega \in \Omega_0(\mathfrak{A}'^{-1})$, then

$$\omega = \left(\sum_{k=-n_j-1}^{\infty} d_{jk} z_j^k \right) dz_j \quad \text{with } d_{j,-1} = 0$$

in terms of the local coordinates z_j. Thus we define a mapping

$$S: \Omega_0(\mathfrak{A}'^{-1}) \to \mathbb{C}^{\deg \mathfrak{A}}$$

by setting

$$S(\omega) = (d_{1,-2}, \ldots, d_{1,-n_1-1}, d_{2,-2}, \ldots, d_{2,-n_2-1}, \ldots, d_{m,-2}, \ldots, d_{m,-n_m-1}).$$

If $\omega \in$ Kernel S, then ω is an abelian differential of the first kind with zero "a" periods, and hence $\omega = 0$ by Proposition III.3.3. Hence S is injective, $\dim \Omega_0(\mathfrak{A}'^{-1}) = \deg \mathfrak{A}$, and every $\omega \in \Omega_0(\mathfrak{A}'^{-1})$ can be written uniquely as

$$\omega = \sum_{j=1}^{m} \sum_{k=-n_j-1}^{-2} d_{jk} \tau_{P_j}^{(-k)}. \tag{4.8.2}$$

We consider the differential operator

$$d: L(\mathfrak{A}^{-1}) \to \Omega_0(\mathfrak{A}'^{-1}).$$

Since Kernel $d = \mathbb{C}$, it is necessary in order to compute $r(\mathfrak{A}^{-1})$ to characterize the image of d. Now $\omega \in dL(\mathfrak{A}^{-1})$ if and only if ω has zero "b" periods. We conclude that

$$\dim \text{Image } d \geq \deg \mathfrak{A} - g \tag{4.8.3}$$

(since each "b" curve imposes precisely one linear condition). Using the "classical" linear algebra equation

$$r(\mathfrak{A}^{-1}) = \dim \text{Image } d + 1, \tag{4.8.4}$$

we obtain from (4.8.3), the *Riemann inequality*,

$$r(\mathfrak{A}^{-1}) \geq \deg \mathfrak{A} - g + 1.$$

To obtain the Riemann–Roch equality we must evaluate dim Image d. We use bilinear relation (3.8.2), to observe that the differential ω of (4.8.2) has zero b_l period if and only if

$$2\pi i \sum_{j=1}^{m} \sum_{k=-n_j-1}^{-2} \frac{1}{k+1} d_{jk} \alpha_{-k-2}^{(l)}(P_j) = 0, \tag{4.8.5}$$

where $\{\zeta_1, \ldots, \zeta_g\}$ is a basis for the holomorphic differentials of the first kind dual to our canonical homology basis and the power series expansion of ζ_l in terms of z_j is given (in analogy to (3.8.1)) by

$$\zeta_l = \left(\sum_{s=0}^{\infty} \alpha_s^{(l)}(P_j) z_j^s \right) dz_j \quad \text{at } P_j.$$

We recognize from (4.8.5) that the image of d is the kernel of a certain operator from $\mathbb{C}^{\deg \mathfrak{A}}$ to $\mathbb{C}^g$. Consider the operator

$$T:\Omega(1) \to \mathbb{C}^{\deg \mathfrak{A}}$$

defined by $T(\omega) = (e_{1,0}, \ldots, e_{1,n_1-1}, e_{2,0}, \ldots, e_{m,0}, \ldots, e_{m,n_m-1})$, where $\omega \in \Omega(1)$ has Taylor series expansion

$$\left(\sum_{k=0}^{\infty} e_{jk} z_j^k\right) dz_j \quad \text{at } P_j.$$

Represent the linear operator T with respect to the basis $\{\zeta_1, \ldots, \zeta_g\}$ of $\Omega(1)$ and the standard basis of $\mathbb{C}^{\deg \mathfrak{A}}$ as the matrix

$$\begin{bmatrix} \alpha_0^{(1)}(P_1) & \alpha_0^{(2)}(P_1) & \cdots & \alpha_0^{(g)}(P_1) \\ \vdots & \vdots & & \vdots \\ \alpha_{n_1-1}^{(1)}(P_1) & & & \alpha_{n_1-1}^{(g)}(P_1) \\ \alpha_0^{(1)}(P_2) & & & \alpha_0^{(g)}(P_2) \\ \vdots & & & \vdots \\ \alpha_0^{(1)}(P_m) & & & \alpha_0^{(g)}(P_m) \\ \vdots & \vdots & & \vdots \\ \alpha_{n_m-1}^{(1)}(P_m) & \alpha_{n_m-1}^{(2)}(P_m) & \cdots & \alpha_{n_m-1}^{(g)}(P_m) \end{bmatrix}$$

Thus we recognize that

$$\text{Image } d = \text{Kernel } {}^tT.$$

We thus conclude that

$$\begin{aligned} \dim \text{Image } d = \dim \text{Kernel } {}^tT &= \deg \mathfrak{A} - \dim \text{Image } {}^tT \\ &= \deg \mathfrak{A} - \dim \text{Image } T \\ &= \deg \mathfrak{A} - (\dim \Omega(1) - \dim \text{Kernel } T). \end{aligned}$$

Since Kernel $T = \Omega(\mathfrak{A})$, and $\dim \Omega(1) = g$, we have shown that

$$\dim \text{Image } d = \deg \mathfrak{A} - g + i(\mathfrak{A}),$$

and (4.8.4) yields (4.8.1). □

III.4.9. We collect some immediate consequences of our theorem.

Corollary 1. *If the genus of M is zero, then M is conformally equivalent to the complex sphere $\mathbb{C} \cup \{\infty\}$.*

PROOF. Consider the point divisor, $P \in M$. Then

$$r(P^{-1}) = 2.$$

Thus, there is a non-constant meromorphic function z in $L(P^{-1})$. Such a function provides an isomorphism between M and $\mathbb{C} \cup \{\infty\}$ by Proposition I.1.6. □

Corollary 2. *The degree of the canonical class Z is $2g - 2$.*

Proof. If $g = 0$, then compute the degree of the divisor dz (which is regular except for a double pole at ∞). Thus we may assume $g > 0$. Since the space of holomorphic abelian differentials has positive dimension, we may choose one such non-trivial differential; say ζ. Since (ζ) is integral

$$r((\zeta)^{-1}) = \deg(\zeta) - g + 1 + i((\zeta)).$$

By Theorem III.4.6 we have $r((\zeta)^{-1}) = i(1)$ and $i((\zeta)) = r(1)$. Thus we have $g = \deg(\zeta) - g + 1 + 1$, and hence the corollary follows. □

Corollary 3. *The Riemann–Roch theorem holds for the divisor $\mathfrak{A}$ provided that*

a. *$\mathfrak{A}$ is equivalent to an integral divisor, or*
b. *$Z/\mathfrak{A}$ is equivalent to an integral divisor for some canonical divisor Z.*

Proof. Statement (a) follows from the trivial observation that all integers appearing in the Riemann–Roch theorem depend only on the divisor class of $\mathfrak{A}$ (by Theorem III.4.6). Thus, to verify (b), we need verify Riemann-Roch for $\mathfrak{A}$ provided we know it for $Z/\mathfrak{A}$. Now

$$\begin{aligned} i(\mathfrak{A}) = r(\mathfrak{A}/Z) &= \deg Z/\mathfrak{A} - g + 1 + i(Z/\mathfrak{A}) \\ &= \deg Z - \deg \mathfrak{A} - g + 1 + r(\mathfrak{A}^{-1}) \\ &= 2g - 2 - \deg \mathfrak{A} - g + 1 + r(\mathfrak{A}^{-1}). \end{aligned}$$

Hence, we have proven the Riemann–Roch theorem for $\mathfrak{A}$. □

III.4.10. To conclude the proof of the Riemann–Roch theorem (for arbitrary divisors), it suffices to study divisors $\mathfrak{A}$ such that neither $\mathfrak{A}$ nor $Z/\mathfrak{A}$ is equivalent to an integral divisor, for all canonical divisors Z.

Proposition. *If $r(\mathfrak{A}^{-1}) > 0$ for $\mathfrak{A} \in \mathrm{Div}(M)$, then $\mathfrak{A}$ is equivalent to an integral divisor.*

Proof. Let $0 \neq f \in L(\mathfrak{A}^{-1})$. Then $(f)\mathfrak{A}$ is integral and equivalent to $\mathfrak{A}$. □

Corollary. *If $i(\mathfrak{A}) > 0$ for $\mathfrak{A} \in \mathrm{Div}(M)$, then $Z/\mathfrak{A}$ is equivalent to an integral divisor.*

Proof. Use $i(\mathfrak{A}) = r(\mathfrak{A}/Z)$. □

III.4.11. If $\mathfrak{A} \in \mathrm{Div}(M)$, and neither $\mathfrak{A}$ nor $Z/\mathfrak{A}$ is equivalent to an integral divisor, then $i(\mathfrak{A}) = 0 = r(\mathfrak{A}^{-1})$. Thus, the Riemann–Roch theorem asserts in this case (to be proven, of course)

$$\deg \mathfrak{A} = g - 1. \tag{4.11.1}$$

Thus verification of (4.11.1) for divisors $\mathfrak{A}$ as above will establish the following theorem.

Theorem. *The Riemann–Roch theorem holds for every divisor on a compact surface M.*

PROOF. We write the above divisor $\mathfrak{A}$ as

$$\mathfrak{A} = \mathfrak{A}_1/\mathfrak{A}_2,$$

with $\mathfrak{A}_j$ $(j = 1,2)$ integral and the pair relatively prime (no points in common). We now have

$$\deg \mathfrak{A} = \deg \mathfrak{A}_1 - \deg \mathfrak{A}_2,$$

and by the Riemann inequality

$$r(\mathfrak{A}_1^{-1}) \geq \deg \mathfrak{A}_1 - g + 1 = \deg \mathfrak{A}_2 + \deg \mathfrak{A} - g + 1.$$

Assume that

$$\deg \mathfrak{A} \geq g.$$

We then have

$$r(\mathfrak{A}_1^{-1}) \geq \deg \mathfrak{A}_2 + 1.$$

Thus, we can find a function $0 \neq f \in L(\mathfrak{A}_1^{-1})$ that vanishes at each point in $\mathfrak{A}_2$ (to the order specified by this divisor). (Vanishing at points of $\mathfrak{A}_2$ imposes $\deg \mathfrak{A}_2$ linear conditions on the vector space $L(\mathfrak{A}_1^{-1})$. If $r(\mathfrak{A}_1^{-1})$ is big enough, linear algebra provides the desired function.) Thus $f \in L(\mathfrak{A}_2/\mathfrak{A}_1) = L(\mathfrak{A}^{-1})$, which contradicts the assumption that $r(\mathfrak{A}^{-1}) = 0$. Hence

$$\deg \mathfrak{A} \leq g - 1.$$

Since $0 = i(\mathfrak{A}) = r(\mathfrak{A}/Z)$, it follows that

$$\deg(Z/\mathfrak{A}) \leq g - 1,$$

or

$$\deg \mathfrak{A} \geq g - 1,$$

concluding the proof of the Riemann–Roch theorem. □

III.4.12. As an immediate application of Corollary 2 to Theorem III.4.8 we give another proof of the Riemann–Hurwitz formula (Theorem I.2.7). Our first proof was topological. This one will be complex analytic. Let f be an analytic map of a compact Riemann surface M of genus g onto a surface N of genus γ. Let $n = \deg f$. Let ω be a meromorphic q-differential on N. We lift ω to a meromorphic q-differential Ω on M as follows: Let z be a local coordinate on M and ζ a local coordinate on N. Assume that in terms of these local coordinates we have

$$\zeta = f(z).$$

If ω is

$$h(\zeta)\, d\zeta^q$$

in terms of ζ, then we set Ω to be

$$h(f(z))f'(z)^q\, dz^q \tag{4.12.1}$$

in terms of z. Note that if z is replaced by z_1, with $z = w(z_1)$, then in terms of z_1 the map f is given by

$$\zeta = (f \circ w)(z_1),$$

and thus we assign to z_1

$$h(f(w(z_1)))f'(w(z_1))^q w'(z_1)^q \, dz_1^q$$

which shows that Ω is indeed a meromorphic q-differential. Without loss of generality, we may assume that for $P \in M$, z vanishes at P, ζ vanishes at $f(P)$,

$$\zeta = z^{b_f(P)+1}.$$

From this and (4.12.1) we see that

$$\operatorname{ord}_P \Omega = (b_f(P) + 1)\operatorname{ord}_{f(P)} \omega + qb_f(P). \tag{4.12.2}$$

Formula (4.12.2) will be very useful in the sequel. For the present we merely use it for $q = 1$. We rewrite it as (for $q = 1$)

$$\operatorname{ord}_P \Omega = (b_f(P) + 1) \operatorname{ord}_{f(P)} \omega + b_f(P). \tag{4.12.3}$$

Let us choose an abelian differential ω that is holomorphic and non-zero at the images of the branch points of f. (With the aid of Riemann–Roch (and allowing lots of poles) the reader should have no trouble producing such an ω. If $\gamma > 0$, ω can be chosen to be holomorphic.) We wish to add (4.12.3) over all $P \in M$. Observe that Corollary 2 of Theorem III.4.8 gives

$$\sum_{P \in M} \operatorname{ord}_P \Omega = 2g - 2,$$

and recall that by definition

$$\sum_{P \in M} b_f(P) = B.$$

We need only analyze

$$\sum_{P \in M} (b_f(P) + 1) \operatorname{ord}_{f(P)} \omega = \sum_{\substack{P \in M \\ b_f(P)=0}} \operatorname{ord}_{f(P)} \omega$$
$$= \sum_{Q \in N} n \operatorname{ord}_Q \omega = n(2\gamma - 2).$$

This last equality is a consequence of the fact that each $Q \in M$ with $\operatorname{ord}_Q \omega \neq 0$ is the image of precisely n points on M.

III.4.13. We discuss some elementary concepts that will turn out to be useful throughout this book. If D is a divisor on M given by the right side of the equality in formula (4.1.1a), then the integer $\alpha(P)$ will be called the *order* or *multiplicity* of D at $P \in M$. The *complete linear series* or *system* of the divisor D, denoted by the symbol $|D|$, is the set of integral divisors equivalent to D.

Proposition. *For every divisor D on M, the points in $|D|$ are in one-to-one canonical correspondence with the points in $\mathbf{P}L(1/D)$, the projective space of the vector space $L(1/D)$.*

PROOF. If $D_1 \in |D|$, then D_1 is integral and $D_1 = D(f)$ for some not identically zero meromorphic function f on M. Since $D_1 \geq 1$, $f \in L(1/D)$. Conversely, for every $f \in L(1/D)$, $D(f)$ is an integral divisor equivalent to D. Two functions f and g in $L(1/D)$ define the same divisor if and only if $f = \lambda g$ for some non-zero complex number λ. Thus $|D|$ is the projectivization of $L(1/D)$. □

A linear subspace of a complete linear series is called a *linear series* or *system*. A linear series of the divisor D is thus of the form $\mathscr{D} = \mathbf{P}V$, where V is a vector subspace of $L(1/D)$; this linear series is said to be a g_d^r if

$$\deg D = d, \qquad \dim V = r + 1.$$

A *base point* of the linear series $\mathscr{D}$ is a point common to all the divisors in the space $\mathscr{D}$.

Assume that P is a base point of a linear series. Using the above notation, we see that this means that $(f)D \geq P$ for all $f \in V$ or that the multiplicity of P in every divisor in $\mathscr{D}$ is greater than minus the multiplicity of P in D. In particular, if we are considering the complete linear series of the divisor D and if the multiplicity of P in D is zero, then P is a base point of $|D|$ if and only if every function in $L(1/D)$ vanishes at P.

III.5. Applications of the Riemann–Roch Theorem

What can we say about the meromorphic functions on the compact Riemann surface M with poles only at one point? What is the lowest degree of such a function? In this section we shall show that on a compact Riemann surface of genus $g \geq 2$, there are finitely many points $P \in M$ (called Weierstrass points) such that there exists on M a meromorphic function f regular on $M \backslash \{P\}$ with $\deg f \leq g$.

We shall see when we study hyperelliptic surfaces (in III.7) and when we study automorphisms of compact surfaces (in Chapter V) that these Weierstrass points carry a lot of information about the Riemann surface.

Throughout this section, M is a compact Riemann surface of genus g (usually positive), and $Z \in \mathrm{Div}(M)$ will denote a canonical divisor.

III.5.1. We recall (to begin) that the Riemann–Roch theorem can be written for $D \in \mathrm{Div}(M)$ as

$$r(D^{-1}) = \deg D - g + 1 + r(D/Z). \tag{5.1.1}$$

Furthermore,

$$r(D) = 0 \text{ provided } \deg D > 0, \tag{5.1.2}$$

and

$$i(D) = 0 \text{ provided } \deg D > 2g - 2. \tag{5.1.2a}$$

If $\deg D = 0$, then

$$r(D) \le 1, \tag{5.1.3}$$

and

$$r(D) = 1 \Leftrightarrow D \text{ is principal.} \tag{5.1.4}$$

Finally, for any $q \in \mathbb{Z}$

$$L(Z^{-q}) \cong \mathscr{H}^q(M), \text{ the vector space of holomorphic } q\text{-differentials.} \tag{5.1.5}$$

To verify (5.1.5) note that we can choose an abelian differential $\omega \neq 0$ such that $(\omega) = Z$, and now observe that $f \in L(Z^{-q})$ if and only if $f\omega^q$ is a holomorphic q-differential. The mapping

$$L(Z^{-q}) \ni f \mapsto f\omega^q \in \mathscr{H}^q(M)$$

establishes a $\mathbb{C}$-linear isomorphism between the spaces involved.

III.5.2. Proposition. *Let $q \in \mathbb{Z}$. The dimension of the space of holomorphic q-differentials on M is given by the following table:*

Genus	*Weight*	*Dimension*
$g = 0$	$q \le 0$	$1 - 2q$
	$q > 0$	0
$g = 1$	all q	1
$g > 1$	$q < 0$	0
	$q = 0$	1
	$q = 1$	g
	$q > 1$	$(2q - 1)(g - 1)$

PROOF. Let $D = Z^q$ in (5.1.1), then

$$r(Z^{-q}) = (2q - 1)(g - 1) + r(Z^{q-1}). \tag{5.2.1}$$

From (5.1.5), the dimension to be computed is $r(Z^{-q})$.

Assume that $g > 1$. If $q < 0$, then

$$\deg Z^{-q} = -q(2g - 2) > 0,$$

and thus $r(Z^{-q}) = 0$ by (5.1.2). We already know that $r(1) = 1$, and that $r(Z^{-1}) = g$. For $q > 1$, $r(Z^{q-1}) = 0$ (by what was said before), and (5.2.1) gives a formula for $r(Z^{-q})$.

Next for $g = 1$, (5.2.1) reads

$$r(Z^{-q}) = r(Z^{q-1}), \tag{5.2.2}$$

and hence gives little information. If $0 \neq \omega$ is a holomorphic q-differential, it must also be free of zeros ($\deg(\omega) = 0$). Thus ω^{-1} is a $(-q)$-differential. Let ω be a non-trivial holomorphic differential. Multiplication by ω establishes an isomorphism of $\mathscr{H}^q(M)$ onto $\mathscr{H}^{q+1}(M)$, for every integer q. From $r(1) = 1$, we conclude that $r(Z^q) = 1$, by induction on q.

Finally, for $g = 0$, deg $Z = -2$. Thus $r(Z^n) = 0$ for $n \leq -1$, and (5.2.1) yields the required results. □

EXERCISE

Establish the above proposition for $g = 0$ and 1 *without* the use of the Riemann–Roch theorem. (For $g = 0$, write any $\omega \in \mathscr{H}^q(M)$ as $\omega = f\,dz^q$ with $f \in \mathscr{K}(M)$. Thus f is a rational function. Describe the singularities of f.)

III.5.3. Theorem (The Weierstrass "gap" Theorem). *Let M have positive genus g, and let $P \in M$ be arbitrary. There are precisely g integers*

$$1 = n_1 < n_2 < \cdots < n_g < 2g \tag{5.3.1}$$

such that there does not exist a function $f \in \mathscr{K}(M)$ holomorphic on $M \backslash \{P\}$ with a pole of order n_j at P.

Remarks

1. The numbers appearing in the list (5.3.1) are called the "gaps" at P. Their complement in the positive integers are called the "non-gaps". The "non-gaps" clearly form an additive semi-group. There are precisely g "non-gaps" in $\{2, \ldots, 2g\}$ with $2g$ always a "non-gap." These are the first g "non-gaps" in the semi-group of "non-gaps."
2. The Weierstrass "gap" theorem trivially holds for $g = 0$. Since on the sphere there is always a function with one (simple) pole, there are no "gaps".
3. The Weierstrass "gap" theorem is a special case of a more general theorem to be stated and proven in the next section.

III.5.4. We stay with the compact Riemann surface M of positive genus g. Let

$$P_1, P_2, P_3, \ldots$$

be a sequence of points on M. Define a sequence of divisors on M by

$$D_0 = 1, \qquad D_{j+1} = D_j P_{j+1}, \qquad j = 0, 1, \ldots.$$

We now pose a sequence of questions.

Question "j" ($j = 1, 2, \ldots$):

Does there exist a meromorphic function f on M with

$$(f) \geq D_j^{-1} \quad \text{and} \quad (f) \not\geq D_{j-1}^{-1}?$$

We can also phrase the question in another way. Does there exist a (non-constant) function $f \in L(D_j^{-1}) \backslash L(D_{j-1}^{-1})$?

Theorem (The Noether "gap" Theorem). *There are precisely g integers n_k satisfying* (5.3.1) *such that the answer to Question "j" is no if and only if j is one of the integers appearing in the list* (5.3.1).

Remark. Taking

$$P = P_1 = P_2 = \cdots = P_j = \cdots$$

we see that the Weierstrass "gap" theorem is a special case of the Noether "gap" theorem. We shall say that j is a "gap" provided the answer to Question "j" is no. When there is need, we will distinguish the "Weierstrass gaps" from the "Noether gaps".

PROOF OF THEOREM. The answer to Question "1" is always no, since $g > 0$. Thus $n_1 = 1$, as asserted. The answer to Question "j" is yes if and only if $r(D_j^{-1}) - r(D_{j-1}^{-1}) = 1$ (the answer is no if and only if $r(D_j^{-1}) - r(D_{j-1}^{-1}) = 0$). From the Riemann–Roch theorem

$$r(D_j^{-1}) - r(D_{j-1}^{-1}) = 1 + i(D_j) - i(D_{j-1}). \tag{5.4.1}$$

Thus for every $k \geq 1$:

$$r(D_k^{-1}) - r(D_0^{-1}) = \sum_{j=1}^{k} (r(D_j^{-1}) - r(D_{j-1}^{-1}))$$

$$= k + \sum_{j=1}^{k} (i(D_j) - i(D_{j-1})) = k + i(D_k) - i(D_0),$$

or

$$r(D_k^{-1}) - 1 = k + i(D_k) - g,$$

and this number *is* the number of "non-gaps" $\leq k$. Thus for $k > 2g - 2$ (thus $\deg D_k > 2g - 2$ and $i(D_k) = 0$)

$$k - (\text{number of "gaps"} \leq k) = k - g.$$

Thus there are precisely g "gaps" and all of them are $\leq 2g - 1$. □

III.5.5. We now begin a more careful study of the Weierstrass "gaps". Let $P \in M$ be arbitrary, and let

$$1 < \alpha_1 < \alpha_2 < \cdots < \alpha_g = 2g$$

be the first g "non-gaps".

Proposition. *For each integer j, $0 < j < g$, we have*

$$\alpha_j + \alpha_{g-j} \geq 2g.$$

PROOF. Suppose that $\alpha_j + \alpha_{g-j} < 2g$. Thus for each $k \leq j$ we would also have $\alpha_k + \alpha_{g-j} < 2g$. Since the sum of "non-gaps" is a "non-gap", we would have at least j "non-gaps" strictly between α_{g-j} and α_g. Thus at least $(g - j) + j + 1 = g + 1$ "non-gaps" $\leq 2g$, contradicting the fact that there are only g such "non-gaps". □

III.5.6. Proposition. *If $\alpha_1 = 2$, then $\alpha_j = 2j$ and $\alpha_j + \alpha_{g-j} = 2g$ for $0 < j < g$.*

PROOF. If $\alpha_1 = 2$, then $2, 4, \ldots, 2g$ are g "non-gaps" $\leq 2g$, and hence these are all the "non-gaps" $\leq 2g$. □

III.5.7. Proposition. *If $\alpha_1 > 2$, then for some j with $0 < j < g$, we have*

$$\alpha_j + \alpha_{g-j} > 2g.$$

PROOF. If $g = 2$, then our assumption implies that $\alpha_1 = 3$ and $\alpha_2 = 4$ and there is nothing to prove. If $g = 3$, then the possible "non-gaps" are $\{3,4,6\}$, $\{3,5,6\}$, and $\{4,5,6\}$, and again there is nothing to prove. So assume that $g \geq 4$. We assume that $\alpha_j + \alpha_{g-j} = 2g$ for all j with $0 < j < g$. For $q \in \mathbb{R}$, let $[q]$ be the greatest integer $\leq q$. Then $\alpha_1, 2\alpha_1, \ldots, [2g/\alpha_1]\alpha_1$ are "non-gaps" $\leq 2g$. If $\alpha_1 > 2$, then the above accounts for at most $\frac{2}{3}g < g$ "non-gaps", and there must be another one $\leq 2g$. Let α be the first "non-gap" not appearing in our previous enumeration. For some integer r, $1 \leq r \leq [2g/\alpha_1] < g - 1$, we must have

$$r\alpha_1 < \alpha < (r+1)\alpha_1.$$

Thus we have "non-gaps"

$$\alpha_1, \qquad \alpha_2 = 2\alpha_1, \qquad \ldots, \qquad \alpha_r = r\alpha_1, \qquad \alpha_{r+1} = \alpha,$$

and by our assumption

$$\alpha_{g-1} = 2g - \alpha_1, \qquad \ldots, \qquad \alpha_{g-r} = 2g - r\alpha_1, \qquad \alpha_{g-(r+1)} = 2g - \alpha.$$

The integers in the last line are all the "non-gaps" which are $\geq \alpha_{g-(r+1)}$ and $< 2g$. It follows that

$$\alpha_1 + \alpha_{g-(r+1)} = \alpha_1 + 2g - \alpha = 2g - (\alpha - \alpha_1) > 2g - r\alpha_1 = \alpha_{g-r}.$$

It therefore follows that there is a "non-gap" $< 2g$, greater than α_{g-r} and not in the list $\alpha_{g-1}, \ldots, \alpha_{g-(r+1)}$. This is an obvious contradiction. □

Corollary. *We have*

$$\sum_{j=1}^{g-1} \alpha_j \geq g(g-1),$$

with equality if and only if $\alpha_1 = 2$.

PROOF. From Proposition III.5.5, $2\sum_{j=1}^{g-1} \alpha_j \geq 2g(g-1)$. Furthermore if $\alpha_1 = 2$, then we have equality in the above by Proposition III.5.6. If $\alpha_1 > 2$, we must have strict inequality by Proposition III.5.7. □

III.5.8. We have seen in III.5.4, that $j \geq 1$ is a "gap" at $P \in M$ if and only if

$$r(P^{-j}) - r(P^{-j+1}) = 0,$$

if and only if

$$i(P^{j-1}) - i(P^j) = 1;$$

that is, if and only if there exists on M an abelian differential of the first kind with a zero of order $j - 1$ at P. Thus the possible orders of zeros of abelian differential of the first kind at P are precisely

$$0 = n_1 - 1 < n_2 - 1 < \cdots < n_g - 1 \leq 2g - 2,$$

where the n_j's are the "gaps" at P (appearing in the list (5.3.1)). The above situation is a special case of a general phenomenon, which we shall proceed to study.

Before proceeding to study the general situation, let us observe that we have established the following basic *Fact. Given a point* P *on a compact Riemann surface* M *of genus* $g > 0$, *then there exists an abelian differential* ω *of the first kind* ($\omega \in \mathscr{H}^1(M)$) *that does not vanish at* P (*that is,* $\operatorname{ord}_P \omega = 0$).

Let A be a finite-dimensional space of holomorphic functions on a domain $D \subset \mathbb{C}$. Assume that $\dim A = n \geq 1$. Let $z \in D$. By a *basis of* A *adapted to* z, we mean a basis $\{\varphi_1, \ldots, \varphi_n\}$ with

$$\operatorname{ord}_z \varphi_1 < \operatorname{ord}_z \varphi_2 < \cdots < \operatorname{ord}_z \varphi_n. \tag{5.8.1}$$

To construct such a basis, let

$$\mu_1 = \min_{\varphi \in A} \{\operatorname{ord}_z \varphi\},$$

and choose $\varphi_1 \in A$ with $\operatorname{ord}_z \varphi_1 = \mu_1$. Then

$$A_1 = \{\varphi \in A; \operatorname{ord}_z \varphi > \mu_1\}$$

is an $(n-1)$-dimensional subspace of A, and we can set

$$\mu_2 = \min_{\varphi \in A_1} \{\operatorname{ord}_z \varphi\}.$$

By induction, we can now construct the basis satisfying (5.8.1). The basis adapted to z is (of course) not unique. We can make it unique as follows: Let $\mu_j = \operatorname{ord}_z \varphi_j$. Consider the Taylor series expansion of φ_j at z (in terms of ζ)

$$\varphi_j(\zeta) = \sum_{k=0}^{\infty} a_{kj}(\zeta - z)^k.$$

We may and *hereafter* do require that

$$a_{\mu_k j} = \begin{cases} 1, & k = j, \\ 0, & k \neq j, \end{cases}$$

where $j, k = 1, \ldots, n$.

Remark. On a Riemann surface "the unique" basis adapted to a point will, of course, depend on the choice of local coordinate.

It is obvious that $\mu_j \geq j - 1$. We define the *weight of* z *with respect to* A by

$$\tau(z) = \sum_{j=1}^{n} (\mu_j - j + 1). \tag{5.8.2}$$

Proposition. *Let* $\{\varphi_1, \ldots, \varphi_n\}$ *be any basis for* A. *Consider the holomorphic function* (*the Wronskian*)

$$\Phi(z) = \det \begin{bmatrix} \varphi_1(z) & \cdots & \varphi_n(z) \\ \varphi_1'(z) & \cdots & \varphi_n'(z) \\ \vdots & & \vdots \\ \varphi_1^{(n-1)}(z) & \cdots & \varphi_n^{(n-1)}(z) \end{bmatrix}. \tag{5.8.3}$$

Then

$$\operatorname{ord}_z \Phi = \tau(z).$$

PROOF. It is easy to see that a change of basis will lead to a non-zero constant multiple of Φ. Hence we may assume, whenever necessary, that the basis used is adapted to the point z. Let us abbreviate equation (5.8.3) by

$$\Phi(z) = \det[\varphi_1(z), \ldots, \varphi_n(z)].$$

In order to prove the proposition we derive some easy properties of the function Φ. First: for every holomorphic function f,

$$\det[f\varphi_1, \ldots, f\varphi_n] = f^n \det[\varphi_1, \ldots, \varphi_n]. \tag{5.8.4}$$

This follows from well-known properties of determinants. Explicitly,

$$\det[f\varphi_1, \ldots, f\varphi_n]$$

$$= \det\begin{bmatrix} f\varphi_1 & \cdots & f\varphi_n \\ f\varphi_1' + f'\varphi_1 & \cdots & f\varphi_n' + f'\varphi_n \\ f\varphi_1'' + 2f'\varphi_1' + f''\varphi_1 & \cdots & f\varphi_n'' + 2f'\varphi_n' + f''\varphi_n \\ \vdots & & \vdots \\ f\varphi_1^{(n-1)} + \cdots + f^{(n-1)}\varphi_1 & \cdots & f\varphi_n^{(n-1)} + \cdots + f^{(n-1)}\varphi_n \end{bmatrix}$$

$$= f \det\begin{bmatrix} \varphi_1 & \cdots & \varphi_n \\ f\varphi_1' + f'\varphi_1 & \cdots & f\varphi_n' + f'\varphi_n \\ \vdots & & \vdots \\ f\varphi_1^{(n-1)} + \cdots + f^{(n-1)}\varphi_1 & \cdots & f\varphi_n^{(n-1)} + \cdots + f^{(n-1)}\varphi_n \end{bmatrix}$$

$$= f \det\begin{bmatrix} \varphi_1 & \cdots & \varphi_n \\ f\varphi_1' & \cdots & f\varphi_n' \\ \vdots & & \vdots \\ f\varphi_1^{(n-1)} + \cdots + f^{(n-1)}\varphi_1 & \cdots & f\varphi_n^{(n-1)} + \cdots + f^{(n-1)}\varphi_n \end{bmatrix},$$

where the last equality arises from multiplying the first row of the determinant by $-f'$ and adding the result to the second row. In a similar fashion we can remove from each column the appropriate multiple of φ_j leaving us with the previous expression equal to

$$f \det\begin{bmatrix} \varphi_1 & \cdots & \varphi_n \\ f\varphi_1' & \cdots & f\varphi_n' \\ f\varphi_1'' + 2f'\varphi_1' & \cdots & f\varphi_n'' + 2f'\varphi_n' \\ \vdots & & \vdots \\ f\varphi_1^{(n-1)} + \cdots + (n-1)f^{(n-2)}\varphi_1' & \cdots & f\varphi_n^{(n-1)} + \cdots + (n-1)f^{(n-2)}\varphi_n' \end{bmatrix}$$

$$= f^2 \det\begin{bmatrix} \varphi_1 & \cdots & \varphi_n \\ \varphi_1' & \cdots & \varphi_n' \\ f\varphi_1'' + 2f'\varphi_1' & \cdots & f\varphi_n'' + 2f'\varphi_n' \\ \vdots & & \vdots \\ f\varphi_1^{(n-1)} + \cdots + (n-1)f^{(n-2)}\varphi_1' & \cdots & f\varphi_n^{(n-1)} + \cdots + (n-1)f^{(n-2)}\varphi_n' \end{bmatrix}$$

We now repeat the same procedure to remove from each column the appropriate multiple of φ_j'. This clearly terminates with the preceding equal to

$$f^n \det[\varphi_1, \ldots, \varphi_n].$$

We now turn our attention to the proof of the proposition which shall be by induction on n. Clearly the result is true for $n = 1$. Let us now assume that the proposition is true for $n = k$. Explicity we are thus assuming that

$$\operatorname{ord}_z \det[\varphi_1, \ldots, \varphi_k] = \sum_{j=1}^{k} (\mu_j - j + 1),$$

where $\mu_j = \operatorname{ord}_z \varphi_j$. Consider now $\det[\varphi_1, \ldots, \varphi_{k+1}]$. It is clear from the preceding remarks that

$$\det[\varphi_1, \ldots, \varphi_{k+1}] = \varphi_1^{k+1} \det[1, \varphi_2/\varphi_1, \ldots, \varphi_{k+1}/\varphi_1].$$

Now the right-hand side is simply $\varphi_1^{k+1} \det[(\varphi_2/\varphi_1)', \ldots, (\varphi_{k+1}/\varphi_1)']$. The induction hypothesis now gives that

$$\begin{aligned} &\operatorname{ord}_z\{\varphi_1^{k+1} \det[(\varphi_2/\varphi_1)', \ldots, (\varphi_{k+1}/\varphi_1)']\} \\ &\quad = (k+1)\mu_1 + \sum_{j=2}^{k+1} \{(\mu_j - \mu_1 - 1) - (j-2)\} \\ &\quad = \mu_1 + \sum_{j=2}^{k+1} (\mu_j - j + 1), \end{aligned}$$

provided that for each j, $\mu_j - (j-1) - \mu_1 \geq 0$. Since the $\{\varphi_j\}$ are a basis adapted to z, this inequality is always satisfied, and we have

$$\operatorname{ord}_z \det[\varphi_1, \ldots, \varphi_{k+1}] = \sum_{j=1}^{k+1} (\mu_j - j + 1).$$

This concludes the proof of the proposition.

Remark. Let $\{\varphi_1, \ldots, \varphi_n\}$ be any set of n holomorphic functions on D. We can define $\Phi = \det[\varphi_1, \ldots, \varphi_n]$. Our argument here shows that Φ is identically zero if and only if the functions $\varphi_1, \ldots, \varphi_n$ are linearly dependent.

Corollary 1. *Let A be a finite-dimensional space of holomorphic functions on a domain $D \subset \mathbb{C}$. The set of $z \in D$ with positive weight with respect to A is discrete.*

Corollary 2. *Under the hypothesis of Corollary* 1, *for an open dense set in D, the basis $\{\varphi_1, \ldots, \varphi_n\}$ of A adapted to z has the property*

$$\operatorname{ord}_z \varphi_j = j - 1.$$

PROOF. By the hypothesis $\operatorname{ord}_z \varphi_j = \mu_j$. It follows from Corollary 1 that $\tau(z) = \sum_{j=1}^{n} (\mu_j - j + 1) = 0$ for an open dense set. Since (as we have previously remarked) $\mu_j \geq j - 1$ we have $\mu_j = j - 1$ for each j, on this open dense set. □

III.5.9. The considerations of the last paragraph apply (of course) to the space $\mathscr{H}^q(M)$ of holomorphic q-differentials ($q \geq 1$) on a compact Riemann surface of genus $g \geq 1$. A point $P \in M$ will be called a *q-Weierstrass point* provided its weight with respect to $\mathscr{H}^q(M)$ is positive. A *1-Weierstrass point* is called simply a *Weierstrass point* or a *classical Weierstrass point*. It is clear from the ideas in the previous paragraph that we have the following

Proposition. *A point P on a Riemann surface M of genus $g \geq 2$ is a q-Weierstrass point if and only if there exists a (not identically zero) holomorphic q-differential on M with a zero of order $\geq \dim \mathscr{H}^q(M)$ at P. For $q = 1$, this condition is equivalent to either (and hence both)*

a. $i(P^g) > 0$, *or*
b. $r(P^{-g}) \geq 2$ *(that is, at least one of the integers $2, \ldots, g$ is not a "gap").*

III.5.10. There are clearly no q-Weierstrass points (for any $q \geq 1$) on a surface of genus 1. We assume thus that $g \geq 2$.

Proposition. *For $g \geq 2$, $q \geq 1$, let $\tau(P)$ be the weight of $P \in M$ with respect to $\mathscr{H}^q(M)$. Let W_q be the Wronskian of a basis for $\mathscr{H}^q(M)$. Set $d = d_q = \dim \mathscr{H}^q(M)$. Then W_q is a (non-trival) holomorphic $m = m_q$-differential where $m = (d/2)(2q - 1 + d)$. Hence*

$$\sum_{P \in M} \tau(P) = (g - 1)d(2q - 1 + d).$$

PROOF. We must merely verify that the determinant Φ defined by (5.8.3) transforms as an m-differential under changes of coordinates. Explicitly, let $\{\zeta_1, \ldots, \zeta_d\}$ be a basis for $\mathscr{H}^q(M)$. Let z and $\tilde{z}$ be local coordinates with $\tilde{z} = f(z)$ on the overlap of their respective domains. Assume that

$$\zeta_j = \varphi_j(z)dz^q = \tilde{\varphi}_j(\tilde{z})d\tilde{z}^q$$

(that is,

$$\tilde{\varphi}_j(f(z))f'(z)^q = \varphi_j(z))$$

in terms of the local coordinates z and $\tilde{z}$. We must show that

$$(\det[\varphi_1, \ldots, \varphi_d])\,dz^m = (\det[\tilde{\varphi}_1, \ldots, \tilde{\varphi}_d])\,d\tilde{z}^m. \tag{5.10.1}$$

But it is easy to verify that

$$\begin{aligned}\det[\varphi_1, \ldots, \varphi_d] &= \det[(\tilde{\varphi}_1 \circ f)(f')^q, \ldots, (\tilde{\varphi}_d \circ f)(f')^q] \\ &= (f')^m(\det[\tilde{\varphi}_1, \ldots, \tilde{\varphi}_d] \circ f),\end{aligned}$$

which is equivalent to (5.10.1). □

Corollary. *For $g \geq 2$ there are q-Weierstrass points for every $q \geq 1$.*

PROOF. Any m_q-differential always has zeros. □

III.5.11. We now finish the study of classical Weierstrass points (the case $q = 1$).

Theorem. *For $g \geq 2$, the weight of a point with respect to the holomorphic abelian differentials is $\leq g(g-1)/2$. This bound is attained only for a point P where the "non-gap" sequence begins with 2.*

PROOF. We have seen that (Proposition III.5.10)

$$\sum_{P \in M} \tau(P) = (g-1)g(g+1). \tag{5.11.1}$$

The above, of course, gives a trivial estimate on $\tau(P)$. We need a better one. Let $2 \leq \alpha_1 < \alpha_2 < \cdots < \alpha_g = 2g$ be the first g "non-gaps" at P. Then let $1 = n_1 < n_2 < \cdots < n_g < 2g$ be the g-"gaps" at P. (That is, the sequence of n_j's is the complement in $\{1, \ldots, 2g\}$ of the sequence of α_j's.) Then (recall III.5.8)

$$\begin{aligned}\tau(P) &= \sum_{j=1}^{g} (n_j - j) = \sum_{j=1}^{2g} j - \sum_{j=1}^{g} \alpha_j - \sum_{j=1}^{g} j \\ &= \sum_{j=g+1}^{2g-1} j - \sum_{j=1}^{g-1} \alpha_j \leq \frac{3g}{2}(g-1) - g(g-1) \\ &= g(g-1)/2,\end{aligned}$$

by the Corollary to Proposition III.5.7, with equality holding if and only if $\alpha_1 = 2$. □

Corollary. *Let W be the number of Weierstrass points on a compact surface of genus $g \geq 2$, then $2g + 2 \leq W \leq g^3 - g$.*

PROOF. The first inequality follows from (5.11.1) and the fact that the maximum weight of a Weierstrass point is $g(g-1)/2 > 0$. The second from the fact that the minimum weight of a Weierstrass point is 1 (so called *simple* Weierstrass points). □

Remark. The first equality is attained if and only if at every Weierstrass point the "gap" sequence is $1, 3, \ldots, 2g - 1$. These are the hyperelliptic surfaces to be studied in III.7. The second equality is attained if and only if the "gap" sequence at each Weierstrass point is $1, 2, \ldots, g-1, g+1$. Existence of such surfaces will be demonstrated in VII.3.9.

III.5.12. In this section we present an interesting application of the theory developed so far, and exhibit a striking difference between open and closed Riemann surfaces. If M is a Riemann surface, then we define the (*first*) *holomorphic de Rham cohomology group* as the vector space of holomorphic differentials on M factored by the subspace of exact holomorphic differentials (the latter are the images of holomorphic functions under the differential operator d). We denote this group by $H^1_{\text{hol}}(M)$. It is obviously a complex vector space. We have seen that if M is a compact surface of genus g, then

$$\dim H^1_{\text{hol}}(M) = g.$$

Theorem. *Let M be a compact Riemann surface of genus $g \geq 0$ and let P_1, $\ldots$, P_k be $k > 0$ distinct points on M. Set $M' = M - \{P_1, \ldots, P_k\}$. Then*

$$\dim H^1_{\text{hol}}(M') = 2g + k - 1.$$

Further, each element of $H^1_{\text{hol}}(M')$ may be represented by an abelian differential of the third kind on M that is regular on M' has a pole of order at most $2g$ at P_1 and at most a simple pole at P_j, $j = 2, \ldots, k$.

Proof. We observe that the rank of $H_1(M')$, the first homology group of M', is $2g + k - 1$. A holomorphic 1-form on M' that has zero periods over a basis for $H_1(M')$ must be exact. Hence we conclude that

$$\dim H^1_{\text{hol}}(M') \leq 2g + k - 1.$$

Since the d operator sends regular functions to regular differentials, meromorphic functions to meromorphic differentials, and functions with essential singularities to differentials with essential singularities, the theorem will be proved if we show that the quotient of the meromorphic 1-forms on M that are regular on M' by the subspace of images under d of the meromorphic functions on M that are regular on M' is exactly (it would suffice to show it is at least) $2g + k - 1$.

We use induction on k. Assume that $k = 1$. For each integer n with $n \geq 2g$, there exists a meromorphic function f on M which is regular on $M - \{P_1\}$ and has a pole of order n at P_1. Therefore, every meromorphic 1-form on M which is regular except possibly at P_1 is equivalent modulo exact forms to one with a pole of order at most $2g$. We compute the dimension of

$$\frac{\Omega(P_1^{-2g})}{dL(P_1^{-2g+1})}.$$

The above dimension equals $i(P_1^{-2g}) - (r(P_1^{-2g+1}) - 1)$ since the kernel of d consists of the constants. The Riemann–Roch theorem shows that the above difference is $2g$. Hence the result is verified for $k = 1$. □

Assume now that $k > 1$ and let $M'' = M - \{P_1, \ldots, P_{k-1}\}$. We assume that the theorem holds on M''. Observe that (by the Noether "gap" theorem, for example) for each positive integer n, there exists a meromorphic function on M which is regular except possibly at P_1 and P_k and which has a pole of order n at P_k. It follows that in passing from M'' to M'

the dimension of the de Rham cohomology group can go up by at most one. To show that it actually increases, we observe that the differential of the third kind $\omega_{P_k P_1}$ is not holomorphic on M'' and represents a nontrivial (since it cannot possibly be exact) class in the cohomology group of M'.

III.5.13. We shall now refine the results of the last section and give an alternate proof of the inequality

$$\dim H^1_{\text{hol}}(M') \geq 2g + k - 1.$$

We single out one of the punctures on M', say P_1, and then construct a unique representative for each cohomology class in $H^1_{\text{hol}}(M')$. We choose holomorphic forms on M' as follows:

(a) a basis for the holomorphic differentials of the first kind on M (for example, a normalized basis $\zeta_1, \ldots, \zeta_g$ dual to some canonical homology basis on the compact surface (without punctures)) as defined by Proposition III.2.8;
(b) for $j = 2, \ldots, k$, we let τ_j be any meromorphic differential of the third kind on M that is regular on M' and has simple poles at both P_1 and P_j (if $k = 1$, then we do not need any differentials of the third kind); and (c) for $j = 1, \ldots, g$, we let θ_j be any meromorphic differential of the second kind on M that is regular on M' and has a pole of order $n_j + 1$ at P_1, where $n_1, \ldots, n_g$ is the "gap" sequence at P_1, see (5.3.1).
(c) We note that $n_g + 1 \leq 2g$ and that we can take for θ_j the differential $\tau_{P_j}^{(n_j+1)}$ defined in III.3.8. We now have the following:

Proposition. *Each element of* $H^1_{\text{hol}}(M')$ *is uniquely represented by a meromorphic differential in the linear span of the* $2g + k - 1$ *linearly independent differentials defined by* (a), (b), *and* (c).

PROOF. Let us write an element in this span as $\zeta + \tau$, where ζ is of the first kind and τ is of the third kind. Assume that this differential is exact and equal to df, then τ must be zero since an exact differential cannot have any non-zero residues. Let $-n = \text{ord}_{P_1} \zeta$. If $n > 0$, then $n > 1$ and $\deg f = -\text{ord}_{P_1} f = n - 1$ contradicting the fact that $n - 1$ is a gap. Thus $n \leq 0$ and ζ is of the first kind (on M). It must hence be the zero differential. □

Remark. The differentials in (a) and (b) span $\Omega(1/P_1 \cdots P_k)$.

Corollary. *If P_1 is not a Weierstrass point on M then each element of* $H^1_{\text{hol}}(M')$ *is uniquely represented by a meromorphic differential in*

$$\Omega(P_1^{-g-1} P_2^{-1} \cdots P_k^{-1}).$$

III.6. Abel's Theorem and the Jacobi Inversion Problem

In this section we determine necessary and sufficient conditions for a divisor of degree zero to be principal (Abel's theorem), and begin the study of the space of positive (integral) divisors on a compact Riemann surface. To each compact surface of positive genus g, we attach a complex torus (of complex dimension g) into which the surface is imbedded. This torus inherits many of the properties of the Riemann surface, and is a tool in the study of the surface and the divisors on it.

The Riemann–Roch theorem showed that every surface of genus 0 is conformally equivalent to the sphere $\mathbb{C} \cup \{\infty\}$ (Corollary 1 in III.4.9). Abel's theorem (Corollary 1 in III.6.4) shows that every surface of genus 1 is a torus ($\mathbb{C}$ modulo a lattice). These are uniformization theorems for compact surfaces of genus $g \leq 1$. For uniformization theorems for surfaces of genus $g \geq 2$, we will have to rely on different methods (involving more analysis and topology). These methods, which will be applicable to all surfaces, will be treated in IV.4 and IV.5.

Throughout this section M represents a compact Riemann surface of genus $g > 0$.

III.6.1. We start with

$$\{a_1, \ldots, a_g, b_1, \ldots, b_g\} = \{a,b\},$$

a canonical homology basis on M, and

$$\{\zeta_1, \ldots, \zeta_g\} = \{\zeta\},$$

the dual basis for $\mathscr{H}^1(M)$; that is,

$$\int_{a_k} \zeta_j = \delta_{jk}, \qquad j, k = 1, 2, \ldots, g.$$

We have seen in III.2, that the matrix Π with entries

$$\pi_{jk} = \int_{b_k} \zeta_j, \qquad j, k = 1, \ldots, g,$$

is symmetric with positive definite imaginary part. Let us denote by $L = L(M)$ the lattice (over $\mathbb{Z}$) generated by the $2g$-columns of the $g \times 2g$ matrix (I,Π). Denote these columns (they are clearly linearly independent over $\mathbb{R}$) by $e^{(1)}, \ldots, e^{(g)}, \pi^{(1)}, \ldots, \pi^{(g)}$. A point of L can be written uniquely as

$$\sum_{j=1}^{g} m_j e^{(j)} + \sum_{j=1}^{g} n_j \pi^{(j)}, \quad \text{with } m_j, n_j \in \mathbb{Z},$$

or

$$Im + \Pi n \quad \text{with } m = {}^t(m_1, \ldots, m_g) \in \mathbb{Z}^g \text{ and } n = {}^t(n_1, \ldots, n_g) \in \mathbb{Z}^g.$$

We shall call $J(M) = \mathbb{C}^g/L(M)$ the *Jacobian variety of* M. It is a compact, commutative, g-dimensional complex Lie group. We define a map

$$\varphi: M \to J(M)$$

by choosing a point $P_0 \in M$ and setting

$$\varphi(P) = \int_{P_0}^{P} {}^t\zeta = {}^t\left(\int_{P_0}^{P} \zeta_1, \ldots, \int_{P_0}^{P} \zeta_g\right).$$

Proposition. *The map* φ *is a well defined holomorphic mapping of* M *into* $J(M)$. *It has maximal rank.*

PROOF. Let c_1 and c_2 be two paths joining P_0 to P, then $c_1 c_2^{-1}$ is homologous to $(a,b)\left[\begin{smallmatrix} m \\ n \end{smallmatrix}\right]$ for some $m, n \in \mathbb{Z}^g$. Thus

$$\int_{c_1} {}^t\zeta - \int_{c_2} {}^t\zeta = Im + \Pi n \in L(M).$$

If z is a local coordinate vanishing at P and $\varphi_1, \ldots, \varphi_g$ are the components of φ (in $\mathbb{C}^g$), then writing $\zeta_j = \eta_j \, dz$, we have

$$\varphi_j(z) = \int_{P_0}^{P} \zeta_j + \int_0^z \eta_j(z)\, dz,$$

and we see that

$$\frac{\partial \varphi_j}{\partial z} = \eta_j(z).$$

Thus φ would not have maximal rank if there were a point at which all the abelian differentials of the first kind vanished. Since this does not occur (recall III.5.8), the rank of φ is constant and equals one. □

A map $\psi : M \to J(M)$ of the form

$$\psi(P) = \varphi(P) + c, \qquad P \in M,$$

with fixed $c \in J(M)$ will be called as *Abel–Jacobi embedding* of M and its Jacobian variety.

III.6.2. Let, for every integer $n \geq 1$, M_n denote the set of integral divisors of degree n. We extend the map φ:

$$\varphi: M_n \to J(M)$$

by setting for

$$D = P_1 \cdots P_n, \qquad \varphi(D) = \sum_{j=1}^{n} \varphi(P_j).$$

Note that (since $\varphi(D) = \varphi(P_0 D)$)

$$\varphi(M_{n+1}) \supset \varphi(M_n) \supset \cdots \supset \varphi(M_1) = \varphi(M).$$

We can also obtain a map that does not depend on the base point P_0.

Let $\text{Div}^{(0)}(M)$ denote the divisors of degree zero of M; define

$$\varphi: \text{Div}(M) \to J(M)$$

by setting

$$\varphi(D) = \sum_{j=1}^{r} \varphi(P_j) - \sum_{j=1}^{s} \varphi(Q_j)$$

for

$$D = P_1 \cdots P_r / Q_1 \cdots Q_s.$$

It is clear that if $r = s$, $\varphi(D)$ is independent of the base point; that is, the map

$$\varphi: \text{Div}^{(0)}(M) \to J(M)$$

is independent of the base point.

III.6.3. Theorem (Abel). *Let $D \in \text{Div}(M)$. A necessary and sufficient condition for D to be the divisor of a meromorphic function is that*

$$\varphi(D) = 0 \bmod (L(M)) \quad \textit{and} \quad \deg D = 0. \tag{6.3.1}$$

PROOF. Assume that f is a meromorphic function. Let $D = (f)$. We have seen (Proposition I.1.6) that $\deg D = 0$. Since for $D = 1$, Abel's theorem trivially holds, we assume that $f \notin \mathbb{C}$. Write

$$D = P_1^{\alpha_1} \cdots P_k^{\alpha_k} / Q_1^{\beta_1} \cdots Q_r^{\beta_r}, \qquad k \geq 1, r \geq 1, \tag{6.3.2}$$

with

$$\begin{aligned} &P_j \neq Q_l, \quad \text{all } j, l; \\ &P_j \neq P_l \quad \text{and} \quad Q_j \neq Q_l, \quad \text{all } j \neq l; \\ &\sum_{j=1}^{k} \alpha_j = \sum_{j=1}^{r} \beta_j \, (\geq 1). \end{aligned} \tag{6.3.3}$$

Without loss of generality, we may assume that none of the points P_j, Q_j lie on the curves representing the canonical homology basis. Recall the normalized abelian differentials τ_{PQ} of the third kind introduced in III.3. Observe that df/f is an abelian differential of the third kind with simple poles and

$$\text{res}_P \frac{df}{f} = \text{ord}_P f, \quad \text{for all } P \in M.$$

Thus

$$\frac{df}{f} - \left(\sum_{j=1}^{k} \alpha_j \tau_{P_j P_0} - \sum_{j=1}^{r} \beta_j \tau_{Q_j P_0} \right)$$

(where P_0 is not any of the P_j or Q_j nor on the curves a_j, b_j) is an abelian differential of the first kind. Hence, we can choose constants c_j, $j = 1, \ldots, g$, such that

$$\frac{df}{f} = \sum_{j=1}^{k} \alpha_j \tau_{P_j P_0} - \sum_{j=1}^{r} \beta_j \tau_{Q_j P_0} + \sum_{j=1}^{g} c_j \zeta_j.$$

It follows that

$$\int_{a_l} \frac{df}{f} = c_l,$$

and by the bilinear relations for the normalized differentials τ_{PQ},

$$\int_{b_l} \frac{df}{f} = 2\pi i \left(\sum_{j=1}^{k} \alpha_j \int_{P_0}^{P_j} \zeta_l - \sum_{j=1}^{r} \beta_j \int_{P_0}^{Q_j} \zeta_l \right) + \sum_{j=1}^{g} c_j \pi_{jl}.$$

Since $df/f = d \log f$, we see that

$$\int_{a_l} \frac{df}{f} = 2\pi i m_l, \qquad \int_{b_l} \frac{df}{f} = 2\pi i n_l,$$

where m_l, n_l are integers. It follows that the lth component of $\varphi(D)$ is

$$\begin{aligned} \sum_{j=1}^{k} \alpha_j \int_{P_0}^{P_j} \zeta_l - \sum_{j=1}^{r} \beta_j \int_{P_0}^{Q_j} \zeta_l &= \frac{1}{2\pi i} \int_{b_l} \frac{df}{f} - \frac{1}{2\pi i} \sum_{j=1}^{g} c_j \pi_{jl} \\ &= n_l - \sum_{j=1}^{g} m_j \pi_{jl}, \end{aligned}$$

and thus $\varphi(D) = 0 \bmod(L(M))$. We have used P_0 as the base point for φ. Recall that since D is of degree 0, $\varphi(D)$ is independent of P_0.

To prove the converse, we let D be given by (6.3.2) subject to (6.3.3). Choose a point Q_0 not equal to P_0, nor any of the P_j nor any of the Q_j nor lying on any of the curves a_j, b_j. Set

$$\begin{aligned} f(P) &= \exp\left(\sum_{j=1}^{k} \alpha_j \int_{Q_0}^{P} \tau_{P_j P_0} - \sum_{j=1}^{r} \beta_j \int_{Q_0}^{P} \tau_{Q_j P_0} + \sum_{j=1}^{g} c_j \int_{Q_0}^{P} \zeta_j \right) \\ &= \exp \int_{Q_0}^{P} \tau, \end{aligned}$$

where the constants $c_1, \ldots, c_g$ are to be determined. It is clear that f is a meromorphic function with $(f) = D$, provided f is single-valued. We compute

$$\int_{a_l} \tau = \sum_{j=1}^{k} \alpha_j \int_{a_l} \tau_{P_j P_0} - \sum_{j=1}^{r} \beta_j \int_{a_l} \tau_{Q_j P_0} + c_l = c_l,$$

and

$$\begin{aligned} \int_{b_l} \tau &= \sum_{j=1}^{k} \alpha_j \int_{b_l} \tau_{P_j P_0} - \sum_{j=1}^{r} \beta_j \int_{b_l} \tau_{Q_j P_0} + \sum_{j=1}^{g} c_j \pi_{jl} \\ &= 2\pi i \sum_{j=1}^{k} \alpha_j \int_{P_0}^{P_j} \zeta_l - 2\pi i \sum_{j=1}^{r} \beta_j \int_{P_0}^{Q_j} \zeta_l + \sum_{j=1}^{g} c_j \pi_{jl}. \end{aligned}$$

For f to be single-valued, we must have that $\int_{a_l} \tau$, $\int_{b_l} \tau$ are of the form $2\pi i n$, $n \in \mathbb{Z}$. Now (6.3.1) yields,

$$\sum_{j=1}^{k} \alpha_j \int_{P_0}^{P_j} \zeta_l - \sum_{j=1}^{r} \beta_j \int_{P_0}^{Q_j} \zeta_l = n_l + \sum_{j=1}^{g} m_j \pi_{jl}$$

with $n_l, m_j \in \mathbb{Z}$, $l = 1, \ldots, g, j = 1, \ldots, g$. This means we can choose paths γ_j and $\tilde{\gamma}_j$ joining P_0 to P_j and Q_j respectively, such that integrating over these paths, leads to the above equation. It is clear that if we choose $c_j = -2\pi i m_j$, f will be single-valued. □

EXERCISE

We outline an alternate proof of the necessity part of Abel's theorem. Let f be a non-constant meromorphic function on M. For each $\alpha \in \mathbb{C} \cup \{\infty\}$, let $f^{-1}(\alpha)$ be viewed as the integral divisor of degree deg f, consisting of the preimage of α. Then $\alpha \mapsto \varphi(f^{-1}(\alpha))$ is a holomorphic mapping of $\mathbb{C} \cup \{\infty\}$ into $J(M)$. Since $\mathbb{C} \cup \{\infty\}$ is simply connected, it lifts to a holomorphic mapping of $\mathbb{C} \cup \{\infty\}$ into $\mathbb{C}^g$. Since $\mathbb{C} \cup \{\infty\}$ is compact this mapping is constant and thus $\varphi(f^{-1}(0)) = \varphi(f^{-1}(\infty))$.

III.6.4. For $g = 1$, $J(M)$ is of course a compact Riemann surface. We have the following

Corollary 1. *If M is of genus* 1, *then*

$$\varphi: M \to J(M)$$

is an isomorphism (conformal homeomorphism).

PROOF. Clearly φ is surjective (since it is not constant). Let $P, Q \in M$, $P \neq Q$. If $\varphi(P) = \varphi(Q)$, then P/Q is principal by Abel's theorem. Thus there is a meromorphic function on M with a single simple pole. This contradiction shows that φ is injective. □

Corollary 2. *If M has genus* >1, *then φ is an injective holomorphic mapping of M onto a proper sub-manifold $\varphi(M)$ of $J(M)$.*

III.6.5. Let D be an integral divisor of degree g on M; that is,

$$D = P_1 \cdots P_g \qquad (P_j \in M, j = 1, \ldots, g). \tag{6.5.1}$$

Then by the Riemann–Roch theorem

$$r(D^{-1}) = 1 + i(D) \geq 1.$$

We call the divisor D *special* provided $r(D^{-1}) > 1$.

Theorem. *Let $D \in \operatorname{Div}(M)$ with $D \geq 1$ and $\deg D = g$. There is an integral divisor D' of degree g close to D such that D' is not special. Further D' may be chosen to consist of g distinct points.*

Remark. Write D as in (6.5.1), and choose a neighborhood U_j of P_j. The condition that D' be close to D is that $D' = P'_1 \cdots P'_g$ with $P'_j \in U_j$.

PROOF OF THEOREM. Note that for divisors of degree g,

$$r(D^{-1}) = 1 \Leftrightarrow i(D) = 0.$$

Assume that D is given by (6.5.1) and define

$$D_j = P_1 \cdots P_j \qquad (j = 1, \ldots, g).$$

Thus $D_g = D$. Set $D_0 = 1$. We prove by induction that for $j \geq 1$, we can find a divisor D'_j of the form $D'_{j-1}P'_j$ ($D'_0 = 1$) with P'_j arbitrarily close to P_j and $i(D'_j) = g - j$, $j = 1, \ldots, g$. The Riemann–Roch theorem (or linear algebra) implies that $i(D'_j) \geq g - j$ (for any integral divisor D'_j of degree j). Note that for arbitrary P_1 (as we have seen before),

$$i(P_1) = r(P_1^{-1}) - 1 + g - 1 = r(P_1^{-1}) + g - 2 = g - 1$$

(since $\mathbb{C} = L(P_1^{-1})$, because a surface of positive genus does not admit a meromorphic function with a single pole). Thus, it suffices to take $P'_1 = P_1$. Assume that for some $1 \leq j < g$, we have constructed a D'_j of the required type. Let $\{\varphi_1, \ldots, \varphi_{g-j}\}$ be a basis for the abelian differentials of the first kind vanishing at $P'_1, \ldots, P'_j$. Look at φ_1. If $\varphi_1(P_{j+1}) \neq 0$, then $i(D'_jP_{j+1}) \leq g - (j + 1)$, and we are done. If $\varphi_1(P_{j+1}) = 0$, then arbitrarily close to P_{j+1} there is a point P'_{j+1} with $\varphi_1(P'_{j+1}) \neq 0$ and once again $i(D'_jP'_{j+1}) \leq g - (j + 1)$. □

III.6.6. We consider now the map

$$\varphi: M_g \to J(M).$$

We will show that this map is always surjective (generalizing the first corollary in III.6.4). To this end let $D_0 = P_1 \cdots P_g$ with $P_j \in M$, $P_j \neq P_k$ for $j \neq k$ be such that $i(D_0) = 0$. Let U_j be a coordinate disk around P_j with local coordinate t_j vanishing at P_j. Let $K = \varphi(P_1 \cdots P_g)$. In terms of the local coordinates t_j we write $\zeta_k = \eta_{kj}(t_j)\,dt_j$ and thus in terms of the local coordinates t_j at P_j the mapping φ in a neighborhood of the origin $(0, \ldots, 0)$ is simply

$$(z_1, \ldots, z_g) \mapsto K + (\varphi_1(z_1, \ldots, z_g), \ldots, \varphi_g(z_1, \ldots, z_g)),$$

where

$$\varphi_l(z_1, \ldots, z_g) = \sum_{j=1}^{g} \int_0^{z_j} \eta_{lj}(t_j)\,dt_j.$$

Thus

$$\frac{\partial \varphi_l}{\partial z_k}(0, \ldots, 0) = \eta_{lk}(0) = \zeta_l(P_k)$$

(the last equality is merely a convenient abbreviation). Thus the Jacobian of the map φ at $P_1 \cdots P_g$ is

$$\begin{bmatrix} \zeta_1(P_1) & \cdots & \zeta_1(P_g) \\ \vdots & & \vdots \\ \zeta_g(P_1) & \cdots & \zeta_g(P_g) \end{bmatrix}.$$

The condition $i(P_1 \cdots P_g) = 0$, however, gives us immediately that the rank of the Jacobian is g. Therefore, the map

$$\varphi : U_1 \times \cdots \times U_g \to \mathbb{C}^g$$

is a homeomorphism in a neighborhood of $(0, \ldots, 0)$ by the inverse function theorem; and thus covers a neighborhood U of K (in $\mathbb{C}^g$ or in $J(M)$). Let $c = {}^t(c_1, \ldots, c_g) \in \mathbb{C}^g$. Then for a sufficiently large integer N, $K + c/N \in U$, and thus there exists a $Q_1, \ldots, Q_g$ such that

$$\varphi(Q_1 \cdots Q_g) = K + c/N$$

or

$$N(\varphi(Q_1 \cdots Q_g) - K) = c.$$

Thus to show that $c \in$ Image φ, it suffices to show that there exists an integral divisor D of degree g such that

$$\varphi(D) = N(\varphi(Q_1 \cdots Q_g) - K).$$

Consider the divisor $(P_1 \cdots P_g)^N/(Q_1 \cdots Q_g)^N P_0^g$. The Riemann–Roch theorem gives

$$r\left(\frac{(P_1 \cdots P_g)^N}{(Q_1 \cdots Q_g)^N P_0^g}\right) = 1 + i\left(\frac{(Q_1 \cdots Q_g)^N P_0^g}{(P_1 \cdots P_g)^N}\right) \geq 1.$$

Hence, there is an $f \in L((P_1 \cdots P_g)^N/(Q_1 \cdots Q_g)^N P_0^g))$; that is, there is an integral divisor D of degree g such that

$$(f) = \frac{D(P_1 \cdots P_g)^N}{(Q_1 \cdots Q_g)^N P_0^g}.$$

By Abel's theorem

$$\varphi(D) + N\varphi(P_1 \cdots P_g) = N\varphi(Q_1 \cdots Q_g).$$

We have solved the Jacobi inversion problem:

Theorem (Jacobi Inversion). *Every point in $J(M)$ is the image of an integral divisor of degree g.*

Corollary. *As a group $J(M)$ is isomorphic to the group of divisors of degree zero modulo its subgroup of principal divisors.*

Remark. We have shown (Corollary 1 of III.6.4) that every surface of genus one can be realized as $\mathbb{C}$ modulo a lattice. This is the uniformization theorem for tori. We shall return to this topic (including uniformization theorems for surfaces of genus > 1) in the next chapter.

III.6.7. Exercise

We have seen that an arbitrary torus M is conformally equivalent to $\mathbb{C}/G$ where G is the group generated by two elements $z \mapsto z + 1$ and $z \mapsto z + \tau$, $\operatorname{Im} \tau > 0$. Further, the origin of $\mathbb{C}/G$ may be made to correspond to any point in M.

(a) Every meromorphic function f on M can be viewed as a doubly periodic function f (this identification should cause no confusion) on $\mathbb{C}$; that is, a meromorphic function f on $\mathbb{C}$ with

$$f(z+1) = f(z) = f(z+\tau), \quad \text{all } z \in \mathbb{C}.$$

The *Weierstrass* $\wp$*-function* is defined by

$$\wp(z) = \frac{1}{z^2} + \sum_{\substack{(n,m) \neq (0,0) \\ (n,m) \in \mathbb{Z}^2}} \left(\frac{1}{(z - n - m\tau)^2} - \frac{1}{(n + m\tau)^2} \right), \; z \in \mathbb{C}.$$

Note that $\wp$ is an even function. Let $P, Q \in M$. Let $f \in L(1/PQ) \setminus \mathbb{C}$. Show that there exists $\alpha \in \text{Aut } M$ (the group of conformal automorphisms of M), $\beta \in \text{Aut}(\mathbb{C} \cup \{\infty\})$, such that

$$f = \beta \circ \wp \circ \alpha.$$

(For a discussion of Aut M, see V.4.)

(b) Show that every meromorphic differential on M is of the form $f(z)\,dz$ where f is a doubly periodic function. If we choose the loop corresponding to $z \mapsto z + 1$ as the "a-curve" and $z \mapsto z + \tau$ as the "b-curve", then we have a canonical homology basis on M. The basis for $\mathscr{H}^1(M)$ dual to this canonical homology basis is then $\{dz\}$.

(c) From what we said above, $\wp(z)\,dz$ is an abelian differential of the third kind with zero residue and singularity $1/z^2$ at the origin. Hence there exists a meromorphic function ζ on $\mathbb{C}$ such that $\zeta' = -\wp$. This function ζ cannot be doubly periodic (why?). However, ζ satisfies for all $z \in \mathbb{C}$

$$\zeta(z+1) = \zeta(z) + \eta_1,$$
$$\zeta(z+\tau) = \zeta(z) + \eta_2,$$

where η_1 and η_2 satisfy Legendre's equation

$$\eta_1\tau - \eta_2 = 2\pi i. \tag{6.7.1}$$

Derive (6.7.1) from (3.8.2).

III.6.8. The following technical and important result is suggested by the development of Section III.6.6.

Proposition. *Let M be a compact Riemann surface of genus $g \geq 2$. For all $q \geq 1$, the differentials with simple zeros are dense and open in $\mathscr{H}^q(M)$.*

PROOF. Let $d = \dim \mathscr{H}^q(M)$ and recall that $1 < d < \infty$. Let ω_1 and ω_2 be two linearly independent elements of $\mathscr{H}^q(M)$ without common zeros. To construct such differentials, let ω_1 be any non-trivial holomorphic q-differential. The set of differentials vanishing at a point in the divisor of ω_1 is a finite union of codimension one subspaces of $\mathscr{H}^q(M)$. We can clearly find an ω_2 in the complement of the finitely many hyperplanes. Then $f = \omega_1/\omega_2$ is a non-construct meromorphic function on the surface of degree $q(2g-2)$. Let α be a complex number which is not a branch value of f. We have excluded a finite (non-empty) set of values. The function f takes on the value α at $q(2g-2)$ distinct points. These points must be zeros of the differential $\omega_1 - \alpha\omega_2$; they must be all the zeros of this differential and all of these must be simple zeros since the degree of the divisor of $\omega_1 - \alpha\omega_2$ is $q(2g-2)$. We have shown that the set of q-differentials with simple zeros is dense. □

Next we show that the differentials with simple zeros are open. Let ω be a differential with simple zeros at $P_1, \ldots, P_{q(2g-2)}$ and let $\omega_1, \ldots, \omega_d$ be a basis for the holomorphic q-differentials. Choose a small disc D_j with center at P_j; we assume that these discs are all disjoint and that the closure of each D_j is contained in the domain of a single local coordinate z_j vanishing at P_j. Let us write in terms of these local coordinates $\omega = \phi_j(z_j)\, dz_j^q$ and $\omega_k = \phi_{kj}(z_j)\, dz_j^q$. Our assumption that ω has a simple zero at each P_j means that $\phi_j(0) = 0$ and that ϕ_j does not vanish on the boundary c_j of D_j; hence the absolute value of ϕ_j has a positive minimum m_j on c_j. Let m_{kj} be the maximum of the absolute value of ϕ_{kj} on c_j. Choose a positive ε such that

$$\varepsilon \sum_{k=1}^{d} m_{kj} < m_j \quad \text{for } j = 1, \ldots, d.$$

Then for $\eta_1, \ldots, \eta_d$ arbitrary complex numbers with $|\eta_k| < \varepsilon$, the differential $\theta = \omega + \sum_{k=1}^{d} \eta_k \omega_k$ has one simple zero inside each disc D_j, $j = 1, \ldots, q(2g-2)$, by Rouche's theorem, since on c_j

$$\left| \sum_{k=1}^{d} \eta_k \phi_{kj} \right| \le \sum_{k=1}^{d} |\eta_k| m_{kj} < \varepsilon \sum_{k=1}^{d} m_{kj} < m_j.$$

Once again these $q(2g-2)$ zeros account for all the points where the differential θ vanishes and each zero is simple.

III.7. Hyperelliptic Riemann Surfaces

In this section we study hyperelliptic Riemann surfaces—the simplest surfaces. These are the two-sheeted (branched) coverings of the sphere. We shall see that there exist hyperelliptic surfaces of each genus g, and that these surfaces are the ones for which the number of Weierstrass points is precisely $2g + 2$. These surfaces thus show that the lower bound obtained in the Corollary to Theorem III.5.11 is sharp.

III.7.1. A compact Riemann surface M is called *hyperelliptic* provided there exists an integral divisor D on M with

$$\deg D = 2, \qquad r(D^{-1}) \ge 2.$$

Equivalently, M is hyperelliptic if and only if M admits a non-constant meromorphic function with precisely 2 poles. If M has such a function, then each ramification point has branch number 1, and hence the genus g of M and the number B of branch points (= ramification points) of f are related by (using Riemann–Hurwitz)

$$B = 2g + 2.$$

Remarks

1. We can hence describe a hyperelliptic surface of genus g as a two-sheeted covering of the sphere branched at $2g + 2$ points.
2. Some authors restrict the term hyperelliptic to surfaces of genus ≥ 2 that satisfy the above condition.

III.7.2. Proposition. *Every surface of genus ≤ 2 is hyperelliptic.*

PROOF. Let D be an integral divisor of degree 2. Riemann–Roch yields

$$r(D^{-1}) = 2 - g + 1 + i(D). \tag{7.2.1}$$

Thus $r(D^{-1}) \geq 2$ for $g \leq 1$, and the only issue is $g = 2$.

First proof for $g = 2$: Let P be a Weierstrass point on a surface of genus 2. Then there is a non-constant $f \in L(P^{-2})$. (This function cannot have degree 1.)

Second proof for $g = 2$: Choose $\omega \neq 0$, an abelian differential of the first kind on M. Then, since $\deg(\omega) = 2$,

$$(\omega) = PQ.$$

Since $i(PQ) = 1$, (7.2.1) yields $r(P^{-1}Q^{-1}) = 2$. □

Remark. Surfaces of genus 1 are also called *elliptic* (tori). Surfaces of genus zero admit, of course, functions of degree 1. Thus, hyperelliptic surfaces are those which admit functions of lowest possible degree.

III.7.3. Let M be a hyperelliptic Riemann surface of genus ≥ 2. Choose a function z of degree 2 on M. Let f be another such function. We claim that f is a Möbius transformation of z. Let the polar divisor of z be P_1Q_1 and the polar divisor of f be P_2Q_2. It suffices to show that $P_1Q_1 \sim P_2Q_2$. For then there is an $h \in \mathscr{K}(M)$, such that multiplication by h establishes an isomorphism between $L(P_1^{-1}Q_1^{-1})$ and $L(P_2^{-1}Q_2^{-1})$. Since $\{1,z\}$ and $\{1,f\}$ are bases for these spaces, there are constants α, β, γ, δ such that

$$1 = \alpha h + \beta hz$$
$$f = \gamma h + \delta hz$$

or

$$f = \frac{\gamma + \delta z}{\alpha + \beta z}.$$

To establish the above equivalence, we observe first that the branch points of f are precisely the Weierstrass points of M. To see this let $P \in M$ be a branch point of f. Then f is locally two-to-one at P. Thus if $f(P) = \infty$, f has a pole of order 2 at P (and no other poles) and then P is a Weierstrass point. If $f(P) \neq \infty$, then

$$\frac{1}{f - f(P)}$$

has a pole of order 2 at P, proving that P is a Weierstrass point on M. It now follows by Proposition III.5.6 that the "gap" sequence at any of the $2g + 2$ branch points of f is

$$1, 3, \ldots, 2g - 1,$$

and thus the weight of any of these points is

$$\sum_{k=1}^{g} (2k - 1) - \sum_{k=1}^{g} k = g^2 - \frac{g(g + 1)}{2} = \frac{1}{2}g(g - 1).$$

Thus these $2g + 2$ points contribute $g(g^2 - 1)$ to the sum of the weights of the Weierstrass points. Since the sum of the weights of all the Weierstrass points is precisely $g(g^2 - 1)$ there are no other Weierstrass points.

Let us choose any Weierstrass point P on M. We claim that the polar divisor of f is equivalent to P^2. If $f(P) = \infty$, there is nothing to prove. Otherwise, look at the function $1/(f - f(P)) = F$. Its polar divisor is P^2 which is equivalent to the polar divisor of f since F is a Möbius transformation of f (that is, $f^{-1}(\infty) \sim f^{-1}(f(P))$).

We have therefore established most of the following

Theorem. *Let M be a hyperelliptic Riemann surface of genus $g \geq 2$. Then the function z of degree 2 on M is unique up to fractional linear transformations. Furthermore, the branch points of z are precisely the Weierstrass points of M. The hyperelliptic surfaces of genus $g \geq 2$ are the only ones with precisely $2g + 2$ Weierstrass points.*

Proof. Only the last statement needs verification. It is clear that if a surface has precisely $2g + 2$ Weierstrass points, then the weight of each such point must be $\frac{1}{2}g(g - 1)$ by III.5.10 and III.5.11, and thus, as we saw there, the "non-gap" sequence must begin with 2. □

Remark. We show next that on a hyperelliptic surface each of the Weierstrass points is also a q-Weierstrass point for every $q > 1$. It follows from the fact that the "gap" sequence is $1, 3, \ldots, 2g - 1$ for each Weierstrass point P, that there is an abelian differential φ of the first kind with divisor P^{2g-2}. Thus φ^q is a holomorphic q differential with divisor $P^{q(2g-2)}$. Since $q(2g - 2) > (2q - 1)(g - 1) - 1$, P must also be a q-Weierstrass point.

III.7.4. We now wish to construct another function on a hyperelliptic surface of genus $g \geq 1$. Let z be a function of degree 2. Let $P_1, \ldots, P_{2g+2}$ be the branch points of z. Without loss of generality we assume that

$$z(P_j) \neq \infty, \qquad j = 1, \ldots, 2g + 2.$$

Consider "the function"

$$w = \sqrt{\prod_{j=1}^{2g+2} (z - z(P_j))}. \tag{7.4.1}$$

Remark. We have introduced above a multivalued function which we will show to be single-valued. Multivalued functions are treated in III.9. In IV.11, we will show how the function w can be obtained without the use of multi-valued functions.

Proposition. *The above defines w as a meromorphic function on M whose divisor is*

$$\frac{P_1 \cdots P_{2g+2}}{Q_1^{g+1} Q_2^{g+1}} \tag{7.4.2}$$

where $Q_1 Q_2$ is the polar divisor of z.

PROOF. Since all the branch points of z have ramification number 2, w locally defines a meromorphic function on M which is two-valued. We must show we can choose a single valued branch. It is convenient at this point to introduce a "concrete" representation of the surface M as a two-sheeted covering of the sphere $\mathbb{C} \cup \{\infty\}$. If $P \in M$, $z(P) \neq \infty$ and P is not a branch point of z, then $z - z(P)$ is a local coordinate vanishing at P. If $z(P) = \infty$ (recall we have assumed that P is not a branch point), then $1/z$ is a local coordinate vanishing at P. If P is a branch point, then either branch of $\sqrt{z - z(P)}$ is a local coordinate vanishing at P. With slight and obvious modification, the above procedure could have been carried out with any meromorphic function on any (compact or not) surface (recall Remark 3 in I.1.6). We define now $e_j = z(P_j)$. Then these e_j are distinct; and $z^{-1}(\alpha)$ consists of precisely two points on M for all $\alpha \in \mathbb{C} \cup \{\infty\} \setminus \{e_1, \ldots, e_{2g+2}\}$, whereas $z^{-1}(e_j)$ consists only of the point P_j. We picture now two copies of the sphere. We label these two copies sheet I and sheet II. On each sheet for each $k = 1, \ldots, g + 1$, we draw a smooth curve called a "cut" joining e_{2k-1} to e_{2k}. We may assume that the e_j's have been ordered so that these cuts do not intersect. Each "cut" is considered to have two banks; an N-bank and an S-bank. We construct a Riemann surface $\tilde{M}$ by joining every S-Bank on sheet I to an N-bank of the corresponding "cut" on sheet II, and then joining the corresponding S-bank on sheet II to the N-bank of the corresponding "cut" on sheet I. It is quite clear that $\tilde{M}$ is a compact Riemann surface and z is a meromorphic function on $\tilde{M}$. Thus $\tilde{M}$ is indeed a concrete model for M. We remark that a simple closed curve around a point e_j, that

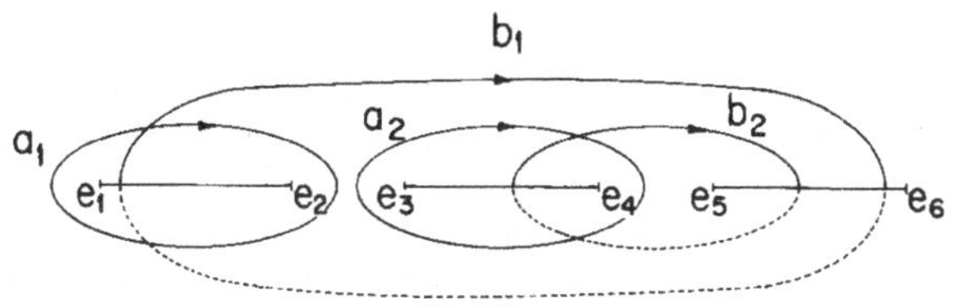

Figure III.3. Cross "cuts" for a hyperelliptic surface of genus 2.

does not go around any e_k with $k \neq j$, may be pictured as beginning in one sheet say at R_1, continuing around until the point, R_2, on the second sheet with $z(R_1) = z(R_2)$ and returning back to R_1. We now construct a canonical homology basis for M using this two sheeted representation. Draw simple smooth closed curves a_k, $k = 1, \ldots, g$, winding once around the "cut" from e_{2k-1} to e_{2k} in one sheet of M oriented as indicated in Figure III.3. This curve a_k exists because we cross between sheets only through the "cuts". Next choose curves b_k, $k = 1, \ldots, g$, starting from a point on the "cut" from e_{2k-1} to e_{2k} going on the first sheet to a point on "cut" from e_{2g+1} to e_{2g+2} and returning on the second sheet (indicated in Figure III.3 by dotted lines) to the original point. The orientation of the b-curves is again illustrated in Figure III.3. Let us stop to analyze what is happening on the surface itself.

The reader should convince himself (or herself) that the picture on the surface (Figure III.4) is actually the lift of the picture in the extended plane via the two-sheeted covering z. (Note that all we are using is that a curve in the plane passing through e_j can be lifted in two ways as a curve in M passing through P_j.) We have actually constructed a canonical homology basis, since by inspection the intersection matrix is of the form

	a	b
a	0	I
b	$-I$	0

We return—after this lengthy digression—to investigate the behavior of our function w. We need only look at what happens in the plane if we continue analytically a branch of $\sqrt{z - e_j}$. The analytic continuation of this function

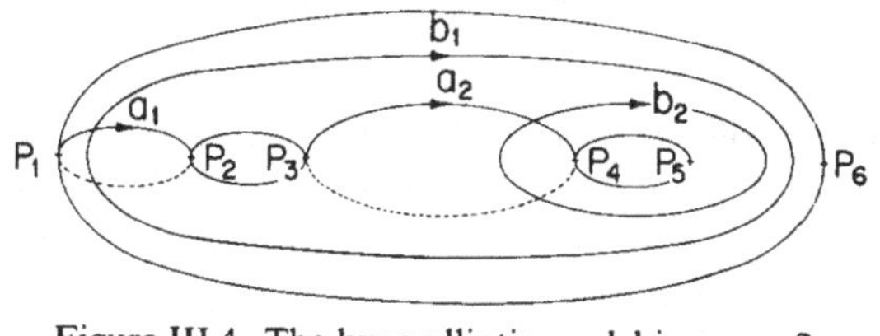

Figure III.4. The hyperelliptic model in genus 2.

element around a closed path changes sign if and only if the path has odd winding number about e_j (if it has even winding number, we return to the original function element). Thus w changes sign if we continue it along a simple closed path in $\mathbb{C} \cup \{\infty\}$ that encloses an odd number of the e_j. If a curve in $\mathbb{C} \cup \{\infty\} \setminus \{e_1, \ldots, e_{2g+2}\}$ begins at z and returns to this point, enclosing an odd number of e_j, then this curve must cross one of the "cuts" and thus its lift to M is a curve joining the points P and Q, $P \neq Q$, on M with $z(P) = z(Q)$. Thus w can be continued analytically along all paths in M. Furthermore, continuation along any closed path in M (which must encircle an even number of e_j's when viewed in the plane) leads back to the original value of w. Thus w is single-valued on M, and for any P, Q on M with $P \neq Q$, we have

$$z(P) = z(Q) \Rightarrow w(P) = -w(Q).$$

The fact that (w) is given by (7.4.2) is an immediate consequence of (7.4.1). □

III.7.5. The above proposition has some immediate consequences.

Corollary 1. *The g differentials*

$$\frac{z^j\,dz}{w}, \qquad j = 0, \ldots, g-1, \tag{7.5.1}$$

form a basis for the abelian differentials of the first kind on M.

PROOF. Without loss of generality $z(P_j) \neq 0$, and

$$(z) = \frac{Q_3 Q_4}{Q_1 Q_2}.$$

It is clear of course that the differentials in (7.5.1) are linearly independent, and all we must show is that they are holomorphic. Since

$$(dz) = \frac{P_1 \cdots P_{2g+2}}{Q_1^2 Q_2^2},$$

we see that

$$\left(\frac{z^j\,dz}{w}\right) = Q_1^{g-j-1} Q_2^{g-j-1} Q_3^j Q_4^j.$$

This divisor is integral as long as $j \leq g-1$ and therefore these differentials are holomorphic. □

Corollary 2. *On a hyperelliptic surface of genus $g \geq 2$ the products of the holomorphic abelian differentials (taken 2 at a time) form a $(2g-1)$-dimensional subspace of the $(3g-3)$-dimensional space of all holomorphic quadratic differentials.*

PROOF. Using the basis constructed in Corollary 1, the span of the products has a basis consisting of

$$\frac{z^j(dz)^2}{w^2}, \qquad j = 0, \ldots, 2g - 2. \qquad (7.5.2) \qquad \square$$

Remarks

1. Note that $2g - 1 = 3g - 3$ if and only if $g = 2$.
2. To obtain a basis for the holomorphic quadratic differentials for a hyperelliptic surface of genus $g > 2$ we must add to the list in (7.5.2), the differentials

$$\frac{z^j(dz)^2}{w}, \qquad j = 0, \ldots, g - 3.$$

EXERCISE

Obtain a basis for the holomorphic q-differentials on a hyperelliptic surface. What is the dimension of the span of the homogeneous polynomials (of degree q in q variables) of the abelian differentials of the first kind?

III.7.6. Recall the injective holomorphic mapping

$$\varphi: M \to J(M)$$

of a Riemann surface M into its Jacobian variety introduced in III.6.1. Let n be a positive integer. Since $J(M)$ is a (commutative) group, we say that a point $e \in J(M)$ is of *order* n, provided $ne = 0$ and $me \neq 0$ for all integers n with $0 < m < n$.

Proposition. *Let M be a hyperelliptic Riemann surface of genus ≥ 2. Choose a Weierstrass point P_0 as a base point for the map φ. Let $P \in M_1$, $P \neq P_0$. Then $\varphi(P)$ is of order 2 whenever P is a Weierstrass point.*

PROOF. We have seen that there is a meromorphic function f on M whose polar divisor is P^2. Since P_0 is a branch point of f, $(f - f(P_0)) = P_0^2/P^2$. Thus $P_0^2 \sim P^2$ and by Abel's theorem,

$$2\varphi(P) = \varphi(P^2) = \varphi(P_0^2) = 0. \qquad \square$$

EXERCISE

Let M be an arbitrary compact Riemann surface of genus $g \geq 1$. Let $\varphi: M \to J(M)$ be the embedding of M into its Jacobian variety with base point P_0. Assume that $\varphi(P)$ has order n, for some $P \in M$, $P \neq P_0$. Show that there exists an $f \in \mathscr{K}(M)$ with $(f) = P^n/P_0^n$. In particular if $g > 1$ and $2 \leq n \leq g$, then both P and P_0 are Weierstrass points on M.

For $g = 1$, return to the Weierstrass $\wp$-function considered in Exercise III.6.7. Show that the 4 branch points of $\wp$ are precisely the 4 half periods of $J(M)$.

III.7.7. If M is an arbitrary Riemann surface, we denote by Aut M the group of conformal automorphisms of M. Let $H \subset \text{Aut } M$ be a finite subgroup. For $P \in M$, set

$$H_P = \{h \in H; h(P) = P\}.$$

Then H_P is clearly a subgroup of H. We claim that H_P is cyclic. This is a consequence of the following

Proposition. *Let $h_1, \ldots, h_n$ be n holomorphic functions defined in a neighborhood of the origin. Assume that $h_j(0) = 0, j = 1, \ldots, n$, and that these n-functions form a group H under composition. Then H is a rotation group (that is, there is a simply connected neighborhood D of the origin and a conformal mapping f of the unit disk Δ onto D such that $f(0) = 0$, $h_j(D) = D$ and $f^{-1} \circ h_j \circ f$ is a rotation for $j = 1, \ldots, n$).*

Remark. The existence of a simply connected D invariant under H implies the rest of the proposition. In this case, choose f to be a Riemann map of Δ onto D with $f(0) = 0$. Then

$$\hat{h}_j = f^{-1} \circ h_j \circ f$$

is a conformal self-mapping of the unit disk that fixes 0, and hence of the form

$$\hat{h}_j(z) = e^{2\pi i\theta_j}z, \qquad 0 \le \theta_j < 2\pi.$$

Choosing the smallest positive θ_j and calling it θ, we see that $\hat{h}(z) = e^{2\pi i\theta}z$ generates the group $f^{-1}Hf$. Thus we also have

Corollary. *The group H_P is cyclic.*

Proof of the Proposition. Let h be a typical element of H. Then $h'(0) \ne 0$, since h is invertible. We claim that there is an $\varepsilon > 0$ such that h maps every disk $\{|z| \le r < \varepsilon\}$ onto a convex region. Such a region is convex if and only if $c = (h\{|z| = r\})$ is a convex curve if and only if the direction of the tangent vector to c is a monotonically increasing function of arg z; that is, if and only if the angle $(\frac{1}{2}\pi + \arg z + \arg h'(z))$ is an increasing function of arg z on $\{|z| = r\}$. A simple calculation shows that the hypothesis $(h'(0) \ne 0)$ guarantees the monotonicity of the direction of the tangent line. The derivative of the above angle with respect to arg z is (recall $\arg h'(z) = \operatorname{Re}(-i \log h')$)

$$1 + \operatorname{Re}\frac{zh''}{h'},$$

and this derivative is positive as long as $|z|$ is small.

Now choose ε so small, so that letting $\Delta_\varepsilon = \{z \in \mathbb{C}; |z| < \varepsilon\}$ we have that

$$h_j(\Delta_\varepsilon) \text{ is convex for } j = 1, \ldots, n.$$

Let $D = \bigcap_{j=1}^{n} h_j(\Delta_\varepsilon)$; D is convex and hence simply connected. Furthermore $h_j D = D$ for $j = 1, \ldots, n$. By the remark preceding the proof, we are done.

Remark. The material of the next chapter (in particular IV.8 and IV.9) will allow us to present at least two other proofs of the corollary to the above proposition.

III.7.8. We continue to use the notation introduced at the beginning of III.7.7. We give a Riemann surface structure to the orbit space M/H as follows. First, we topologize M/H such that the natural projection

$$\pi: M \to M/H$$

from a point onto its orbit is continuous. It is straightforward to check that this makes M/H into a Hausdorff space and π an open mapping. We introduce next a complex structure on M/H. If $P \in M$, and H_P is trivial, then any local coordinate at P serves as a local coordinate at $\pi(P)$ on M/H. In general, choose a neighborhood U of P in M so that H_P fixes U and so that in terms of some local coordinate vanishing at P, the action of the generator of H_P on U is given by

$$z \mapsto e^{2\pi i/k} z.$$

Then z^k is a local coordinate on M/H vanishing at $\pi(P)$.

Remark. The Riemann–Hurwitz relation allows us to compute the genus of M/H in terms of the genus of M and the branch points of π (= fixed points of elements of H). We will use this fact in V.1.

III.7.9. Proposition. *Let M be a compact Riemann surface of genus g. Then M is hyperelliptic if and only if there exists a conformal involution J ($J \in \text{Aut}\, M$ with $J^2 = 1$) on M that fixes $2g + 2$ points.*

Proof. Assume M is hyperelliptic. Let z be a function of degree 2 on M. For $P \in M$ set $J(P)$ to be the unique point Q such that $z(P) = z(Q)$ and $Q \neq P$ if such a point exists and $J(P) = P$ otherwise. It is clear that J is conformal and that if ζ is the local coordinate, $\zeta = \sqrt{z - z(P_j)}$ in a neighborhood of a branch point P_j of z, then $J(\zeta) = -\zeta$. It should be obvious to the reader why J will also be called the *sheet interchange* or the *hyperelliptic involution*. The fixed points of J are clearly the $2g + 2$ branch points of z.

Conversely, let J be a conformal involution of M with $2g + 2$ fixed points. Consider the group of order 2, $\langle J \rangle$, generated by J, and the two-sheeted covering

$$M \to M/\langle J \rangle$$

which is branched at the $2g + 2$ fixed points of J. Riemann–Hurwitz implies that $M/\langle J \rangle$ has genus 0, and thus M has a meromorphic function of degree 2. □

Corollary 1. *If $g \geq 2$, then the fixed points of the hyperelliptic involution are the Weierstrass points.*

Corollary 2. *If $g \geq 2$, then the hyperelliptic involution is the unique involution with $2g + 2$ fixed points.*

PROOF. Let $\tilde{J}$ be another involution with $2g + 2$ fixed points. Then, as we have seen, the fixed points must be the Weierstrass points. Also if z is a function of degree 2 on M, so is $z \circ \tilde{J}$. Thus by Theorem III.7.3, there is a Möbius transformation A such that $z \circ \tilde{J} = A \circ z$. Let $P_1, \ldots, P_{2g+2}$ be the Weierstrass points on M. Then $z(P_j) = z(\tilde{J}(P_j)) = A(z(P_j))$. Thus A fixes $z(P_j)$, $2g + 2$ distinct complex numbers (or ∞), and must be the identity. Hence $\tilde{J}$ is the sheet interchange. □

Corollary 3. *The hyperelliptic involution J on a (hyperelliptic) surface M of genus $g \geq 2$ is in the center of* Aut M.

PROOF. Let $h \in \operatorname{Aut} M$. Then $h \circ J \circ h^{-1}$ is an involution and fixes the $2g + 2$ points $h(P_j)$. Thus it is the hyperelliptic involution. Hence $h \circ J \circ h^{-1} = J$, or h commutes with J. □

III.7.10. Proposition. *On a hyperelliptic surface of genus g any function of degree $\leq g$ must be of even degree.*

PROOF. Clearly the proposition has content only for $g \geq 3$. Let f be a meromorphic function with polar divisor D with $\deg D \leq g$. Riemann–Roch says

$$2 \leq r(D^{-1}) = \deg D - g + 1 + i(D).$$

Thus $i(D) \geq 1$ and there is a holomorphic abelian differential ω such that $f\omega$ is also a holomorphic abelian differential. Using the basis for abelian differentials of the first kind introduced in Corollary 1 to Proposition III.7.4, we see that

$$\omega = \sum_{j=1}^{g} \alpha_j \frac{z^{j-1}\,dz}{w},$$

$$f\omega = \sum_{j=1}^{g} \beta_j \frac{z^{j-1}\,dz}{w},$$

with $\alpha_j, \beta_j \in \mathbb{C}$. Thus

$$f = \frac{\sum_{j=1}^{g} \beta_j z^{j-1}}{\sum_{j=1}^{g} \alpha_j z^{j-1}}$$

is a rational function of z (a function of degree 2), and must be of even degree. □

III.7.11. Proposition. *Let M be a hyperelliptic Riemann surface of genus ≥ 2. Let $T \in \operatorname{Aut} M$. Assume $T \notin \langle J \rangle$, where J is the hyperelliptic involution. Then T has at most four fixed points.*

PROOF. Let z be a function with two poles on M. For $T \in \text{Aut } M$, $z \circ T$ is also a function with two poles. Thus there is a Mobius transformation $A = \left[\begin{smallmatrix} a & b \\ c & d \end{smallmatrix}\right] \neq 1$ such that

$$z \circ T = \frac{az + b}{cz + d} = A \circ z.$$

We thus obtain an anti-homomorphism

$$\text{Aut } M \to SL(2,\mathbb{C})/\pm I.$$

The kernel of this anti-homomorphism is $\langle J \rangle$. If P is a fixed point of T, then

$$z(P) = z(T(P)) = A(z(P)),$$

and $z(P)$ is a fixed point of A. Since $A \neq 1$, A can have at most 2 fixed points and T can have at most 4. □

Corollary. *If T fixes a Weierstrass point, then T has at most 2 other fixed points.*

PROOF. Without loss of generality, we may assume the Weierstrass point is a pole of order 2 of z. Thus the A above fixes ∞, and must be affine (of the form $\left[\begin{smallmatrix} a & b \\ 0 & 1 \end{smallmatrix}\right]$). Since such an A has at most one other fixed point, T can have at most two other fixed points. □

III.8. Special Divisors on Compact Surfaces

Throughout this section, M is a compact Riemann surface of positive genus g. As before, $\text{Div}(M)$ denotes the group of divisors on M and Z denotes a canonical divisor (usually integral).

In this section we use the Clifford index (an integer invariant of a divisor class) to characterize hyperelliptic surfaces and to show (among other things) that every surface of genus 4 can be represented as either a two- or three-sheeted cover of the sphere (this is an improvement over the Weierstrass "gap" theorem).

The object is to represent a compact Riemann surface of genus g as a branched m-sheeted covering of the sphere, with m as small as possible. The methods of this section give sharp answers only for small g.

III.8.1. Let $D \in \text{Div}(M)$. We define the *Clifford index* of D by

$$c(D) = \deg D - 2r(D^{-1}) + 2.$$

The fact that we have introduced a useful concept is not at all clear. Clifford's theorem (III.8.4) will convince the reader of the usefulness of this definition.

We are interested in *special* divisors; that is, integral divisors D such that there exists an integral divisor D^* with

$$DD^* = Z.$$

We call D^* a *complementary divisor* of D.

It should be noted that for an integral divisor D, we have:

(a) D is special if and only if $i(D) > 0$;
(b) D is special whenever $\deg D \leq g - 1$ (because by Riemann–Roch, $i(D) \geq r(D^{-1}) \geq 1$); and
(c) if $\deg D = g$, then D is special if and only if $r(D^{-1}) \geq 2$, if and only if $L(D^{-1})$ contains a non-constant function (again by Riemann–Roch as in III.6.5).

Trivial (but Important) Remark. From the definition of $c(D)$ it follows that $c(D)$ and $\deg D$ have the same parity (both are even or both are odd).

III.8.2. It is clear that the Clifford index depends only on the divisor class. More is true.

Proposition. *For* $D \in \text{Div}(M)$, $c(D) = c(Z/D)$.

PROOF. We compute

$$\begin{aligned} c(Z/D) &= \deg(Z/D) - 2r(D/Z) + 2 \\ &= 2g - 2 - \deg D - 2i(D) + 2 \\ &= 2g - 2 - \deg D - 2(r(D^{-1}) - \deg D + g - 1) + 2 \\ &= \deg D - 2r(D^{-1}) + 2 = c(D). \end{aligned}$$ □

Remark. Again, from Riemann–Roch,

$$\begin{aligned} c(D) &= \deg D - 2(\deg D - g + 1 + i(D)) + 2 \\ &= -\deg D + 2g - 2i(D). \end{aligned} \qquad (8.2.1)$$

Thus, the proposition can be restated as

$$2i(D) + \deg D = 2i(Z/D) + \deg(Z/D).$$

or

$$i(D) - i(Z/D) = (g - 1) - \deg D.$$

In particular, if $\deg D = g - 1$, then $i(D) = i(Z/D)$.

III.8.3. Let D_1 and D_2 be integral divisors. The *greatest common divisor* (gcd) of D_1 and D_2 is the unique integral divisor D satisfying the following two properties:

a. $D \leq D_1$, $D \leq D_2$, and
b. whenever $\tilde{D}$ is an integral divisor with $\tilde{D} \leq D_1$ and $\tilde{D} \leq D_2$, then $\tilde{D} \leq D$.

If the divisor D_j is given by

$$D_j = \prod_{P \in M} P^{\alpha_j(P)} \qquad (\alpha_j(P) \geq 0 \text{ for all } P \in M \text{ and } \alpha_j(P) > 0 \text{ for only finitely many } P \in M), \quad (8.3.1)$$

then

$$\gcd(D_1,D_2) = (D_1,D_2) = \prod_{P \in M} P^{\min\{\alpha_1(P),\alpha_2(P)\}}.$$

Similarly, the *least common multiple* (lcm) of D_1 and D_2 is the unique integral divisor D satisfying:

a. $D \geq D_1$ and $D \geq D_2$, and
b. whenever $\tilde{D}$ is an integral divisor with $\tilde{D} \geq D_1$ and $\tilde{D} \geq D_2$, then $\tilde{D} \geq D$.

If D_j is given by (8.3.1), then

$$\operatorname{lcm}(D_1,D_2) = \prod_{P \in M} P^{\max\{\alpha_1(P),\alpha_2(P)\}}.$$

Furthermore,

$$\operatorname{lcm}(D_1,D_2)\gcd(D_1,D_2) = D_1 D_2.$$

Proposition. *Let D_1, D_2 be integral divisors. Set $D = (D_1,D_2)$. Then*

$$r(D_1^{-1}) + r(D_2^{-1}) - r(D^{-1}) \leq r\left(\frac{D}{D_1 D_2}\right).$$

PROOF. Observe that

$$L(D_j^{-1}) \subset L(DD_1^{-1}D_2^{-1}).$$

This inclusion follows from the fact that D_1D_2/D is integral and is $\geq D_j$, for $j = 1, 2$. Thus

$$L(D_1^{-1}) \vee L(D_2^{-1}) \subset L(DD_1^{-1}D_2^{-1}), \quad (8.3.2)$$

where $\vee$ denotes linear span. Next we show

$$L(D_1^{-1}) \cap L(D_2^{-1}) = L(D^{-1}). \quad (8.3.3)$$

To verify (8.3.3) assume D_j is given by (8.3.1). If $f \in L(D_1^{-1}) \cap L(D_2^{-1})$ has a pole at P of order $\alpha \geq 1$, then

$$\alpha \leq \alpha_j(P), \qquad j = 1, 2.$$

Thus also

$$\alpha \leq \min\{\alpha_1(P),\alpha_2(P)\},$$

and $f \in L(D^{-1})$. We have established

$$L(D_1^{-1}) \cap L(D_2^{-1}) \subset L(D^{-1}).$$

The reverse inclusion follows from the fact that $D_j \geq D$, $j = 1, 2$. This verifies (8.3.3).

Finally, using (8.3.2) and (8.3.3) and a little linear algebra,

$$\begin{aligned} r(D_1^{-1})+r(D_2^{-1})-r(D^{-1}) &= \dim L(D_1^{-1})+\dim L(D_2^{-1})-\dim(L(D_1^{-1})\cap L(D_2^{-1})) \\ &= \dim(L(D_1^{-1}) \vee L(D_2^{-1})) \leq \dim L(DD_1^{-1}D_2^{-1}) \\ &= r(DD_1^{-1}D_2^{-1}). \end{aligned}$$

□

Corollary 1. *Under the hypothesis of the proposition,*

$$c(D_1) + c(D_2) \geq c(D) + c(D_1 D_2 D^{-1}). \tag{8.3.4}$$

PROOF. The proof is by direct computation. □

Corollary 2. *If D is a special divisor with complementary divisor D^*, then*

$$c(D) \geq c((D,D^*)).$$

PROOF. By Corollary 1 and Proposition III.8.2,

$$2c(D) = c(D) + c(D^*) \geq c((D,D^*)) + c\left(\frac{DD^*}{(D,D^*)}\right)$$

$$= 2c((D,D^*)).$$ □

III.8.4. Theorem (Clifford). *Let D be a special divisor on M. We have:*

a. $c(D) \geq 0$.
b. *If* $\deg D = 0$ *or* $\deg D = 2g - 2$, *then* $c(D) = 0$.
c. *If* $c(D) = 0$, *then* $\deg D = 0$ *or* $\deg D = 2g - 2$ *unless the Riemann surface M is hyperelliptic.*

PROOF. Since $c(D) = c(D^*)$ and $\deg D + \deg D^* = 2g - 2$, where D^* is a complementary divisor of D, the theorem need be verified only for special divisors of degree $\leq g - 1$.

If $\deg D = 0$, then $r(D^{-1}) = 1$ and $c(D) = 0$. Thus part (b) is verified. Next, if $\deg D = 1$, then also $c(D) = 1$. We proceed by induction. Suppose now that $1 < \deg D \leq g - 1$, and $c(D) < 0$. Thus

$$r(D^{-1}) > \tfrac{1}{2} \deg D + 1 \geq 2.$$

Thus there is a non-constant function in $L(D^{-1})$. Let D^* be a complementary divisor. Replacing D by an equivalent divisor (it has the same Clifford index as D), we may assume that $(D,D^*) \neq D$. (This last assertion is of critical importance. It will be used in ever more sophisticated disguises. We shall hence verify it in detail. We must show that there is $P \in M$ which appears in D^* with lower multiplicity than in an integral divisor equivalent to D. Let ω be an abelian differential of the first kind such that $(\omega) = DD^*$. Let $f \in L(D^{-1})\backslash\mathbb{C}$. Then for all $c \in \mathbb{C}$, $(f - c)D$ is integral and equivalent to D. By properly choosing c (for example, $f^{-1}(c)$ should contain a point not in D^*), $(f - c)D$ will contain a point not in D^*. Now $(f - c)D$ and D^* are still complementary divisors since $(f - c)DD^* = ((f - c)\omega)$.) By Corollary 2 to our previous proposition, $c((D,D^*)) < 0$. But $\deg(D,D^*) < \deg D$ and (D,D^*) is a special divisor. By repeating the procedure we ultimately arrive at a divisor of degree zero or 1, which is a contradiction. This establishes (a).

It remains to verify (c). We have seen that if $0 < \deg D < 2g - 2$ with $c(D) = 0$, then $1 < \deg D < 2g - 3$, and

$$r(D^{-1}) = \tfrac{1}{2} \deg D + 1. \tag{8.4.1}$$

If $\deg D = 2$, then $r(D^{-1}) = 2$ and the surface is hyperelliptic. Now since $\deg D$ is even, we may assume that $\deg D \geq 4$. Choose D from its equivalence class so that

$$1 \neq (D,D^*) \neq D. \tag{8.4.2}$$

This can be done since $r(D^{-1}) \geq 3$. All we want is a function in $L(D^{-1})$ that vanishes at some point of D^* and at one point not in D^*. Let f be such a function. Then $(f)D$ is integral, equivalent to D, and satisfies (8.4.2). We have produced a special divisor (D,D^*) with $0 < \deg(D,D^*) < \deg D$ and $c((D,D^*)) = 0$. Thus $2 \leq \deg(D,D^*) \leq \deg D - 2$, and we can arrive at a special divisor of degree 2, with Clifford index zero. □

III.8.5. We now derive some consequences of Clifford's theorem. If $D \in \operatorname{Div}(M)$ is arbitrary with $0 \leq \deg D \leq 2g - 2$, then $c(D) \geq \deg D$ unless $r(D^{-1}) \geq 2$. In the latter case (since there is a non-constant function f in $L(D^{-1})$) we have $D' = (f)D$ is integral and equivalent to D. Since equivalent divisors have the same Clifford index, we have almost obtained

Corollary 1. *Let D be a divisor on M with $0 \leq \deg D \leq 2g - 2$. Then $c(D) \geq 0$. Equality occurs if and only if D is principal or canonical, unless M is a hyperelliptic Riemann surface.*

Proof. The remarks preceding the statement of the corollary show that unless $r(D^{-1}) \geq 2$, $c(D) \geq \deg D$. If $r(D^{-1}) \geq 2$, we have D equivalent to an integral divisor D' of the same degree. Proposition III.8.2 allows us to assume with no loss of generality that $\deg D \leq g - 1$. Now D' is a special divisor. The fact that $c(D') = c(D) \geq 0$ follows from Clifford's theorem. If D is neither principal nor canonical (and since as already stated we may restrict our attention to divisors of degree $\leq g - 1$, we are only interested in the case of D not principal), we have $c(D) = 0$ gives $r(D^{-1}) = 1 + \frac{1}{2}\deg D$. If $\deg D > 0$ we have once again $r(D^{-1}) \geq 2$, and as before D is equivalent to a special divisor and Clifford's theorem implies that M is hyperelliptic. If $\deg D = 0$, we have $r(D^{-1}) = 1$ and D is principal. □

Corollary 2. *If D is a divisor on M with $0 \leq \deg D \leq 2g - 2$, then*

$$i(D) \leq g - \frac{\deg D}{2}.$$

Equality implies that D is principal or canonical, unless M is hyperelliptic.

Proof. In view of (8.2.1), this is a restatement of Corollary 1. □

Corollary 3. *Let M be a compact Riemann surface of genus $g > 4$. Let D_1 and D_2 be two inequivalent integral divisors of degree 3 such that $r(D_1^{-1}) = 2 = r(D_2^{-1})$. Then M is hyperelliptic.*

Proof. Choose non-constant functions $f_j \in L(D_j^{-1})$, $j = 1, 2$. We may assume that each f_j is of degree 3 as otherwise there is nothing to prove. Since D_1

and D_2 are inequivalent, $f_1 \neq cf_2$ for any $c \in \mathbb{C}$. As a matter of fact, we have that $f_1 \neq A \circ f_2$ for any Möbius transformation A (of course, $A \circ f_2 = (af_2 + b)/(cf_2 + d)$ where $a, b, c, d \in \mathbb{C}$, $ad - bc \neq 0$). For if $f_1 = A \circ f_2$, then the divisor of poles of f_1 would be equivalent to the divisor of poles of f_2 (since the polar divisor of f_2 is equivalent to the polar divisor of $A \circ f_2$). This would contradict the fact that f_j is of degree 3, and that the two divisors D_j are inequivalent. Thus, we have 4 linearly independent functions in $L(D_1^{-1}D_2^{-1})$, namely: $1, f_1, f_2, f_1 f_2$. For if

$$c_0 + c_1 f_1 + c_2 f_2 + c_3 f_1 f_2 = 0$$

with $c_j \in \mathbb{C}$, then f_1 is a Möbius transformation of f_2. Thus $r(1/D_1D_2) \geq 4 = \frac{1}{2} \deg D_1D_2 + 1$, and $c(D_1D_2) = 0$. Since $\deg D_1D_2 = 6 < 2g - 2$, Clifford's theorem implies that M is hyperelliptic. □

Remark. Since M is hyperelliptic Proposition III.7.10 implies that M has no functions of degree three. Hence the functions f_1 and f_2 are really functions of degree 2, and thus by Theorem III.7.3, f_1 is indeed a Möbius transformation of f_2.

Furthermore, the above remark and the proof of Corollary 3 yield

Corollary 4. *If a surface of genus $g > 4$ admits a function f of degree 3, then any other function of degree 3 must be a fractional linear transformation of f. Further, on this surface we cannot find a function of degree 2.*

III.8.6. What happens in genus 4? We prove a special case of a more general (see the Corollary to Theorem III.8.13) result.

Proposition. *Every surface M of genus 4 has a special divisor D of degree 3 with $r(D^{-1}) = 2$.*

PROOF. Let $\{\zeta_1, \ldots, \zeta_4\}$ be a basis for the abelian differentials of the first kind on M. Then

$$\zeta_1^2, \zeta_1\zeta_2, \ldots, \zeta_1\zeta_4, \zeta_2^2, \zeta_2\zeta_3, \ldots, \zeta_3\zeta_4, \zeta_4^2$$

are 10 holomorphic quadratic differentials on M. They are linearly dependent since the dimension of the space of holomorphic quadratic differentials on M is 9. Hence, there are constants a_{jk} $(1 \leq k \leq 4,\ 1 \leq j \leq k)$ such that

$$\sum_{j \leq k} a_{jk}\zeta_j\zeta_k = 0.$$

We write this dependence relation in matrix form

$$(\zeta_1, \zeta_2, \zeta_3, \zeta_4) \begin{bmatrix} a_{11} & \frac{1}{2}a_{12} & \frac{1}{2}a_{13} & \frac{1}{2}a_{14} \\ \frac{1}{2}a_{12} & a_{22} & \frac{1}{2}a_{23} & \frac{1}{2}a_{24} \\ \frac{1}{2}a_{13} & \frac{1}{2}a_{23} & a_{33} & \frac{1}{2}a_{34} \\ \frac{1}{2}a_{14} & \frac{1}{2}a_{24} & \frac{1}{2}a_{34} & a_{44} \end{bmatrix} \begin{bmatrix} \zeta_1 \\ \zeta_2 \\ \zeta_3 \\ \zeta_4 \end{bmatrix} = 0. \tag{8.6.1}$$

Let Ξ be the column vector of differentials appearing in (8.6.1), $\Xi = {}^t(\zeta_1, \ldots, \zeta_4)$. We rewrite (8.6.1) as

$$ {}^t\Xi A \Xi = 0. \tag{8.6.2}$$

Since A is symmetric, there exists a non-singular matrix B such that ${}^tBAB$ is a diagonal matrix with 1 on the first l (≤ 4) diagonal entries and zeros on the rest of the diagonal. Thus, we rewrite (8.6.2) with respect to $B^{-1}\Xi$, a new basis for abelian differentials of the first kind, as

$$ {}^t(B^{-1}\Xi){}^tBAB(B^{-1}\Xi) = 0. \tag{8.6.3}$$

From now on we replace A by ${}^tBAB$ and Ξ by $B^{-1}\Xi$.

We claim now that A has rank 3 or 4. It clearly has positive rank. It cannot have rank 1, since in this case (8.6.3) reads $\zeta_1^2 = 0$. Similarly, it cannot have rank 2, since in this case (8.6.3) reads

$$\zeta_1^2 + \zeta_2^2 = 0,$$

which implies that $(\zeta_1) = (\zeta_2)$ and hence that ζ_1 and ζ_2 are dependent. Thus, the relation (8.6.3) is of the form

$$\zeta_1^2 + \zeta_2^2 + \zeta_3^2 = 0 \quad \text{or} \quad \zeta_1^2 + \zeta_2^2 + \zeta_3^2 + \zeta_4^2 = 0.$$

Another change of basis (for example, $\omega_1 = \zeta_1 + i\zeta_2$, $\omega_2 = \zeta_1 - i\zeta_2$, $\omega_3 = \zeta_3 + i\zeta_4$, $-\omega_4 = \zeta_3 - i\zeta_4$, in the second case) leads to the simpler relations

$$\zeta_1^2 = \zeta_3\zeta_4 \quad \text{or} \quad \zeta_1\zeta_2 = \zeta_3\zeta_4. \tag{8.6.4}$$

Write

$$(\zeta_1) = P_1 \cdots P_6.$$

Thus either ζ_3 or ζ_4 (say ζ_3) must vanish at at least 3 of the zeros of ζ_1. Thus ζ_1/ζ_3 is a non-constant function with at most three poles. This function gives rise to a special divisor D of degree 3 with $r(D^{-1}) \geq 2$. Theorem III.8.4 shows that $r(D^{-1}) \leq 2$. Hence $r(D^{-1}) = 2$. □

III.8.7. Theorem. *Let M be a compact Riemann surface of genus 4. Then one and only one of the following holds:*

a. *M is hyperelliptic.*
b. *M has a function f of degree 3 such that $(f) = \mathfrak{A}/D$ with D, $\mathfrak{A}$ integral and $D^2 \sim Z$. Any other function of degree 3 on M is a fractional linear transformation of f.*
c. *M has exactly two functions of degree 3 that are not Möbius transformations of each other.*

PROOF. Proposition III.7.10 implies that (a) cannot occur simultaneously with (b) or (c). We have already shown that every compact surface of genus 4 admits a non-constant function of degree ≤ 3. Note that for any divisor D of degree 3 on M (of genus 4), we have

$$r(D^{-1}) = i(D). \tag{8.7.1}$$

Suppose we are in case (b). Let us assume there is an integral divisor $D_1 \neq D$ with $\deg D_1 = 3$ and $r(D_1^{-1}) = 2$. Let $f_1 \in L(D_1^{-1})\backslash\mathbb{C}$. If D_1 is equivalent to D, then f_1 is a Möbius transformation of f. To see this, let $h \in \mathscr{K}(M)$ be such that $(h)D_1 = D$. Then $L(D^{-1}) = \langle 1, f\rangle$ and $L(D_1^{-1}) = \langle h, fh\rangle = \langle 1, f_1\rangle$. Thus, there are constants α, β, γ, δ such that

$$1 = \alpha h + \beta f h$$

$$f_1 = \gamma h + \delta f h$$

or

$$f_1 = \frac{\gamma + \delta f}{\alpha + \beta f}.$$

Thus we assume that f_1 is not a Möbius transformation composed with f, which implies that D_1 is not equivalent to D. We therefore have $1, f, f_1, ff_1$ are 4 linearly independent functions in $L(D^{-1}D_1^{-1})$. Then $r(D^{-1}D_1^{-1}) \geq 4$ and hence

$$i(DD_1) = r(D^{-1}D_1^{-1}) - 6 + 4 - 1 \geq 1.$$

Because $\deg DD_1 = 6$, DD_1 must be canonical, and hence $i(DD_1) = 1$. Since also $i(D^2) = 1$ by hypothesis, we conclude D_1 is equivalent to D. This contradiction establishes the uniqueness of f up to Möbius transformations.

We consider case (c). We assume that for no function f of degree 3 is it true that the square of the polar divisor D of f is canonical. Let $(f) = \mathfrak{A}/D$. Choose D_1, integral of degree 3, such that $DD_1 \sim Z$. Note that D and D_1 are inequivalent (otherwise $D^2 \sim Z$). Since $r(D^{-1}) = 2$, we conclude from (8.7.1) that there is a divisor D_2 of degree 3 such that $D_2 \neq D_1$ and $DD_2 \sim Z$. Choose holomorphic abelian differentials ω_j such that $(\omega_j) = DD_j$ $(j = 1,2)$ and set $f_1 = \omega_2/\omega_1$. Thus, since the polar divisor D_1 of f_1 is not linearly equivalent to D, we have produced a function of degree 3 which is not a Möbius transformation of f.

If $\mathfrak{A}$ is the polar divisor of an arbitrary function h of degree 3, and h is not a Möbius transformation of f, then $1, f, h, fh$ are 4 linearly independent functions in $L(1/D\mathfrak{A})$. Thus $r(1/D\mathfrak{A}) \geq 4$ and by Riemann–Roch we must have equality, and also conclude that $D\mathfrak{A}$ is canonical. Hence $D\mathfrak{A}$ is the divisor of the differential $c_1\omega_1 + c_2\omega_2$ for some constants c_1, c_2 (not both zero). Say $c_2 \neq 0$. Then $\omega_1/(c_1\omega_1 + c_2\omega_2)$ is a meromorphic function with divisor $D_1/\mathfrak{A}$. Hence h is a Möbius transformation of f_1. □

Remark. Cases (b) and (c) correspond to the relations of rank 3 and 4 of (8.6.4) respectively. In case (c) we have $(f) = \mathfrak{A}/D$, $(f_1) = \mathfrak{A}_1/D_1 = D_2/D_1$. Consider the identity

$$(DD_1)(\mathfrak{A}\mathfrak{A}_1) = (\mathfrak{A}D_1)(D\mathfrak{A}_1).$$

There are holomorphic differentials ζ_j, $j = 1, \ldots, 4$, satisfying

$$(\zeta_1) = DD_1, \qquad (\zeta_4) = (f_1\zeta_1) = D\mathfrak{A}_1,$$

$$(\zeta_2) = (f\zeta_4) = \frac{\mathfrak{A}}{D} D\mathfrak{A}_1 = \mathfrak{A}\mathfrak{A}_1,$$

$$(\zeta_3) = (f_1^{-1}f\zeta_4) = \frac{D_1}{\mathfrak{A}_1}\frac{\mathfrak{A}}{D} D\mathfrak{A}_1 = D_1\mathfrak{A}.$$

Thus, by multiplying these differentials by constants, we obtain,

$$\zeta_1\zeta_2 = \zeta_3\zeta_4.$$

The four differentials we have produced are linearly independent, since otherwise the functions f and f_1 would be related by a Möbius transformation. This follows from the fact that $\zeta_1/\zeta_4 = f_1^{-1}$, $\zeta_2/\zeta_4 = f$, $\zeta_3/\zeta_4 = f_1^{-1}f$. Thus, we have a relation of rank 4. In case (b), we start from the identity

$$(DD_1)^2 = D^2D_1^2,$$

where $D_1 \neq D$ is chosen to be equivalent to D. We know there are differentials ζ_j satisfying

$$(\zeta_2) = D^2, \qquad (\zeta_1) = DD_1, \qquad (\zeta_3) = (\zeta_1 f_1) = DD_1\frac{D_1}{D} = D_1^2,$$

(where $(f_1) = D_1/D$). Thus, we obtain the relation of rank 3 (after adjusting constants)

$$\zeta_1^2 = \zeta_2\zeta_3.$$

Again, the three differentials in question are independent, since $D_1 \neq D$. Thus if $a_1\zeta_1 + a_2\zeta_2 + a_3\zeta_3 = 0$, choosing a point $P \in D$, $P \notin D_1$, gives $a_3 = 0$. Similarly, choosing a point $P \in D_1$, $P \notin D$, gives $a_2 = 0$.

III.8.8. Proposition. *Let B be the polar divisor of a meromorphic function on M. Let A be an arbitrary divisor on M. Then*

$$2r(A^{-1}) \leq r(A^{-1}B^{-1}) + r(BA^{-1}). \tag{8.8.1}$$

Proof. If $B = 1$, the result reduces to a trivial equality. So assume B is not the unit divisor. Thus, there is a non-constant function f on M with polar divisor B. We can now find integral divisors B' and B'' such that

$$B' \sim B \sim B'',$$

and such that the above three divisors have no points in common (for example, $B' = f^{-1}(0)$, $B'' = f^{-1}(1)$).

We claim that

$$L(A^{-1}) \cap L(B''A^{-1}(B')^{-1}) = L(B''A^{-1}). \tag{8.8.2}$$

It is clear that

$$L(B''A^{-1}) \subset L(A^{-1}), \qquad L(B''A^{-1}) \subset L(B''A^{-1}(B')^{-1}),$$

and thus $L(B''A^{-1})$ is contained in the intersection.

Conversely, suppose $f \in L(A^{-1}) \cap L(B''A^{-1}(B')^{-1})$. Thus $(f)A = I_1$ and $(f)AB'/B'' = I_2$ with I_1 and I_2 integral divisors. It follows therefore that $I_1 = B''I_2/B'$. Since B'' and B' have no points in common, I_1 can be integral only if I_2 is a multiple of B'. Thus $I_2 = B'I_3$ and $I_1 = B''I_3$, and in particular $(f)A/B'' = I_1/B'' = I_3$ or $f \in L(B''A^{-1})$. This concludes the proof of (8.8.2).

We observe next that

$$L(A^{-1}) \subset L(A^{-1}(B')^{-1}), \qquad L(B''A^{-1}(B')^{-1}) \subset L(A^{-1}(B')^{-1}).$$

It thus follows that

$$\begin{aligned} r(A^{-1}) + r(B''A^{-1}(B')^{-1}) - r(B''A^{-1}) &= \dim(L(A^{-1}) \vee L(B''A^{-1}(B')^{-1})) \\ &\leq r(A^{-1}(B')^{-1}). \end{aligned}$$

Since $B \sim B' \sim B''$, (8.8.1) follows from the above. □

Corollary 1. *Under the hypothesis of the proposition,*

$$2c(A) \geq c(AB) + c(AB^{-1}).$$

Corollary 2. *Let M be a compact Riemann surface of genus $g \geq 4$. Let A and B be inequivalent integral divisors with*

$$3 \leq \deg B \leq \deg A \leq g - 1,$$

and

$$c(A) = 1 = c(B).$$

Then unless $B = A^*$ (a complementary divisor of A) M is hyperelliptic.

PROOF. From the definition of Clifford index,

$$2r(A^{-1}) = 1 + \deg A \geq 4.$$

Thus $r(A^{-1}) \geq 2$ and $r(B^{-1}) \geq 2$. There are now two possibilities.

Case I: B is not the polar divisor of a function. Then there is at least one $P \in B$ such that $r(PB^{-1}) = r(B^{-1})$. Let $B = B'P$ and observe that

$$2 \leq \deg B' \leq g - 2,$$

and

$$\begin{aligned} c(B') &= \deg B' - 2r((B')^{-1}) + 2 \\ &= \deg B - 1 - 2r(B^{-1}) + 2 = c(B) - 1 = 0. \end{aligned}$$

Hence the surface is hyperelliptic by Clifford's theorem.

Case II: B is the polar divisor of a function. In this case we apply Corollary 1 and Corollary 1 to Clifford's theorem (III.8.5), to obtain

$$2 = 2c(A) \geq c(AB) + c(AB^{-1}) \geq 0.$$

Recall that the degree and Clifford index of a divisor have the same parity. Thus, A, B have odd degree and AB and AB^{-1} have even degree and also even Clifford index. Thus either $c(AB) = 0$ or $c(AB^{-1}) = 0$. If $c(AB) = 0$, then by Clifford's theorem (Corollary 1 in III.8.5), M is hyperelliptic unless $AB \sim Z$. If $c(AB^{-1}) = 0$, then M is hyperelliptic unless AB^{-1} is principal. In this latter case $A \sim B$, contrary to hypothesis. □

III.8.9. Proposition. *Let A and B be integral divisors of the same degree $\leq g - 1$ with $r(A^{-1}) = r(B^{-1}) = t + 1$, $t \geq 1$. Then,*

$$c(AB) \leq \deg(A,B) - c((A,B)) + 2(c(A) - s), \tag{8.9.1}$$

whenever s is defined by

$$r(DA^{-1}B^{-1}) = r(A^{-1}B^{-1}) - s, \tag{8.9.2}$$

for some integral divisor D of degree $s \leq t$ with $(A,B)/D$ integral.

PROOF. We write

$$A = A'D, \qquad B = B'D,$$

with A', B' integral divisors. Clearly $L((A,B)/AB) \subset L(D/AB)$. Hence $r((A,B)/AB) \leq r(D/AB)$, and thus (8.9.2) implies

$$r\left(\frac{1}{AB}\right) - s \geq r\left(\frac{(A,B)}{AB}\right).$$

Translating the above inequality to Clifford indices, we obtain

$$\deg(AB) - c(AB) - 2s \geq \deg AB - \deg(A,B) - c\left(\frac{AB}{(A,B)}\right),$$

or

$$c(AB) \leq \deg(A,B) + c\left(\frac{AB}{(A,B)}\right) - 2s.$$

We now use Corollary 1 to Proposition III.8.3 in the form $2c(A) \geq c((A,B)) + c(AB/(A,B))$ to obtain (8.9.1). □

III.8.10. Let A', B' be two integral divisors of the same degree. Assume that $r(A'^{-1}) = r(B'^{-1}) = t + 1$, with $t \geq 1$. For almost all (to be defined in the proof of the assertion) integral divisors D of positive degree $s \leq t$, it is possible to find integral divisors A, B satisfying

$$A \sim A', \qquad B \sim B' \tag{8.10.1}$$

$$(A,B)/D \text{ is integral,} \tag{8.10.2}$$

and

$$r(A^{-1}B^{-1}) - s = r(DA^{-1}B^{-1}). \tag{8.10.3}$$

To verify the above claim, we begin by showing that for every $s \leq t + 1$ there is an integral divisor $D = P_1 \cdots P_s$ such that $r(1/A'B') - s = r(D/A'B')$. This is clearly equivalent to showing that the matrix $(f_k(P_j))$, $k = 1, \ldots, d$, $j = 1, \ldots, s$, with $f_1, \ldots, f_d$ a basis for $L(1/A'B')$, has rank s. Note first that $d \geq t + 1 \geq s$.

We choose P_1 such that P_1 does not appear in $A'B'$ and such that $f_1(P_1) \neq 0$. Consider now the meromorphic function of P:

$$\det\begin{bmatrix} f_1(P_1) & f_2(P_1) \\ f_1(P) & f_2(P) \end{bmatrix}.$$

Since $\{f_1, f_2\}$ are linearly independent, the determinant is not identically zero, and thus we can find P_2 such that P_2 does not appear in $A'B'$ and such that the determinant does not vanish at P_2. Hence

$$\operatorname{rank}\begin{bmatrix} f_1(P_1) & \cdots & f_d(P_1) \\ f_1(P_2) & \cdots & f_d(P_2) \end{bmatrix} = 2.$$

Having now chosen $P_1, \ldots, P_{k-1}$ such that no P_l appears in $A'B'$ and such that $\det(f_m(P_j))$, $m = 1, \ldots, k-1, j = 1, \ldots, k-1$ $(k \leq s)$ does not vanish, we consider

$$\det\begin{bmatrix} f_1(P_1) & \cdots & f_k(P_1) \\ \vdots & & \vdots \\ f_1(P_{k-1}) & \cdots & f_k(P_{k-1}) \\ f_1(P) & \cdots & f_k(P) \end{bmatrix}.$$

The linear independence of $\{f_1, \ldots, f_k\}$ assures us that the determinant does not vanish identically so that we can choose P_k not to appear in $A'B'$ and such that $\det(f_m(P_j))$, $m = 1, \ldots, k, j = 1, \ldots, k$ does not vanish. Hence

$$\operatorname{rank}\begin{bmatrix} f_1(P_1) & \cdots & f_d(P_1) \\ \vdots & & \vdots \\ f_1(P_k) & \cdots & f_d(P_k) \end{bmatrix} = k.$$

This verifies the existence of a divisor D of degree s with the required properties and in fact, shows that almost all divisors D of degree s would work.

Finally, we need show the existence of divisors $A \sim A'$ and $B \sim B'$ such that $(A,B)/D$ is integral. To this end consider $L(D/A')$. Clearly (for $s \leq t$) $r(D/A') \geq t + 1 - s \geq 1$, and thus there is a non-constant function $f \in L(D/A')$ such that $(f) = DI_1/A'$ and we may take $A = DI_1$. Similarly there is an integral divisor I_2 such that B may be chosen as DI_2.

III.8.11. Theorem. *Let M be a compact Riemann surface of genus $g \geq 4$. Assume that M is not hyperelliptic. Let $\mathfrak{A}$ be an integral divisor of degree $\leq g - 1$ with $c(\mathfrak{A}) = 1$. Then $r(\mathfrak{A}^{-1}) \leq 2$ (and thus $\deg \mathfrak{A} \leq 3$) except possibly if $g = 6$. In this case it is possible that $r(\mathfrak{A}^{-1}) = 3$ and $\mathfrak{A}^2$ is canonical.*

PROOF. Suppose that $r(\mathfrak{A}^{-1}) > 2$. Let $s = r(\mathfrak{A}^{-1}) - 2$. Choose integral divisors A, B, D with $\deg D = s$ such that they satisfy (8.10.1), (8.10.2), and (8.10.3) with $A' = \mathfrak{A} = B'$. We use now (8.3.4) to obtain

$$c\left(\frac{AB}{(A,B)}\right) + c((A,B)) \leq c(A) + c(B) = 2.$$

From (8.10.2) we see that

$$1 \leq s = \deg D \leq \deg(A,B) \leq \deg A \leq g - 1, \tag{8.11.1}$$

and hence,

$$1 \leq \deg\left(\frac{AB}{(A,B)}\right) \leq 2g - 3.$$

Thus, by Clifford's theorem,

$$c\left(\frac{AB}{(A,B)}\right) = 1 = c((A,B)).$$

We now consider cases:

Case I: $r(1/(A,B)) = 1$.

From the definition of Clifford index,

$$1 = c((A,B)) = \deg(A,B).$$

But Proposition III.8.9 implies that

$$c(\mathfrak{A}^2) = c(AB) \leq \deg(A,B) - c((A,B)) + 2(1 - s) = 2(1 - s).$$

Since $s \geq 1$, we see by Clifford's theorem that $s = 1$, $\mathfrak{A}^2 \sim Z$, and $r(\mathfrak{A}^{-1}) = 3$. Furthermore,

$$1 = c(\mathfrak{A}) = \deg \mathfrak{A} - 4,$$

shows that $\deg Z = 2 \deg \mathfrak{A} = 10$ or $g = 6$.

Case II: $r(1/(A,B)) \geq 2$ (thus $\deg(A,B) \geq 3$).

In this case we may assume that (A,B) is the polar divisor of a function. (Note first that if $P \in M$ appears in the divisor (A,B), then

$$r\left(\frac{P}{(A,B)}\right) = r\left(\frac{1}{(A,B)}\right) - 1.$$

Otherwise,

$$c\left(\frac{(A,B)}{P}\right) = \deg(A,B) - 1 - 2r\left(\frac{1}{(A,B)}\right) + 2 = c((A,B)) - 1 = 0,$$

contradicting that M is not hyperelliptic. A function belonging to

$$L\left(\frac{1}{(A,B)}\right)\setminus \bigcup_{P\in(A,B)} L\left(\frac{P}{(A,B)}\right)$$

(this is non-empty since every term in the finite union is of codimension 1 in $L(1/(A,B))$ will necessarily have polar divisor (A,B).) We thus apply Corollary 1 to Proposition III.8.8,

$$2 = 2c(\mathfrak{A}) \geq c(\mathfrak{A}(A,B)) + c\left(\frac{\mathfrak{A}}{(A,B)}\right). \tag{8.11.2}$$

Since $\mathfrak{A}$ and (A,B) have odd Clifford index, they also have odd degree. Hence $\mathfrak{A}(A,B)$ and $\mathfrak{A}/(A,B)$ have even degree and even Clifford index. Thus one of the terms on the right hand side of (8.11.2) must be zero. Since neither $\mathfrak{A}(A,B)$ nor $\mathfrak{A}/(A,B)$ is principal or canonical we are done. (If, for example, $\mathfrak{A}/(A,B)$ were principal, then $A \sim B \sim \mathfrak{A} \sim (A,B)$. Thus $A = (A,B) = B$, which clearly can be avoided from the beginning because $r(D/\mathfrak{A}) \geq r(1/\mathfrak{A}) - s = 2$.) □

III.8.12. To change the pace slightly, we prove a result in linear algebra. For the proposition of this section, we will need some elementary results from algebraic geometry.

To begin with, the set of $r \times r$ symmetric matrices can be viewed in a natural way as a vector space of dimension $\frac{1}{2}r(r+1)$ which we identify with $\mathbb{C}^{(1/2)r(r+1)}$.

Proposition. *The set V_ρ of $r \times r$ symmetric matrices of rank $\leq \rho$ is an irreducible homogeneous, algebraic subvariety of $\mathbb{C}^{(1/2)r(r+1)}$ of dimension $(1/2)\rho(\rho+1) + \rho(r-\rho)$.*

PROOF. The points in V_ρ are those $r \times r$ symmetric matrices which satisfy the homogeneous equations of degree $\rho + 1$ obtained by equating all $(\rho+1) \times (\rho+1)$ subdeterminants to zero. Thus V_ρ is a homogeneous algebraic subvariety. Furthermore, for every $\tilde{T} \in V_\rho$ there exists an $r \times r$ matrix T such that

$$\tilde{T} = {}^tTE_\rho T, \tag{8.12.1}$$

where E_ρ is the diagonal matrix with ones along the first ρ diagonal entries and zeros on the remaining $r - \rho$ diagonal entries. Thus V_ρ is connected and irreducible. To compute the dimension of V_ρ, we may consider only the matrices in V_ρ of rank precisely ρ. Consider such a

$$\tilde{T} = \begin{bmatrix} \tilde{A} & \tilde{B} \\ \tilde{C} & \tilde{D} \end{bmatrix} \in V_\rho$$

with $\tilde{A}$ a $\rho \times \rho$ non-singular symmetric matrix and thus $\tilde{B}$ is a $\rho \times (r-\rho)$

matrix ... etc. By (8.12.1) we see that there exists an $r \times r$ matrix

$$T = \begin{bmatrix} A & B \\ C & D \end{bmatrix}$$

such that

$$\begin{bmatrix} {}^tA & {}^tC \\ {}^tB & {}^tD \end{bmatrix} \begin{bmatrix} 1 & 0 \\ 0 & 0 \end{bmatrix} \begin{bmatrix} A & B \\ C & D \end{bmatrix} = \begin{bmatrix} {}^tAA & {}^tAB \\ {}^tBA & {}^tBB \end{bmatrix} = \begin{bmatrix} \tilde{A} & \tilde{B} \\ \tilde{C} & \tilde{D} \end{bmatrix}.$$

From the above it follows that A is non-singular (${}^tAA = \tilde{A}$ implies $(\det A)^2 = \det \tilde{A} \neq 0$). We claim that $\tilde{A}$, $\tilde{B}$ uniquely determine $\tilde{C}$, $\tilde{D}$. It is clear that $\tilde{C} = {}^t\tilde{B}$. Further, since $\tilde{B} = {}^tAB$ and A is non-singular $\tilde{D} = {}^tBB = {}^t\tilde{B}({}^tAA)^{-1}\tilde{B} = {}^t\tilde{B}\tilde{A}^{-1}\tilde{B}$. Conversely, given a non-singular symmetric $\tilde{A}$ as above, it can be written as ${}^tAA$ for some A. Also given an arbitrary $\tilde{B}$ as above, we can define a matrix $\tilde{T}$ of the above form. Thus we see that $\dim V_\rho$ is equal to the sum of the dimension of all possible $\tilde{A}$ and the dimension of all possible $\tilde{B}$. □

Corollary. *Let V be a k-dimensional vector space of symmetric $r \times r$ matrices. A sufficient condition for $V \cap V_\rho$ to contain a non-trivial matrix is that*

$$\frac{r(r+1)}{2} < k + \frac{1}{2}\rho(\rho + 1) + \rho(r - \rho).$$

Proof. The vector space V is clearly an irreducible homogeneous subvariety of $\mathbb{C}^{(1/2)r(r+1)}$. So is V_ρ. The intersection of two such varieties is never empty and in our case has dimension

$$\geq k + \tfrac{1}{2}\rho(\rho + 1)\rho(r - \rho) - \tfrac{1}{2}r(r + 1) \geq 1. \qquad \square$$

III.8.13. Theorem. *Let M be a compact Riemann surface of genus g and let A be an integral divisor of degree $n \geq g$. If $r(A^{-1}) > \frac{1}{4}(2n + 7 - g)$, then there is an integral divisor B on M with* $\deg B \leq \frac{1}{2}n$ *and $r(B^{-1}) \geq 2$.*

Proof. From the Riemann-Roch theorem,

$$r(A^{-2}) = 2n - g + 1.$$

Let $r = r(A^{-1})$. Let $\{f_1, \ldots, f_r\}$ be a basis for $L(A^{-1})$. Then $f_j f_k \in L(A^{-2})$ for $1 \leq j \leq k \leq r$, and these $\frac{1}{2}r(r + 1)$ elements are dependent (because $\frac{1}{2}r(r + 1) \geq \frac{1}{32}(2n + 8 - g)(2n + 12 - g) > 2n - g + 1$), and satisfy at least $k = \frac{1}{2}r(r + 1) - (2n - g + 1)$ linearly independent symmetric relations of the form

$$\sum_{j,k=1}^{r} a_{jk} f_j f_k = 0.$$

By the corollary to the previous proposition, a sufficient condition for the existence of a non-trivial rank ≤ 4 relation among these is that

$$\tfrac{1}{2}r(r+1) < k + 10 + 4(r-4),$$

which is precisely the condition imposed on $r = r(A^{-1})$. By a change of basis in $L(A^{-1})$, we may assume the relation is of the form (it cannot be of rank ≤ 2)

$$f_1 f_2 = f_3 f_4 \quad \text{or} \quad f_1^2 = f_3 f_4.$$

From here it is easy to get the desired conclusion. We use a variation of a previous argument. Let $A = P_1 \cdots P_n$. Assume (in case of the relation of rank 3 set $f_2 = f_1$)

$$(f_j) = \frac{Q_{j1} \cdots Q_{jn}}{P_1 \cdots P_n}, \qquad j = 1, \ldots, 4$$

(we are not assuming that $Q_{jl} \neq P_k$ for all j, l, k). Then

$$Q_{11} \cdots Q_{1n} Q_{21} \cdots Q_{2n} = Q_{31} \cdots Q_{3n} Q_{41} \cdots Q_{4n}.$$

Half the $\{Q_{1k}; k = 1, \ldots, n\}$ occur in $\{Q_{3k}\}$ or $\{Q_{4k}\}$. Say in the $\{Q_{3k}\}$, then

$$\left(\frac{f_1}{f_3}\right) = \frac{Q_{11}Q_{12} \cdots Q_{1n}}{Q_{31}Q_{32} \cdots Q_{3n}},$$

and $f = f_1/f_3$ has not more than $n/2$ zeros and is not constant. We can set $B =$ the divisor of zeros of f, and observe that $1/f \in L(B^{-1})$. □

Corollary. *Let M be a surface of genus $g \geq 4$. Then M carries a non-constant meromorphic function of degree $\leq \frac{1}{4}(3g+1)$.*

Proof. Let D be a special divisor of degree n with $\frac{3}{2}g \leq n \leq 2g - 2$. The Riemann–Roch theorem implies that

$$r(D^{-1}) = n - g + 1 + i(D) \geq n - g + 2.$$

The assumption that $n \geq \frac{3}{2}g$ implies $n - g + 2 > \frac{1}{4}(2n - g + 7)$ and allows us to apply the theorem. We conclude that there exists an integral divisor B of degree $\leq \frac{1}{2}n$, with $L(B^{-1}) \geq 2$. Choosing the integer n as low as possible, the Corollary follows. □

Remark. The above result generalizes Proposition III.8.6. The inequality is not sharp. A surface of even genus g is always an $n \leq \frac{1}{2}(g+2)$ sheeted cover of the sphere; the corresponding bound for odd genus is $\frac{1}{2}(g+3)$. Our methods do not produce these sharp results.

III.8.14. We consider a special case of the preceding theorem. Let Z be an integral canonical divisor (then $r(Z^{-1}) = g$). Note $r(Z^{-1}) > \frac{1}{4}(2 \deg Z + 7 - g)$ if and only if $g > 3$. Hence we obtain, in this case, from

Theorem III.8.13, an integral divisor B of degree $\leq g - 1$ with $r(1/B) \geq 2$. Thus, also $i(B) \geq 2$. (The above also follows from the corollary to Theorem III.8.13.)

We have produced the divisor B as a consequence of a quadratic relation of rank ≤ 4 among products of meromorphic functions. It could have equivalently been produced as a consequence of a quadratic relation of rank ≤ 4 among products of abelian differentials of the first kind. Now, there are $\frac{1}{2}g(g+1)$ such products and $3g - 3$ linearly independent holomorphic quadratic differentials. Thus there are at least $\frac{1}{2}g(g+1) - (3g-3) = \frac{1}{2}(g-2)(g-3)$ linearly independent symmetric relations

$$\sum_{j,k=1}^{g} a_{jk}\varphi_j\varphi_k = 0$$

among products of a basis $\{\varphi_1, \ldots, \varphi_g\}$ of abelian differentials of the first kind. The space of symmetric $g \times g$ matrices of rank ≤ 4 has dimension $4g - 6$, and thus the dimension of the "space of relations of rank ≤ 4" is at least

$$(4g-6) + \tfrac{1}{2}(g-2)(g-3) - \tfrac{1}{2}g(g+1) = (4g-6) - (3g-3) = g - 3.$$

Thus (another loose statement), the dimension of the space of integral divisors of degree $\leq g - 1$ and index of specialty ≥ 2 is $\geq g - 3$. The proofs of these assertions involve new ideas, and will be presented in III.11.

III.8.15. The next lemma is both technically very useful, and explains what it means for a reciprocal of an integral divisor to have positive dimension. Roughly it says that for an integral divisor D, $L(1/D)$ has dimension $\geq s$ if and only if D has $s - 1$ "free points", and gives a precise meaning to this statement. Note that since D is integral, $s \geq 1$.

Remark. The lemma could have been established at the end of III.4. We have delayed its appearance because its proof uses techniques of this section. In fact, we have already used and proved part of the lemma in III.8.10.

Lemma. *Let D be an integral divisor on M. A necessary and sufficient condition for $r(1/D) \geq s$ is: given any integral divisor D' of degree $\leq s - 1$, there is an integral divisor D'' such that $D'D'' \sim D$. Further, for sufficiency it suffices to assume that D' is restricted to any open subset U of $M_{(s-1)}$, the $(s-1)$-symmetric product of M.*

PROOF. To prove the sufficiency, we assume $r(1/D) = d$ and that $\{f_1, \ldots, f_d\}$ is a basis for $L(1/D)$. As in III.8.10, we construct a divisor $D' = Q_1 \cdots Q_{s-1}$ such that rank $(f_k(Q_j))$, $k = 1, \ldots, d$, $j = 1, \ldots, s-1$, is precisely $\min\{d, s-1\}$ and such that none of the points Q_j appear in D. If $d \leq s - 1$ we would have that the only function which vanished at $Q_1, \ldots, Q_{s-1}$ in

$L(1/D)$ would be the zero function contradicting the fact that we can choose an integral divisor D'' such that $D'D'' \sim D$. Hence $d \geq s$.

To show necessity assume that $r(D^{-1}) = s \geq 1$. Note that for arbitrary divisor $\mathfrak{A}$ and arbitrary point $P \in M$, $r(P\mathfrak{A}^{-1}) \geq r(\mathfrak{A}^{-1}) - 1$. Now let D' be integral of degree $\leq s - 1$. By the above remark, $r(D'/D) \geq s - (s - 1) = 1$. Now choose $f \in L(D'/D)$. Thus there exists a divisor D'' such that $D'D'' \sim D$. □

III.9. Multivalued Functions

We have on several occasions referred to certain functions as being (perhaps) multivalued, and then proceeded to show that they were single-valued. Examples of this occurred in the proof of the Riemann–Roch theorem (in a slightly disguised form), in the proof of the sufficiency part of Abel's theorem (Theorem III.6.3), and in the section on hyperelliptic surfaces. In this section we give a precise meaning to the term multivalued function and generalize Abel's theorem and the Riemann–Roch theorem to include multivalued functions. Multivalued functions will also be treated in IV.4 and IV.11. Analytic continuation (= multivalued functions) is one of the motivating elements in the development of Riemann surface theory. Riemann surfaces are the objects on which multivalued functions become single-valued. This aspect of multivalued functions will be explored in IV.11.

III.9.1. Let M be a Riemann surface. By a *function element* (f,U) on M, we mean an open set $U \subset M$ and a meromorphic function

$$f : U \to \mathbb{C} \cup \{\infty\}.$$

Two function elements (f,U) and (g,V) are said to be *equivalent* at $P \in U \cap V$, provided there is an open set W with $P \in W \subset U \cap V$ and

$$f|W = g|W.$$

The equivalence class of (f,U) at $P \in U$ is called the *germ* of (f,U) at P and will be denoted by (f,P). It is obvious that the set of germs of all function elements at a point $P \in M$ is in bijective correspondence with the set of convergent Laurent series at P (in terms of some local parameter) with finite singular parts. We topologize the set of germs as follows. Let (f,P) be a germ. Assume it is the equivalence class of the function element (f,U) with $P \in U$. By a neighborhood of (f,P), we shall mean the set of germs (f,Q) with $Q \in U$. It is obvious that each connected component $\mathscr{F}$, of the set of germs, is a Riemann surface equipped with two holomorphic maps

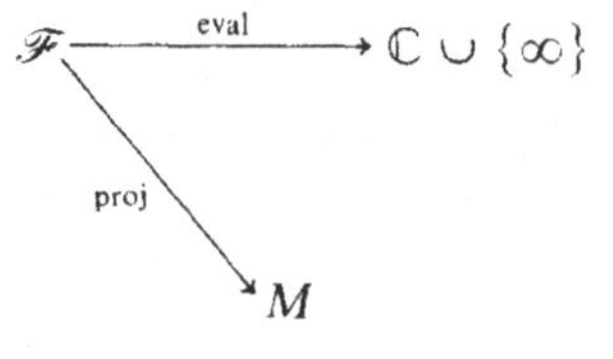

where

$$\text{eval}(f,P) = f(P)$$
$$\text{proj}(f,P) = P.$$

The verification of the above claims is routine, and hence left to the reader. In this connection the reader should see also IV.11.

III.9.2. We shall be interested exclusively in a restricted class of components $\mathscr{F}$ as above. Namely, we require that

i. proj be surjective, and that
ii. for every path $c:I \to M$, and every $f \in \mathscr{F}$ with $\text{proj}(f) = c(0)$, there exists a (necessarily) unique path $\tilde{c}:I \to \mathscr{F}$ with $\tilde{c}(0) = f$ and $c = \text{proj} \circ \tilde{c}$.

We shall call $\tilde{c}$ *the analytic continuation of* $\tilde{c}(0)$ *along* c.

Note that by the Monodromy theorem $\tilde{c}(1)$ depends only on the homotopy class of the path c and the point $\tilde{c}(0)$. (In the language of I.2.4, we are considering only those components $\mathscr{F}$ which are smooth unlimited covering manifolds of M.)

III.9.3. Let M be a compact Riemann surface and $\pi_1(M)$, its fundamental group. By a *character* χ on $\pi_1(M)$ we mean a homomorphism of $\pi_1(M)$ into the multiplicative subgroup of $\mathbb{C}$, $\mathbb{C}^* = \mathbb{C}\backslash\{0\}$. Since the range is commutative, a character χ is actually a homomorphism

$$\chi: H_1(M) \to \mathbb{C}^*.$$

The character χ is *normalized* if it takes values in the unit circle $\{z \in \mathbb{C}; |z| = 1\}$. If M has positive genus g, and $\{\aleph_1, \ldots, \aleph_{2g}\} = \{a_1, \ldots, a_g, b_1, \ldots, b_g\}$ is a canonical homology basis on M, then χ is determined uniquely by its values on a canonical homology basis, and these values may be arbitrarily assigned.

The set of characters on $\pi_1(M)$ forms an abelian group (under the obvious multiplication) that will be denoted by Char M.

III.9.4. We are finally ready to define a *multiplicative multivalued function belonging to a character* χ on M. By this we mean a component $\mathscr{F}$ of the germs of meromorphic functions on M satisfying the two properties listed in III.9.2 and the following additional property:

iii. the continuation of any $f \in \mathscr{F}$ along a closed curve c leads to a point f_1 with

$$\text{eval } f_1 = \chi(c) \text{ eval } f.$$

By varying the curve c in its homotopy class, it is easy to see that the above property holds or an open set; that is, as germs.

III.9.5. We examine more closely a multiplicative function $\mathscr{F}$ belonging to a character χ. Let c be a closed path in M and $\tilde{c}$ a path in $\mathscr{F}$ lying above it ($\text{proj} \circ \tilde{c} = c$). For each point $t \in [0,1]$, the point $\tilde{c}(t)$ is represented by a function element $(f_t, U(t))$ with $U(t)$ open in M and $c(t) \in U(t)$. By the continuity of the map $\tilde{c}: I \to \mathscr{F}$ and the compactness of I, we can find a subdivision

$$0 = t_0 < t_1 < t_2 < \cdots < t_n < t_{n+1} = 1$$

and function elements (f_0, U_0), (f_1, U_1), ..., (f_n, U_n) such that $\tilde{c}(t)$ is the equivalence class of (f_j, U_j) for $t \in [t_j, t_{j+1}]$, $j = 0, \ldots, n$. Furthermore, if c is a closed path, $c(0) = c(1)$, we may assume without loss of generality that $U_0 = U_n$. We say that the function element (f_n, U_n) has been obtained from the function element (f_0, U_0) by *analytic continuation along the curve* c. Since $\mathscr{F}$ belongs to the character χ, we see that

$$f_n = \chi(c) f_0.$$

Thus, we may view a multiplicative function $\mathscr{F}$ belonging to a character χ as a collection of function elements (f, U) on M with the properties

(i) given two elements (f_1, U_1) and (f_2, U_2) in $\mathscr{F}$, then (f_2, U_2) has been obtained by analytic continuation of (f_1, U_1) along some curve c on M, and

(ii) continuation of a function element (f, U) in $\mathscr{F}$ along the closed curve c leads to the function element $(\chi(c) f, U)$.

III.9.6. To define multiplicative differentials belonging to a character, we proceed as follows: Consider the triples (ω, U, z) where U is an open set in M, z is a local coordinate on U, and ω is a meromorphic function of z. If (ω_1, V, ζ) is another such triple and $P \in U \cap V$, then the two triples are said to be *equivalent* at P provided there is an open set $W \subset U \cap V$, with $P \in W$, and for all $Q \in W$

$$\omega_1(\zeta(z)) \frac{d\zeta}{dz} = \omega(z), \qquad z = z(Q).$$

Repeating the previous arguments, with this equivalence relation, one arrives at *multiplicative (or Prym) differentials belonging to a character* χ. To fix ideas, this involves a collection of triples (ω, U, z). If (ω_0, U_0, z_0) and (ω_n, U_n, z_n) are two such triples, then we can find a chain of triples

$$(\omega_j, U_j, z_j), \qquad j = 0, \ldots, n,$$

such that

$$U_j \cap U_{j+1} \neq \varnothing, \qquad j = 0, \ldots, n-1,$$

and

$$\omega_j(z_j(z_{j+1})) \frac{dz_j}{dz_{j+1}} = \omega_{j+1}(z_{j+1}), \; z_{j+1} = z_{j+1}(Q), \qquad Q \in U_j \cap U_{j+1}.$$

Furthermore, if we continue an element (ω, U, z) along a closed curve c to the element (ω_1, U, z), then

$$\omega_1(z) = \chi(c)\omega(z), \qquad z = z(Q), \; Q \in U.$$

III.9.7. It is quite clear that if f_j $(j = 1, 2)$ is a multiplicative function belonging to the character χ_j, then $f_1 f_2$ is a multiplicative function belonging to the character $\chi_1\chi_2$ (f_1/f_2 is a multiplicative function belonging to $\chi_1\chi_2^{-1}$ provided $f_2 \neq 0$). Similarly, if ω_j ($j = 3,4$) is a multiplicative differential belonging to the character χ_j, then $f_1\omega_3$ is a multiplicative differential belonging to the character $\chi_1\chi_3$ and ω_3/ω_4 is a multiplicative function belonging to the character $\chi_3\chi_4^{-1}$, provided $\omega_4 \neq 0$.

Proposition. *If $f \neq 0$ is a multiplicative function belonging to the character χ, then df is a multiplicative differential belonging to the same character and df/f is an abelian differential.*

Proof. We need only assure the reader that df is exactly what one expects it to be. If f is represented by (f,U) on the domain of the local parameter z, then df is represented by $(f'(z),U,z)$. □

III.9.8. It is quite obvious that a multiplicative function (and differential) has a well-defined order at every point, and thus we can assign to it a divisor.

Corollary 1. *If $f \neq 0$ is a multiplicative function, then* $\deg(f) = 0$.

Proof. The order of f at P, $\operatorname{ord}_P f$, just as in the ordinary case, is given by the residue at P of df/f. Since df/f is an abelian differential, the sum of its residues is zero. □

Corollary 2. *If $\omega \neq 0$ is a multiplicative differential, then* $\deg(\omega) = 2g - 2$.

Proof. Choose an abelian differential ω_1 on M, $\omega_1 \neq 0$. Then ω/ω_1 is a multiplicative function belonging to the same character as ω. Since $\deg(\omega_1) = 2g - 2$, Corollary 1 yields Corollary 2. □

III.9.9. If f is a multiplicative function without zeros and poles, then $df/f = d(\log f)$ is a holomorphic abelian differential. Letting $\{\zeta_1, \ldots, \zeta_g\}$ be a basis for the abelian differentials of the first kind on M dual to the canonical homology basis, we see that

$$\frac{df}{f} = d \log f = \sum_{j=1}^{g} c_j\zeta_j,$$

and thus

$$f(P) = f(P_0) \exp \int_{P_0}^{P} \sum_{j=1}^{g} c_j\zeta_j, \quad \text{with } c_j \in \mathbb{C}. \tag{9.9.1}$$

The character χ of the function f is then given by

$$\chi(a_k) = \exp c_k, \qquad k = 1, \ldots, g, \tag{9.9.2}$$

$$\chi(b_k) = \exp\left(\sum_{j=1}^{g} c_j\pi_{jk}\right), \qquad k = 1, \ldots, g, \qquad \left(\pi_{jk} = \int_{b_k} \zeta_j\right). \tag{9.9.3}$$

We shall call a character χ as above *inessential*, and f as above a *unit*.

Proposition. *If χ is an arbitrary character, then there exists a unique inessential character χ_1 such that χ/χ_1 is normalized.*

PROOF. Assume

$$\chi(a_k) = e^{s_k + it_k}, \qquad s_k, t_k \in \mathbb{R},$$
$$\chi(b_k) = e^{u_k + iv_k}, \qquad u_k, v_k \in \mathbb{R},$$

for $k = 1, \ldots, g$. To construct an inessential character χ_1 with $|\chi_1(c)| = |\chi(c)|$ for all $c \in H_1(M)$, we choose constants $c_k = \alpha_k + i\beta_k$ so that (compare with 9.9.1, 9.9.2, and 9.9.3),

$$|e^{c_k}| = e^{\alpha_k} = e^{s_k} \quad \text{or} \quad \alpha_k = s_k,$$

and

$$\left| e^{\sum_{j=1}^{g} c_j \pi_{jk}} \right| = e^{\sum_{j=1}^{g} \operatorname{Re} c_j \pi_{jk}} = e^{u_k} \quad \text{or} \quad \sum_{j=1}^{g} \operatorname{Re} c_j \pi_{jk} = u_k. \tag{9.9.4}$$

To see that this choice is indeed possible and in fact unique, recall that we can write the matrix $\Pi = (\pi_{jk})$ as

$$\Pi = X + iY$$

with Y positive definite, and thus non-singular. To solve (9.9.4) we write $c = {}^t(c_1, \ldots, c_g)$, $\alpha = {}^t(\alpha_1, \ldots, \alpha_g)$, etc., ..., and note that we want to solve

$$\operatorname{Re}[(X + iY)(\alpha + i\beta)] = u.$$

Since we have already chosen $\alpha = s$ the equation we wish to solve is

$$\operatorname{Re}[(X + iY)(s + i\beta)] = u,$$

or what amounts to the same thing

$$Y\beta = -u + Xs.$$

Since Y is non-singular there is a unique β which solves this system of equations. □

Corollary. *A normalized inessential character is trivial.*

III.9.10. Theorem. *Every divisor D of degree zero is the divisor of a unique (up to a multiplicative constant) multiplicative function belonging to a unique normalized character.*

PROOF. Let $D = P_1 \cdots P_r/Q_1 \cdots Q_r$ with $P_j \neq Q_k$ all $j, k = 1, \ldots, r > 0$. If f_j is a multiplicative function belonging to the normalized character χ_j ($j = 1,2$), and $(f_j) = D$, then f_1/f_2 is a multiplicative function without zeros and poles. Thus, $\chi_1\chi_2^{-1}$ is inessential and normalized; hence trivial. Thus it suffices to prove existence for the divisor $D = P_1/Q_1$, with $P_1, Q_1 \in M$, $P_1 \neq Q_1$. Recall the normalized abelian differential $\tau_{P_1Q_1}$ introduced in

III.3.6. Define

$$f(P) = \exp \int_{P_0}^{P} \tau_{P_1Q_1}.$$

The character to which f belongs is not necessarily normalized. But the arguments in III.9.9 showed how to get around this obstacle. □

Corollary. *Every divisor D of degree $2g - 2$ is the divisor of a unique (up to a multiplicative constant) multiplicative differential belonging to a unique normalized character.*

Proof. Let Z be a canonical divisor, and apply the theorem to the divisor of degree zero D/Z. □

III.9.11. Theorem. *Every character χ is the character of a multiplicative function (that does not vanish identically).*

Proof. Let $\{a_1, \ldots, a_g, b_1, \ldots, b_g\}$ be a canonical homology basis on M. It is obvious that χ may be replaced (without loss of generality) by $\chi\chi_1$ with χ_1 inessential (a unit always exists with character χ_1). Since the value of an inessential character can be prescribed arbitrarily on the "a" periods (see III.9.9), we may assume

$$\chi(a_j) = 1, \qquad j = 1, \ldots, g. \tag{9.11.1}$$

Furthermore, writing

$$\chi(b_j) = \exp(\beta_j), \qquad j = 1, \ldots, g,$$

and fixing $\varepsilon > 0$, we may assume in addition

$$|\beta_j| < \varepsilon, \qquad j = 1, \ldots, g. \tag{9.11.2}$$

(For if χ is arbitrary and satisfies (9.11.1), then choose an integer $N > 0$ so that

$$|\beta_j/N| < \varepsilon, \qquad j = 1, \ldots, g.$$

Set

$$\hat{\chi}(a_j) = 1, \qquad \hat{\chi}(b_j) = e^{\beta_j/N}, \qquad j = 1, \ldots, g.$$

If f is a function belonging to the character $\hat{\chi}$, then f^N is a function belonging to the character χ.) Let us now fix a single parameter disc U on M with local coordinate z. We may assume U is equivalent via z to the unit disc. Choose a point $(z_1, \ldots, z_g) \in U^g$ such that the divisor $D = P_1 \cdots P_g$ ($P_j = z^{-1}(z_j)$) is not special. (This is, as we have previously seen, always possible.) We use once again the normalized abelian differential τ_{PQ} of the third kind (introduced in III.3).

We define

$$f(P) = \exp \sum_{j=1}^{g} \int_{P_0}^{P} \tau_{\hat{z}_j z_j},$$

where $\hat{z} = (\hat{z}_1, \ldots, \hat{z}_g) \in U^g$ is a variable point, and $P_0 \notin U$. For fixed $\hat{z}$ we get a multiplicative function with character χ_f satisfying (9.11.1). This is, of course, a consequence of the fact that $\int_{a_k} \tau_{PQ} = 0$, for $k = 1, \ldots, g$. Furthermore,

$$\chi_f(b_k) = \exp \sum_{j=1}^{g} \int_{b_k} \tau_{\hat{z}_j z_j} = \exp \sum_{j=1}^{g} 2\pi i \int_{z_j}^{\hat{z}_j} \zeta_k \tag{9.11.3}$$

by the bilinear relation (3.6.3), where $\{\zeta_1, \ldots, \zeta_g\}$ is the normalized basis for the abelian differentials of the first kind dual to the given holomogy basis.

Now (9.11.3) obviously defines a holomorphic mapping $\Phi: U^g \to \mathbb{C}^g$

$$\Phi = (\Phi_1, \ldots, \Phi_g)$$

$$\Phi_k(\hat{z}) = \exp \sum_{j=1}^{g} 2\pi i \int_{z_j}^{\hat{z}_j} \zeta_k$$

$$\Phi(z_1, \ldots, z_g) = (1, \ldots, 1).$$

The theorem will be established if we show that Φ covers a neighborhood in $\mathbb{C}^g$ of $(1, \ldots, 1)$. By the inverse function theorem it suffices to show that the Jacobian of Φ at $(z_1, \ldots, z_g)$ is non-singular. Write $2\pi i \zeta_j(z) = \varphi_j(z)\,dz$ in terms of the local coordinate z, and note that

$$\left.\frac{\partial \Phi_k}{\partial z_j}\right|_{z_j} = \varphi_k(z_j), \qquad j, k = 1, \ldots, g.$$

Thus, we have to show that the matrix

$$\begin{bmatrix} \varphi_1(z_1) & \varphi_1(z_2) & \cdots & \varphi_1(z_g) \\ \varphi_2(z_1) & & \cdots & \varphi_2(z_g) \\ \vdots & & & \vdots \\ \varphi_g(z_1) & & \cdots & \varphi_g(z_g) \end{bmatrix}$$

is non-singular. If this matrix were singular, then a non-trivial linear combination of the rows would be the zero vector in $\mathbb{C}^g$. That is (since $\zeta_1, \ldots, \zeta_g$ are linearly independent), there would be a non-zero abelian differential of the first kind vanishing at $P_1, \ldots, P_g$, contradicting the assumption that the divisor D is not special. □

Remark. The reader should notice the similarity between the proof of the above theorem and the proof of the Jacobi inversion theorem.

III.9.12. Let $\mathfrak{A}$ be an arbitrary divisor on M. In analogy to our work in III.4, we define for an arbitrary character χ,

$L_\chi(\mathfrak{A}) = \{$multiplicative functions f belonging to the character χ such that $(f) \geq \mathfrak{A}\}$,

$r_\chi(\mathfrak{A}) = \dim L_\chi(\mathfrak{A})$,

$\Omega_\chi(\mathfrak{A}) = \{$multiplicative differentials ω on M belonging to the character χ such that $(\omega) \geq \mathfrak{A}\}$, and

$i_\chi(\mathfrak{A}) = \dim \Omega_\chi(\mathfrak{A})$.

Theorem (Riemann–Roch). *Let M be a compact surface of genus g, and χ a character on M. Then for every divisor $\mathfrak{A}$ on M, we have*

$$r_\chi(\mathfrak{A}^{-1}) = \deg \mathfrak{A} - g + 1 + i_{\chi^{-1}}(\mathfrak{A}). \tag{9.12.1}$$

PROOF. Choose a multiplicative function $0 \neq f$ belonging to the character χ. (Theorem III.9.11 gives us the existence of such an f.) Then

$$h \in L_\chi(\mathfrak{A}^{-1}) \Leftrightarrow h/f \in L(\mathfrak{A}^{-1}(f)^{-1}), \tag{9.12.2}$$

and

$$\omega \in \Omega_{\chi^{-1}}(\mathfrak{A}) \Leftrightarrow \omega f \in \Omega(\mathfrak{A}(f)). \tag{9.12.3}$$

Thus

$$r_\chi(\mathfrak{A}^{-1}) = r((\mathfrak{A}(f))^{-1}) \quad \text{and} \quad i_{\chi^{-1}}(\mathfrak{A}) = i(\mathfrak{A}(f)). \tag{9.12.4}$$

We apply now the standard Riemann–Roch theorem (4.11) to the divisor $\mathfrak{A}(f)$ and obtain

$$r((\mathfrak{A}(f))^{-1}) = \deg((\mathfrak{A}(f))) - g + 1 + i(\mathfrak{A}(f)),$$

which is equivalent to (9.12.1) in view of (9.12.4) and the fact that

$$\deg(\mathfrak{A}(f)) = \deg \mathfrak{A} + \deg(f) = \deg \mathfrak{A}.$$

Remark. The isomorphisms established in (9.12.2) and (9.12.3) are also useful in their own right.

III.9.13 An important class of characters are the so-called nth-integer characteristics. Let $\{a_1, \ldots, a_g, b_1, \ldots, b_g\}$ be a canonical homology basis on M. Consider an integer $n \geq 2$ and a $2 \times g$ matrix

$$\begin{bmatrix} \varepsilon \\ \varepsilon' \end{bmatrix}_n = \begin{bmatrix} \varepsilon_1/n, \ldots, \varepsilon_g/n \\ \varepsilon'_1/n, \ldots, \varepsilon'_g/n \end{bmatrix}$$

with $\varepsilon_j, \varepsilon'_j$ integers between 0 and $n - 1$. The symbol $[{}^{\varepsilon}_{\varepsilon'}]_n$ will be called an nth-*integer characteristic*. It determines a normalized character χ on M via

$$\chi(a_j) = \exp(2\pi i \varepsilon_j/n),$$
$$\chi(b_j) = \exp(2\pi i \varepsilon'_j/n), \qquad j = 1, \ldots, g.$$

It is clear that for every multiplicative function f belonging to such a character χ, $|f|$ is a (single-valued) function on M, and f itself lifts to a function on an n-sheeted covering surface of M. Furthermore, we can construct such an f by taking nth roots of a meromorphic function h on M provided (h) is an nth power of a divisor on M (that is, provided the order of zeros and poles of h are integral multiples of n).

III.9.14. Proposition. *For $\chi \in \operatorname{Char} M$,*

$$i_\chi(1) = \begin{cases} g & \text{if } \chi \text{ is inessential,} \\ g-1 & \text{if } \chi \text{ is essential.} \end{cases}$$

PROOF. The Riemann–Roch theorem says:

$$r_{\chi^{-1}}(1) = -g + 1 + i_\chi(1).$$

Thus, $i_\chi(1) \geq g - 1$, and $i_\chi(1) \leq g$. Furthermore, $i_\chi(1) = g$ if and only if there is an $f \in L_{\chi^{-1}}(1)$, $f \neq 0$. Since such an f must be a unit if it is not identically zero, we are done.

III.9.15. We end this section with the statement of Abel's theorem for multiplicative functions belonging to a character χ. The proof is omitted since it is exactly the same as the proof already studied (in III.6.3).

Theorem (Abel). *Let $D \in \operatorname{Div}(M)$, $\chi \in \operatorname{Char} M$. A necessary and sufficient condition for D to be the divisor of a multiplicative function belonging to the character χ is*

$$\varphi(D) = \frac{1}{2\pi i} \sum_{j=1}^{g} \log \chi(b_j) e^{(j)} - \frac{1}{2\pi i} \sum_{j=1}^{g} \log \chi(a_j) \pi^{(j)} \,(\operatorname{mod} L(M)),$$

and

$$\deg D = 0.$$

III.9.16. Exercise

Define a mapping

$$\varphi : \operatorname{Char} M \to J(M)$$

as follows. For $\chi \in \operatorname{Char} M$, select a non-constant meromorphic function f belonging to the character χ. Let $\varphi(\chi) = \varphi((f))$, where, as before, $\varphi((f))$ is the image of the divisor (f) in the Jacobian variety. Show that:

(1) The mapping φ is a well defined group homomorphism.

(2) The mapping φ is surjective.

(3) $\chi \in \operatorname{Kernel} \varphi$ if and only if χ is an inessential character. Conclude that as groups

$$J(M) \cong \operatorname{Char} M/(\text{Inessential characters on } M).$$

(4) Obtain an alternate proof of the Jacobi inversion theorem as follows. First show that every $e \in J(M)$ is the image of a divisor of degree zero. (We established this fact by the use of the implicit function theorem.) Thus every $e \in J(M)$ is the image of some $\chi \in \operatorname{Char} M$. Now use Theorem III.9.12 (Riemann–Roch) to show that $r_\chi(1/P_0^g) = 1$. Thus there is a multiplicative function belonging to the character χ with $\leq g$ poles. Jacobi inversion follows from this observation. The same method (see III.11) can also determine the dimension of the space of divisors of degree g that have image e.

Remark. The reader should at this point review the remark at the end of III.9.11.

III.9.17. Exercise

(1) Let $P \in M$ and let z be a local coordinate vanishing at P. Let $n \in \mathbb{Z}$, $n \geq 1$. By a *principal part (of a meromorphic function) at* P we mean a rational function of z of the form

$$f(z) = \sum_{k=-n}^{-1} a_k z^k.$$

(2) Let $\{P_1, \ldots, P_m\}$ be m distinct points and z_j a local coordinate vanishing at P_j. Let f_j be a principal part of a meromorphic function at P_j (in terms of the local coordinate z_j). The collection $F = \{f_1, \ldots, f_m\}$ will be called a *system of principal parts at* $\{P_1, \ldots, P_m\}$.

(3) Let F be a system of principal parts at $\{P_1, \ldots, P_m\}$. Let φ be an abelian differential on M that is regular at P_j for $j = 1, \ldots, m$. Then

$$F(\varphi) = \sum_{j=1}^{m} \operatorname{Res}_{P_j} f_j \varphi$$

is a well-defined linear functional on the space of all such φ. In particular, F *induces a linear function* on $\mathscr{H}^1(M)$, the vector space of holomorphic differentials on M.

(4) Let $\chi \in \operatorname{Hom}(H_1(M), \mathbb{C})$; that is, χ is a homomorphism from the first homology group of M into the complex numbers. By an *additive multivalued function belonging to* χ we mean a component $\mathscr{F}$ of the germs of meromorphic functions on M satisfying the two properties listed in III.9.2, and (iii)$'$ the continuation of any $f \in \mathscr{F}$ along a closed curve c leads to $f_1 \in \mathscr{F}$ with

$$\operatorname{eval} f_1 = \operatorname{eval} f + \chi(c).$$

Show that for every additive function $\mathscr{F}$, $d\mathscr{F}$ is a (well-defined) holomorphic differential. Further $\mathscr{F}$ defines a system of principal parts and this induces a linear functional on $\mathscr{H}^1(M)$.

(5) Using the normalized abelian differentials of the second kind introduced in III.3, show that
   (a) Every system of principal parts is the system of principal parts of an additive function.
   (b) For every $\chi \in \operatorname{Hom}(H_1(M), \mathbb{C})$ there is an additive function belonging to χ.
   (c) Let $\{a_1, \ldots, a_g, b_1, \ldots, b_g\}$ be a canonical homology basis on M. A homomorphism $\chi \in \operatorname{Hom}(H_1(M), \mathbb{C})$ is called *normalized* if $\chi(a_j) = 0$, $j = 1, \ldots, g$. Show that every system of principal parts belongs to a unique normalized homomorphism.

(6) Prove that a system F of principal parts is the system of principal parts of a (single-valued) meromorphic function on M if and only if F induces the zero linear functional on $\mathscr{H}^1(M)$.

(7) Use the notation of III.3, and show that

$$\frac{\alpha_{n-2}}{n-1} = \int_Q^P \tau_R^{(n)},$$

where

$$\tau_{PQ} = \sum_{j=0}^{\infty} (\alpha_j z^j)\, dz \quad \text{at } P,$$

and

$$R, P, Q, \text{ are three distinct points on } M.$$

(8) Let F be a system of principal parts at $\{P_1, \ldots, P_m\}$. Let Q be arbitrary but distinct from P_j, $j = 1, \ldots, m$. Define for $P \neq P_j$, $P \neq Q$,

$$E(P) = -F(\tau_{PQ}),$$

Show that E agrees on $M \setminus \{P_1, \ldots, P_m, Q\}$ with the unique (up to additive constant) additive function $\mathscr{F}$ with F as its system of principal parts and belonging to a normalized homomorphism.

III.10. Projective Imbeddings

Throughout this section, M is a compact Riemann surface of genus $g \geq 2$, q is an integer ≥ 1, and $d = \dim \mathscr{H}^q(M)$ ($= g$ for $q = 1$, $= (2q - 1)(g - 1)$ for $q \geq 2$). We show that every such surface M can be realized as a submanifold of three-dimensional complex projective space.

III.10.1. We have seen that for each $P \in M$, there is a $\omega \in \mathscr{H}^1(M)$ with $\omega(P) \neq 0$ (thus also $\omega^q \in \mathscr{H}^q(M)$ with $\omega^q(P) \neq 0$). Let $\{\zeta_1, \ldots, \zeta_d\}$ be a basis for $\mathscr{H}^q(M)$. The preceding remark shows that we have a well-defined holomorphic mapping, called the *q-canonical mapping*, of M into complex projective space

$$\theta : M \to \mathbb{P}^{d-1},$$

where $\theta(P)$ is given by $(\zeta_1(P), \ldots, \zeta_d(P))$ in homogeneous coordinates. We are using here our usual conventions: Let z be any local coordinate on M. Express the differential

$$\zeta_j = \varphi_j(z)\, dz^q$$

in terms of this local coordinate. Then

$$\theta(z) = (\varphi_1(z), \ldots, \varphi_d(z))$$

in terms of the local coordinate z. The image of $P \in M$ in $\mathbb{P}^{d-1}$ is clearly independent of the choice of local coordinate used, and a change of basis of $\mathscr{H}^q(M)$ leads to projective transformation of $\mathbb{P}^{d-1}$. Hence the map θ is canonical, up to the natural self-maps of projective space.

III.10.2. Theorem. *The q-canonical holomorphic mapping $\theta : M \to \mathbb{P}^{d-1}$ is injective (and of maximal rank) unless $q = 1$ and M is hyperelliptic or $g = 2 = q$. In these exceptional cases θ is two-to-one.*

Proof. Assume that for $P \in M$, 2 is a "q-gap" (that is, there exists a holomorphic q-differential with a simple zero at P). Then by choosing a basis for

$\mathscr{H}^q(M)$ adapted to P, we see that (z = local coordinate vanishing at P)

$$\varphi_1(z) = (1 + O(|z|))\,dz^q, \qquad z \to 0,$$
$$\varphi_2(z) = (z + O(|z|^2))\,dz^q, \qquad z \to 0.$$

Thus

$$\frac{\varphi_2}{\varphi_1}(z) = z + O(|z|^2), \qquad z \to 0,$$

and we conclude that θ has a non-vanishing differential at P ($z = 0$) and is, hence, of maximal rank at P.

Now assume there exist P and $Q \in M$, $P \neq Q$, with $\theta(P) = \theta(Q)$. By choosing a basis adapted to P as above, we see that in homogeneous coordinates

$$\theta(P) = (1,\underbrace{0,\ldots,0}_{(d-1)\text{-times}}).$$

Thus, if $\theta(P) = \theta(Q)$, we must have,

$$\theta(Q) = (\lambda,\underbrace{0,\ldots,0}_{(d-1)\text{-times}}), \qquad \lambda \neq 0.$$

In particular, we see that every holomorphic q-differential that vanishes at P also vanishes at Q, or

$$L(Z^{-q}P) = L(Z^{-q}PQ), \tag{10.2.1}$$

since $L(Z^{-q}P)$ and $L(Z^{-q}PQ)$ are isomorphic to the vector spaces of holomorphic q-differentials which vanish at P and P and Q, respectively. Hence

$$r(Z^{-q}P) = r(Z^{-q}PQ) = d - 1. \tag{10.2.2}$$

Let us assume $q = 1$, and use (10.2.2) and Riemann–Roch to compute

$$r\left(\frac{1}{PQ}\right) = 2 - g + 1 + r(PQ/Z) = 2.$$

Thus M must be hyperelliptic. Next we assume $q > 1$, and compute

$$\begin{aligned} r(Z^{-q}PQ) &= q(2g - 2) - 2 - g + 1 + r(Z^{q-1}/PQ) \\ &= (2q - 1)(g - 1) - 2 + r(Z^{q-1}/PQ). \end{aligned}$$

Now for (10.2.2) to hold we must have that

$$r(Z^{q-1}/PQ) = 1;$$

which implies that

$$\deg(Z^{q-1}/PQ) \leq 0.$$

This last statement is equivalent to

$$(q - 1)(g - 1) \leq 1.$$

Since $g \geq 2$ and $q \geq 2$, this is only possible for $g = 2 = q$. Note that the above argument also establishes that for each $P \in M$, 2 is a "q-gap" except if $g = 2 = q$ or $q = 1$ and M is hyperelliptic.

III.10.3. To study the excluded cases, we represent a hyperelliptic surface M by

$$w^2 = (z - e_1) \cdots (z - e_{2g+2})$$

with distinct e_j. We have seen that a basis for $\mathscr{H}^1(M)$ is then

$$\left\{\frac{dz}{w}, \frac{zdz}{w}, \ldots, \frac{z^{g-1}\,dz}{w}\right\}.$$

Thus in affine coordinates

$$\theta(P) = (1, z(P), \ldots, z(P)^{g-1}).$$

Thus θ is clearly two-to-one in this case (since z is two-to-one), and not of maximal rank at the Weierstrass points $z^{-1}(e_j)$.

For the other excluded case ($q = 2$ and $g = 2$), M is hyperelliptic again. A basis for $\mathscr{H}^2(M)$ is

$$\left\{\frac{dz^2}{w^2}, z\frac{dz^2}{w^2}, z^2\frac{dz^2}{w^2}\right\}.$$

Again

$$\theta(P) = (1, z(P), z(P)^2)$$

is independent of w, and the 2-canonical map (for genus 2) is two-to-one, and not of maximal rank at the Weierstrass points. □

III.10.4. Since the image of a compact manifold under an analytic mapping of maximal rank is a sub-manifold, we have obtained

Corollary 1. *Every Riemann surface of genus ≥ 2 is a submanifold of a complex projective space.*

Corollary 2

a. *Every non-hyperelliptic surface of genus 3 is a submanifold of $\mathbb{P}^2$.*
b. *Every other surface of genus $g \geq 2$ is a submanifold of $\mathbb{P}^3$.*

PROOF. Part (a) has already been proven. We have also seen that every surface is a submanifold of $\mathbb{P}^{d-1}$ for large d. Now assume that a Riemann surface M can be realized as a submanifold of $\mathbb{P}^k$ for $k > 3$. Call a point $x \in \mathbb{P}^k \backslash M$ *good*, provided for all $P \in M$, the (projective) line joining x to P is neither tangent to M at P, nor does it intersect M in another point. Say that we can find a good point x. We may assume that

$$x = (1, \underbrace{0, \ldots, 0}_{k\text{-times}}).$$

A line L through x may be represented thus by any other point y on it. Since $y \neq x$, $y = (\lambda_0, \lambda_1, \ldots, \lambda_k)$, where $(\lambda_1, \ldots, \lambda_k) = y(L)$, determines a well defined (depending only on L) point in $\mathbb{P}^{k-1}$. Now the mapping

$$M \ni P \mapsto y(L) \in \mathbb{P}^{k-1},$$

where L is the line joining x to P, is a one-to-one holomorphic mapping of maximal rank. Thus we are reduced to finding good points. The tangent lines to M form a two-dimensional subspace of $\mathbb{P}^k$, and the lines through two points of M form a three dimensional subspace of $\mathbb{P}^k$. Thus a good point (a point not in the union of these two subspaces) can certainly be found provided $\mathbb{P}^k$ has dimension ≥ 4. □

We also saw in the above proof that the embedding of M as a submanifold of $\mathbb{P}^3$ can be achieved by using 3 or 4 linearly independent holomorphic q-differentials ($q = 1$ except in the few exceptional cases). We will show in IV.11 that every two meromorphic functions on M are algebraically dependent. Hence we will also have

Corollary 3. *Every compact Riemann surface of genus $g \geq 2$ can be realized as an algebraic submanifold of $\mathbb{P}^3$.*

III.10.5. Let S be a non-singular curve of genus g in $\mathbb{P}^{g-1}$. The curve S is called *canonical* if it is the image of a compact Riemann surface of genus g under the 1-canonical mapping. Let $\theta: M \to \mathbb{P}^{g-1}$ be the canonical mapping and assume that M is not hyperelliptic. Let $(z_1, \ldots, z_g)$ be homogeneous coordinates on $\mathbb{P}^{g-1}$ and let $(\alpha_1, \ldots, \alpha_g) \in \mathbb{C}^g - \{0\}$. The hyperplane $0 = \sum_{i=1}^{g} \alpha_i z_i$ in $\mathbb{P}^{g-1}$ intersects $\theta(M)$ at P if and only if P is a zero of the non-trivial holomorphic differential $\sum_{i=1}^{g} \alpha_i \zeta_i$. We have established the necessity part of the following

Theorem. *Let S be a non-singular curve of genus g in $\mathbb{P}^{g-1}$. Then S is canonical if and only if S is non-degenerate (that is, the curve does not lie in any hyperplane in projective space) and of degree $2g - 2$ (that is, the curve intersects every hyperplane in $2g - 2$ points, counting multiplicities).*

PROOF. The theorem is the geometric interpretation of the following observation. Let D be a divisor of degree $2g - 2$ on a compact Riemann surface of genus g. From Riemann–Roch

$$r(1/D) = g - 1 + i(D).$$

Now $i(D) = 1$ if and only if D is canonical (otherwise $i(D) = 0$). Thus $r(1/D) = g - 1$ whenever D is not canonical

Necessity has already been shown. For sufficiency, assume that S is a non-degenerate curve of degree $2g - 2$ in $\mathbb{P}^{g-1}$. For $j = 1, \ldots, g$, the hyperplane $\{z_j = 0\}$ intersects S in $2g - 2$ points in the divisor

$D_j = P_1^{(j)} \cdots P_{2g-2}^{(j)}$. Then $f_j = z_j/z_1$ is a meromorphic function on S with $(f_j) = D_j/D_1$. We claim that the functions $f_1, \ldots, f_g$ are linearly independent. If not, there exist constants $\alpha_1, \ldots, \alpha_g$, not all zero, such that

$$\sum_{j=1}^{g} \alpha_j f_j(P) = 0 \quad \text{all } P \in S.$$

Thus S is contained in the hyperplane $\sum_{j=1}^{g} \alpha_j z_j = 0$ contrary to the non-degeneracy assumption. Since the divisor D_j is canonical, there exists a holomorphic differential ω_j on S with $(\omega_j) = D_j$. Hence for $j = 1, \ldots, g$, there exists a non-zero complex number λ_j such that $f_j = \lambda_j(\omega_j/\omega_1)$; of course, $\lambda_1 = 1$. Thus the mapping of S into projective space using the functions $\{1, f_2, \ldots, f_g\}$ differs from the map using the (linearly independent) differentials $\{\omega_1, \ldots, \omega_g\}$ by the projective transformation $\operatorname{diag}(1, \lambda_2, \ldots, \lambda_g)$. The first mapping is the identity and the second is canonical. □

III.10.6. We will study an interesting map $\theta: M \to \mathbb{P}^g$. Here M is a compact hyperelliptic Riemann surface of genus $g \geq 2$. Let $P_1, \ldots, P_{2g+2}$ be the Weierstrass points on M. Let us choose the function z of degree 2 on M with $(z) = P_1^2/P_{2g+2}^2$. The map θ into projective space is obtained using the $(g+1)$-dimensional space of meromorphic differentials $\Omega(1/P_1P_2)$. If we view M as the algebraic curve $w^2 = \prod_{j=1}^{2g+1}(z - e_j)$, then $e_j = z(P_j)$, $j = 1, \ldots, 2g+1$, $e_{2g+2} = z(P_{2g+2}) = \infty$ ($e_1 = 0$, of course). A basis for $\Omega(1/P_1P_2)$ is

$$\left\{\omega_0 = \frac{dz}{w}, \omega_1 = \frac{z\,dz}{w}, \ldots, \omega_{g-1} = \frac{z^{g-1}\,dz}{w}, \omega_g = \frac{dz}{(z-e_1)(z-e_2)}\right\};$$

we note from III.7.5 that

$$\left(\frac{z^j\,dz}{w}\right) = P_1^{2j}P_{2g+2}^{2g-2j-2}, \qquad j = 0, \ldots, g-1,$$

and

$$\left(\frac{dz}{(z-e_1)(z-e_2)}\right) = \frac{P_3P_4 \cdots P_{2g+2}}{P_1P_2}.$$

In affine coordinates

$$\theta(P) = \left(1, z(P), \ldots, z(P)^{g-1}, \frac{w(P)}{(z(P)-e_1)(z(P)-e_2)}\right),$$

for $P \neq P_j$, $j = 1, \ldots, 2g+2$. Thus for P and Q not Weierstrass points $\theta(P) = \theta(Q)$ if and only if $P = Q$ (for $z(P) = z(Q)$ and $P \neq Q$ implies $w(P) = -w(Q)$). The map θ is also of maximal rank at the non-Weierstrass points. Further, none of the components of $\theta(P)$ vanish.

Let ζ be a local coordinate vanishing at P_1. There exist holomorphic and non-vanishing functions $\phi_0, \ldots, \phi_g$ with

$$\frac{z^j\,dz}{w} = \zeta^{2j}\phi_j(\zeta)\,d\zeta \quad \text{near } P_1 \text{ for } j = 0, \ldots, g-1,$$

and

$$\frac{dz}{(z-e_1)(z-e_2)} = \zeta^{-1}\phi_g(\zeta)\,d\zeta \quad \text{near } P_1.$$

Thus for $|\zeta|$ small,

$$\theta(\zeta) = (\zeta\phi_0(\zeta), \zeta^3\phi_1(\zeta), \ldots, \zeta^{2g-1}\phi_{g-1}(\zeta), \phi_g(\zeta));$$

in particular,

$$\theta(P_1) = (0, \ldots, 0, \lambda_1) \quad \text{for some non-zero complex number } \lambda_1.$$

Similarly,

$$\theta(P_2) = (0, \ldots, 0, \lambda_2) \quad \text{for some non-zero complex number } \lambda_2,$$

and we see that

$$\theta(P_1) = \theta(P_2).$$

It is clear that θ is of maximal rank at both P_1 and P_2. At P_j, $j = 3, \ldots, 2g+2$, θ is also of maximal rank and the g-th component of $\theta(P_j)$ vanishes. It is also easy to see that $\theta(P_j) \neq \theta(P_k)$ for $3 \leq j < k \leq 2g+2$, because the canonical map already discriminates between these points.

If $\sum_{j=0}^{g} \alpha_j z_j = 0$ is a hyperplane on $\mathbb{P}^g$, then $\theta(M)$ intersects this hyperplane in the zeros of the abelian differential $\sum_{j=0}^{g} \alpha_j \omega_j$. This intersection will (generically) consist of $2g$ points (provided $\alpha_g \neq 0$). Thus $\theta(M)$ is a curve of genus g and degree $2g$ in $\mathbb{P}^g$ with a single double point (the image of the points P_1 and P_2).

Consider the case $g = 2$. In this situation, the defining equation for the curve $\theta(M)$ is obtained by letting $W = w/z(z-1)$ (without loss of generality $e_1 = 0$, $e_2 = 1$, $e_6 = \infty$) as

$$W^2 = \frac{(z-e_3)(z-e_4)(z-e_5)}{z(z-1)}$$

or in homogeneous coordinates as

$$Z(Z-Y)W^2 = Y(Z-e_3Y)(Z-e_4Y)(Z-e_5Y).$$

Exercise

Carry through the above analysis for non-hyperelliptic surfaces; also for hyperelliptic surfaces but do not assume that P_1 and P_2 are Weierstrass points.

III.11. More on the Jacobian Variety

This section is devoted to a closer study of various spaces of divisors on a compact Riemann surface M, and the images of these spaces in the Jacobian variety $J(M)$. We show that $J(M)$ satisfies a universal mapping property (Proposition III.11.7). The technical side involves the calculation of dimensions of subvarieties of $J(M)$. The most important consequence is Noether's theorem (III.11.20). Throughout the section, we assume that the reader is familiar with elementary properties of complex manifolds of dimension > 1.

III.11.1. By a *complex torus* T, we mean the quotient space, $T = \mathbb{C}^n/G$, where G is a group of translations generated by $2n$ $\mathbb{R}$-linearly independent vectors in $\mathbb{C}^n$. A torus T is thus a group (under addition modulo G) and a complex analytic manifold with the natural projection

$$\rho : \mathbb{C}^n \to T,$$

a holomorphic local homeomorphism.

We introduce now some notation that we will follow throughout this section. If $u \in T$, then $\hat{u} \in \mathbb{C}^n$ will denote a point with $\rho(\hat{u}) = u$. The generators of G will be denoted by the column vectors $\Pi^{(1)}, \ldots, \Pi^{(2n)} \in \mathbb{C}^n$. The j-th component of the vector $\Pi^{(k)}$ will be denoted by π_{jk}. The $n \times 2n$ matrix $(\pi_{jk}) = \Pi$ will be called the *period matrix* of T.

Our first observation is that the $2n \times 2n$ matrix $\left[\begin{smallmatrix}\Pi\\ \bar{\Pi}\end{smallmatrix}\right]$ is non-singular. To see this, assume that there exists a vector $c \in \mathbb{C}^{2n}$ such that $\left[\begin{smallmatrix}\Pi\\ \bar{\Pi}\end{smallmatrix}\right]c = \left[\begin{smallmatrix}\Pi c\\ \bar{\Pi} c\end{smallmatrix}\right] = 0$. Then $\Pi(c + \bar{c}) = 0 = \Pi(c - \bar{c})$. Thus $\operatorname{Re} c = 0 = \operatorname{Im} c$, by the $\mathbb{R}$-linear independence of the columns of Π. Thus $c = 0$.

Hence we have for ${}^t x$, ${}^t y \in \mathbb{C}^n$, ${}^t c \in \mathbb{C}^{2n}$ the equation $(x,y)\left[\begin{smallmatrix}\Pi\\ \bar{\Pi}\end{smallmatrix}\right] = c$ has a unique solution. In particular if $c \in \mathbb{R}^{2n} \subset \mathbb{C}^{2n}$, then

$$x\Pi + y\bar{\Pi} = c = \bar{c} = \bar{x}\bar{\Pi} + \bar{y}\Pi,$$

or

$$(x - \bar{y})\Pi + (y - \bar{x})\bar{\Pi} = 0.$$

Thus $x = \bar{y}$, and

$$c = 2\operatorname{Re}(x\Pi). \tag{11.1.1}$$

III.11.2. Since G acts fixed point freely on $\mathbb{C}^n$, $\pi_1(T) \cong G$, and since G is abelian $H_1(T,\mathbb{Z}) \cong G$. As a matter of fact, the paths corresponding to the columns of Π (that is,

$$t \mapsto t\Pi^{(k)}, \qquad t \in [0,1],\ k = 1, \ldots, 2n)$$

project to a basis of $H_1(T,\mathbb{Z})$. We observe next that the holomorphic 1-differentials, $d\hat{u}_j$ are invariant under G, and hence project to holomorphic 1-differentials du_j on T.

Remark. Let V be a complex manifold of dimension m. By definition, a holomorphic 1-form ω on V is one that can be written locally as dF for some locally defined holomorphic function F on V, or in terms of local coordinates, $z = (z_1, \ldots, z_m)$, as

$$\omega = \sum_{j=1}^{m} f_j(z)\, dz_j,$$

with f_j holomorphic. In particular, ω is closed. Since

$$d\omega = \sum_{j=1}^{m} \sum_{k=1}^{m} \frac{\partial f_j}{\partial z_k}\, dz_k \wedge dz_j = 0,$$

we also have

$$\frac{\partial f_j}{\partial z_k} = \frac{\partial f_k}{\partial z_j}, \quad \text{all } k, j = 1, \ldots, m.$$

Lemma. *The projections to T of the differentials $d\hat{u}_1, \ldots, d\hat{u}_n$ form a basis for the holomorphic 1-differentials on T, that will be denoted by $\{du_1, \ldots, du_n\}$.*

PROOF. Let $\{a_1, \ldots, a_{2n}\}$ be a basis for $H_1(T, \mathbb{Z})$ corresponding to the generators of G. Then

$$\int_{a_k} du_j = \pi_{jk}, \qquad j = 1, \ldots, n,\ k = 1, \ldots, 2n.$$

Thus, letting $c \in \mathbb{R}^{2n}$, it follows by (11.1.1) that there is a unique vector ${}^t x \in \mathbb{C}^n$ such that

$$\operatorname{Re} \int_{a_k} \sum_{j=1}^{n} x_j\, du_j = c_k, \qquad k = 1, \ldots, 2n. \tag{11.2.1}$$

Now (11.2.1) shows that the differentials $du_1, du_2, \ldots, du_n$ are linearly independent over $\mathbb{C}$. Furthermore, given any holomorphic 1-differential δ on T, there exists a differential $\omega = \sum_{j=1}^{n} x_j\, du_j$ such that

$$\operatorname{Re} \int_{a_k} \delta = \operatorname{Re} \int_{a_k} \omega, \qquad k = 1, \ldots, 2n. \tag{11.2.2}$$

Define a function F on T by

$$F(P) = \operatorname{Re} \int_0^P (\delta - \omega), \qquad P \in T.$$

By (11.2.2), F is well-defined. Since $\delta - \omega$ is holomorphic, F is locally the real part of a holomorphic function. Thus F satisfies the maximum (and minimum) principle. Since T is compact, F is constant and thus $= 0$. By the Cauchy–Riemann equations

$$\operatorname{Im} \int_0^P (\delta - \omega) = 0,$$

and thus $\delta = \omega$. □

Definition. By du we will denote the column vector of 1-differentials ${}^t\{du_1, \ldots, du_n\}$.

Remark. The notation of this section involves some abuse of language. The closed differentials du_j are *not* exact (the $d\hat{u}_j$ are exact).

III.11.3. Let V be any connected compact m-dimensional complex analytic manifold, and let $\Phi: V \to T$ be a holomorphic mapping. Let $z = (z_1, \ldots, z_n)$ be a local coordinate at a point $\Phi(P) \in T$. If ω is a holomorphic 1-form on T, then locally

$$\omega = \sum_{j=1}^{n} f_j(z)\, dz_j.$$

Let $\zeta = (\zeta_1, \ldots, \zeta_m)$ be a local coordinate at P on V. The pullback of ω via Φ, $\Phi^*\omega$, is the holomorphic 1-form defined in terms of the local coordinate ζ by

$$\Phi^*\omega = \sum_{j=1}^{m} g_j(\zeta)\, d\zeta_j,$$

where we write $z = h(\zeta) = (h_1(\zeta), \ldots, h_n(\zeta))$, and

$$g_j(\zeta) = \sum_{k=1}^{n} f_k(h(\zeta)) \frac{\partial h_k}{\partial \zeta_j}, \qquad j = 1, \ldots, m.$$

Let $\delta_1, \ldots, \delta_n$ denote the pullbacks of $du_1, \ldots, du_n$ via Φ. Let δ be the column vector formed by the δ_j, and ρ be, as before, the projection from $\mathbb{C}^n \to T$.

Proposition. *Let $P_0 \in V$. Then*

$$\Phi(P) = \Phi(P_0) + \rho\left(\int_{P_0}^{P} \delta\right). \tag{11.3.1}$$

PROOF. Since $\delta = \Phi^* du$, and

$$\int_{P_0}^{P} \delta = \int_{\Phi(P_0)}^{\Phi(P)} du = \Phi(P) - \Phi(P_0) \tag{11.3.2}$$

modulo periods, the result is clear. ☐

III.11.4. Corollary. *Let $\Phi_j: V \to T$ be holomorphic mappings for $j = 0,1$. Assume Φ_0 is homotopic to Φ_1. Then there exists a $c_0 \in T$ such that*

$$\Phi_0(P) = \Phi_1(P) + c_0, \quad \textit{all } P \in V.$$

PROOF. Since Φ_0 is homotopic to Φ_1, there exists a continuous function

$$\Phi: V \times I \to T$$

($I = [0,1]$) such that

$$\Phi(\cdot,0) = \Phi_0, \qquad \Phi(\cdot,1) = \Phi_1.$$

Thus for every closed curve a in V, $\Phi_0(a)$ is homotopic to $\Phi_1(a)$. Let $\delta^{(j)} = \Phi_j{}^*du$. From (11.3.2) we see that $\delta_k^{(0)} = \delta_k^{(1)}$ (since they have the same periods over every closed curve a on V). Thus Φ_0 and Φ_1 satisfy (11.3.1) with the same δ. □

III.11.5. Let $T_1 = \mathbb{C}^m/G_1$ and $T_2 = \mathbb{C}^n/G_2$ be two complex tori. Let $\Phi: T_1 \to T_2$ be a holomorphic mapping. As above, let $\delta = \Phi^*\, du$. Then there exists an $n \times m$ matrix A such $\delta = A\; dv$ where $dv = {}^t\{dv_1, \ldots, dv_m\}$ is a basis for the holomorphic 1-forms on T_1. Hence, as a consequence of the previous proposition, the map Φ can be written in the form

$$\Phi(P) = \rho_2 \circ A \circ \rho_1^{-1}(P) + c_0, \qquad P \in T_1,$$

for some $c_0 \in T_2$. (Since A is an $n \times m$ matrix it also represents a linear transformation from $\mathbb{C}^m$ into $\mathbb{C}^n$.)

We have thus established the following

Proposition. *The only holomorphic maps of a complex torus into a complex torus are the group homomorphisms composed with translations.*

III.11.6. By an *underlying real structure* for the complex torus $T = \mathbb{C}^n/G$ we mean the real torus $\mathbb{R}^{2n}/\mathbb{Z}^{2n}$ together with the map $\mathbb{R}^{2n}/\mathbb{Z}^{2n} \to T$ induced by the linear map

$$\mathbb{R}^{2n} \ni x \mapsto \Pi x \in \mathbb{C}^n.$$

We have seen (as a consequence of Proposition III.11.5) that any endomorphism of T is induced by a linear transformation $A: \mathbb{C}^n \to \mathbb{C}^n$ that preserves periods. Thus if A represents the matrix of this linear transformation with respect to the canonical basis for $\mathbb{C}^n$, then there exists a $2n \times 2n$ integral matrix M such that $A\Pi = \Pi M$. The matrix M now induces an endomorphism of the underlying real structure. The endomorphism A is completely determined by M (since $\left[\frac{\Pi}{\bar{\Pi}}\right]$ is nonsingular), and the following diagrams commute:

$$\begin{array}{ccc} \mathbb{R}^{2n} & \xrightarrow{\ \Pi\ } & \mathbb{C}^n \\ \downarrow{\scriptstyle M} & & \downarrow{\scriptstyle A} \\ \mathbb{R}^{2n} & \xrightarrow{\ \Pi\ } & \mathbb{C}^n \end{array} \qquad \begin{array}{ccc} \mathbb{R}^{2n}/\mathbb{Z}^{2n} & \xrightarrow{\ \Pi\ } & T \\ \downarrow{\scriptstyle M} & & \downarrow{\scriptstyle A}, \\ \mathbb{R}^{2n}/\mathbb{Z}^{2n} & \xrightarrow{\ \Pi\ } & T \end{array}$$

which is simply another way of saying $A\Pi = \Pi M$.

It follows immediately from the previous remarks that a holomorphic injective map of one complex torus into another torus of the same dimension is necessarily biholomorphic. If Π_1 and Π_2 are the period matrices of the two tori, the map can be represented by a matrix $A: \mathbb{C}^n \to \mathbb{C}^n$ such that $A\Pi_1 = \Pi_2 M$ for some M as above. It is necessarily the case that both A and M are non-singular. Thus we have established the following

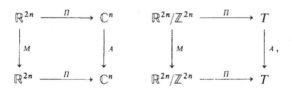

Proposition. *Two complex tori* $T_1 = \mathbb{C}^n/G_1$ *and* $T_2 = C^m/G_2$ *are holomorphically equivalent if and only if* $n = m$ *and there are matrices* $A \in GL(n,\mathbb{C})$, $M \in GL(2n,\mathbb{Z})$ *such that* $A\Pi_1 = \Pi_2 M$ *where* Π_j *is a period matrix of* T_j, $j = 1, 2$.

Let A represent an endomorphism of a complex torus

$$A:\mathbb{C}^n/G \to \mathbb{C}^n/G.$$

Let Π_1, Π_2 be two period matrices for this torus. Then for some $M_j \in GL(2n,\mathbb{Z})$,

$$A\Pi_1 = \Pi_1 M_1 \quad \text{and} \quad A\Pi_2 = \Pi_2 M_2.$$

Further,

$$\Pi_2 = \Pi_1 M, \qquad M \in GL(2n,\mathbb{Z}).$$

Thus

$$\Pi_1 M_1 M = A\Pi_1 M = A\Pi_2 = \Pi_2 M_2 = \Pi_1 M M_2,$$

and hence (since $\Pi_1 : \mathbb{R}^{2n} \to \mathbb{C}^n$ is an isomorphism)

$$M_1 M = M M_2.$$

It thus follows, since M is non-singular, that trace M_1 = trace M_2.

Corollary. *The trace of the endomorphism* A *is well defined by setting it equal to trace* M_1 *(or* M_2*).*

III.11.7. We return now to the situation of III.6. Let M be a compact Riemann surface of genus $g > 0$, ${}^t\{a,b\}$ a canonical homology basis on M, and $\{\zeta\}$ the dual basis for $\mathscr{H}^1(M)$. As before Π denotes the period matrix of the surface. (Here Π is a $g \times g$ matrix.) Let $J(M)$ be the Jacobian variety of M; it is, of course, a complex torus (with period matrix (I,Π)). Let φ be the mapping of M into $J(M)$ with base point P_0 previously defined. Note that $\varphi^* du_j = \zeta_j, j = 1, \ldots, g$.

Now let $\Phi: M \to T$ be any holomorphic mapping of M into a complex torus $T = \mathbb{C}^n/G$. Let $\{dv_1, \ldots, dv_n\}$ be a basis for the holomorphic differentials on T. Let $\delta_j = \Phi^* dv_j \in \mathscr{H}^1(M)$. Since $\{\zeta\}$ is a basis for $\mathscr{H}^1(M)$, there are unique complex numbers a_{jk} such that

$$\delta_j = \sum_{k=1}^{g} a_{jk}\zeta_k, \qquad j = 1, \ldots, n,$$

(or $\delta = A\zeta$, $\delta = {}^t\{\delta_1, \ldots, \delta_n\}$, $A = (a_{jk})$). Then

$$\mathbb{C}^g \ni \hat{u} \mapsto \rho(A\hat{u}) + \Phi(P_0) \in T$$

defines a unique mapping

$$\psi: J(M) \to T.$$

It now follows from (11.3.1) that

$$\Phi = \psi \circ \varphi. \tag{11.7.1}$$

We have established the following

Proposition. *Let $\Phi: M \to T$ be a holomorphic mapping of a Riemann surface M into a complex torus T, then there exists a unique holomorphic mapping $\psi: J(M) \to T$ such that* (11.7.1) *holds.*

The above proposition shows that $J(M)$ is determined by M up to a canonical isomorphism. The mapping φ is, of course, determined up to an additive constant by the canonical homology basis on M.

III.11.8. We have seen in III.6, that the mapping φ extends to divisors on M. Let W_n be the image in $J(M)$ of the integral divisors of degree n. Set by definition $W_0 = \{0\}$. It is then clear that

$$W_n \subset W_{n+1},$$

(because $\varphi(D) = \varphi(DP_0)$ for every divisor D) and $W_g = J(M)$ (Jacobi inversion theorem).

Let W_n^r be the set of points in $J(M)$ which are images of integral divisors D that satisfy the two conditions $\deg D = n$ and $r(D^{-1}) \geq r + 1$. Thus W_n^r consists of the complete g_n^r's as defined in III.4.13. Denote by K the image under φ of the canonical (integral) divisors. By Abel's theorem, K consists of one point.

Proposition. $\{K\} = W_{2g-2}^{g-1}$.

Proof. Let D be an integral with $\deg D = 2g - 2$. Then $r(D^{-1}) \geq g$ if and only if $i(D) \geq 1$ if and only if D is canonical. □

III.11.9. Let M be a Riemann surface and $n > 0$ an integer. The n-fold Cartesian product, M^n, is naturally a complex manifold of dimension n. By M_n we denote the set of integral divisors of degree n on M. We topologize and give a complex structure to M_n as follows: Let $\mathcal{S}_n$ be the symmetric group on n letters. An element $\sigma \in \mathcal{S}_n$ acts on M^n by:

$$\sigma(P_1, \ldots, P_n) = (P_{\sigma(1)}, \ldots, P_{\sigma(n)}).$$

As point sets M_n and $M^n/\mathcal{S}_n$ can clearly be identified. We shall show that much more is true. A function f on M_n is said to be *continuous* (*holomorphic*) provided $f \circ p$ is continuous (holomorphic) on M^n, where

$$p: M^n \to M_n$$

is the canonical projection. This gives a natural topology and complex structure to M_n. To see that with this structure M_n is a complex manifold, let $(P_1, \ldots, P_n) \in M^n$ and let z_j be a local coordinate at P_j. Define the elementary symmetric functions of z_j:

$$\zeta_1 = (-1) \sum_j z_j,$$

$$\zeta_2 = (-1)^2 \sum_{j<k} z_j z_k,$$

$$\vdots \qquad \vdots$$

$$\zeta_n = (-1)^n z_1 \cdots z_n.$$

Notice that ζ_j is (locally) a holomorphic function on M_n and that the ordered set $\{\zeta_1, \ldots, \zeta_n\}$ determines the unordered set $\{z_1, \ldots, z_n\}$ uniquely as the roots of the polynomial $z^n + \zeta_1 z^{n-1} + \cdots + \zeta_n = 0$. Hence $\zeta = (\zeta_1, \ldots, \zeta_n)$ is a local homeomorphism of a neighborhood of $p(P_1, \ldots, P_n) \in M_n$ and thus ζ serves as a local coordinate at $p(P_1, \ldots, P_n) \in M_n$.

Another useful set of local coordinates on M_n is obtained by considering the elementary symmetric functions of the second kind:

$$t_k = \sum_{j=1}^{n} z_j^k, \qquad k = 1, \ldots, n.$$

An easy induction argument shows that for each k, $1 \leq k \leq n$, $\{t_1, \ldots, t_k\}$ uniquely determines and is uniquely determined by $\{\zeta_1, \ldots, \zeta_k\}$.

We remark that with the choice of local coordinates made, p is obviously a holomorphic map (between complex manifolds) and that M_n is compact if and only if M is. We have established the following

Proposition. *If M is a compact Riemann surface, then M_n can be given a unique n-dimensional complex structure so that the natural projection $p: M^n \to M_n$ is holomorphic. Further, if V is a complex manifold, and the diagram*

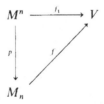

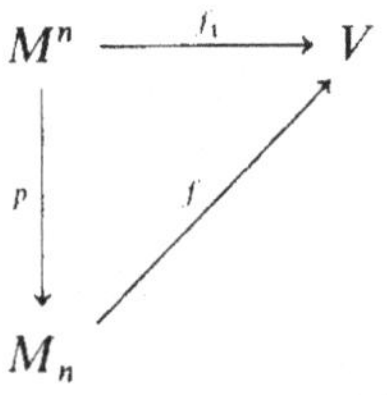

commutes, then f is holomorphic if and only if f_1 is.

Remarks

1. Let f be a holomorphic function in a neighborhood of 0 in $\mathbb{C}$. Thus

$$f(z) = \sum_{k=0}^{\infty} a_k z^k.$$

For $(z_1, \ldots, z_n) \in \mathbb{C}^n$ with $|z_j|$ sufficiently small $F(z_1, \ldots, z_n) = \sum_{j=1}^{n} f(z_j)$ defines a symmetric holomorphic function in a neighborhood of the origin *in* $\mathbb{C}^n$. On the quotient space $\mathbb{C}^n/\mathcal{S}_n$, we can write

$$F(z_1, \ldots, z_n) = \sum_{k=0}^{\infty} \sum_{j=1}^{n} a_k z_j^k,$$

and conclude that in terms of the local coordinate $(t_1, \ldots, t_n)$ on $\mathbb{C}^n/\mathcal{S}_n$,

$$\left.\frac{\partial F}{\partial t_j}\right|_{t=(0,\ldots,0)} = \left.\frac{1}{j!}\frac{d^j f}{dz^j}\right|_{z=0} = a_j.$$

2. If $D = P_1 \cdots P_n \in M_n$ and the P_j are distinct, we can identify a neighborhood of D in M_n with a neighborhood of $(P_1, \ldots, P_n) \in M^n$. Thus the symmetric group and symmetric functions enter only at those points $D \in M_n$ which contain non-distinct entries. It is also clear from this remark that in general we can describe a neighborhood of D by considering blocks consisting of the distinct points of D listed according to their multiplicities.

III.11.10. Let δ be a holomorphic 1-differential on $J(M)$. Let $a \in J(M)$ and set $\varphi_a(P) = \varphi(P) + a$, $P \in M$. Then $\delta' = \varphi_a^*\delta$ is a holomorphic 1-differential on M, and is independent of a. Thus there is a canonical identification via φ of the space of holomorphic 1-differentials on M and $J(M)$.

Let $Q_1, \ldots, Q_s$ be distinct points on M and let z_k be a local coordinate vanishing at Q_k. Assume $s \le n$, and let $m_1, \ldots, m_s$ be positive integers with

$$\sum_{j=1}^{s} m_j = n.$$

Consider the map

$$\varphi: M_n \to W_n$$

in a neighborhood of the n-tuple A made up of each Q_k appearing m_k times. Let δ'' be δ pulled back to M_n. A neighborhood of the point A consists of points

$$(Q_1^1, Q_2^1, \ldots, Q_{m_1}^1, Q_1^2, \ldots, Q_{m_2}^2, \ldots, Q_1^s, \ldots, Q_{m_s}^s),$$

with Q_j^k near Q_k. Consider the point

$$\tilde{A} = (\underbrace{Q_1, \ldots, Q_1}_{m_1\text{-times}}, \underbrace{Q_2, \ldots, Q_2}_{m_2\text{-times}}, Q_3, \ldots, \underbrace{Q_s, \ldots, Q_s}_{m_s\text{-times}}) \in M^n.$$

A local coordinate (on M^n) vanishing at this point is therefore given by

$$(z_{11}, \ldots, z_{m_1 1}, z_{12}, \ldots, z_{m_2 2}, \ldots, z_{m_s s}),$$

where

$$z_{jk}(Q) = z_k(Q), \qquad j = 1, \ldots, m_k, \qquad k = 1, \ldots, s.$$

Let $h_k(z_k)$ be a local primitive of δ' near Q_k on M. Without loss of generality $h_k(0) = 0$. The sum

$$\sum_{k=1}^{s} m_k h_k(z_k)$$

defines a function on M_n whose differential is δ''. (To see this recall the definition of the pullback of a differential. The differential δ can be written as df with f a linear function on $\mathbb{C}^g$. Now h_k is simply $f \circ \varphi$. The differential of $f \circ \varphi_n$ is δ'', where φ_n is φ viewed as a map of M_n into $J(M)$. Since f is linear, $f(\varphi_n(P_1 \cdots P_n)) = f(\varphi(P_1) + \cdots + \varphi(P_n)) = f(\varphi(P_1)) + \cdots + f(\varphi(P_n))$.) Assume now that δ'' vanishes at the point A. Since the pullback of δ'' to M^n vanishes at $\tilde{A}$, we must have $dh_k/dz_k = 0$ at $z_k = 0$ for $k = 1, \ldots, s$. We need more accurate information about the differential δ' in the case that at least one $m_k > 1$. In this case, the function

$$h_k(z_{1k}) + \cdots + h_k(z_{m_k k}) \tag{11.10.1}$$

can be expressed as a power series in the elementary symmetric functions of the second kind of the m_k variables z_{jk}. If we write

$$h_k(z_k) = \sum_{v=1}^{\infty} a_v z_k^v,$$

we see that the coefficient of the l-th symmetric function of the second kind in (11.10.1) is precisely a_l. We see therefore that

$$\delta'' = d\left(\sum_{k=1}^{s} m_k h_k\right),$$

vanishes at $A = Q_1^{m_1} \cdots Q_s^{m_s}$ if and only if dh_k vanishes at zero to order $\geq m_k$. Thus the differential δ' vanishes at Q_k to order $\geq m_k$. We have thus proved the following.

Lemma. *If δ is a holomorphic 1-differential on $J(M)$, whose pullback to M_n via the map $\varphi: M_n \to J(M)$ vanishes at a preimage of a point $u \in W_n$, corresponding to an integral divisor D of degree n on M, then the holomorphic differential δ' on M which corresponds to δ under the canonical map $\varphi: M \to J(M)$ has the property that $(\delta')/D$ is integral.*

III.11.11. Before proceeding we define a filtration of M_n that is analogous to the filtration of W_n by W_n^r described in III.11.8. Let

$$M_n^r = \{D \in M_n;\ r(D^{-1}) \geq r + 1\}.$$

(We abbreviate M_n^0 by M_n). It is obvious that

$$M_n^r = \varphi^{-1}(W_n^r).$$

Proposition

a. *Let $D \in M_n$. The Jacobian of the mapping $\varphi: M_n \to W_n \subset J(M)$ at D has rank equal to $n + 1 - r(D^{-1})$.*
b. *The fiber $\varphi^{-1}(\varphi(D))$ of the mapping is an analytic subvariety of M_n of dimension $v = r(D^{-1}) - 1$, which can be represented as a one-to-one analytic image of $\mathbb{P}^v$.*

c. *The mapping* $\varphi: M_n \to W_n$ *establishes an analytic isomorphism between* $M_n \backslash M_n^1$ *and* $W_n \backslash W_n^1$.

PROOF. Let us examine the differential $d\varphi_D$ of the analytic mapping $\varphi: M_n \to J(M)$ at a point $D = Q_1^{m_1} \cdots Q_s^{m_s}$ (where $m_1 + \cdots + m_s = n$). It is, of course, a linear mapping between tangent spaces

$$d\varphi_D : T_D(M_n) \to T_{\varphi(D)}(J(M)).$$

Let us denote by z_k a local coordinate vanishing at Q_k and let ζ_j, the j-th normalized differential of first kind, have the power series expansion

$$\left(\sum_{l=0}^{\infty} a_l^{jk} z_k^l\right) dz_k$$

in a neighborhood of Q_k. We must explain which coordinates we are using. We, of course, use the usual coordinates on $J(M)$ coming from $\mathbb{C}^g$. At the point $D \in M_n$, we use modified coordinates of the second kind. The modification is obtained by bunching together only those coordinates that arise from a single Q_k. Thus if our coordinates in a neighborhood of D in M^n are

$$(\underbrace{z_{11}, \ldots, z_{1m_1}}_{\substack{\text{corresponding}\\ \text{to } z_1}}, \underbrace{z_{21}, \ldots, z_{2m_2}}_{\substack{\text{corresponding}\\ \text{to } z_2}}, \ldots, \underbrace{z_{sm_s}}_{\substack{\text{corresponding}\\ \text{to } z_s}}),$$

the related coordinates on M_n are

$$(t_{11}, \ldots, t_{1m_1}, t_{21}, \ldots, t_{2m_2}, \ldots, t_{sm_s}),$$

where

$$t_{pq} = \sum_{l=1}^{m_p} z_{pl}^q, \qquad q = 1, \ldots, m_p,\ p = 1, \ldots, s.$$

The matrix which represents $d\varphi_D$ is then

$$\begin{bmatrix} a_0^{11} & \tfrac{1}{2}a_1^{11} & \cdots & \dfrac{1}{m_1} a_{m_1-1}^{11} & a_0^{12} & \cdots & \dfrac{1}{m_2} a_{m_2-1}^{12} & \cdots & a_0^{1s} & \cdots & \dfrac{1}{m_s} a_{m_s-1}^{1s} \\ \vdots & & & & & & & & & & \\ a_0^{g1} & \tfrac{1}{2}a_1^{g1} & \cdots & \dfrac{1}{m_1} a_{m_1-1}^{11} & a_0^{g2} & \cdots & \dfrac{1}{m_2} a_{m_2-1}^{g2} & \cdots & a_0^{gs} & \cdots & \dfrac{1}{m_s} a_{m_s-1}^{gs} \end{bmatrix},$$

or interpreting the number a_l^{jk} in the obvious manner the matrix can be written (up to certain obvious constants) as

$$\begin{bmatrix} \zeta_1(Q_1) & \cdots & \dfrac{1}{m_1!}\zeta_1^{(m_1-1)}(Q_1) & \cdots & \zeta_1(Q_s) & \cdots & \dfrac{1}{m_s!}\zeta_1^{(m_s-1)}(Q_s) \\ \vdots & & & & & & \vdots \\ \zeta_g(Q_1) & \cdots & \dfrac{1}{m_1!}\zeta_g^{(m_1-1)}(Q_1) & \cdots & \zeta_g(Q_s) & \cdots & \dfrac{1}{m_s!}\zeta_g^{(m_s-1)}(Q_s) \end{bmatrix}.$$

The rank of the above matrix is $g - i(D)$, which by Riemann–Roch equals $n + 1 - r(D^{-1})$. Thus (a) holds.

Let $D_j \in M_n \backslash M_n^1$, $j = 1, 2$. If $\varphi(D_1) = \varphi(D_2)$, then by Abel's theorem D_1/D_2 is principal. Hence $D_1 = D_2$ unless $r(D_2^{-1}) \geq 2$. The latter would imply that $D_2 \in M_n^1$. We have thus established (c).

Since $\varphi: M_n \to J(M)$ is analytic, the fiber $\varphi^{-1}(\varphi(D))$ is a subvariety of M_n. Thus it suffices, in order to establish (b), to produce a one-to-one surjective holomorphic mapping $\Psi: \mathbb{P}^\nu \to \varphi^{-1}(\varphi(D))$. By Abel's theorem, $\varphi^{-1}(\varphi(D))$ consists of integral divisors of degree n equivalent to the divisor D. Let D' be an arbitrary integral divisor of degree n equivalent to D, then D'/D is principal and we conclude that every such D' is of the form $(f)D$ for some $f \in L(D^{-1})$. We have reinterpreted Proposition III.4.13. We now produce an analytic mapping from $\mathbb{P}L(D^{-1})$ to $\varphi^{-1}(\varphi(D))$.

Let $\{f_0 = 1, f_1, \ldots, f_\nu\}$ be a basis for $L(D^{-1})$. Send the point $0 \neq (c_0, c_1, \ldots, c_\nu) = c$ onto the divisor $\Psi(c) = D(f)$ with $f = \sum_{j=0}^{\nu} c_j f_j$. Note that if $\Psi(c) = \Psi(\tilde{c})$, then $(f)/(\tilde{f})$ is the unit divisor and $\tilde{f}$ is a constant multiple of f. Thus we have produced a well defined one-to-one surjective mapping

$$\Psi: \mathbb{P}^\nu \to \varphi^{-1}(\varphi(D)) \subset M_n.$$

To show that the mapping Ψ is holomorphic, we write

$$D = Q_1^{m_1} \cdots Q_s^{m_s}, \qquad Q_i \in M,\ m_i > 0,\ i = 1, 2, \ldots, s,\ m_1 + \cdots + m_s = n.$$

Fix a point $c^o = (c_0^o, \ldots, c_\nu^o) \in \mathbb{C}^{\nu+1}$ and let $c = (c_0, \ldots, c_\nu) \in \mathbb{C}^{\nu+1}$ be sufficiently close (to be specified below) to c^o. We let $f^o = \sum_{j=0}^{\nu} c_j^o f_j$ and $f = \sum_{j=0}^{\nu} c_j f_j$. Then

$$D(f^o) = P_1^{\mu_1} \cdots P_r^{\mu_r}, \qquad P_j \in M,\ \mu_j > 0,\ j = 1, 2, \ldots, r,\ \mu_1 + \cdots + \mu_r = n.$$

Let U_j be an open neighborhood of P_j, $j = 1, \ldots, r$, and z_j a local coordinate vanishing at P_j that is defined in a neighborhood of the closure of U_j We assume that the closures of these open sets are disjoint. Further, the points Q_i that do not equal any of the P_j are not in any of the closures of the U_j and if $Q_i = P_j$ for some pair i and j, then we assume that no other Q_k, $k \neq i$, is in the closure of U_j. By shrinking the sets U_j further we may assume that whenever $P_j = Q_i$, we can and do choose a holomorphic function h_j on U_j that has a zero of order m_i at P_j and is non-zero on $U_j - \{P_j\}$. For $P_j \notin \{Q_1, \ldots, Q_s\}$, we set $h_j = 1$.

We are again using, as local coordinates on M_n, the modified elementary symmetric functions of the second kind:

$$t_{pq} = \sum_{l=1}^{\mu_p} z_{pl}^q, \qquad q = 1, \ldots, \mu_p,\ p = 1, \ldots, r,$$

where $z_{pl} = z_p$, $l = 1, \ldots, \mu_p$, $p = 1, \ldots, r$. We observe that by definition, $t_{pq}(\Psi(c))$ is the sum of the qth powers of the values of the local coordinate z_p at points in the open set U_p that appear in divisor $(f)D$ (with proper multiplicities). Note that $t_{pq}(\Psi(c^o)) = 0$. We consider the holomorphic function $h_j f^o$ on the set U_j. It vanishes to order μ_j at P_j and has no other zeros on U_j. By Rouche's theorem (see III.6.8 for a similar argument), the function $h_j f$ will have μ_j zeros (counting multiplicities) in U_j if $|c - c^o|$ is sufficiently small. The product of these n zeros (as we sum over the j's) forms the divisor $(f)D$, since D is integral of degree n. The residue theorem yields

$$t_{pq}(\Psi(c)) = \sum_{l=1}^{\mu_p} z_{pl}^q(\Psi(c)) = \frac{1}{2\pi i}\int_{\delta U_p} \frac{(h_p(z_p)f(z_p))'}{h_p(z_p)f(z_p)} z_p^q \, dz_p$$

for $q = 1, \ldots, \mu_p$, $p = 1, \ldots, r$. We have shown that

$$t_{pq}(\Psi(c))$$

is a complex analytic function of c.

Remarks

1. We saw in the proof that the rank of the Jacobian of φ at $D \in M_n$ can be written as $g - i(D)$.
2. A "generic" (see III.6.5) divisor $D \in M_n$, $n \le g$, has index of specialty $i(D) = g - n$. Thus W_n for $n \le g$ has dimension precisely n (as expected), by Remark (1).

III.11.12. By the inverse function theorem it follows that at the image of an integral divisor D of degree $n < g$ with $r(D^{-1}) = 1$, there are local coordinates $z_1, \ldots, z_g$ for $J(M)$ so that the points of W_n are given by the equations $z_{n+1} = \cdots = z_g = 0$. We say in this case that W_n is *regularly embedded* at $\varphi(D)$.

Conversely, if W_n is regularly embedded at a point $u \in J(M)$, then there are $g - n$ linearly independent holomorphic differentials on $J(M)$ whose pull-backs to M_n vanish at the pre-image of u on M_n. We have seen (Lemma III.11.10) that such a holomorphic differential corresponds to a differential δ on M such that (δ) is a multiple of D for all $D \in \varphi^{-1}(u)$. Now if $u \in W_n^1$, then for any $Q \in M$ there is a divisor D of degree n containing Q such that $\varphi(D) = u$. Hence the differential δ must vanish identically on M. But then it vanishes identically on $J(M)$. Now if $g - n \ge 1$, and W_n is regularly im-bedded at u, there is at least one differential on $J(M)$ that vanishes on W_n and is not identically zero. We have established the

Proposition. *If $n \le g - 1$, then $u \in W_n$ is a singularity of W_n if and only if $u \in W_n^1$.*

Remarks

1. The above considerations allow us to describe more intrinsically the tangent space to $J(M)$. Let $u_1, \ldots, u_g$ be the canonical coordinates on $\mathbb{C}^g$. Then these also serve as local coordinates at an arbitrary point $x \in J(M) = \mathbb{C}^g/G$. In terms of these coordinates, a natural basis for the (complex) tangent space $T_x(J(M))$ of the manifold $J(M)$ at the point x is given by the vectors $\partial/\partial u_1, \ldots, \partial/\partial u_g$. On the cotangent space $T_x^*(J(M))$, the

covectors $du_1, \ldots, du_g$ provide a basis. The dual pairing

$$T_x(J(M)) \times T_x^*(J(M)) \to \mathbb{C}, \tag{11.12.1}$$

is then given by

$$\left(\sum_{j=1}^{g} a_j \frac{\partial}{\partial u_j}\right) \times \left(\sum_{j=1}^{g} b_j\, du_j\right) = \sum_{j=1}^{g} a_j b_j,$$

here $a_j, b_j \in \mathbb{C}$. Of course, every cotangent vector may also be viewed as a holomorphic 1-form on $J(M)$ and conversely. As we have already seen the holomorphic 1-forms on $J(M)$ restrict to holomorphic 1-forms on $W_1 \subset J(M)$ and pull back via an isomorphism to elements of $\mathscr{H}^1(M)$. Thus the cotangent space to the complex manifold $J(M)$ at any point x is naturally identified with the abelian differentials of the first kind on M.

We return briefly to the differential $d\varphi_D$ discussed in III.11.11. The image of $d\varphi_D$ is spanned by the column vectors of the matrix (11.11.1). The linear subspace of $T^*_{\varphi(D)}(J(M))$ dual to the image of $d\varphi_D$ consists (by definition) of those cotangent vectors $(\sum_{j=1}^{g} b_j\, du_j)$ that annihilate the image of $d\varphi_D$ under the pairing (11.12.1); in other words, of those covectors $(b_1\, du_1 + \cdots + b_g\, du_g)$ which are annihilated by the transpose of the above matrix. Under the natural identification of $T^*_{\varphi(D)}(J(M))$ with $\mathscr{H}^1(M)$, these covectors correspond to those $\omega \in \mathscr{H}^1(M)$ with $(\omega) \geq D$.

2. If $V \subset J(M)$ is any analytic subvariety, then $R(V)$ denotes the *regular points* of V; that is, those points $v \in V$ at which V is an analytic submanifold of $J(M)$. The remaining points are called *singular*, and the set of singular points is denoted by $S(V)$. To any point $x \in V$, there is associated a linear subspace $T_x^*(V) \subset T_x^*(J(M))$ spanned by those covectors of the form df_x, where f is any analytic function in an open neighborhood of the point x in $J(M)$ which vanishes (identically) on V. The natural dual to the subspace $T_x^*(V)$ is a linear subspace $T_x(V) \subset T_x(J(M))$ called the *tangent space* to the variety $V \subset J(M)$ at the point x, and $\dim T_x(V)$ is called the *embedding dimension* of the variety V at the point x. The embedding dimension is the dimension of the smallest submanifold of $J(M)$ which contains the intersection of V with a small neighborhood of x in $J(M)$. It is thus clear that the embedding dimension of $x \in R(V)$ is precisely the local dimension of x at V. The proposition preceding these remarks has shown that W_n has embedding dimension g precisely at the points of W_n^1 for $n \leq g - 1$. (Hence for $n \geq g$ as well.)

III.11.13. We introduce now some useful operations on subsets of $J(M)$. If $S \subset J(M)$ and $a \in J(M)$, then we set

$$S + a = \{s + a; s \in S\},$$
$$-S = \{-s; s \in S\}.$$

If $S, T \subset J(M)$, then we set

$$S \oplus T = \cup\{S + a; a \in T\} = \{s + t; s \in S, t \in T\},$$
$$S \ominus T = \cap\{S - a; a \in T\}.$$

A subvariety S of $J(M)$ is *irreducible* if every meromorphic function on $J(M)$ that vanishes on an open subset of S vanishes on S. Proposition III.11.11 has as an immediate consequence the following

Corollary. *For each $n \leq g$, W_n is an irreducible subvariety of $J(M)$ of dimension n.*

III.11.14. Proposition. *For any $a \in J(M)$,*

$$-W_{g-1} - a = W_{g-1} - a - K.$$

PROOF. For any integral divisor D of degree $g - 1$, we can find an integral divisor D' of the same degree such that DD' is canonical. Thus

$$\varphi(D) + \varphi(D') = K,$$

showing that

$$W_{g-1} = K - W_{g-1}. \qquad \square$$

III.11.15. Proposition. *Let $0 \leq r \leq t \leq g - 1$. Let $a \in J(M)$, $b \in J(M)$. Then*

$$(W_r + a) \subset (W_t + b) \Leftrightarrow a \in (W_{t-r} + b) \Leftrightarrow b \in (-W_{t-r} + a).$$

PROOF. First note that $a \in (W_{t-r} + b) \Leftrightarrow a = x + b$, $x \in W_{t-r} \Leftrightarrow b = a - x$, $x \in W_{t-r} \Leftrightarrow b \in (-W_{t-r} + a)$. Hence the last equivalence is trivial. Also if $a \in (W_{t-r} + b)$, it is easy to see that $(W_r + a) \subset (W_t + b)$.

Assume now that $(W_r + a) \subset (W_t + b)$. Thus for every integral divisor D of degree $\leq r$ there is a $D' \in M_t$ such that

$$\varphi(D) + a = \varphi(D') + b. \tag{11.15.1}$$

We need show that there is an $A \in M_{t-r}$ such that $a = \varphi(A) + b$. Now the Jacobi inversion theorem implies that there is an integral divisor B such that $a = \varphi(B) + b$, and (11.15.1) implies that $\deg B \leq t$. Let A be a divisor of minimal degree such that $a = \varphi(A) + b$. It follows that A is unique, since $r(A^{-1})$ is necessarily equal to one. If $r(A^{-1}) > 1$, then A would be equivalent to a divisor which contains P_0. Thus $\varphi(A) = \varphi(P_0 A') = \varphi(A')$; contradicting minimality of the degree of A. We now wish to show that $\deg A \leq t - r$. We now have for every $D \in M_r$, there is a $D' \in M_t$ such that

$$\varphi(D) + \varphi(A) = \varphi(D'), \quad \text{where } \deg A = s \leq t. \tag{11.15.2}$$

To show that $s \leq t - r$, we assume that $s > t - r$. We can now choose a divisor D of degree $t + 1 - s \leq r$ such that $P_0 \notin D$ and such that $r(A^{-1}D^{-1}) = 1$. (We are here using the facts that $\deg A + \deg D \leq g$ and $r(A^{-1}) = 1$.) Now, (11.15.2), $r(A^{-1}D^{-1}) = 1$, and Abel's theorem give us $AD = D'P_0$; which is a contradiction since $P_0 \notin A$, $P_0 \notin D$. Hence we conclude $s \leq t - r$ and thus $a \in W_{t-r} + b$. $\square$

Remark. The proposition shows that the subvarieties $W_t \subset J(M)$ for $t < g$ are as far from being translation invariant as possible. Specifically an inclusion $W_t + u \subset W_t$ is equivalent to $u \in W_0 = \{0\}$.

III.11.16. Proposition

a. *For any* $r, t \geq 0$ *and any* $a, b \in J(M)$, *we have*

$$(W_r + a) + (W_t + b) = (W_{r+t} + a + b).$$

b. *For* $0 \leq r \leq t \leq g - 1$ *and any* $a, b \in J(M)$, $(W_t + a) \ominus (W_r + b) = (W_{t-r} + a - b)$.

PROOF. Part (a) is a triviality. To prove (b) we use Proposition III.11.15 to obtain $u \in (W_{t-r} + (a - b)) \Leftrightarrow (b - a) \in (W_{t-r} - u) \Leftrightarrow (W_r + (b - a)) \subset (W_t - u) \Leftrightarrow (W_r + b) \subset (W_t + a - u)$. Assume now that $u \in (W_{t-r} + a - b)$ [and thus $(W_r + b) \subset (W_t + a - u)$] and $v \in (W_r + b)$. Thus $v \in (W_t + a - u)$, or $u \in W_t + a - v$ for all $v \in (W_r + b)$. Hence $u \in (W_t + a) \ominus (W_r + b)$. Conversely, if $u \in (W_t + a) \ominus (W_r + b)$ then $u \in W_t + a - v$ for all $v \in W_r + b$ or $v \in (W_t + a - u)$. Hence $(W_r + b) \subset (W_t + a - u)$ or $u \in (W_{t-r} + a - b)$. □

Remarks

1. The condition that $t \leq g - 1$ in (b) is necessary because $W_t = J(M)$ for $t \geq g$. Hence for any $S \subset J(M)$, $W_t \ominus S = J(M)$ for $t \geq g$.
2. A special case of the proposition is also worthy of mention here:

$$W_1 = W_{g-1} \ominus W_{g-2} = \bigcap_{u \in W_{g-2}} W_{g-1} - u.$$

Thus $W_1 \cong M$ can be recovered from W_{g-2} and W_{g-1}. Hence M is determined by W_{g-1} and W_{g-2}.

III.11.17. Proposition. *For any* $n > 0$ *and any* $r \geq 0$,

$$W_n^r = W_{n-r} \ominus (-W_r), \quad \textit{whenever } r \leq n, \textit{ and} \tag{11.17.1}$$

$$W_n^r = \varnothing \quad \textit{whenever } r \geq n. \tag{11.17.2}$$

PROOF. Let $x \in W_n^r$. Then $x = \varphi(D)$, $D \in M_n$, and $r(D^{-1}) \geq r + 1$. The Riemann–Roch theorem gives $r(D^{-1}) = n - g + 1 + i(D)$. This is clearly impossible when $r \geq n$.

We have previously seen (Lemma III.8.15) that $u \in W_n^r$ if and only if for every $v \in W_r$ there is a $v' \in W_{n-r}$ such that $u = v + v'$. Hence

$$W_n^r = \bigcap_{v \in W_r} (W_{n-r} + v). \tag{11.17.3}$$ □

Remarks

1. Formula (11.17.3) can be reformulated as $u \in W_n^r \Leftrightarrow -(W_r - u) \subset W_{n-r}$. Clifford's theorem gives a necessary condition that W_n^r not be empty.

The condition is that $2r \leq n$ or that $r \leq n - r$. Hence the above gives a geometric interpretation of this theorem.

2. It is an immediate consequence of the proposition (in the form of Equation (11.17.3)) that the subsets W_n^r are complex analytic subvarieties of $J(M)$.
3. If we define W_n to be the empty set for $n < 0$, then (11.17.2) becomes a special case of (11.17.1), and (11.17.3) is always valid. We shall adopt this convention.

III.11.18. Proposition. *Let $1 \leq n \leq g - 1$ and let $x = \varphi(Q)$ with $Q \neq P_0$. Then*

$$W_n \cap (W_n + x) = (W_{n-1} + x) \cup W_{n+1}^1. \tag{11.18.1}$$

PROOF. Assume $u \in W_n \cap (W_n + x)$. Then

$$u = \varphi(Q_1 \cdots Q_n) + \varphi(Q) = \varphi(P_1 \cdots P_n).$$

Thus $QQ_1 \cdots Q_n \sim P_0 P_1 \cdots P_n$. If $\sim$ can be replaced by $=$, then Q appears among $P_1, \ldots, P_n$ and we may assume $Q = P_n$. Thus

$$u = \varphi(P_1 \cdots P_{n-1}) + \varphi(Q) \in (W_{n-1} + x).$$

If the two divisors are not identical, then $r(P_0^{-1} \cdots P_n^{-1}) \geq 2$ and $u \in W_{n+1}^1$. The reverse inclusion is trivial. □

Remark. A non-empty component of W_{n+1}^1 cannot be a subset of $W_{n-1} + x$ for all $x \in W_1$. For if V is a component of W_{n+1}^1 and $V \subset W_{n-1} + x$ for all $x \in W_1$, then

$$V \subset \bigcap_{x \in W_1} (W_{n-1} + x) = W_n^1.$$

Now $V \subset W_n^1$ certainly implies that $V + W_1 \subset W_{n+1}^1$. Since $V + W_1$ is an irreducible subvariety of $J(M)$ contained in W_{n+1}^1 (it is the image of the irreducible subvariety $V \times W_1$ under the mapping $(x,y) \mapsto x + y$) that contains V, it must agree with V. Thus in particular we have W_1 invariant under translations by $v \in V$. This contradicts the remark following Proposition III.11.15.

III.11.19. Theorem

a. *Let r be the dimension of a non-empty component of W_{n+1}^1 with $1 \leq n \leq g - 1$. Then*

$$2n - g \leq r \leq n - 1.$$

b. *A component of W_{n+1}^1, $1 \leq n \leq g - 1$, has dimension $n - 1$ if and only if M is hyperelliptic.*

PROOF. Let V be a component of W_{n+1}^1. By the above remark we can choose an $x \in W_1$ such that $V \not\subset (W_{n-1} + x)$. By (11.18.1), V must be a component of the intersection $W_n \cap (W_n + x)$. Hence $\dim V \geq 2n - g$. Since W_n is irreducible, and $W_n + x \neq W_n$, for each component V of the intersection, we must have $\dim V \leq n - 1$.

Assume now that M is hyperelliptic. Then W_2^1 is non-empty and consists of a single point x. Hence $x + W_{n-1}$ is a component of W_{n+1}^1 of dimension $n - 1$.

Conversely, assume W_{n+1}^1 has a component V of dimension $n - 1$. If $n = 1$, then W_2^1 is not empty and M is hyperelliptic. Thus assume $n \geq 2$. Let us choose a point $x \in W_1$, and a point $u \in R(V) \subset W_n \cap (W_n + x)$ that is a manifold point of the two varieties of the intersection. (This is possible since V has dimension $n - 1$ and each of the two singular sets W_n^1, $(W_n^1 + x)$ have dimensions at most $n - 2$.) The tangent space to W_{n+1}^1 at u has dimension $n - 1$. Thus the dual cotangent space has dimension $g - n + 1$. Now the point $u \in W_n \cap (W_n + x)$ is given by

$$u = \varphi(Q_1 \cdots Q_n) + \varphi(Q) = \varphi(P_1 \cdots P_n) + \varphi(P_0),$$

where $x = \varphi(Q)$. Assume now that $P_1 \cdots P_n P_0$ is the polar divisor of a function. We can then choose Q so that neither Q nor any of the Q_j appear among the P_k. The space of differentials vanishing at u on W_{n+1}^1 is the span of those vanishing at u on W_n and on $W_n + x$, which are, respectively those vanishing at $P_1, \ldots, P_n$ and $Q_1, \ldots, Q_n$. However, $r(P_1^{-1} \cdots P_n^{-1} P_0^{-1}) = r(P_1^{-1} \cdots P_n^{-1}) + 1$ implies that $i(P_1 \cdots P_n P_0) = i(P_1 \cdots P_n)$. Thus every differential vanishing at $P_1, \ldots, P_n$ also vanishes at P_0. Similarly, every differential vanishing at $Q_1, \ldots, Q_n$ also vanishes at Q. Each of these spaces are of dimension $g - n$. Their span must have dimension $g - n + 1$. Thus their intersection has dimension

$$2(g - n) - (g - n + 1) = g - n - 1.$$

The intersection is $\Omega(Q_1 \cdots Q_n Q P_1 \cdots P_n P_0) = \Omega(D)$. Now

$$r(D^{-1}) = 2n + 2 - g + 1 + (g - n - 1) = n + 2,$$

and hence

$$c(D) = 2n + 2 - 2(n + 2) + 2 = 0.$$

By Clifford's theorem, M is hyperelliptic unless $D \sim Z$.

If $D \sim Z$, then $2n + 2 = 2g - 2$ or $n = g - 2$. In this case $2u = K$, which has only finitely many solutions in $J(M)$. But $n \geq 2$ implies ($g \geq 4$ and) $\dim V = n - 1 \geq 1$. Thus V is a continuum and u could be chosen so that $2u \neq K$ (from the beginning).

There remains the possibility that $D = P_1 \cdots P_n P_0$ is not the polar divisor of a function for every $D \in \varphi^{-1}(V)$ with D containing P_0. Note that every $D' \in M_{n+1}^1$ is equivalent to a $D \in M_{n+1}^1$ that contains P_0. Thus for every $D \in \varphi^{-1}(V)$ there is a $P \in D$, such that $r(PD^{-1}) = r(D^{-1})$; showing that every $u \in V$ can be written in the form $u = v + y$ with $v \in W_n^1$ and $y \in W_1$.

Thus W_n^1 must have a component of dimension $\geq n-2$. By induction this implies the hyperellipticity of M. $\square$

III.11.20. An application of the above results on the dimension of W_{n+1}^1 yields the following.

Theorem (Noether). *On a non-hyperelliptic surface of genus $g \geq 3$, for every $q \geq 2$, the q-fold products of the abelian differentials of the first kind span the space of holomorphic q-differentials.*

Before we prove Noether's theorem we recall (see III.6.8) the following useful observation: we can always find two elements of $\mathscr{H}^1(M)$ that have no common zeros. Choose any $\omega_1 \in \mathscr{H}^1(M)$ and let $P_1, \ldots, P_r$ be its distinct zeros. If the result were not true, every other element of $\mathscr{H}^1(M)$ would have to vanish at one or more of the points $P_1, \ldots, P_r$. Consider, however, $\Omega(P_1) \cup \cdots \cup \Omega(P_r)$. Each set is $g-1$ dimensional, so that the union is not all of $\mathscr{H}^1(M)$. Thus there is an $\omega_2 \in \mathscr{H}^1(M)$ which does not vanish at any of the points $P_1, \ldots, P_r$.

The preceding implies that the function $f = \omega_1/\omega_2$ gives rise to a $(2g-2)$-sheeted cover of the sphere with the property that for each $z \in \mathbb{C} \cup \{\infty\}$, $f^{-1}(z)$ is a canonical integral divisor. Since the function f is branched over only finitely many points we can even assume that the differentials ω_1 and ω_2 have only simple zeros.

A similar argument applied to $\Omega(P_0)$ shows that we can always find two elements $\omega_1, \omega_2 \in \mathscr{H}^1(M)$ with precisely one common zero at P_0 (and no other common zeros), provided M is not hyperelliptic. We shall use these results in the proof of Noether's theorem.

Proof of Theorem. Assume that M is a non-hyperelliptic surface. Thus there exists on M an integral divisor D of degree $g-2$, such that $i(D) = 2$ and for each $Q \in M$ we have $i(DQ) = 1$. Assume, on the contrary, that for every $D \in M_{g-2}$, there is a $Q \in M$ such that $i(DQ) \geq 2$. Thus for all $v \in W_{g-2}$, there is an $x \in W_1$ such that $v + x \in W_{g-1}^1$, or $v \in W_{g-1}^1 - x \subset W_{g-1}^1 \oplus (-W_1)$. But then $W_{g-2} \subset W_{g-1}^1 \oplus (-W_1)$, and it must be the case that $\dim W_{g-1}^1 \geq g-3$. Theorem III.11.19 implies that $\dim W_{g-1}^1 \leq g-3$, and that M is hyperelliptic.

Let ω_1, ω_2 span $\Omega(D)$ and let ω_3 be any holomorphic differential which does not vanish at any of the zeros of $\omega_1\omega_2$. Our assumptions on D give that ω_1/ω_2 is a meromorphic function on M with precisely g poles and $i((\omega_2)/D) = 1$. The function ω_1/ω_2 has precisely g poles because ω_1 and ω_2 have no common zeros other than in D. The assertion on the index of $(\omega_2)/D$ follows from the equality $i((\omega_2)/D) = r(D^{-1}) = -1 + i(D)$. Let $\{\theta_1, \ldots, \theta_g\}$ be a basis for $\mathscr{H}^1(M)$ and consider the $2g$ elements of $\mathscr{H}^2(M)$:

$$\omega_1\theta_1, \ldots, \omega_1\theta_g; \omega_2\theta_1, \ldots, \omega_2\theta_g.$$

Note that the first g products are linearly independent and so are the last g products. Let A_1 and A_2 denote the subspaces of $\mathscr{H}^2(M)$ spanned by these products. Now $\dim(A_1 \cap A_2) = \dim A_1 + \dim A_2 - \dim(A_1 \vee A_2)$, and $\dim(A_1 \cap A_2) = 1$. To verify the last assertion write $(\omega_j) = DD_j, j = 1, 2$, and note that D_1 and D_2 are relatively prime integral divisors of degree g (with $(\omega_1/\omega_2) = D_1/D_2$). Now if $\eta \in A_1 \cap A_2$, then $\eta = \zeta_j\omega_j$, $j = 1, 2$, with $\zeta_j \in \mathscr{H}^1(M)$ (these ζ_j need not be normalized). Thus $\zeta_2/\zeta_1 = \omega_1/\omega_2$. In particular, $(\zeta_1) = \tilde{D}D_2$ and $(\zeta_2) = \tilde{D}D_1$ for some $\tilde{D} \in M_{g-2}$. Since $i(D_1) = 1 = i(D_2)$, we must have ζ_1/ω_2 and $\zeta_2/\omega_1 \in \mathbb{C}$ and $A_1 \cap A_2$ is spanned by $\omega_1\omega_2$. Thus $\dim(A_1 \vee A_2) = 2g - 1$, and the $2g$ products span a subspace of $\mathscr{H}^2(M)$ of dimension $2g - 1$.

We now adjoin $\omega_3\theta_1, \ldots, \omega_3\theta_g$ to our list, and denote the subspace of $\mathscr{H}^2(M)$ spanned by these g linearly independent products by A_3. Once again

$$\begin{aligned}\dim((A_1 \vee A_2) \cap A_3) &= \dim(A_1 \vee A_2) + \dim A_3 - \dim(A_1 \vee A_2 \vee A_3)\\ &= 2g - 1 + g - \dim(A_1 \vee A_2 \vee A_3).\end{aligned}$$

We show that $\dim((A_1 \vee A_2) \cap A_3) = 2$, and in fact the space in question is spanned by $\omega_1\omega_3, \omega_2\omega_3$. Note that if $\eta \in (A_1 \vee A_2) \cap A_3$, then $\eta = \zeta_3\omega_3 = \zeta_1\omega_1 + \zeta_2\omega_2$. Since the right-hand side vanishes at D, it follows that $(\zeta_3) = D\tilde{D}$. Thus $\zeta_3 = x_1\omega_1 + x_2\omega_2$ (with $x_j \in \mathbb{C}$) and the dimension is 2, as claimed. Thus $\dim(A_1 \vee A_2 \vee A_3) = 3g - 3 = \dim \mathscr{H}^2(M)$. This concludes the proof for $q = 2$.

For the case $q = 3$, we start by choosing two elements $\omega_1, \omega_2 \in \mathscr{H}^1(M)$ such that ω_1 and ω_2 have precisely one common zero at P_0, and $\omega_3 \in \mathscr{H}^1(M)$ such that ω_3 does not vanish at the zeros of $\omega_1\omega_2$.

Let $\{f_1, \ldots, f_{3g-3}\}$ be a basis of $\mathscr{H}^2(M)$ with each f_j a 2-fold product of elements of $\mathscr{H}^1(M)$. Let A_j be the $(3g - 3)$-dimensional space spanned by $\omega_j f_1, \ldots, \omega_j f_{3g-3}$. As before

$$\dim(A_1 \cap A_2) = \dim A_1 + \dim A_2 - \dim(A_1 \vee A_2).$$

If $\omega_1\eta_1 = \omega_2\eta_2 \in A_1 \cap A_2$, then η_2/ω_1 is an abelian differential with at most a single pole at P_0 and therefore holomorphic at P_0. Hence η_2 can be written as $\eta_2 = \omega_1\omega$ with $\omega \in \mathscr{H}^1(M)$. This shows that $\dim(A_1 \cap A_2) = g$, and thus that $\dim(A_1 \vee A_2) = 5g - 6$. Clearly ω_3^3 does not vanish at P_0, while every element in $A_1 \vee A_2$ does. Hence ω_3^3 is not in $A_1 \vee A_2$. Adjoining it to our list, we have a space of dimension $5g - 5 = \dim \mathscr{H}^3(M)$, spanned by three-fold products. This concludes the proof for $q = 3$.

For $q \geq 4$ we now can proceed by induction. We let $\omega_1, \omega_2 \in \mathscr{H}^1(M)$ be such that ω_1 and ω_2 have no common zeros. Let $\{f_1, \ldots, f_{(2m-1)(g-1)}\}$ be a basis for $\mathscr{H}^m(M)$, with $m \geq 3$, composed of m-fold products of elements of $\mathscr{H}^1(M)$. Let A_j denote the space spanned by $\omega_j f_1, \ldots, \omega_j f_{(2m-1)(g-1)}$. Then $\dim(A_1 \cap A_2) = 2(2m - 1)(g - 1) - \dim(A_1 \vee A_2)$. Clearly however $\dim(A_1 \cap A_2) = (2m - 3)(g - 1)$, and thus $\dim(A_1 \vee A_2) = (2m + 1)(g - 1) = \dim \mathscr{H}^{m+1}(M)$. □

EXERCISE

Recall the exercises of III.7.5 and complete the above discussion for hyperelliptic surfaces (including the case $g = 2$).

III.12. Torelli's Theorem

This section contains a proof of Torelli's theorem which states that the conformal equivalence class of a compact Riemann surface is determined by its period matrix. If the surface is of genus 0, there is of course nothing to prove. For surfaces of genus 1, the result is a consequence of Abel's theorem (see III.6.4) which shows that each torus is its own Jacobian variety. Thus we are left with the case of surfaces of genus $g > 1$. We will use the notation of the previous section and show that $W_1 \subset J(M)$ is determined up to a translation and reflection (this is the map that sends a point in the Jacobian variety to its inverse) by W_{g-1}, the so-called *canonical polarization of* $J(M)$. Of course, W_1 is isomorphic to the surface M (Abel's theorem again) and we shall show in Chapter VI (Theorem VI.3.1) that W_{g-1} is determined by the period matrix of M. We follow the arguments of Henrik H. Martens, which show that Torelli's theorem is a combinatorial consequence of the Riemann–Roch theorem and Abel's theorem.

III.12.1. We start with a number of observations.

(a) For every $a \in J(M)$, we have

$$-(W_{g-1} + a) + K = W_{g-1} - a.$$

This is a restatement of Proposition III.11.14.

(b) For every integer r with $0 \leq r \leq g - 1$ and all a and b in $J(M)$, we have

$$(W_r + a) \subset (W_{g-1} + b) \Leftrightarrow a \in (W_{g-1-r} + b).$$

This result is a special case of Proposition III.11.15 for the case $t = g - 1$.

(c) For every integer r as above

$$W_{g-1-r} = W_{g-1} \ominus W_r$$

and

$$-W_{g-1-r} + K = W_{g-1} \ominus (-W_r).$$

The first statement is Proposition III.11.16(b) for $t = g - 1$. The second formula is a consequence of

$$\begin{aligned} W_{g-1} \ominus (-W_r) &= \bigcap \{W_{g-1} + u; u \in W_r\} = \bigcap \{-W_{g-1} + u + K; u \in W_r\} \\ &= -\bigcap \{W_{g-1} - u; u \in W_r\} + K \\ &= -(W_{g-1} \ominus W_r) + K = -W_{g-1-r} + K. \end{aligned}$$

We now establish the following

Lemma. *Let r be an integer with $0 \le r \le g-2$. Let a and b be points in $J(M)$ that are related by the equation $b = a + x - y$ where $x \in W_1$ and $y \in W_{g-1-r}$. Then either $(W_{r+1} + a) \subset (W_{g-1} + b)$ or else*

$$(W_{r+1} + a) \cap (W_{g-1} + b) = (W_r + a + x) \cup S,$$

where

$$S = (W_{r+1} + a) \cap (-W_{g-2} - y + a + K).$$

PROOF. By assumption $x = \varphi(R)$, $y = \varphi(\hat{R})$, and $\varphi(R) + a = \varphi(\hat{R}) + b$, where R and $\hat{R}$ are integral divisors of degree 1 and $g-1-r$, respectively. If R is a point of $\hat{R}$, then our equation reduces to $a = \varphi(R') + b$, where $R' = \hat{R}/R$ is an integral divisor of degree $g-2-r$. In this case, $a \in W_{g-2-r} + b$ and $W_{r+1} + a \subset W_{g-1} + b$, by (b) above. Hence we assume that R is not a point of $\hat{R}$.

Let $u \in (W_{r+1} + a) \cap (W_{g-1} + b)$. Then there exist integral divisors D and $\hat{D}$ of degree $r+1$ and $g-1$, respectively, such that $u = \varphi(D) + a = \varphi(\hat{D}) + b$. Hence $D\hat{R}$ is equivalent to $\hat{D}R$. If $D\hat{R} = \hat{D}R$, then R must be a point of D and $u = \varphi(D) + a = \varphi(D') + \varphi(R) + a$, where D' is an integral divisor of degree r and thus $u \in W_r + a + x$. If $D\hat{R} \ne \hat{D}R$, then $r(1/D\hat{R}) \ge 2$ and hence given any point $Q \in M$, there exists an integral divisor $\hat{Q}$ of degree $g-1$ such that $D\hat{R}$ is equivalent to $Q\hat{Q}$. Then

$$u = \varphi(D) + a = \varphi(\hat{Q}) + \varphi(Q) - \varphi(\hat{R}) + a;$$

that is,

$$u \in \bigcap \{W_{g-1} + a - y + v; v \in W_1\} = -(W_{g-2} + y - a) + K.$$

We have used part two of (c). Since

$$-(W_{g-2} + y - a) + K \subset -(W_{g-1} + y - a - x) + K = W_{g-1} + b,$$

the reverse inclusion is trivial. □

III.12.2. We will need two more facts. The first observation involves reflection in the Jacobian variety. We change the canonical homology basis on the surface M

$$\text{from} \quad \{a_1, \ldots, a_g, b_1, \ldots, b_g\} \quad \text{to} \; \{-a_1, \ldots, -a_g, -b_1, \ldots, -b_g\}.$$

This changes the elements in the dual basis for the holomorphic differentials to their negatives while leaving unaltered both the period matrix and the Jacobian variety of the surface. Our reflection changes W_r into $-W_r$ for each positive integer r.

Our second observation involves the intersection of the curve in the Jacobian (that is, W_1) with the polarization divisor (that is, translates of W_{g-1}). The needed fact is best established using the results of Chapter VI on the Riemann theta function. Let b be a point in $J(M)$ and assume that W_1 is not contained in $W_{g-1} + b$. Then the intersection $W_1 \cap (W_{g-1} + b)$ consists of g points (counting multiplicity) $\varphi(P_1), \ldots, \varphi(P_g)$ with $P_i \in M$, $i = 1, \ldots, g$ and there exists a point $c \in J(M)$ that is independent of b such that

$$\varphi(P_1 \cdots P_g) = b + c.$$

To verify this assertion, we let κ be the vector of Riemann constants (see VI.2.4.1). By Theorem VI.2.4, the function on $\mathbf{C}^g$

$$z \mapsto \theta(z - b + \kappa)$$

vanishes precisely on $W_{g-1} + b$ and by Theorem VI.3.2, the multivalued function on M

$$P \mapsto \theta(\varphi(P) - b + \kappa)$$

either vanishes identically or has precisely g zeros $P_1, \ldots, P_g$ on M that satisfy

$$\varphi(P_1 \cdots P_g) + \kappa = b - \kappa.$$

The images under φ of the zeros of the multivalued function are precisely the points in $W_1 \cap (W_{g-1} + b)$. Thus we have verified our claim with

$$c = -2\kappa = K$$

by Theorem VI.3.6.

III.12.3. Torelli's Theorem. *Let $\varphi : M \to J(M)$ be an Abel–Jacobi embedding of a compact Riemann surface M of genus $g \geq 2$ into its Jacobian variety $J(M)$. Then $W_1 = \varphi(M)$ is determined up to translation and reflection by the canonical polarization of $J(M)$; that is, by the class of translates of W_{g-1}.*

PROOF. By translation, if necessary, we may normalize φ so that $\varphi(P_o) = 0$ for some base point $P_o \in M$. Let N be a second Riemann surface of genus g with the same Jacobian variety as M and let $\psi : N \to J(M)$ be an Abel–Jacobi embedding of N into its Jacobian variety with some normalization $\psi(Q_o) = 0$ for a base point $Q_o \in N$. Let V_r denote the image under ψ of the set of integral divisors of degree r on N. We must show that if V_{g-1} is a translate of W_{g-1} then V_1 is a translate of either W_1 or of $-W_1$.

Let r be the smallest integer such that

$$V_1 \subset (W_{r+1} + a) \quad \text{or} \quad V_1 \subset -(W_{r+1} + a) + K \quad \text{for some } a \in J(M).$$

The theorem will be proved if we can show that $r = 0$. Assume for contradiction that $r \geq 1$. Clearly, $r < g - 1$. By changing the canonical homology basis on N to inverses as in III.12.2, if necessary, we may assume that $V_1 \subset W_{r+1} + a$ for some given $a \in J(M)$. Choose $x \in W_1$ and $y \in W_{g-1-r}$. Set $b = a + x - y$. Then unless $W_{r+1} + a \subset W_{g-1} + b$, we have by the assumed inclusion and the previous lemma (in the notation of that lemma)

$$\begin{aligned} V_1 \cap (W_{g-1} + b) &= V_1 \cap (W_{g-1} + b) \cap (W_{r+1} + a) \\ &= (V_1 \cap (W_r + a + x)) \cup (V_1 \cap S). \end{aligned}$$

For given a, $W_r + a + x$ depends only on the choice of x and S only on the choice of y.

We show first, that for a fixed x, $V_1 \not\subset (W_{g-1} + b)$ for almost all choices of y, and hence also $(W_{r+1} + a) \not\subset (W_{g-1} + b)$ for the same y. As y varies over W_{g-1-r}, $-b$ varies over $W_{g-1-r} - a - x$. By assumption, there exists a $k \in J(M)$ such that $V_{g-1} + k = W_{g-1}$. Hence $V_1 \subset (W_{g-1} + b)$ if and only if $V_1 \subset (V_{g-1} + b + k)$ if and only if $-b \in (V_{g-2} + k)$ by III.12.1(b). Thus the set of b for which $V_1 \subset (W_{g-1} + b)$ is precisely the set of b with $-b \in (V_{g-2} + k) \cap (W_{g-1-r} - a - x)$. Now, if $V_1 \subset (W_{g-1} + b)$ for all $-b \in (W_{g-1-r} - a - x)$, then $V_1 \subset W_{g-1} \ominus (W_{g-1-r} - a - x) = W_r + a + x$ by part one of III.12.1(c). This contradicts the minimality assumption on r. Hence $(W_{g-1-r} - a - x) \not\subset (V_{g-2} + k)$ and the intersection of these two sets is a lower dimensional subset of $(W_{g-1-r} - a - x)$.

We now consider again the intersection

$$V_1 \cap (W_{g-1} + b) = (V_1 \cap (W_r + a + x)) \cup (V_1 \cap S).$$

By III.12.2, if $V_1 \not\subset (W_{g-1} + b)$, then there is a unique integral divisor $D(b)$ of degree g on N such that

$$\psi(D(b)) = b + c, \tag{$*$}$$

where $c \in J(M)$ is independent of b and the points in $D(b)$ are mapped by ψ onto the intersection $V_1 \cap (W_{g-1} + b)$.

We show next that $V_1 \cap (W_r + a + x)$ contains at most one point. If not, then as $-b$ varies over almost all points of $(W_{g-1-r} - a - x)$ for fixed x, the divisor $D(b)$ will contain at least two fixed points and hence $\psi(D(b))$ varies over a translate of V_{g-2}. By the last displayed equation $(*)$, we would have an inclusion of $-W_{g-1-r} + K$ in a translate of V_{g-2}; say $(-W_{g-1-r} + K) \subset (V_{g-2} + d)$ for some $d \in J(M)$. Then

$$(V_{g-1} + k) \ominus (V_{g-2} + d) \subset W_{g-1} \ominus (-W_{g-1-r} + K)$$

and, using III.12.1(c), we get an inclusion of V_1 in a translate of $-W_r + K$. This again contradicts the minimality or r.

Keeping y fixed and varying x, we see from $(*)$ that $V_1 \cap (W_r + a + x)$ must contain at least one and hence exactly one point. By the preceding argument this point occurs in the divisor $D(b)$ with multiplicity 1 for almost all choices of y.

It is now easily seen that we can find x and x' in W_1 and a $y \in W_{g-1-r}$ such that $D(a + x - y) = Q\hat{D}$ and $D(a + x' - y) = Q'\hat{D}$, where Q and Q' are points of N and $\hat{D}$ is an integral divisor on N of degree $g - 1$ not containing either Q or Q'. By $(*)$, $\varphi(Q) - \varphi(Q') = x - x'$ and hence W_1 has two distinct points in common with some translate of V_1. Now, if $x, x' \in W_1$, then $(W_{g-1} - x) \cap (W_{g-1} - x') = W_{g-2} \cup (-W_{g-2} - x - x' + K)$ by the last lemma. By III.12.1(c), we now have an inclusion of some translate of V_{g-2} in W_{g-2} or in $-W_{g-2} + K$, hence by III.12.1(c) again, we get an inclusion of some translate of V_1 in W_1 or in $-W_1 + K$. We have completed the proof of the theorem. □

CHAPTER IV

Uniformization

This chapter has two purposes. The first and by far the most important is to prove the uniformization theorem for Riemann surfaces. This theorem describes all simply connected Riemann surfaces and hence with the help of topology, all Riemann surfaces.

The second purpose is to give different proofs for the existence of meromorphic functions on Riemann surfaces. These proofs will not need the topological facts we assumed in Chapter II (triangulability of surfaces). As a matter of fact, all the topology can be quickly recovered from the complex structure.

This chapter also contains a discussion of the exceptional surfaces (those surfaces with abelian fundamental groups), an alternate proof of the Riemann–Roch theorem, and a treatment of analytic continuation (algebraic functions on compact surfaces).

IV.1. More on Harmonic Functions (A Quick Review)

In this paragraph we establish some of the basic properties of harmonic functions. The material presented here is probably familiar to most readers.

IV.1.1. We begin by posing a problem that will motivate the presentation of this section. Details will only be sketched. For more see Ahlfors' book *Complex Analysis.*

Dirichlet Problem. Let D be a region on a Riemann surface M with boundary δD. Let f be a continuous function on δD. Does there exist a continuous function F defined on $D \cup \delta D$ such that

i. $F|D$ is harmonic, and
ii. $F|\delta D = f$?

IV.1.2. Let g be a real-valued harmonic function in $\{|z| < \rho\}$. There exists, of course, a holomorphic function f defined on $\{|z| < \rho\}$ with $g = \text{Re } f$. We may define f by

$$\text{Im } f(z) = \int_0^z *dg,$$

and observe that the integral is independent of the path (because $\{|z| < \rho\}$ is simply connected). The Taylor series expansion of f about the origin is

$$f(re^{i\theta}) = \sum_{n=0}^{\infty} a_n r^n e^{in\theta}, \qquad 0 \le r < \rho.$$

Without loss of generality $a_0 \in \mathbb{R}$. Hence

$$g(re^{i\theta}) = \frac{1}{2}(f(re^{i\theta}) + \overline{f(re^{i\theta})})$$

$$= a_0 + \frac{1}{2}\sum_{n \ge 1} r^n(a_n e^{in\theta} + \bar{a}_n e^{-in\theta}).$$

Multiplying the above by $e^{-in\theta}$ and integrating, we get

$$a_0 = \frac{1}{2\pi}\int_0^{2\pi} g(re^{i\theta})\, d\theta,$$

$$a_n = \frac{1}{\pi}\int_0^{2\pi} \frac{g(re^{i\theta})}{(re^{i\theta})^n}\, d\theta, \qquad n \ge 1.$$

Thus for $|z| < r$, we have

$$f(z) = \frac{1}{2\pi}\int_0^{2\pi} g(re^{i\theta})\left[1 + 2\sum_{n \ge 1}\left(\frac{z}{re^{i\theta}}\right)^n\right] d\theta$$

$$= \frac{1}{2\pi}\int_0^{2\pi} g(re^{i\theta})\frac{re^{i\theta} + z}{re^{i\theta} - z}\, d\theta, \tag{1.2.1}$$

and

$$g(z) = \text{Re } f(z) = \frac{1}{2\pi}\int_0^{2\pi} g(re^{i\theta}) \text{ Re } \frac{re^{i\theta} + z}{re^{i\theta} - z}\, d\theta$$

$$= \frac{1}{2\pi}\int_0^{2\pi} g(re^{i\theta})\frac{r^2 - |z|^2}{|re^{i\theta} - z|^2}\, d\theta. \tag{1.2.2}$$

The expression $(r^2 - |z|^2)/|re^{i\theta} - z|^2$ is known as the Poisson kernel (for the disc of radius r about the origin). It has the following important properties:

$$\frac{1}{2\pi}\int_0^{2\pi} \frac{r^2 - |z|^2}{|re^{i\theta} - z|^2}\, d\theta = 1, \qquad |z| < r. \tag{1.2.3}$$

$$\frac{r^2 - |z|^2}{|re^{i\theta} - z|^2} > 0 \quad \text{for } |z| < r. \tag{1.2.4}$$

$$\lim_{\substack{z \to re^{i\theta_0} \\ |z|<r}} \frac{1}{2\pi} \int_{|\theta-\theta_0|>\eta} \frac{r^2 - |z|^2}{|re^{i\theta} - z|^2}\, d\theta = 0 \quad \text{for } 0 < \eta < \pi. \tag{1.2.5}$$

From the reproducing formula (1.2.1) we also get a formula for the harmonic conjugate of g that vanishes at $z = 0$, namely,

$$\begin{aligned} \operatorname{Im} f(z) &= \frac{1}{2\pi} \int_0^{2\pi} g(re^{i\theta}) \operatorname{Im} \frac{re^{i\theta} + z}{re^{i\theta} - z}\, d\theta \\ &= \frac{1}{2\pi i} \int_0^{2\pi} g(re^{i\theta}) \frac{re^{-i\theta} z - re^{i\theta}\bar{z}}{|re^{i\theta} - z|^2}\, d\theta. \end{aligned} \tag{1.2.6}$$

IV.1.3. Theorem. *The Dirichlet problem for the disc has a unique solution; that is, given a continuous function f defined on $\{|z| = r\}$, there exists a continuous function F on $\{|z| \le r\}$ such that F is harmonic in $\{|z| < r\}$ and $F(re^{i\theta}) = f(re^{i\theta})$, $0 \le \theta \le 2\pi$.*

Proof. Without loss of generality f is real-valued. Since real harmonic functions satisfy the maximum and minimum principle, uniqueness is obvious. For existence, one sets for $|z| < r$

$$F(z) = \frac{1}{2\pi} \int_0^{2\pi} \frac{r^2 - |z|^2}{|re^{i\theta} - z|^2} f(re^{i\theta})\, d\theta \tag{1.3.1}$$

and uses the properties of the Poisson kernel (1.2.3)–(1.2.5). □

Corollary. *If f is a continuous function on a domain $D \subset \mathbb{C}$, and f satisfies the mean-value property in D, then f is harmonic in D.*

Proof. Again without loss of generality f is real-valued. Solve the Dirichlet problem for $f|\{|z - z_0| = r\}$ with $\{|z - z_0| \le r\} \subset D$. Call the solution F. Then $F - f$ satisfies both minimum and maximum principles, since the mean value property is all that one needs to prove these principles. Hence $F = f$ on $\{|z - z_0| \le r\}$, and f is harmonic. □

IV.1.4. For $|z| < r$, it is easy to see that

$$\frac{r - |z|}{r + |z|} \le \frac{r^2 - |z|^2}{|re^{i\theta} - z|^2} \le \frac{r + |z|}{r - |z|}.$$

These estimates on the Poisson kernel imply almost immediately

Harnack's Inequality. *Let D be a domain in $\mathbb{C}$ and $D_1 \subset\subset D$ (that is, D_1 is a relatively compact subdomain of D). Let u be a positive harmonic function on D. Then there exists a constant $c = c(D_1, D)$ that depends only on D_1 and D (not on*

u) such that

$$\frac{1}{c} \leq \frac{u(z_1)}{u(z_2)} \leq c, \quad \text{all } z_1, z_2 \in D_1.$$

IV.1.5. Because harmonicity is a local property and we have the Poisson reproducing formula (1.2.2) for harmonic functions, we can establish the following

Proposition. *If $\{u_n\}$ is a sequence of harmonic functions on a Riemann surface M and $\{u_n\}$ converges to u uniformly on compact subsets of M, then u is also harmonic.*

This can be seen from the fact that the limit function is continuous and necessarily has the mean-value property.

IV.1.6. Harnack's Principle. *Consider a sequence of real valued harmonic functions $\{u_n\}$ each defined on a domain D_n on the Riemann surface M. Assume that each $P_0 \in M$ has a neighborhood U such that $U \subset D_n$ for all but finitely many n. Further assume that*

$$u_n(P) \leq u_{n+1}(P), \quad P \in U, n \text{ large}.$$

Then either

i. $\lim_{n\to\infty} u_n = +\infty$, *uniformly on compact subsets of M, or*
ii. $\lim_{n\to\infty} u_n = u$, *uniformly on compact subsets of M with u a harmonic function.*

Outline of Proof. Without loss of generality we may assume u_n are positive harmonic functions. Define

$$u(P) = \lim_n u_n(P).$$

Harnack's inequalities show that the sets

$$\{P \in M; u(P) = +\infty\}$$
$$\{P \in M; u(P) < +\infty\}$$

are both open. Hence one of them is empty. The same inequalities show that the convergence is locally uniform. Thus the result follows from the previous proposition. □

IV.1.7. Theorem. *Let u be a harmonic function on $\{0 < |z| < 1\}$. Then there exist constants α, β such that for $0 < r < 1$,*

$$\int_0^{2\pi} u(re^{i\theta})\, d\theta = \alpha \log r + \beta. \tag{1.7.1}$$

Proof. Recall Formula I (4.4.2). For u_1, u_2 harmonic and $0 < \rho < r$ (and the usual counterclockwise orientation for the circles) we have:

$$\left(\int_{|z|=r} - \int_{|z|=\rho}\right)(u_1 {}^*du_2 - u_2 {}^*du_1) = \iint_{\rho<|z|<r} u_1 \Delta u_2 - u_2 \Delta u_1 = 0.$$

Now we let $u = u_2$, and set

$$u_1(z) = \log|z|$$

(recall that on $|z| = r$, $(\partial u_j/\partial r)\, r\, d\theta = {}^*du_j$), to obtain

$$-\beta = \int_0^{2\pi} \log r \frac{\partial u}{\partial r} r\, d\theta - \int_0^{2\pi} u(re^{i\theta}) \frac{1}{r} r\, d\theta$$

($-\beta$ is, of course, the value of the right-hand side of the above equation for $r = \rho$, which we take to be a fixed value). Thus

$$\int_0^{2\pi} u(re^{i\theta})\, d\theta = \beta + \log r \int_{|z|=r} {}^*du = \beta + \alpha \log r.$$

The last equality holds because $\int_{|z|=r} {}^*du$ is independent of r. This is simply Cauchy's theorem for $du + i{}^*du$ is a holomorphic differential in $\{0 < |z| < 1\}$ —alternatively, because ${}^*du$ is closed. □

Remark. The above also shows how to evaluate α:

$$\alpha = \int_{|z|=r} {}^*du.$$

Furthermore, we may view (1.7.1) as a formula for computing β, especially when we know that $\alpha = 0$.

Corollary 1. *If u is harmonic and bounded in $\{0 < |z| < 1\}$, then $\alpha = 0$.*

Proof. If $M = \sup|u|$ on $0 < |z| < 1$, then

$$\left|\int_0^{2\pi} u(re^{i\theta})\, d\theta\right| \le M2\pi.$$ □

Corollary 2. *If u is harmonic and bounded in $\{0 < |z| < 1\}$, then u can be extended as a harmonic function to $\{|z| < 1\}$.*

Proof. Since $\alpha = 0$, $\int_{|z|=r} {}^*du = 0$. Thus (we assume u is real, this involves no loss of generality) there exists an analytic function f on $\{0 < |z| < 1\}$, with $u = \operatorname{Re} f$ on $\{0 < |z| < 1\}$. Set $F = \exp f$. Since $|F| = \exp u$, and u is bounded, so is $|F|$. By the Riemann removable singularity theorem F can be extended to $|z| < 1$. Since a pole or an essential singularity of f is an essential singularity for F, f has a removable singularity at $z = 0$. □

Remark. As a consequence of the preceding, we have the following: The Dirichlet problem does *not* have a solution for $D = \{0 < |z| < 1\}$, $f(0) = 1$, $f(z) = 0$ for $|z| = 1$.

IV.2. Subharmonic Functions and Perron's Method

The linear functions $f(x) = ax + b$ in one (real) variable satisfy the mean-value property. The harmonic functions in two variables are the natural generalization. Similarly, both classes of functions satisfy the (appropriate) Laplace equations. Subharmonic functions are the natural generalization of convex functions. Throughout this section, M is a Riemann surface. *All functions considered will be real-valued.*

In this section we establish Perron's principle which gives sufficient conditions for the supremum of a family of subharmonic functions to be a harmonic function. Using this principle, the Dirichlet problem is solved.

IV.2.1. A continuous function u on M is called *subharmonic* on M if and only if for every domain D on M and every harmonic function h on D with $u \leq h$ on D we have $u \equiv h$ on D or $u < h$ on D. The function u is called *superharmonic* if and only if $-u$ is subharmonic. Obviously every harmonic function is both subharmonic and superharmonic.

Proposition. *A continuous function u is subharmonic on M if and only if for every domain $D \subset M$ and every harmonic function h on M, $u + h$ has no maximum in D unless $u + h$ is constant.*

PROOF. Say u is subharmonic and $u + h \leq H$ with H constant and H assumed by $u + h$ at some point in D, then $u \leq H - h$ in D. Since $H - h$ is harmonic in D and equality holds at least one point in D, $u = H - h$. Thus, $u + h = H$, and $u + h$ is constant.

Conversely, say $u \leq h$ on D. Then $u + (-h) \leq 0$. Then either $u + (-h) < 0$ on D or $u + (-h)$ equals zero at a point in D. In the latter case, the hypothesis implies $u + (-h)$ is constant (and $= 0$). Thus u is subharmonic. □

Corollary. *Subharmonicity is a local property (that is, a function subharmonic in a neighborhood of every point is subharmonic).*

IV.2.2. A *conformal disc* $K \subset M$ is an open set (K) whose closure ($\mathrm{Cl}\,K$) is in a single coordinate patch (with local coordinate z) such that $z(\mathrm{Cl}\,K)$ is a closed disc in $\mathbb{C}$ of radius > 1, and center $z = 0$.

Let u be a continuous function on $M (u \in C(M))$. Fix a conformal disc K on M. We define a new function $u^{(K)}$ on M as follows:

$$u^{(K)} \in C(M),$$
$$u^{(K)} | M \backslash K = u | M \backslash K,$$
$$u^{(K)} \text{ is harmonic in } K.$$

The solution of the Dirichlet problem for the disc gives the existence and uniqueness of $u^{(K)}$. Furthermore,

$$C(M) \ni u \mapsto u^{(K)} \in C(M)$$

defines an $\mathbb{R}$-linear operator (to be called *harmonization*) on $C(M)$. Also, u is harmonic if and only if $u = u^{(K)}$ for all conformal discs K on M.

Proposition. *Let $u \in C(M)$. The function u is subharmonic if and only if $u \leq u^{(K)}$ for every conformal disc K on M.*

PROOF. Assume u is subharmonic. Consider

$$u - u^{(K)}.$$

This function vanishes on $M \setminus K$. It has no maximum in K(unless it is constant). Its maximum must be on δK. Thus $u - u^{(K)} \leq 0$.

To prove the converse, let h be harmonic on D. We show that the maximum principle holds for $u + h$. Assume $u + h$ achieves a maximum H on D, a domain in M. Set

$$D_H = \{P \in D; u(P) + h(P) = H\}.$$

Then D_H is non-empty and closed in D. Pick a conformal disc $K \subset D$ around $P \in D_H$ with the local coordinate z. Now for $0 < r < 1$, $K_r = z^{-1}(\{|z| < r\})$ is also a conformal disc, and

$$\begin{aligned} H &= u(P) + h(P) = u(0) + h(0) \leq u^{(K)}(0) + h(0) \\ &= \frac{1}{2\pi} \int_0^{2\pi} (u(re^{i\theta}) + h(re^{i\theta}))\, d\theta \leq H. \end{aligned}$$

Thus $u(re^{i\theta}) + h(re^{i\theta}) = H$ all r, $0 \leq r \leq 1$, and all θ, $0 \leq \theta \leq 2\pi$. Hence D_H is open in D. Since we have taken D to be connected (without loss of generality), $D_H = D$, and $u + h$ is constant in D. Hence u is subharmonic by Proposition IV.2.1. □

Corollary (of Proof). *Let $u \in C(M)$. Then u is subharmonic if and only if*

$$u(0) \leq \frac{1}{2\pi} \int_0^{2\pi} u(e^{i\theta}) d\theta$$

for every conformal disc on M.

Corollary (of Corollary). *Let D be a domain on M. Assume that $u \in C(\text{Cl } D)$ is subharmonic and non-negative on D and identically zero on δD. Extend u to be zero on $M \setminus D$. Then u is subharmonic on M.*

IV.2.3. Proposition. *Let u, v be subharmonic functions on M and $c \in \mathbb{R}$, $c \geq 0$. Let K be a conformal disc on M. Then cu, $u + v$, $\max\{u,v\}$, and $u^{(K)}$ are all subharmonic.*

PROOF. That cu and $u + v$ are subharmonic follows immediately from the above corollaries. Next assume that $\max\{u,v\}(P_0) = u(P_0)$. Letting z be a local coordinate vanishing at P_0, corresponding to a conformal disc, we see that

$$\max\{u,v\}(P_0) = u(P_0) \le \frac{1}{2\pi}\int_0^{2\pi} u(e^{i\theta})\,d\theta$$

$$\le \frac{1}{2\pi}\int_0^{2\pi} \max\{u,v\}(e^{i\theta})\,d\theta,$$

showing that $\max\{u,v\}$ is also subharmonic. Finally, $u^{(K)}$ is clearly subharmonic on $M\backslash\delta K$. Thus, let $P_0 \in \delta K$. Using the same notation as above, we see that

$$u^{(K)}(0) = u(0) \le \frac{1}{2\pi}\int_0^{2\pi} u(e^{i\theta})\,d\theta \le \frac{1}{2\pi}\int_0^{2\pi} u^{(K)}(e^{i\theta})\,d\theta,$$

proving that $u^{(K)}$ is subharmonic. □

IV.2.4. Proposition. *Let $u \in C(M)$. Then u is harmonic if and only if u is subharmonic and superharmonic.*

PROOF. If u is subharmonic, then $u \le u^{(K)}$ for all conformal discs K. If u is superharmonic then $u \ge u^{(K)}$. Thus, if u is both, $u = u^{(K)}$. Since harmonization does not affect such a function u, it must be harmonic. The converse is, of course, as previously remarked, trivial. □

IV.2.5. Proposition. *Let $u \in C^2(M)$. Then u is subharmonic if and only if $\Delta u \ge 0$.*

PROOF. Since subharmonicity is a local property, it involves no loss of generality to assume that M is the unit disc $\{z = x + iy; |z| < 1\}$. Furthermore, $\Delta u \ge 0$ is a well-defined concept on any Riemann surface (because we are interested only in complex analytic coordinate changes). We view Δ as an operator from functions to functions:

$$\Delta u = \frac{\partial^2 u}{\partial x^2} + \frac{\partial^2 u}{\partial y^2}.$$

Say $\Delta u > 0$. Thus u has no maximum on M. (If u had a relative maximum at P_0, then

$$\frac{\partial^2 u}{\partial x^2}(P_0) \le 0, \qquad \frac{\partial^2 u}{\partial y^2}(P_0) \le 0 \quad \text{or} \quad \Delta u(P_0) \le 0.)$$

If h is harmonic on M, then $\Delta(u + h) > 0$, and, as seen above, this implies that $u + h$ has no maximum on M. Thus u is subharmonic.

Suppose now $\Delta u \ge 0$. Let $\varepsilon > 0$ be arbitrary. Set

$$v(x,y) = u(x,y) + \varepsilon(x^2 + y^2).$$

Then $v \in C^2(M)$ and $\Delta v = \Delta u + 4\varepsilon > 0$. Thus, by the first part, v is subharmonic. Hence, for every K, a conformal disc on M, $v \le v^{(K)}$, or

$$u^{(K)} + \varepsilon \ge u^{(K)} + \varepsilon(x^2 + y^2)^{(K)} \ge u + \varepsilon(x^2 + y^2),$$

or (by letting ε approach zero) $u \le u^{(K)}$; that is, u is subharmonic.

Conversely, say u is subharmonic and $\Delta u < 0$ at some point. Then also in a neighborhood D of this point. By the above, u is superharmonic in D. Hence, we conclude that such a u must be harmonic in D, and we arrive at the contradiction $\Delta u = 0$ in D. □

IV.2.6. A family $\mathscr{F}$ of subharmonic functions on a Riemann suface M is called a *Perron family* (on M) provided:

$$\mathscr{F} \text{ is non-empty,} \tag{2.6.1}$$

$$\text{for every conformal disc } K \subset M \text{ and every } u \in \mathscr{F},$$

$$\text{there is a } v \in \mathscr{F} \text{ such that } v|K \text{ is harmonic and } v \ge u, \tag{2.6.2}$$

and

$$\text{for every } u_1 \in \mathscr{F} \text{ and every } u_2 \in \mathscr{F}, \text{ there is a}$$
$$v \in \mathscr{F} \text{ such that } v \ge \max\{u_1,u_2\}. \tag{2.6.3}$$

Remarks

1. In most applications the functions v satisfying (2.6.2) and (2.6.3) will be $u^{(K)}$ and $\max\{u_1,u_2\}$, respectively.
2. If $\mathscr{F}$ is a Perron family on M, if K is a conformal disc in M, and if $u_j \in \mathscr{F}$, $j = 1, \ldots, n$, then there is a $v \in \mathscr{F}$ such that $v|K$ is harmonic and $v \ge u_j$, $j = 1, \ldots, n$.

Theorem (Perron's Principle). *Let $\mathscr{F}$ be a Perron family and define*

$$u(P) = \sup_{v \in \mathscr{F}} v(P), \qquad P \in M.$$

Then either $u \equiv +\infty$ or u is harmonic.

PROOF. Cover M by a family of discs $\{D_\alpha\}$. If we have the theorem for discs, we have it for all of M. We claim that if u is harmonic on one disc in the cover, say D_1, then u is harmonic on all discs. Let D_* be another such disc. Since M is connected we can find a chain of discs

$$D_2, D_3, \ldots, D_{n+1} = D_*$$

with

$$D_j \cap D_{j+1} \text{ non-empty for } j = 1, \ldots, n. \tag{2.6.4}$$

By the theorem for discs $u|D_j$ is either harmonic or $\equiv +\infty$. Since we have (2.6.4), it is impossible for $u|D_1$ to be harmonic and $u|D_{n+1}$ to be $\equiv +\infty$. Thus we may assume

$$K = \{z \in \mathbb{C}; |z| < 1\} \subset \{z \in \mathbb{C}; |z| < r\} = M, \qquad r > 1,$$

and prove the theorem for K. Let $\{z_j\}_{j=1}^{\infty}$ be a dense set of points in M. (We cannot at this point choose a countable dense set in our original M, since we do not know yet that every Riemann surface is second countable.) For each j, choose a sequence $v_{jk} \in \mathscr{F}$ such that

$$u(z_j) = \lim_{k\to\infty} v_{jk}(z_j).$$

Choose any $v_1 \in \mathscr{F}$ such that v_1 is harmonic on K and $v_1 \geq v_{11}$. Having chosen

$$\{v_1, \ldots, v_n\} \subset \mathscr{F},$$

choose $v_{n+1} \in \mathscr{F}$ such that

$$v_{n+1}\big|K \text{ is harmonic,}$$

$$v_{n+1} \geq v_n,$$

and

$$v_{n+1} \geq v_{ml}, \quad \text{all } m \leq n+1, \text{ all } l \leq n+1.$$

We now observe that

$$v_n(z_j) \geq v_{jk}(z_j), \quad \text{for } n \geq k \geq j,$$

and thus

$$\lim_{n\to\infty} v_n(z_j) = \sup_n v_n(z_j) = u(z_j).$$

Assume $u \not\equiv +\infty$. Without loss of generality we assume that $u(z_1) < +\infty$. By Harnack's principle

$$W = \lim_k v_k \tag{2.6.5}$$

is harmonic in K. We must verify $W = u$. Since $v_k \in \mathscr{F}$, $v_k \leq u$, and hence $W \leq u$ by (2.6.5). Further $W = u$ on a dense set (we do not know yet however that u is even continuous). Thus $W \geq v$ on a dense set for all $v \in \mathscr{F}$. Since all $v \in \mathscr{F}$ are continuous, $W \geq v$ for all $v \in \mathscr{F}$. Thus, $W \geq u$ on K. □

Remark. The above proof also showed how to obtain the function u. Let K be an arbitrary compact subset of M. Cover K by finitely many discs $\{K_j; j = 1, \ldots, n\}$. For each j, there is a sequence of increasing harmonic functions $\{v_{jk}\}$ such that

$$u = \lim_k v_{jk} \text{ on } K_j.$$

Choose $v_k \in \mathscr{F}$ such that $v_k \geq \max\{v_{jk}; j = 1, \ldots, n\}$. Then

$$u = \lim_k v_k$$

uniformly on K. In general, the functions v_k are only subharmonic on K (that is, not always harmonic).

IV.2.7. We now return to the Dirichlet problem introduced in IV.1.1. We take a region D with boundary δD on a Riemann surface M.

Let $P \in \delta D$. We shall say that β is a *barrier* at P provided there exists a neighborhood N of P such that

$$\beta \in C(\mathrm{Cl}(D \cap N)), \tag{2.7.1}$$

$$\beta \text{ is superharmonic on } \mathrm{Int}(D \cap N), \tag{2.7.2}$$

$$\beta(P) = 0, \tag{2.7.3}$$

and

$$\beta(Q) > 0 \quad \text{for } Q \neq P, Q \in \mathrm{Cl}(D \cap N). \tag{2.7.4}$$

Remark. If $P \in \delta D$ can be reached by an analytic arc (=image of straight line under holomorphic mapping) with no points in common with Cl D, then there exists a barrier at P.

PROOF. This is a local problem and so we may assume that P can be reached by a straight line with no points in common with D and $D \subset \mathbb{C}$. Furthermore, we may assume $P = 0$, and the line segment is $y = 0$, $x \leq 0$. Choose a single-valued branch of $\sqrt{z}$ in the complement of this segment and set $\beta(z) = \mathrm{Re}\sqrt{z}$. Writing $z = re^{i\theta}$, we see that $\beta(z) = r^{1/2}\cos\theta/2$ with $-\pi < \theta < \pi$, and is thus a barrier at 0. □

We need a slight (free) improvement. Let β be a barrier at $P \in \delta D$ with N (as in the definition) a relatively compact neighborhood of P. Let us choose any smaller neighborhood N_0 of P with Cl $N_0 \subset$ Int N. Set

$$m = \min\{\beta(Q); Q \in (\mathrm{Cl}(N \backslash N_0) \cap \mathrm{Cl}\, D)\} > 0.$$

Then set

$$\tilde{\beta}(Q) = \begin{cases} \min\{m, \beta(Q)\}, & Q \in N \cap D, \\ m, & Q \in \mathrm{Cl}(D \backslash N). \end{cases}$$

Then $\tilde{\beta}$ is continuous on D, $\tilde{\beta} \geq 0$, $\tilde{\beta}(Q) = 0$ if and only if $Q = P$, and $\tilde{\beta}$ is superharmonic on D. Further $\tilde{\beta}/m$ is again a barrier at P with $\tilde{\beta}/m = 1$ outside N. (It is defined on all of Cl D.) We shall call $\tilde{\beta}/m$ a *normalized* barrier at P.

IV.2.8. Let $D \subset M$. A point $P \in \delta D$ is called a *regular* point (for the Dirichlet problem) if there exists a barrier at P. A solution u to the Dirichlet problem for a bounded $f \in C(\delta D)$ is called *proper* provided

$$\inf\{f(P); P \in \delta D\} \leq u(Q) \leq \sup\{f(P); P \in \delta D\}, \quad \text{all } Q \in \mathrm{Cl}\, D.$$

Theorem. *The following are equivalent for* $D \subset M$:

a. *There exists a proper solution for every bounded* $f \in C(\delta D)$.
b. *Every point of* δD *is regular.*

PROOF. (a) ⇒ (b): Let $P \in \delta D$. It is easy to construct an $f \in C(\delta D)$ with $0 \leq f \leq 1$ and $f(Q) = 0$ if and only if $Q = P$. Let u be a proper solution to the Dirichlet problem with boundary value f. Then $0 \le u \le 1$. We claim $u > 0$ in D. Otherwise, $u \equiv 0$ in D (by the minimum principle) and $f \equiv 0$. Since u is harmonic, it is superharmonic, and thus a barrier at P.

(b) $\Rightarrow$ (a): Let

$$\begin{aligned}\mathscr{F} = \{&v \in C(\mathrm{Cl}\, D);\ v \text{ is subharmonic in } D,\\ &m_0 = \inf f \le v \le \sup f = m_1, \text{ and } v(Q) \le f(Q)\\ &\text{for all } Q \in \delta D\}.\end{aligned}$$

Clearly the function which is identically m_0 is in $\mathscr{F}$. If $u_1, u_2 \in \mathscr{F}$, then $\max\{u_1,u_2\} \in \mathscr{F}$. Also for every $u \in \mathscr{F}$ and every conformal disc K in D, $u^{(K)} \in \mathscr{F}$. Thus $\mathscr{F}$ is a Perron family on D. Let

$$u(Q) = \sup_{v \in \mathscr{F}} v(Q), \qquad Q \in D.$$

Then u is harmonic in D and $m_0 \le u \le m_1$.

We verify two statements for $P \in \delta D$:

1. $\liminf_{Q \to P} u(Q) \ge f(P)$; and
2. $\limsup_{Q \to P} u(Q) \le f(P)$.

PROOF OF (1). If $f(P) = m_0$, there is nothing to prove. So assume $f(P) > m_0$. Choose $\varepsilon > 0$ such that $f(P) - \varepsilon > m_0$. There is then a neighborhood $N(P)$ such that

$$f(Q) \ge f(P) - \varepsilon \quad \text{for all } Q \in N(P) \cap \delta D.$$

Let β be a normalized barrier at P which is $\equiv 1$ outside $N(P)$. Set

$$w(Q) = -(f(P) - m_0 - \varepsilon)\beta(Q) + f(P) - \varepsilon, \qquad Q \in \mathrm{Cl}\, D.$$

Clearly, $w \in C(\mathrm{Cl}\, D)$ and w is subharmonic. For $Q \in \mathrm{Cl}\, D$,

$$w(Q) \le f(P) - \varepsilon < m_1,$$

and

$$\begin{aligned} w(Q) &= m_0\beta(Q) + (f(P) - \varepsilon)(1 - \beta(Q))\\ &\ge m_0\beta(Q) + m_0(1 - \beta(Q)) = m_0.\end{aligned}$$

Finally, for $Q \in \delta D$,

$$w(Q) = m_0 \le f(Q), \qquad Q \notin N(P),$$

and

$$w(Q) \le f(P) - \varepsilon \le f(Q), \qquad Q \in N(P).$$

We have shown that $w \in \mathscr{F}$. Hence, $w(Q) \le u(Q)$, all $Q \in D$, and thus

$$\liminf_{Q \to P} u(Q) \ge w(P) = f(P) - \varepsilon.$$

Since ε is arbitrary, (1) follows. □

PROOF OF (2). There is nothing to verify if $f(P) = m_1$. Assume $f(P) < m_1$. Choose $\varepsilon > 0$ so that $f(P) + \varepsilon < m_1$. Choose $N = N(P)$, a relatively compact neighborhood of P, such that

$$f(Q) \leq f(P) + \varepsilon \quad \text{for all } Q \in N(P) \cap \delta D.$$

Let $v \in \mathcal{F}$ be arbitrary. We claim

$$v(Q) - (m_1 - f(P) - \varepsilon)\beta(Q) \leq f(P) + \varepsilon, \qquad Q \in N(P) \cap D. \quad (2.8.1)$$

(Here β is again a normalized barrier at P that is $\equiv 1$ outside $N(P)$.) Since the function on the left of the inequality is subharmonic, it suffices to check the inequality on $\delta(N \cap D)$. If $Q \in \delta(N \cap D)$, then either (i) $Q \in \delta N \cap \operatorname{Cl} D$ or (ii) $Q \in \operatorname{Cl} N \cap \delta D$. In case (i), the left-hand side of (2.8.1) satisfies

$$= v(Q) - m_1 + f(P) + \varepsilon \leq f(P) + \varepsilon.$$

In case (ii) we have the estimates (for the same quantity)

$$\leq v(Q) \leq f(Q) \leq f(P) + \varepsilon.$$

We have thus verified (2.8.1). Thus for $Q \in N(P) \cap D$

$$v(Q) \leq f(P) + \varepsilon + (m_1 - f(P) - \varepsilon)\beta(Q),$$

and hence also

$$u(Q) \leq f(P) + \varepsilon + (m_1 - f(P) - \varepsilon)\beta(Q).$$

From this last inequality we conclude

$$\limsup_{Q \to P} u(Q) \leq f(P) + \varepsilon.$$

Since ε can be chosen arbitrarily small, we have (2). □

IV.3. A Classification of Riemann Surfaces

In this section we partition the family of all Riemann surfaces into three mutually exclusive classes: compact (=elliptic), parabolic, and hyperbolic. The partition depends on the existence or non-existence of certain subharmonic functions. It will turn out (next section) that each of these classes contains precisely one simply connected Riemann surface. Perhaps of equal importance is the fact that this classification also enables us to construct non-constant meromorphic functions on each Riemann surface. The constructions in this section differ in a few important respects from the constructions in II.4. We do not need here the topological facts that were previously used (triangulability of surfaces and existence of partitions of unity), and we get sharper information about the meromorphic functions that we construct (see, for example, Theorem IV.3.11).

IV.3.1. Lemma. *Let K be a compact connected region on a Riemann surface M, with $K \neq M$. There exists a domain D on M such that*

$$K \subset D, \tag{3.1.1}$$

$$\text{Cl } D \textit{ is compact}, \tag{3.1.2}$$

and

$$\delta D \textit{ consists of finitely many closed analytic curves.} \tag{3.1.3}$$

PROOF. If M is compact the lemma is trivial. Choose $P \notin K$ and a small disc U about P in $M \backslash K$. In the general case, every $P \in K$ is included in a conformal disc. Finitely many such discs will cover K. Let D_1 be the union of these discs. By changing the radii of the discs, if necessary, we may assume that the boundaries of the discs intersect locally only in pairs and non-tangentially. Delete from D_1 a small disc U in $D_1 \backslash K$ and call the resulting domain D_2. Solve the Dirichlet problem with boundary values 1 on δU, 0 on δD_1 (note that $D_2 = D_1 \backslash \text{Cl } U$ has a regular boundary). Let D be the component containing K of

$$\{P \in D_2; u(P) > \varepsilon > 0\},$$

where $\varepsilon < \min\{u(P); P \in K\}$. The critical points of the harmonic function u (the points with $du = 0$) form a discrete set. By changing ε we eliminate them from δD. Thus, the domain D satisfies (3.1.1)–(3.1.3). □

IV.3.2. Let M be a Riemann surface. We will call M *elliptic* if and only if M is compact (=closed). We will call M *parabolic* if and only if M is not compact and M does not carry a negative non-constant subharmonic function. We will call M *hyperbolic* if and only if M does carry a negative non-constant subharmonic function.

Remark. It is obvious that a hyperbolic surface cannot be compact (by the maximum principle for subharmonic functions), and thus we have divided Riemann surfaces into three mutually exclusive families.

IV.3.3. Subharmonic functions on a parabolic surface satisfy a strong maximum principle.

Theorem. *Let D be an open set on a parabolic surface M. Let $u \in C(\text{Cl } D)$. Assume u is subharmonic in D. Furthermore, assume there exist $m_1, m_2 \in \mathbb{R}$ such that $u \leq m_1$ on δD and $u \leq m_2$ on $\text{Cl } D$. Then $u \leq m_1$ on $\text{Cl } D$.*

PROOF. Assume $m_2 > m_1$ (otherwise there is nothing to prove). Choose ε, $0 < \varepsilon < m_2 - m_1$. Define

$$v = \begin{cases} \max\{u, m_1 + \varepsilon\} - m_2 & \text{in } D, \\ m_1 + \varepsilon - m_2 & \text{in } M \backslash D. \end{cases}$$

It is clear that v is subharmonic on M, $v < 0$. Thus v is constant; that is,

$$\max\{u, m_1 + \varepsilon\} = m_1 + \varepsilon \quad \text{in } D.$$

Hence

$$u \leq m_1 + \varepsilon,$$

and since ε is arbitrary, we are done. □

Corollary 1. *Let u be a bounded real-valued harmonic function on an open set D on a parabolic surface M. Assume that $u \in C(\text{Cl } D)$, with $m_0 = \inf\{u(P); P \in \delta D\}$ and $m_1 = \sup\{u(P); P \in \delta D\}$. Then $m_0 \leq u \leq m_1$ (on Cl D).*

Corollary 2. *On a parabolic Riemann surface, the Dirichlet problem has at most one bounded solution.*

IV.3.4. On hyperbolic surfaces we do not, in general, have uniqueness of solutions to the Dirichlet problem. To see this we first establish the following

Theorem. *Let M be a hyperbolic Riemann surface and K a compact subset with $\delta(M \backslash K)$ regular and $M \backslash K$ connected. Then there exist a function $\omega \in C(\text{Cl}(M \backslash K))$ such that*

i. *ω is harmonic on $M \backslash K$,*
ii. *$\omega = 1$ on $\delta(M \backslash K)$, and*
iii. *$0 < \omega < 1$ on $M \backslash K$.*

Remark. We will call the smallest ω as above, *the harmonic measure of K*.

Proof of Theorem. Let ψ_0 be a non-constant superharmonic function on M with $\psi_0 > 0$. Let $m_0 = \min(\psi_0 | K)$ and $\psi_1 = \psi_0 / m_0$. Then ψ_1 is superharmonic on M, non-constant, $\psi_1 > 0$ and $\psi_1 | K \geq 1$.

We claim there exists $P_0 \in \delta K$ such that $\psi_1(P_0) = 1$. Clearly, since K is compact we can find such a $P_0 \in K$. If $P_0 \in \text{Int } K$, then ψ_1 is constant on the component of K containing P_0, and thus also on the boundary of that component. Thus we can find a $P_0 \in \delta K$ at which ψ_1 has the value 1.

There exists a $Q_0 \in M \backslash K$ with $\psi_1(Q_0) < 1$. Otherwise $\psi_1 \geq 1$ and since $\psi_1(P_0) = 1$, this would mean that ψ_1 is constant.

Finally, we set $\psi = \min\{1, \psi_1\}$, and note that ψ is superharmonic on M, $0 < \psi \leq 1$, $\psi(Q_0) < 1$, $\psi | K = 1$.

We define

$$\mathscr{F} = \{v \in C(\text{Cl}(M \backslash K)); v \text{ is subharmonic on } M \backslash K \text{ and } v \leq \psi | (M \backslash K)\}.$$

The family $\mathscr{F}$ is clearly closed under formation of maxima and under harmonization. Thus $\mathscr{F}$ is a Perron family on $M \backslash K$ provided it is not empty. Clearly, $-\psi \in \mathscr{F}$. However, we have to show that $\mathscr{F}$ has more interesting functions.

Choose a domain $D \supsetneqq K$ with Cl D compact, and δD consisting of (finitely many) Jordan curves (this is possible by Lemma IV.3.1). Let v_0 be the solution to the Dirichlet problem on $D \backslash K$ with boundary values 1 on

δK and 0 on δD. Clearly, $0 \le v_0 \le 1$ on $\mathrm{Cl}(D \backslash K)$. Extend v_0 to be zero on $M \backslash D$ and observe that v_0 is subharmonic on M (by the Corollary to Proposition IV.2.2).

We show that $v_0 \in \mathscr{F}$. We must verify that $v_0 \le \psi$ on $M \backslash K$. Note that $v_0 - \psi$ is subharmonic on M, and $v_0 - \psi \le 0$ on δD, and also on $M \backslash D$. By the maximum principle $v_0 - \psi \le 0$ on D because $\mathrm{Cl}\, D$ is compact. Define

$$\omega(P) = \sup_{v \in \mathscr{F}} v(P), \qquad P \in \mathrm{Cl}(M \backslash K),$$

then

$$v_0 \le \omega \le \psi,$$

and in particular ω is harmonic on $M \backslash K$, and $\omega = 1$ on δK (and also $\omega \in C(\mathrm{Cl}(M \backslash K))$). The function ω is non-constant since $\omega(Q_0) \le \psi(Q_0) < 1$. Furthermore, $0 \le \omega \le 1$ on $\mathrm{Cl}(M \backslash K)$ and thus $0 < \omega < 1$ on $M \backslash K$. □

IV.3.5. Important Addition. We can obtain the harmonic measure of K with slight modification of the above argument. Set

$$\mathscr{F}_1 = \{v \in \mathscr{F};\ v \text{ has compact support}\}$$

and define

$$\omega_1(P) = \sup_{v \in \mathscr{F}_1} v(P), \qquad P \in \mathrm{Cl}(M \backslash K).$$

Note that $\mathscr{F}_1$ contains v_0 and is hence non-empty. Furthermore, ω_1 has all the properties of ω. We claim that $\omega_1 \le \tilde{\omega}$ for all $\tilde{\omega}$ enjoying the properties of the theorem. Consider

$$\tilde{\omega} - v, \qquad v \in \mathscr{F}_1.$$

Assume that $v | M \backslash K' = 0$ for some compact set K'. Thus

$$\tilde{\omega} - v \ge 0 \quad \text{on } \delta(K' \backslash K)$$

(since $\tilde{\omega} \ge 0$ on $\delta K'$, $v = 0$ on $\delta K'$; $\tilde{\omega} \ge v$ on δK). Hence $\tilde{\omega} \ge v$ on $(K' \backslash K) \cup (M \backslash K')$, and since v is arbitrary $\tilde{\omega} \ge \omega_1$ on $M \backslash K$.

Corollary. *Let M be a hyperbolic Riemann surface and D a domain in M with regular boundary such that $M \backslash D$ is compact. Then we have non-uniqueness of solutions for the Dirichlet problem for D.*

Proof. If u is any solution to a Dirichlet problem, then so is $(1 - \omega) + u$, where ω is the harmonic measure for $M \backslash D$. □

Remark. Any D as above cannot have compact closure in M. If it did, we would have a unique solution to the Dirichlet problem for D.

IV.3.6. Let M be a Riemann surface and $P \in M$. A function g is called a *Green's function for M with singularity at P* provided:

$$g \text{ is harmonic in } M \backslash \{P\}. \tag{3.6.1}$$

$$g > 0 \text{ in } M\backslash\{P\}. \tag{3.6.2}$$

If z is a local parameter vanishing at P, then

$$g(z) + \log|z| \text{ is harmonic in a neighborhood of P.} \tag{3.6.3}$$

$$\text{If } \hat{g} \text{ is another function satisfying (3.6.1)–(3.6.3), then } \hat{g} \geq g. \tag{3.6.4}$$

Remark. Condition (3.6.3) is independent of the choice of local parameter vanishing at P (that is, the condition makes sense on M). For if ζ is another such parameter, then (with $a_1 \neq 0$)

$$\begin{aligned}\zeta(z) &= a_1 z + a_2 z^2 + \cdots \\ &= a_1 z\left(1 + \frac{a_2}{a_1} z + \cdots\right) = a_1 z f(z),\end{aligned}$$

where f is holomorphic near $z = 0$ and $f(0) \neq 0$. We write $f(z) = e^{h(z)}$, with h holomorphic near $z = 0$, and conclude that

$$\log|\zeta| = \log|a_1| + \log|z| + \operatorname{Re} h(z).$$

Since $\operatorname{Re} h(z)$ is harmonic, we see that $g(z) + \log|z|$ is harmonic if and only if $g(\zeta) + \log|\zeta|$ is.

A Classical Example. Let D be a domain in $\mathbb{C}$ with $\operatorname{Cl} D$ compact and δD regular (for example δD consisting of finitely many analytic arcs). Let $z_0 \in D$. By solving the Dirichlet problem we can find a function $\gamma \in C(\operatorname{Cl} D)$ such that γ is harmonic in D and

$$\gamma(z) = \log|z - z_0|, \qquad z \in \delta D.$$

Set

$$g(z) = -\log|z - z_0| + \gamma(z), \qquad z \in D.$$

Show that g is the Green's function for D with singularity at z_0. (The proof of this assertion follows the proof of Lemma IV.3.8.) Formulate and prove a general theorem so that this example and Lemma IV.3.8 become special cases of the theorem.

Exercise

Let $D \subset \mathbb{C}$ be a domain bounded by finitely many disjoint Jordan curves $C_1, \ldots, C_n$. The *harmonic measure of* $C_j (j = 1, \ldots, n)$ *with respect to* D is classically defined as the unique harmonic function ω_j in D that is continuous on $\operatorname{Cl} D$ and has boundary values δ_{jk}, $k = 1, \ldots, n$. Show that this definition makes sense and relate this concept to the one introduced in IV.3.4. Show that $\omega_1 + \cdots + \omega_n = 1$ and discuss the properties of the $(n-1)$ by $(n-1)$ "period" matrix $\int_{c_j} *d\omega_k$.

IV.3.7. Theorem. *Let M be a Riemann surface. There exists a Green's function on M (with singularity at some point $P \in M$) if and only if M is hyperbolic.*

Proof. Let g be a Green's function on M with singularity at P. Let $m > 0$. Set $f = \min\{m,g\}$. Then f is clearly superharmonic in $M\backslash\{P\}$. Since $g(Q) \to +\infty$ as $Q \to P$, $f \equiv m$ in a neighborhood of P. Thus f is superharmonic on M. Also $f > 0$. If f were constant, then $g \geq m$, and thus $g - m$ would be another candidate for the Green's function with $g - m < g$. This contradiction shows that f cannot be constant and hence M is hyperbolic.

Conversely, assume that M is hyperbolic. Let $P \in M$. Let K be a conformal disc about P with local coordinate z. Set

$$\begin{aligned}\mathscr{F} = \{v;\ & v \text{ is subharmonic in } M\backslash\{P\},\\ & v \geq 0,\ v \text{ has compact support, and}\\ & v(z) + \log|z| \text{ is subharmonic in } |z| < 1\}.\end{aligned}$$

To note that $\mathscr{F}$ is non-empty, we define

$$v_0(z) = \begin{cases} -\log|z|, & 0 < |z| < 1\\ 0, & |z| \geq 1,\end{cases}$$

and observe that $v_0 \in \mathscr{F}$. It is easy to see that $\mathscr{F}$ is a Perron family (on $M\backslash\{P\}$).

We establish now the following

Lemma. *Outside every neighborhood of P, $\mathscr{F}$ is uniformly bounded.*

Proof of Lemma. Choose r, $0 < r < 1$. Let ω_r be the harmonic measure of $\{|z| \leq r\}$. (It is only here that we use the fact that M is hyperbolic.) Thus ω_r is harmonic on $|z| > r$, $\omega_r = 1$ on $|z| = r$, and $0 < \omega_r < 1$ for $|z| > r$. Let $\lambda_r = \max\{\omega_r(z); |z| = 1\}$. Thus $0 < \lambda_r < 1$. For $u \in \mathscr{F}$, let $u_r = \max\{u(z); z = r\}$. We claim

$$u_r\omega_r - u \geq 0 \quad \text{for } |z| \geq r. \tag{3.7.1}$$

Clearly $u_r\omega_r - u$ is superharmonic on $|z| > r$, and $u_r\omega_r - u \geq 0$ on $|z| = r$. Let $\mathscr{K}$ be the support of u. Then $u = 0$ on $\delta\mathscr{K}$, and thus (3.7.1) holds on $\delta\mathscr{K}$. Hence (3.7.1) holds on $\mathscr{K}\backslash\{|z| \leq r\}$ by the minimum principle for superharmonic functions. Inequality (3.7.1) obviously holds on $M\backslash\mathscr{K}$.

Finally, $u(z) + \log|z|$ is subharmonic in $|z| < 1$ and continuous on $|z| \leq 1$. Thus

$$u_r + \log r = \max_{|z|=r} u + \log r \leq \max_{|z|=1} u + \log 1 = \max_{|z|=1} u \leq u_r\lambda_r$$

(the last inequality is a consequence of (3.7.1)). Thus

$$u_r \leq \frac{-\log r}{1 - \lambda_r}.$$

Since $u = 0$ off a compact set $\mathscr{K}$, we conclude that

$$\max\{u(z); |z| \geq r\} \leq \frac{\log r}{\lambda_r - 1}. \tag{3.7.2}$$ □

Conclusion of Proof of Theorem. Set

$$g(Q) = \sup_{u \in \mathscr{F}} u(Q), \qquad Q \in M \backslash \{P\}.$$

Inequality (3.7.2) implies that $g(Q) < \infty$ for $Q \neq P$. Thus, by Perron's principle, g is harmonic in $M \backslash \{P\}$. The fact that $g > 0$ on $M \backslash \{P\}$ follows from the corresponding fact that $u > 0$ on $M \backslash \{P\}$ for all $u \in \mathscr{F}$. Since $u > 0$ implies $g \geq 0$ on $M \backslash \{P\}$. If $g(Q) = 0$ for some $Q \in M \backslash \{P\}$, then g is constant. Thus $g > 0$ on $M \backslash \{P\}$. Next we show (3.6.3). Note that for $|z| < r$, $u \in \mathscr{F}$,

$$u(z) + \log|z| \leq u_r + \log r \leq \frac{\log r}{\lambda_r - 1} + \log r = \frac{\lambda_r \log r}{\lambda_r - 1}.$$

Thus also

$$g(z) + \log|z| \leq \frac{\lambda_r \log r}{\lambda_r - 1} \quad \text{for } |z| \leq r.$$

Hence $z = 0$ is a removable singularity for the bounded harmonic function $g(z) + \log|z|$.

To finish the proof that g is the Green's function, let $\hat{g}$ be a competing function satisfying (3.6.1)–(3.6.3). Let $u \in \mathscr{F}$. Then the function $\hat{g} - u$ is superharmonic on M. Since u has compact support, say K', $\hat{g} - u \geq 0$ on $M \backslash K'$ and by the minimum principle for superharmonic functions also on K'. Thus $\hat{g} \geq u$ on $M \backslash \{P\}$. Thus $\tilde{g} \geq g$. (In fact, either $\tilde{g} > g$ or $\tilde{g} = g$ on $M \backslash \{P\}$.) □

Remark. We have shown a little more than claimed in the theorem:

Existence of Green's function at one point

⇓

Hyperbolic

⇓

Existence of harmonic measures

⇓

Existence of Green's function at every point.

We hence define $g(P,Q)$ as the value at Q of the Green's function (on the hyperbolic surface M) with singularity at P.

IV.3.8. Lemma. *Let M be a hyperbolic Riemann surface. Let D be a domain on M with* Cl *D compact and δD regular for the Dirichlet problem. Let $P \in D$. Let u be the unique harmonic function on D with $u(Q) = g(P,Q)$ on δD. Then $g_D(P,Q) = g(P,Q) - u(Q)$, $Q \in D$ defines the Green's function for D with singularity at P.*

Proof. The function g_D is harmonic in $D \backslash \{P\}$. Take a small disc Δ about P so that $g - u$ is positive in its interior. Thus by the minimum principle for harmonic functions, $g - u \geq 0$ on $D \backslash \{P\}$ since we also have these estimates on $\delta(D \backslash \Delta)$. Condition (3.6.3) is trivially satisfied. Let $\hat{g}$ be another candidate for the Green's function, and let $Q_0 \in D \setminus \{P\}$. It suffices to show that

$$\hat{g}(Q_0) \geq g_D(P,Q_0) - \varepsilon$$

for all $\varepsilon > 0$. For each $Q \in \delta D$, there exists a neighborhood U_Q of Q in $\mathrm{Cl}\, D \setminus \{Q_0\}$ such that $g_D < \varepsilon$ on U_Q. Set

$$U = \bigcup_{Q \in \delta D} U_Q.$$

Then on $D \setminus U$ we have $\hat{g} - g_D \geq -\varepsilon$ since we have this estimate on $\delta(D \setminus U) = \delta U$. In particular, $\hat{g}(Q_0) - g_D(P,Q_0) \geq -\varepsilon$. (Intuitively g_D is necessarily the right choice since it has the smallest possible value on δD, namely 0.) □

IV.3.9. Lemma. *In addition to the hypothesis of the previous lemma, assume that δD consists of closed analytic arcs, and $P, Q \in D$. We have*

$$g_D(P,Q) = g_D(Q,P), \quad \textit{all } P, Q \in D.$$

PROOF. Let U_1, U_2 be two small disjoint discs about P and Q respectively. Let $\tilde{D} = D \setminus (U_1 \cup U_2)$. For $R \in D \setminus \{P,Q\}$ set

$$u(R) = g_D(P,R)$$
$$v(R) = g_D(Q,R).$$

Then by I (4.4.2),

$$0 = \iint_{\tilde{D}} (u\,\Delta v - v\,\Delta u) = \int_{\delta \tilde{D}} (u\,{*dv} - v\,{*du})$$
$$= -\int_{\delta U_1 + \delta U_2} (u\,{*dv} - v\,{*du}). \qquad (3.9.1)$$

Note that by the reflection principle for harmonic functions, we may assume that u and v are C^2 in a neighborhood of the closure of D. We introduce now a conformal disc (U_1) at P with local coordinate $z = re^{i\theta}$. In terms of this local coordinate, we write

$$u(z) = \tilde{u}(z) - \log r$$

with $\tilde{u}$ harmonic in $|z| \leq 1$. Thus

$$\int_{\delta U_1} u\,{*dv} - v\,{*du} = \int_{\delta U_1} (\tilde{u} - \log r)\,{*dv} - v\,{*d(\tilde{u} - \log r)}$$
$$= \int_{\delta U_1} \tilde{u}\,{*dv} - v\,{*d\tilde{u}} - \int_{\delta U_1} (\log r)\,{*dv} - v\,{*d(\log r)}$$
$$= \iint_{U_1} \tilde{u}\,\Delta v - v\,\Delta\tilde{u} - \int_{\delta U_1} (\log r)\,{*dv} - v(re^{i\theta})\frac{1}{r} r\,d\theta$$
$$= \int_0^{2\pi} v(re^{i\theta})\,d\theta = 2\pi v(0) = 2\pi g_D(Q,P).$$

Similarly,

$$\int_{\delta U_2} u\,{*dv} - v\,{*du} = -2\pi g_D(P,Q).$$

□

IV.3.10. Theorem. *If M is hyperbolic, then*

$$g(P,Q) = g(Q,P), \quad \text{all } P, Q \in M.$$

PROOF. Fix a point $P \in M$. Consider the collection $\mathscr{D}$ of all domains $D \subset M$ such that

1. $P \in D$,
2. Cl D is compact, and
3. δD is analytic.

Note that if $D_j \in \mathscr{D}$ $(j = 1,2)$, then there is also a $D \in \mathscr{D}$ with $D_1 \cup D_2 \subset D$. Extend each of the Green's functions $g_D(P,\cdot)$ to be identically zero outside D. Let

$$\mathscr{F} = \{g_D(P,\cdot);\ D \in \mathscr{D}\}.$$

The last corollary in IV.2.2 shows that the functions in $\mathscr{F}$ are subharmonic on $M\backslash\{P\}$.

Let K be any conformal disc on $M\backslash\{P\}$. Let $D \in \mathscr{D}$. Choose $D^* \in \mathscr{D}$ such that $D^* \supset D \cup K$. Thus $g_{D^*}(P,\cdot) \geq g_D(P,\cdot)$ and $g_{D^*}(P,\cdot)|\text{Int } K$ is harmonic. Similarly, if $D_1, D_2 \in \mathscr{D}$, we can choose $D \in \mathscr{D}$ with $D \supset D_1 \cup D_2$ and observe that $g_D(P,\cdot) \geq g_{D_j}(P,\cdot)$, $j = 1, 2$. Thus the family $\mathscr{F}$ is a Perron family. By Perron's Principle

$$\gamma(Q) = \sup_{D \in \mathscr{D}} \{g_D(P,Q)\}, \qquad Q \in M\backslash\{P\}$$

is harmonic on $M\backslash\{P\}$, since

$$g_D(P,\cdot) \leq g(P,\cdot)$$

shows that $\gamma(Q) < +\infty$, all $Q \in M\backslash\{P\}$. The last inequality also shows that

$$\gamma \leq g(P,\cdot) \quad \text{on } M\backslash\{P\}.$$

Since γ is a competitor for the Green's function (it clearly satisfies (3.6.2) and (3.6.3)), $\gamma \geq g(P,\cdot)$ on $M\backslash\{P\}$. We have thus shown that

$$\gamma(Q) = g(P,Q) = \sup_{D \in \mathscr{D}} \{g_D(P,Q)\}, \qquad Q \in M.$$

Hence for $Q, P \in D$

$$g_D(P,Q) = g_D(Q,P) \leq g(Q,P).$$

Now fixing P and taking the supremum over $D \in \mathscr{D}$, we see that

$$g(P,Q) \leq g(Q,P).$$

Reversing the roles of P and Q gives the opposite inequality. Thus the proof of the theorem is complete.

EXERCISE

Show that the Green's function is a conformal invariant; that is, if

$$f: M \to N$$

is a conformal mapping and g is the Green's function on N with singularity at $f(P)$ for some $P \in M$, then $g \circ f$ is the Green's function on M with singularity at P.

DIGRESSION. Let D be a domain on a Riemann surface M with δD consisting of finitely many analytic arcs. Let us generalize the Dirichlet problem and (recall II.4.5) our earlier discussion of the Dirichlet principle. Let Ω be a 2-form which is C^2 on a neighborhood of the closure of D. (We shall show in IV.8 that there is a C^∞ 2-form Ω_0 on M that never vanishes. Thus $\Omega = F\Omega_0$ for some function $F \in C^2(\mathrm{Cl}\, D)$.) Let $f \in C(\delta D)$. We want to solve the boundary-value problem for $u \in C^2(\mathrm{Cl}\, D)$:

$$\Delta u = \Omega,$$
$$u|\delta D = f.$$

Let $P_0 \in D$ be arbitrary and assume our problem has a solution u. Let us take a conformal disc z centered about P_0 and let

$$D_r = D\backslash\{|z| \le r\}, \qquad 0 < r < 1.$$

Apply Formula I (4.4.2) on D_r with $\varphi = u$ and $\psi = g(P_0,\cdot)$, the Green's function for the domain D. Recall that $u = f$ on δD and that $\Delta g(P_0,\cdot) = 0$ on $D\backslash\{P_0\}$ while $g(P,\cdot) = 0$ on δD. Thus

$$\int_{\delta D} f * dg(P_0,\cdot) - \int_{|z|=r} (u(z) * dg(P_0,z) - g(P_0,z) * du(z)) = -\iint_{D_r} g(P_0,\cdot)\Omega.$$

Letting $r \to 0$, we see that

$$2\pi u(P_0) = -\iint_D g(P_0,\cdot)\Omega - \int_{\delta D} f * dg(P_0,\cdot). \tag{3.10.1}$$

Conversely, it can be shown that Formula (3.10.1) does indeed provide a solution to our problem.

IV.3.11. Theorem. *Let M be a non-hyperbolic Riemann surface. Let D be a domain on M with $P \in D$. Let f be a holomorphic function on $D\backslash\{P\}$. Then there exists a unique harmonic function u on $M\backslash\{P\}$ such that*

$u - \operatorname{Re} f$ is harmonic in D and vanishes at P, and (3.11.1)

u is bounded outside every neighborhood of P. (3.11.2)

Remarks

1. The theorem should be contrasted with Theorem II. 4.1 (and its companions). What is important for us is not the more general singularity at P that we can produce, but the boundedness statement (3.11.2).

2. If f has no singularity at P, then existence is trivial. We set $u = (\text{Re } f)(P)$.
3. The uniqueness part of the theorem is also quite easy to establish. If u_1 and u_2 satisfy the conditions for u, then $u_1 - u_2$ is a bounded harmonic function on M and hence constant. Since $(u_1 - u_2)(P) = 0$, $u_1 = u_2$.

Before obtaining existence, we must establish some lemmas.

IV.3.12. Lemma. *Let $D \subset \mathbb{C}$. Let u be harmonic in D and $|u| \leq m$. Let $P \in D$, then*

$$\|(\text{grad } u)_P\| \leq \frac{cm}{|P - \delta D|},$$

with c a universal constant.

PROOF. First: by grad u we mean the vector (u_x, u_y), and by its norm we mean $\max\{|u_x|, |u_y|\}$. Choose $r > 0$ such that the closed disc of radius r about P is contained in D. The function u_x is harmonic in D. Thus (without loss of generality $P = 0$), by the mean-value property for harmonic functions

$$u_x(0) = \frac{1}{2\pi} \int_0^{2\pi} u_x(\rho e^{i\theta})\, d\theta, \qquad 0 < \rho < r.$$

Multiplying both sides by ρ and integrating from 0 to r we obtain the "areal mean-value property"

$$u_x(0) = \frac{1}{\pi r^2} \iint_{|z|<r} u_x\, dx\, dy.$$

Thus

$$\begin{aligned} |u_x(0)| &= \frac{1}{\pi r^2} \left| \iint_{|z|<r} u_x\, dx\, dy \right| \\ &= \frac{1}{\pi r^2} \left| \int_{-r}^{r} (u(\sqrt{r^2 - y^2}, y) - u(-\sqrt{r^2 - y^2}, y))\, dy \right| \\ &\leq \frac{2m}{\pi r^2} 2r = \frac{4m}{\pi r}. \end{aligned}$$

From the above we see that $c = 4/\pi$. □

IV.3.13. Lemma. *Let $\{u_j\}$ be a sequence of harmonic functions on a domain $D \subset \mathbb{C}$. If $\{u_j\}$ converges uniformly on compact subsets to a function u, then $\{u_{j,x}\}$ converges uniformly on compact subsets to u_x.*

PROOF. Since the result is local in nature, it involves no loss of generality to assume that D is the unit disc and to show uniform convergence on a smaller disc. Now choose analytic functions f_j such that

$$u_j = \text{Re } f_j \quad \text{and} \quad f_j(0) = u_j(0).$$

Formula (1.2.1) shows that f_j converges uniformly to f (similarly constructed). Thus f'_j converges uniformly to f' (on any compact disc) and since

$$f'_j = u_{j,x} - iu_{j,y}$$

we are done. □

EXERCISE

Reprove the above lemma using the methods and result of Lemma IV.3.12.

IV.3.14. Lemma. *Let $\{u_j\}$ be a uniformly bounded sequence of harmonic functions on a Riemann surface M. Then there exists a subsequence converging uniformly on compact subsets of M.*

PROOF. If D is a closed disc contained in M, then the sequence $\{f_j\}$ constructed in the proof of the previous lemma is uniformly bounded on the closure of any smaller disc D_0 and thus contains a subsequence converging uniformly on D_0. The same holds for the sequence $\{u_j\}$—the real part of the sequence $\{f_j\}$. If M were second countable (it is—but we have not yet established this) the general result follows by a "diagonalization" procedure to be described in detail in IV.3.16. For the present we can use t' e lemma only for surfaces we know to be second countable. □

IV.3.15. Lemma. *Let M be a non-hyperbolic Riemann surface, and K a conformal disc. Let u be a bounded harmonic function on $M \backslash K$. Assume that u is C^1 on $\mathrm{Cl}(M \backslash K)$. Then*

$$\int_{\delta K} {*du} = 0.$$

PROOF. Without loss of generality we may assume (by adding a constant) that $u \geq 0$ on $M \backslash K$. Let z be the local coordinate corresponding to K so that $\delta K = \{|z| = 1\}$. Since

$$\int_{|z|=1} {*du} = \lim_{\rho \to 1^+} \int_{|z|=\rho} {*du},$$

it suffices to assume that u is harmonic in $\{|z| \geq r\}$, $r < 1$ and to show that

$$\int_{|z|=1} {*du} = 0.$$

Let $\mathscr{D}$ be the collection of all domains D such that $K \subset\subset D \subset\subset M$ and δD is analytic. Let u_D be the solution to the Dirichlet problem on $\mathrm{Cl}(D \backslash K_r)$ with $u_D | \delta K_r = u | \delta K_r$ and $u_D | \delta D = 0$, where $K_r = \{|z| < r\}$. Extend u_D to be zero outside D. Let

$$\mathscr{F} = \{u_D; D \in \mathscr{D}\},$$

and

$$\gamma = \sup_{D \in \mathscr{D}} u_D.$$

The reasoning in Theorem IV.3.10 shows $\mathscr{F}$ is a Perron family on $M\backslash\mathrm{Cl}\ K_r$. For every $D \in \mathscr{D}$, $u_D \leq u$; hence $\gamma \leq u$. Now $u - \gamma$ is a bounded harmonic function on $M\backslash\mathrm{Cl}\ K_r$. By Theorem IV.3.3, the maximum and minimum of this function occur on δK_r. Thus $u = \gamma$. We have shown that

$$u = \lim_{D \in \mathscr{D}} u_D = \sup_{D \in \mathscr{D}} u_D \quad \text{on } \{|z| > r\}.$$

Let ω_D be the solution to the Dirichlet problem on $\mathrm{Cl}(D\backslash K_r)$ with $\omega_D|\delta K_r = 1$ and $\omega_D|\delta D = 0$. Then (as a special case of the previous argument)

$$\lim_{D \in \mathscr{D}} \omega_D = \sup_{D \in \mathscr{D}} \omega_D = 1 \quad \text{on } \{|z| > r\}.$$

Now note that by the reflection principle, u_D and ω_D have harmonic extensions (which we do not use) across δD. It thus follows that $*du_D$ and $*d\omega_D$ are defined and smooth on $\mathrm{Cl}(D\backslash K_r)$.

Now use the remark following Perron's principle (IV.2.6), and choose sequences of domains $\{D_j^{(1)}\} \subset \mathscr{D}$, $\{D_j^{(2)}\} \subset \mathscr{D}$, such that

$$u = \lim_j u_{D_j^{(1)}}, \qquad 1 = \lim_j \omega_{D_j^{(2)}},$$

uniformly on a neighborhood of $\delta K = \{|z| = 1\}$. Choose $D_j \in \mathscr{D}$ such that $D_j \supset D_j^{(1)} \cup D_j^{(2)}$. Then $u = \lim_j u_{D_j}$, $1 = \lim_j \omega_{D_j}$ uniformly on a neighborhood of δK. Further, the functions involved are all harmonic in this neighborhood. Finally,

$$\begin{aligned} 0 &= \iint_{D_j\backslash K} (\omega_{D_j} \Delta u_{D_j} - u_{D_j} \Delta \omega_{D_j}) = \int_{\delta(D_j\backslash K)} (\omega_{D_j} {*du_{D_j}} - u_{D_j} {*d\omega_{D_j}}) \\ &= -\int_{\delta K} ({*du_{D_j}} - u\, {*d\omega_{D_j}}). \end{aligned}$$

By Lemma IV.3.13. this last integral converges to $-\int_{\delta K} *du$. □

IV.3.16. Proof of Theorem IV.3.11. Choose a local parameter z vanishing at P. We may assume that $\{|z| < R\} \subset D$ with $R > 1$. Let $0 < \rho < 1$. Let u^ρ be the solution to the Dirichlet problem on $M\backslash\{|z| < \rho\}$ with $u^\rho|\{|z| = \rho\} = \mathrm{Re} f|\{|z| = \rho\}$. We claim that there is a constant $c(r)$—independent of ρ—such that for $0 < \rho < r < 1$,

$$|u^\rho| \leq c(r) \quad \text{for } |z| \geq r. \tag{3.16.1}$$

It suffices to verify (3.16.1) on $\{|z| = r\}$.

Now, Lemma IV.3.15 implies that

$$\int_{|z|=t} *du^\rho = 0, \quad \text{for } \rho \leq t \leq R.$$

Thus we can define a holomorphic function F^ρ on $\{\rho < |z| < R\}$ by choosing z_0 with $\rho < |z_0| < R$ and setting

$$F^\rho(z) = u^\rho(z_0) + \int_{z_0}^{z} (du^\rho + i\, {*du^\rho}).$$

We know that $\operatorname{Re} F^\rho(z) = u^\rho(z)$, $\rho < |z| < R$. Now both F^ρ and f have Laurent series expansions on $\rho < |z| < R$, say

$$F^\rho(z) - f(z) = \sum_{n=-\infty}^{\infty} a_n z^n.$$

Thus, taking real parts, we obtain (after changing the names of the constants in the Fourier series) for $z = te^{i\theta}$

$$u^\rho(te^{i\theta}) - \operatorname{Re} f(te^{i\theta}) = \frac{\alpha_0^\rho}{2} + \sum_{n=1}^{\infty} (\alpha_n^\rho t^n + \alpha_{-n}^\rho t^{-n}) \cos n\theta$$
$$+ \sum_{n=1}^{\infty} (\beta_n^\rho t^n + \beta_{-n}^\rho t^{-n}) \sin n\theta. \tag{3.16.2}$$

Multiplying (3.16.2) by $\cos k\theta$ or $\sin k\theta$ and integrating from 0 to 2π, we get (for $\rho \le t \le R$)

$$\frac{1}{\pi}\int_0^{2\pi} (u^\rho(te^{i\theta}) - \operatorname{Re} f(te^{i\theta}))d\theta = \alpha_0^\rho.$$

$$\frac{1}{\pi}\int_0^{2\pi} (u^\rho(te^{i\theta}) - \operatorname{Re} f(te^{i\theta})) \cos k\theta\, d\theta = \alpha_k^\rho t^k + \alpha_{-k}^\rho t^{-k}, \qquad k = 1, 2, \ldots,$$

$$\frac{1}{\pi}\int_0^{2\pi} (u^\rho(te^{i\theta}) - \operatorname{Re} f(te^{i\theta})) \sin k\theta\, d\theta = \beta_k^\rho t^k + \beta_{-k}^\rho t^{-k}, \qquad k = 1, 2, \ldots.$$

(3.16.3)

We first use each of the above equations for $t = \rho$ ($u^\rho(\rho e^{i\theta}) = \operatorname{Re} f(\rho e^{i\theta})$) to obtain ($k = 1,2,3, \ldots$)

$$\alpha_0^\rho = 0,$$
$$\alpha_{-k}^\rho = -\rho^{2k}\alpha_k^\rho, \tag{3.16.4}$$
$$\beta_{-k}^\rho = -\rho^{2k}\beta_k^\rho.$$

We let

$$m = \sup\{|f(z)|;\ |z| = 1\},$$
$$m(\rho) = \sup\{|u^\rho(z)|;\ |z| = 1\}.$$

Using Equation (3.16.3) for $t = 1$, we obtain

$$|\alpha_k^\rho + \alpha_{-k}^\rho| \le 2(m(\rho) + m),$$
$$|\beta_k^\rho + \beta_{-k}^\rho| \le 2(m(\rho) + m).$$

Combining the above result with (3.16.4) we have:

$$|\alpha_k^\rho|\,|1 - \rho^{2k}| \le 2(m(\rho) + m),$$
$$|\beta_k^\rho|\,|1 - \rho^{2k}| \le 2(m(\rho) + m);$$

and for $\rho < \frac{1}{2}$:

$$|\alpha_k^\rho| \le 4(m(\rho) + m),$$
$$|\beta_k^\rho| \le 4(m(\rho) + m).$$

Also with the same restrictions on ρ:

$$|\alpha^{\rho}_{-k}| \le 4(m(\rho) + m)\rho^{2k},$$
$$|\beta^{\rho}_{-k}| \le 4(m(\rho) + m)\rho^{2k}.$$

Thus (using (3.16.2))

$$\max_{|z|=r} u^{\rho} = c(r,\rho) \le \max_{|z|=r} |\operatorname{Re} f| + 8 \sum_{n=1}^{\infty} (m(\rho) + m)\left(r^n + \frac{\rho^{2n}}{r^n}\right).$$

Since $\rho^2/r < r$, we conclude that

$$c(r,\rho) \le \max_{|z|=r} |\operatorname{Re} f| + 16(m(\rho) + m)\frac{r}{1-r}. \tag{3.16.5}$$

To finish verifying our claim (that $c(r,\rho) = c(r)$) we must show that $m(\rho)$ is bounded independently of ρ. Now

$$m(\rho) = \max_{|z|=1} |u^{\rho}| \le \max_{|z|=r} |u^{\rho}| = c(r,\rho)$$

(by the maximum principle).

Thus by (3.16.5)

$$m(\rho) \le \max_{|z|=r} |\operatorname{Re} f| + 16m(\rho)\frac{r}{1-r} + 16m\frac{r}{1-r}.$$

Choose r small so that $r/(1-r) = q < \frac{1}{16}$ (for example, $r < \frac{1}{32}$). Hence, we have

$$m(\rho) \le \frac{1}{1-16q}\left(\max_{|z|=r} |\operatorname{Re} f| + 16m\frac{r}{1-r}\right) \le c(r).$$

We have now verified (3.16.1) as well as the following estimates (by adjusting $c(r)$)

$$|\alpha^{\rho}_{k}| \le c(r), \qquad |\alpha^{\rho}_{-k}| \le c(r)\rho^{2k}, \qquad k = 1, 2, 3, \ldots.$$

We now let A be the annulus

$$A = \{r < |z| < 1\}, \qquad \rho < r < \tfrac{1}{32}.$$

The set of harmonic functions

$$\{u^{\rho}; \rho < r\}$$

is uniformly bounded in A (actually in $\{|z| \ge r\}$). Thus we can choose a sequence converging uniformly on A. By the maximum principle this sequence converges uniformly on $\{|z| \ge r\}$ and the limit function is harmonic.

Let us choose now a decreasing sequence of radii $r_j \to 0$ and the corresponding annuli A_j. Let $\{\rho_1, \rho_2, \ldots\}$ be a sequence with

$$\rho_1 > \rho_2 > \cdots,$$
$$\rho_j < r_j, \qquad j = 1, 2, \ldots,$$

and

$$\lim_{k} u^{\rho_k} = u \quad \text{on } \{|z| \ge r_1\}.$$

We relabel $\{\rho_1,\rho_2,\rho_3,\ldots\}$ as $\{\rho_{11},\rho_{12},\rho_{13},\ldots\}$. Now we can choose a subsequence $\{\rho_{21},\rho_{22},\rho_{23},\ldots\}$ of $\{\rho_{11},\rho_{12},\rho_{13},\ldots\}$ so that $u^{\rho_{2k}}$ converges uniformly to u on $\{|z| \geq r_2\}$. By induction the sequence $\{\rho_{j-1,1},\rho_{j-1,2},\rho_{j-1,3},\ldots\}$ satisfies

$$\rho_{j-1,1} > \rho_{j-1,2} > \cdots,$$

$$\rho_{j-1,k} < r_k, \qquad k = 1, 2, \ldots,$$

and

$$\lim_k u^{\rho_{j-1,k}} = u \quad \text{on } \{|z| \geq r_{j-1}\}.$$

So we can choose a subsequence

$$\{\rho_{j1},\rho_{j2},\rho_{j3},\ldots\} \text{ of } \{\rho_{j-1,1},\rho_{j-1,2},\rho_{j-1,3},\ldots\}$$

so that

$$\lim_k u^{\rho_{jk}} = u \quad \text{on } \{|z| \geq r_j\}.$$

We now let $\mu_k = \rho_{kk}$ (the "diagonalization" procedure) and observe that μ_k is a decreasing sequence of positive real numbers with $\mu_k < r_k$ such that

$$\lim_k u^{\mu_k} = u \quad \text{uniformly on } \{|z| \geq r\} \text{ for all } r > 0.$$

Note that

$$(u - \operatorname{Re} f)(te^{i\theta}) = \sum_{n=1}^{\infty} (\alpha_n \cos n\theta + \beta_n \sin n\theta)t^n$$

with

$$\alpha_n = \lim_{\rho\to 0} \alpha_n^\rho, \qquad \beta_n = \lim_{\rho\to 0} \beta_n^\rho,$$

$$\left(\lim_{\rho\to 0} \alpha_{-n}^\rho = 0 = \lim_{\rho\to 0} \beta_{-n}^\rho\right) \quad \text{for } n = 1, 2, \ldots.$$

Thus, in particular, taking subsequences was completely unnecessary by the uniqueness part of our theorem. □

IV.3.17. As we saw in II.5, existence of harmonic functions already implies the existence of meromorphic functions.

Theorem. *Every Riemann surface M carries non-constant meromorphic functions.*

Proof. Let $P_j \in M$ for $j = 1, 2$ with $P_1 \neq P_2$. Let z_j be a local coordinate that vanishes at P_j. We consider two cases:

M is hyperbolic. Let u_j be the Green's function with singularity at P_j. Let $z = x + iy$ be an arbitrary local coordinate on M. Then

$$\varphi(z) = \frac{u_{1x} - iu_{1y}}{u_{2x} - iu_{2y}}$$

defines a meromorphic function with a pole (of order ≥ 1) at P_1 and a zero (of order ≥ 1) at P_2.

M is elliptic or parabolic. Choose u_j harmonic on $M \backslash \{P_j\}$ such that

$$u_j = \operatorname{Re} \frac{1}{z_j} + \gamma_j(z_j)$$

with γ_j harmonic in a neighborhood of zero. Define φ as before. This time $\operatorname{ord}_{P_1} \varphi \leq -2$ and $\operatorname{ord}_{P_2} \varphi \geq 2$. □

IV.3.18. Theorem IV.3.17 has many consequences.

Corollary 1. *On every Riemann surface M we can introduce a C^∞-Riemannian metric consistent with the conformal structure.*

Proof. Let f be any non-constant meromorphic function on M. Let $\{P_1, P_2, \ldots\}$ be the set of poles and critical values (those $P \in M$ with $df(P) = 0$) of f. Let z_j be a local coordinate vanishing at P_j with the sets $\{|z_j| < 1\}$ all disjoint. Let $0 < r_1 < r_2 < 1$ and let ω_j be a C^∞ function with $0 \leq \omega_j \leq 1$,

$$\omega_j = 1 \quad \text{on } \{|z_j| \leq r_1\} \quad \text{and} \quad \omega_j = 0 \quad \text{on } \{|z_j| \geq r_2\}.$$

Define arc length ds by

$$ds^2 = |df|^2 \left(1 - \sum_j \omega_j\right) + \sum_j \omega_j |dz_j|^2. \qquad \square$$

Remark. We shall (see IV.8) be able to do much better. We shall show that the metric may actually be chosen to have constant curvature.

Corollary 2. *Every Riemann surface is metrizeable.*

Corollary 3. *Every Riemann surface has a countable basis for its topology.*

Corollary 4. *Every Riemann surface may be triangulated.*

While it is possible to prove the last two corollaries at this point, we shall delay the proof until after IV.5 when we will be able to give shorter proofs.

IV.4. The Uniformization Theorem for Simply Connected Surfaces

The Riemann mapping theorem classifies the simply connected domains in the complex sphere. It is rather surprising that there are no other simply connected Riemann surfaces. Our development does not require the Riemann mapping theorem, which will be a consequence of our main result.

IV.4.1. Our aim is to prove the following

Theorem. *Let M be a homologically trivial ($H_1(M) = \{0\}$) Riemann surface. Then M is conformally equivalent to one and only one of the following three:*

a. $\mathbb{C} \cup \{\infty\}$,
b. $\mathbb{C}$,
c. $\Delta = \{z \in \mathbb{C}; |z| < 1\}$.

These correspond to the surface M being

a. *elliptic,*
b. *parabolic,*
c. *hyperbolic.*

Of course, $\mathbb{C} \cup \{\infty\}$ is not even topologically equivalent to either of the other two, and $\mathbb{C}$ and Δ are conformally distinct by Liouville's theorem. The fact that these three cases are mutually exclusive will also follow from our proof of the theorem, and the fact that the types (elliptic, parabolic, hyperbolic) are conformal invariants.

IV.4.2. We recall the machinery introduced in III.9. Let M be an arbitrary Riemann surface. Let $P \in M$. Denote by $\mathcal{O}_P(M)$ the germs of holomorphic functions at $P \in M$. Let

$$\mathcal{O}(M) = \bigcup_{P \in M} \mathcal{O}_P(M)$$

denote the *sheaf of germs of holomorphic functions.* This space is given a topology and conformal structure as in III.9. The natural projection

$$\text{proj} = \pi : \mathcal{O}(M) \to M$$

is holomorphic and locally univalent. Similarly, we define the *sheaf of germs of meromorphic functions on M, $\mathcal{M}(M)$.*

Theorem. *Let σ be a discrete set on a Riemann surface M. Let u be a harmonic function on $M \backslash \sigma$ and assume that for each $P \in M$, there is a neighborhood $N(P)$ and a meromorphic function f_P defined on $N(P)$ with either*

a. $\log |f_P| = u$ *on* $N(P) \cap (M \backslash \sigma)$, *or*
b. $\operatorname{Re} f_P = u$ *on* $N(P) \cap (M \backslash \sigma)$.

Take any f_{P_0} as above, and let $\varphi \in \mathcal{M}_{P_0}$ be the germ of f_{P_0} at $P_0 \in M$. Then

i. *φ can be continued analytically along any path in M beginning at P_0, and*
ii. *the continuation of φ along any closed path (beginning and ending at P_0) depends only on the homology class of the path.*

Remark. The hypothesis ((a) or (b)) is automatically satisfied on $M \backslash \sigma$.

Proof of Theorem (The proof is almost shorter than the statement of the theorem and has implicitly been established in the section on multivalued functions.) Let $c:[0,1] \to M$ be a continuous path with $c(0) = P_0$. Let

$$\theta = \{t \in [0,1];\ \varphi \text{ can be continued along } c_t = c|[0,t]\}.$$

It is easy to see that θ is non-empty ($0 \in \theta$), open (without any hypothesis), and closed (by use of (a) or (b)). Note also that the analytic continuation satisfies (a) or (b).

Let c be a closed path beginning at P_0. Let φ_1 be the continuation of φ along c. Then, assuming hypothesis (a), we have

$$\log|\varphi| = u = \log|\varphi_1| \quad \text{near } P_0.$$

Thus there is a $\chi(c) \in \mathbb{C}$, $|\chi(c)| = 1$, such that

$$\varphi_1 = \chi(c)\varphi \quad \text{near } P_0. \tag{4.2.1}$$

Since φ_1 depends only on the homotopy class of c, χ determines a homomorphism

$$\chi:\pi_1(M,P_0) \to S^1$$

from the fundamental group of M (based at P_0) into the unit circle (viewed as a multiplicative subgroup of $\mathbb{C}$). Since S^1 is a commutative group, the kernel of χ contains all commutators and thus χ determines a homomorphism from the first homology group

$$\chi: H_1(M) \to S^1. \tag{4.2.2}$$

In case (b), (4.2.1) is replaced by

$$\varphi_1 = \varphi + i\chi(c)$$

with $\chi(c) \in \mathbb{R}$, and (4.2.2) is replaced by a homomorphism

$$\chi:H_1(M) \to \mathbb{R}.$$

□

Corollary. *Let M be a Riemann surface with $H_1(M) = \{0\}$. Let σ be a discrete subset of M, and u a harmonic function on $M\backslash\sigma$. If u is locally the real part (respectively, the log modulus) of a meromorphic function on M, then u is globally so.*

IV.4.3. As an application of the above corollary we establish the following

Proposition. *Let D be a domain in the extended plane $\mathbb{C} \cup \{\infty\} = S^2$ with $H_1(D) = \{0\}$. Then D is hyperbolic unless $S^2\backslash D$ contains at most one point.*

Remarks

1. The two sphere S^2 is of course elliptic. If $S^2\backslash D$ is a point, then (via a Möbius transformation) $D \cong \mathbb{C}$. The plane $\mathbb{C}$ is parabolic. (Prove this directly. This is also a consequence of Theorem IV.4.5.)

2. The proposition can be viewed as a consequence of the Riemann mapping theorem. It can also be obtained by examining any of the classical proofs of the Riemann mapping theorem. The proposition together with Theorem IV.4.4, imply the Riemann mapping theorem.

PROOF OF PROPOSITION. Say $S^2 \backslash D$ contains 2 points. By replacing D by a conformally equivalent domain, we may assume that $S^2 \backslash D$ contains the points $0, \infty$.

We now construct a single valued branch of $\sqrt{z}$ on D. Observe that $\frac{1}{2} \log|z|$ is a harmonic function on D. Thus there exists a holomorphic function f on D with

$$\log|f(z)| = \tfrac{1}{2} \log|z|, \qquad z \in D.$$

Thus

$$\log|f(z)|^2 = \log|z|, \qquad z \in D,$$

and there is a $\theta \in \mathbb{R}$

$$f(z)^2 = e^{i\theta} z, \qquad z \in D.$$

Set

$$g(z) = e^{-i\theta/2} f(z), \qquad z \in D.$$

Then

$$g(z)^2 = z, \qquad z \in D.$$

Let $a \in D$ and $g(a) = b$. We claim that g does not take on the value $-b$ on D. For if $g(z) = -b$, then $z = g(z)^2 = b^2 = g(a)^2 = a$. This contradiction shows that g misses a ball about $-b$ (using the same argument—since g takes on all values in a ball about b). We define a bounded holomorphic function

$$h(z) = \frac{1}{g(z) + b}, \qquad z \in D.$$

Since D carries a non-constant bounded analytic function, it is hyperbolic. □

IV.4.4. Theorem. *If M is a hyperbolic Riemann surface with $H_1(M) = \{0\}$, then M is conformally equivalent to the unit disc Δ.*

PROOF. Choose a point $P \in M$. Let $g(P, \cdot)$ be the Green's function on M with singularity at P. The function $g(P, \cdot)$ satisfies the hypothesis (a) of Theorem IV.4.2. The only issue is at P. Let z be a local coordinate vanishing at P. Then $g(P,z) + \log|z| = v(z)$ is harmonic in a neighborhood of $z = 0$. Choose a harmonic conjugate v^* of v, write $f = e^{v + iv^*}$. Then

$$g(P,z) + \log|z| = \operatorname{Re} \log f(z) = \log|f(z)|, \quad \text{near } 0,$$

or

$$g(P,z) = \log\left|\frac{f(z)}{z}\right|, \quad \text{near } 0.$$

By Corollary IV.4.2, (applied to $-g(P,\cdot)$) there exists a meromorphic function, $f(P,\cdot)$ on M such that

$$\log|f(P,Q)| = -g(P,Q), \quad \text{all } Q \in M.$$

The function $f(P,\cdot)$ is holomorphic: For $Q \neq P$, $g(P,Q) > 0$, thus $|f(P,Q)| < 1$. By the Riemann removable singularity theorem, $f(P,\cdot)$ extends to an analytic function on M – that is, to the point P. Furthermore,

$$f(P,P) = 0, \qquad f(P,Q) \neq 0 \quad \text{for } Q \neq P. \tag{4.4.1}$$

We want to show that $f(P,\cdot)$ is one-to-one and onto.

Let R, S, $T \in M$ with R and S fixed and T variable. Set

$$\varphi(T) = \frac{f(R,S) - f(R,T)}{1 - \overline{f(R,S)}f(R,T)}.$$

Since φ is $f(R,\cdot)$ followed by a Möbius transformation that fixes the unit disc, φ is holomorphic, $\varphi(S) = 0$, and

$$|\varphi(T)| < 1, \qquad T \in M.$$

Let ζ be a local parameter vanishing at S. We may assume that ζ maps a neighborhood of S onto a neighborhood of the closed unit disc. Then

$$\varphi(\zeta) = a\zeta^n(1 + a_1\zeta + a_2\zeta^2 + \ldots), \qquad n \geq 1, a \neq 0.$$

Set

$$u(T) = -\frac{1}{n}\log|\varphi(T)|.$$

Then u is harmonic and >0 except at the (isolated) zeros of φ. Also

$$-\frac{1}{n}\log|\varphi(\zeta)| + \log|\zeta|$$

is harmonic for $|\zeta|$ small. Let $v \in \mathscr{F}$, where $\mathscr{F}$ is the family of subharmonic functions defined in the proof of Theorem IV.3.7, with P replaced by S (and z by ζ). Let D be the support of v, and D_0 be D with small discs about the (finite number of) singularities of u deleted such that $u - v \geq 0$ on $(D \backslash D_0)$. By the maximum principle $u \geq v$ on D_0. Hence, on $M\backslash\{S\}$. We conclude that

$$-\frac{1}{n}\log|\varphi(T)| = u(T) \geq \sup_{v \in \mathscr{F}} v(T) = g(S,T)$$
$$= -\log|f(S,T)|,$$

or

$$|\varphi(T)| \leq |\varphi(T)|^{1/n} \leq |f(S,T)|. \tag{4.4.2}$$

Setting $T = R$, we get (by (4.4.2))

$$|f(R,S)| \leq |f(S,R)|$$

or (since R and S are arbitrary)

$$|f(R,S)| = |f(S,R)|.$$

Remark. The above equality also follows from the symmetry of the Green's function (Theorem IV.3.10).

Continuation of Proof of Theorem. We consider once again a holomorphic function of T, mapping M into the closed unit disc, namely $h(T) = \varphi(T)/f(S,T)$. By (4.4.2),

$$|h(T)| \leq 1,$$

and since

$$|h(R)| = \left|\frac{\varphi(R)}{f(S,R)}\right| = \left|\frac{f(R,S) - f(R,R)}{1 - \overline{f(R,S)}f(R,R)}\frac{1}{f(S,R)}\right|$$

$$= \left|\frac{f(R,S)}{f(S,R)}\right| = 1,$$

we conclude that

$$\varphi(T) = \chi f(S,T), \qquad \chi \in \mathbb{C}, |\chi| = 1.$$

We rewrite the above equation as

$$\chi f(S,T) = \frac{f(R,S) - f(R,T)}{1 - \overline{f(R,S)}f(R,T)},$$

and deduce that

$$f(R,S) = f(R,T) \Leftrightarrow f(S,T) = 0 \Leftrightarrow S = T.$$

We have shown that $f(P,\cdot)$ is an injective holomorphic function of M into Δ.

Remark. If we are willing to use the Riemann mapping theorem, there is nothing more to prove.

Conclusion of Proof of Theorem. We shall show that $f(P,\cdot)$ is onto. Let $a^2 \in \Delta$ with $a^2 \notin \text{Image } f(P,\cdot)$. Since $f(P,P) = 0$, $a^2 \neq 0$. Let us abbreviate $f(P,\cdot)$ by f. Since $f(M)$ is homologically trivial, we can take a square root (as in IV.4.3) of a non-zero holomorphic function defined on $f(M)$. Let

$$h(z) = \frac{\sqrt{\dfrac{z - a^2}{1 - \bar{a}^2 z}} - ia}{1 + i\bar{a}\sqrt{\dfrac{z - a^2}{1 - \bar{a}^2 z}}}, \qquad z \in f(M),$$

where we choose that branch of the square root with $\sqrt{-a^2} = ia$. Now

$$h(0) = 0, \qquad |h'(0)| = \frac{1 + |a|^2}{2|a|} > 1,$$
$$h(z) < 1 \quad \text{for } z \in f(M). \tag{4.4.3}$$

We now define

$$F(Q) = h(f(Q)), \qquad Q \in M.$$

Then F is a holomorphic function of M into Δ with a simple zero at P and no other zeros. Thus $-\log|F|$ is a competing function for the Green's function and we have

$$-\log|F(Q)| \geq g(P,Q) = -\log|f(P,Q)|,$$

or

$$|F(Q)| \leq |f(Q)|.$$

Hence we conclude that

$$\left|\frac{h(z)}{z}\right| \leq 1 \quad \text{for } |z| \text{ small},$$

and

$$|h'(0)| \leq 1$$

contradicting (4.4.3). □

IV.4.5. Theorem. *Let M be a Riemann surface with $H_1(M) = \{0\}$. If M is compact (respectively, parabolic), then M is conformally equivalent to the complex sphere, $\mathbb{C} \cup \{\infty\} \cong S^2$ (the complex plane, $\mathbb{C}$).*

Before proceeding to the proof of the theorem, we must establish some preliminary results.

IV.4.6. Lemma. *Let D be a relatively compact domain on a Riemann surface M and $P_0 \in D$. Let f be a meromorphic function on* Cl D *whose only singularity is a simple pole at P_0. Then there is a neighborhood N of P_0 such that for any $Q_0 \in N$,*

$$f(Q) \neq f(Q_0), \quad \text{all } Q \in D\backslash\{Q_0\}.$$

PROOF. There is a neighborhood N_0 of P_0 such that $f|N_0$ is injective. Let $m_0 = \max_{P \in \delta(D \backslash \mathrm{Cl}\, N_0)}\{|f(P)|\}$ and let $m_1 > m_0$ be so large so that $\{|z| \geq m_1\}$ is contained in $f(N_0)$. Let $N = f^{-1}(\{z \in \mathbb{C}, |z| > m_1\}) \subset N_0$. Then

$$|f(Q)| < m_1 \quad \text{on } D\backslash \mathrm{Cl}\, N$$
$$|f(Q)| \geq m_1 \quad \text{on } \mathrm{Cl}\, N.$$

□

Corollary. *Let M be a parabolic or compact Riemann surface. Let $P_0 \in M$. Assume that $f \in \mathscr{K}(M)$ with $\mathrm{ord}_{P_0} f = -1$, f is holomorphic on $M\backslash\{P_0\}$ and f is bounded outside some relatively compact (therefore, outside every) neighborhood of P_0. Then there exists a neighborhood N of P_0 such that for any $Q_0 \in N$,*

$$f(Q) \neq f(Q_0), \quad \textit{all } Q \in M\backslash\{Q_0\}.$$

PROOF. This is actually a corollary of the proof of the lemma, using the fact that under the hypothesis, a bounded analytic function assumes its bound on the boundary (Theorem IV.3.3). □

IV.4.7. Lemma. *Let M be a parabolic or compact Riemann surface with $H_1(M) = \{0\}$. Let $P_0 \in M$. There exists a function $f \in \mathscr{K}(M)$ with $\operatorname{ord}_{P_0} f = -1$ that is bounded outside every neighborhood of P_0.*

PROOF. Let z be a conformal disc at P_0. By Theorem IV.3.11, there exists a function u that is harmonic and bounded outside every neighborhood of P_0 and such that

$$u(z) - \operatorname{Re}\frac{1}{z}$$

is harmonic in a neighborhood of $z = 0$, and vanishes at $z = 0$. By the corollary to Theorem IV.4.2, there exists a function f holomorphic on $M\backslash\{P_0\}$ with

$$f = u + iv,$$

$f(z) - 1/z$ holomorphic near $z = 0$ and $(f(z) - 1/z)$ vanishes at 0 (for some harmonic function v on $M\backslash\{P_0\}$). Similarly, there exists a holomorphic function $\tilde{f}$ on $M\backslash\{P_0\}$

$$\tilde{f} = \tilde{u} + i\tilde{v},$$

$|\tilde{u}|$ bounded outside every neighborhood of P_0, and $\tilde{f}(z) - i/z$ is holomorphic near $z = 0$ and it vanishes at zero. We want to prove that $|v|$ is bounded outside every neighborhood of P_0. We shall show that $\tilde{f} = if$ (thus $v = \tilde{u}$), which will conclude the proof.

Choose $m > 0$ so that

$$|u(z)| < m, \qquad |\tilde{u}(z)| < m \quad \text{on } \{|z| = 1\}.$$

Thus, also on $\{|z| \geq 1\}$. It involves no loss of generality to assume that f and $\tilde{f}$ are one-to-one on $\{|z| \leq 1\}$. Choose Q_0 with $0 \neq z(Q_0) = z_0$ and $|z_0| < 1$ such that $|u(Q_0)| > 2m$ and $|\tilde{u}(Q_0)| > 2m$.

Define

$$g(Q) = \frac{1}{f(Q) - f(Q_0)}, \qquad \tilde{g}(Q) = \frac{1}{\tilde{f}(Q) - \tilde{f}(Q_0)}, \qquad Q \in M.$$

The functions g and $\tilde{g}$ are holomorphic on $M\backslash\{Q_0\}$. Furthermore $\operatorname{ord}_{Q_0} g = -1 = \operatorname{ord}_{Q_0} \tilde{g}$. We claim that these two functions are bounded outside every neighborhood of Q_0. It suffices to show that they are bounded outside $\{|z| < 1\}$. Now for $|z(Q)| \geq 1$,

$$|g(Q)|^2 = \frac{1}{(u(Q) - u(Q_0))^2 + (v(Q) - v(Q_0))^2} \leq \frac{1}{m^2},$$

and similarly

$$|\tilde{g}(Q)|^2 \leq \frac{1}{m^2}.$$

The Laurent series expansions of g and $\tilde{g}$ in terms of $z - z_0$ are

$$g(z) = \frac{c}{(z - z_0)} + a_0 + a_1(z - z_0) + \cdots$$

$$\tilde{g}(z) = \frac{\tilde{c}}{(z - z_0)} + \tilde{a}_0 + \tilde{a}_1(z - z_0) + \cdots.$$

Thus $\tilde{c}g - c\tilde{g}$ is a bounded holomorphic function on M and thus constant. In particular $\tilde{f}$ is a Möbius transformation of f. Thus

$$\tilde{f}(Q) = \frac{\alpha f(Q) + \beta}{\gamma f(Q) + \delta}, \qquad \alpha\delta - \beta\gamma \neq 0.$$

Setting $Q = P_0$, we see that $\gamma = 0$ (we may thus take $\delta = 1$) and thus (because we know the singularity of f and $\tilde{f}$ at P_0) $\tilde{f} = if$. □

IV.4.8. Proof of Theorem IV.4.5. Let $P \in M$. A function $g \in \mathscr{K}(M)$ will be called *admissible* at P provided

$$g \in \mathscr{H}(M \backslash \{P\}), \tag{4.8.1}$$

$$\operatorname{ord}_P g = -1, \text{ and} \tag{4.8.2}$$

$$g \text{ is bounded outside every neighborhood of } P. \tag{4.8.3}$$

We have seen that there exist functions admissible at every point P and that any two such functions are related by a Möbius transformation.

Assertion: Given f admissible at P and g admissible at $Q \in M$, then there exists a Möbius transformation L such that $g = L \circ f$.

PROOF OF ASSERTION. Fix $P \in M$, and a function f admissible at P. Let $\Sigma \subset M$ be defined by

$$\Sigma = \{Q \in M; g \text{ admissible at } Q \Rightarrow g = L \circ f, \text{ for some Möbius transformation } L\}.$$

The set Σ is non-empty because $P \in \Sigma$. Let $Q_0 \in \Sigma$. Let g_0 be admissible at Q_0. Thus $g_0 = L_0 \circ f$, for some Möbius transformation L_0. Choose a a neighborhood N of Q_0 such that every value g_0 taken on in N is not assumed in $M \backslash N$. Let $Q_1 \in N$, $Q_1 \neq Q_0$. Then

$$g_1(Q) = \frac{1}{g_0(Q) - g_0(Q_1)} = (L_1 \circ g_0)(Q), \qquad Q \in M,$$

is admissible at Q_1. If g is admissible at Q_1, then $g = L_2 \circ g_1 = L_2 \circ L_1 \circ g_0 = L_2 \circ L_1 \circ L_0 \circ f$. Thus Σ is open. Similarly (by exactly the same argument), Σ is closed in M, and thus $\Sigma = M$.

Assertion: Every admissible function is univalent.

Proof of Assertion. Let f be admissible at P_0. Thus $f(P_0) = \infty$. Assume $f(P_1) = f(P_2)$. Choose g admissible at P_1 and a Möbius transformation L such that $f = L \circ g$. Thus $L(g(P_1)) = L(g(P_2))$. Since L is univalent, $\infty = g(P_1) = g(P_2)$. Thus $P_2 = P_1$ since g is by hypothesis admissible at P_1.

Conclusion of Proof. Choose an admissible

$$f: M \to \mathbb{C} \cup \{\infty\}.$$

If M is compact, then f is also surjective (and conversely). Thus if M is parabolic, $f(M)$ omits at least one point. Following f by a Möbius transformation L, we may assume $L \circ f(M) \subset \mathbb{C}$. If $L \circ f(M) \subsetneqq \mathbb{C}$, then $L \circ f(M)$ (and hence M) would be hyperbolic. □

Remark. We have now established Theorem IV.4.1. This is a major result in this subject.

IV.5. Uniformization of Arbitrary Riemann Surfaces

In this section we introduce the concept of a Kleinian group and show how each Riemann surface can be represented by a special Kleinian group — known as a Fuchsian group. These "uniformizations" will involve only fixed point free groups. More general uniformizations will be treated in IV.9.

IV.5.1. Let G be a subgroup of $PSL(2,\mathbb{C})$ (that is, G is a group of Möbius transformations). Thus G acts as a group of biholomorphic automorphisms of the extended plane $\mathbb{C} \cup \{\infty\}$. Let $z_0 \in \mathbb{C} \cup \{\infty\}$. We shall say that G acts *(properly) discontinuously at* z_0 provided

$$\text{the isotropy subgroup of } G \text{ at } z_0,\quad G_{z_0} = \{g \in G; g(z_0) = z_0\}, \text{ is finite,} \tag{5.1.1}$$

and

$$\text{there exists a neighborhood } U \text{ of } z_0 \text{ such that}$$

$$g(U) = U \quad \text{for } g \in G_{z_0},$$

and

$$g(U) \cap U = \varnothing \quad \text{for } g \in G \backslash G_{z_0}. \tag{5.1.2}$$

Denote by $\Omega(G)$ $(=\Omega)$ the *region of discontinuity* of G; that is, the set of points $z_0 \in \mathbb{C} \cup \{\infty\}$ such that G acts discontinuously at z_0. Set

$$\Lambda = \Lambda(G) = \mathbb{C} \cup \{\infty\} \backslash \Omega(G),$$

and call $\Lambda(G)$ the *limit set* of G.

It is immediately clear from the definitions that Ω is an open G-invariant ($G\Omega = \Omega$) subset of $\hat{\mathbb{C}}$. If $\Omega \neq \emptyset$, we call G a *Kleinian* group.

From now on we assume that G is a Kleinian group. Since Ω is open in $\mathbb{C} \cup \{\infty\}$, it can be written as a union of at most countably many components. We say that two components D and $\tilde{D}$ are *equivalent* if there is a $g \in G$ such that $gD = \tilde{D}$. If Ω contains a component Δ which is G-invariant, then G is called a *function* group. In general, let

$$\Omega_1, \Omega_2, \ldots$$

be a maximal set of non-equivalent components of Ω. It is clear that (as point sets) the orbit spaces

$$\Omega/G \quad \text{and} \quad \bigoplus_j \Omega_j/G_j$$

are isomorphic, where (the *stabilizer* of Ω_j)

$$G_j = \{g \in G; g\Omega_j = \Omega_j\}.$$

It follows from our work in III.7.7 that each Ω_j/G_j is a Riemann surface and that the canonical projection $\Omega_j \to \Omega_j/G_j$ is holomorphic. Thus Ω/G is a (perhaps countable) union of Riemann surfaces with the projection

$$\pi: \Omega \to \Omega/G$$

holomorphic.

IV.5.2. Proposition. *If G is a Kleinian group, then it is finite or countable.*

PROOF. Choose $z \in \Omega(G)$ such that G_z is trivial. Then

$$\{g(z); g \in G\}$$

is a discrete set in Ω. Hence finite or countable. But this set has the same cardinality as the group G. □

IV.5.3. Since $SL(2,\mathbb{C})$ inherits a topology from its imbedding into $\mathbb{C}^4$, $PSL(2,\mathbb{C}) \cong SL(2,\mathbb{C})/\pm I$ is a topological group. A group $G \subset PSL(2,\mathbb{C})$ is called *discrete* if it is a discrete subset of the topological space $PSL(2,\mathbb{C})$. It is obvious that

Proposition. *Every Kleinian group is discrete.*

The Picard group

$$G = \left\{z \mapsto \frac{az + b}{cz + d}; ad - bc = 1 \text{ and } a,b,c,d \in \mathbb{Z}[i]\right\}$$

shows that the converse is not true.

IV.5.4. A Kleinian group G is called *Fuchsian* if there is a disc (or half plane) that is invariant under G.

Theorem. *Let* $G \subset \operatorname{Aut} \Delta$, $\Delta =$ *unit disc. The following conditions are equivalent*:

a. $\Delta \subset \Omega(G)$,
b. $\Delta \cap \Omega(G) \neq \varnothing$,
c. G *is discrete.*

PROOF. (a) $\Rightarrow$ (b): This implication is trivial.

(b) $\Rightarrow$ (c): Assume that G is not discrete. There then exists a distinct sequence $\{g_n\} \subset G$ such that

$$g_n \to g$$

with $g \in \operatorname{Aut} \Delta$. Thus also

$$g_n^{-1} \to g^{-1}$$

(as is easily verified using the matrix representation of elements of $PL(2,\mathbb{C})$). Consider

$$h_n = g_{n+1}^{-1} \circ g_n, \qquad n = 1, 2, \ldots .$$

The sequence $\{h_n\}$ contains an infinite distinct subsequence and $h_n \to 1$. Thus $h_n(z) \to z$, all $z \in \Delta$, and $\Delta \cap \Omega(G) = \varnothing$.

(c) $\Rightarrow$ (a): If the group G is not discontinuous at some point $z_0 \in \Delta$, then there is an infinite sequence of points $\{z_n\}$, $n = 1, 2, \ldots$, in Δ equivalent under G to z_0 and which converges to z_0. Choose $g_n \in G$ with $g_n(z_n) = z_0$. Consider the element $A_n \in \operatorname{Aut} \Delta$ defined by

$$A_n(z) = \frac{z - z_n}{1 - \bar{z}_n z}, \qquad z \in \Delta,\ n = 0, 1, \ldots,$$

and set for $n = 1, 2, 3, \ldots$

$$C_n = A_{n+1} \circ g_{n+1}^{-1} \circ g_n \circ A_n^{-1}.$$

Since $C_n(0) = 0$, we conclude from Schwarz's lemma that

$$C_n(z) = \lambda_n z, \qquad |\lambda_n| = 1.$$

Thus there is a subsequence—which may be taken as the entire sequence—of $\{C_n\}$ that converges to C_0 (where $C_0(z) = \lambda_0 z$). Since $A_n \to A_0$, we see that $h_n = g_{n+1}^{-1} \circ g_n \to A_0^{-1} \circ C_0 \circ A_0$. Since the $\{z_n\}$ are distinct, so are the $\{g_n\}$, and hence also the $\{h_n\}$. □

Corollary. *If* G *is a Fuchsian group with invariant disc* D, *then* $\Lambda(G) \subset \delta D$.

PROOF. The exterior of D is also a disc invariant under G. Since G is a Kleinian group, it must be discrete. Hence, both D and the exterior of D are subsets of $\Omega(G)$. □

Before proceeding to our main result, we need a

Definition. A Kleinian group G with $\Lambda(G)$ consisting of two or less points is called an *elementary* group.

IV.5.5. Let M be an arbitrary Riemann surface. We let $\tilde{M}$ be its universal covering space. Let

$$\pi:\tilde{M} \to M$$

be the canonical projection, and G the covering group; that is, the group of topological automorphisms g of $\tilde{M}$ for which the following diagram commutes:

$$\begin{array}{ccc} \tilde{M} & \xrightarrow{\;g\;} & \tilde{M} \\ & {\scriptstyle\pi}\searrow \quad \swarrow{\scriptstyle\pi} & \\ & M & \end{array}.$$

Of course, since π is a normal covering, $G \cong \pi_1(M)$. Furthermore, G acts properly discontinuously and fixed point freely on $\tilde{M}$. Since π is a local homeomorphism, it introduces via the analytic structure on M, an analytic structure on $\tilde{M}$. With this analytic structure, G becomes a group of conformal automorphisms of $\tilde{M}$; that is, a subgroup of Aut $\tilde{M}$.

IV.5.6. Theorem IV.4.1 gave us all the candidates for $\tilde{M}$; that is, all the simply connected Riemann surfaces: $\mathbb{C} \cup \{\infty\}$, $\mathbb{C}$ or $\Delta \cong U = \{z \in \mathbb{C};\ \text{Im}\, z > 0\}$. Each of these domains has the property that its group of conformal automorphisms is a group of Möbius transformations $z \mapsto (az + b)/(cz + d)$. In fact:

$$\begin{aligned} \text{Aut}(\mathbb{C} \cup \{\infty\}) &\cong PSL(2,\mathbb{C}), \\ \text{Aut}\ \mathbb{C} &\cong P\Delta(2,\mathbb{C}), \\ \text{Aut}\ U &\cong PSL(2,\mathbb{R}). \end{aligned}$$

(By $P\Delta(2,\mathbb{C})$ we mean the projective group of 2×2 upper triangular complex matrices of non-zero determinant.) We have hence established the following general uniformization

Theorem. *Every Riemann surface M is conformally equivalent to D/G with $D = \mathbb{C} \cup \{\infty\}$, $\mathbb{C}$, or U and G a freely acting discontinuous group of Möbius transformations that preserve D. Furthermore, $G \cong \pi_1(M)$.*

Remarks

1. We will see in IV.6 that for most Riemann surfaces M, $\tilde{M} \cong U$.
2. At this point, it is quite easy to give alternate proofs of Corollaries 2–4 in IV.3.18. A stronger form of Corollary 1 will be established in IV.8.

IV.5.7. If $\tilde{M} = U$, then the group G is, of course, a Fuchsian group. If $\tilde{M} = \mathbb{C}$ or $\mathbb{C} \cup \{\infty\}$, then G can have at most one limit point (G acts discontinuously on $\mathbb{C}$). Thus we have established the following

Theorem. *Every Riemann surface can be represented as a domain in the plane factored by a fixed point free Fuchsian or elementary group.*

IV.5.8. A Riemann surface M will be called *prolongable* if there exists a Riemann surface M' and a one-to-one holomorphic mapping $f: M \to M'$ such that $M' \backslash f(M)$ has a non-empty interior.

Compact surfaces and those obtained from compact surfaces by omitting finitely many points are clearly not prolongable. We shall see (trivially) in the next section that those Riemann surfaces that can be represented as $\mathbb{C}/G$ with G a fixed point free elementary group are not prolongable. For the Fuchsian case, we need the following

Definition. Let G be a Fuchsian group leaving invariant the interior of the circle C. The group G is called of the *second kind* if $\Lambda(G) \subsetneq C$. If $\Lambda(G) = C$, G is called of the *first kind.*

Theorem. *Let G be a fixed point free Fuchsian group acting on U. Then U/G is prolongable if G is ofthe second kind.*

PROOF. If G is of the second kind, then $\Omega(G)$ is connected and the complement of U/G in Ω/G certainly contains L/G, where L is the lower half plane. □

Remark. The theorem thus shows that compact surfaces are uniformized by groups of the first kind.

EXERCISE

Formulate and prove a converse to Theorem IV.5.8.

IV.6. The Exceptional Riemann Surfaces

The title of this section is explained by Theorem IV.6.1. The same surfaces will reappear in V.4.

IV.6.1. The fundamental groups of most Riemann surfaces are not commutative. The exceptions are listed in the following

Theorem

a. *The only simply connected Riemann surfaces are the ones conformally equivalent to $\mathbb{C} \cup \{\infty\}$, $\mathbb{C}$, or $\Delta = \{z \in \mathbb{C}, |z| < 1\}$.*
b. *The only surfaces with $\pi_1(M) \cong \mathbb{Z}$ are (conformally equivalent to) $\mathbb{C}^* = \mathbb{C}\backslash\{0\}$, $\Delta^* = \Delta\backslash\{0\}$, or $\Delta_r = \{z \in \mathbb{C}; r < |z| < 1\}$, $0 < r < 1$.*
c. *The only surfaces with $\pi_1(M) \cong \mathbb{Z} \oplus \mathbb{Z}$ (the commutative free group on two generators) are the tori $\mathbb{C}/G$, where G is generated by $z \mapsto z + 1$ and $z \mapsto z + \tau$, $\operatorname{Im} \tau > 0$.*
d. *For all other surfaces M, $\pi_1(M)$ is not abelian.*

The remainder of this section is devoted to the proof of this theorem.

IV.6.2. Let A be a Möbius transformation, $A \neq 1$. Then A has one or two fixed points. If A has one fixed point, it is called *parabolic*. We write $A(z) = (az + b)/(cz + d)$, $ad - bc = 1$. Then

$$trace^2 A = (a + d)^2$$

is well defined. It is easy to check that A is parabolic if and only if $\text{trace}^2 A = 4$. The element A is called *elliptic* if $\text{trace}^2 A \in \mathbb{R}$ and $0 \leq \text{trace}^2 A < 4$. It is called *loxodromic* if $\text{trace}^2 A \notin [0,4] \subset \mathbb{R}$. A loxodromic element A with $\text{trace}^2 A > 4$ (we are assuming here that $\text{trace}^2 A \in \mathbb{R}$) is called *hyperbolic*.

Let A be parabolic with fixed point $z_0 \in \mathbb{C} \cup \{\infty\}$. Choose C Möbius such that $C(z_0) = \infty$. Thus $C \circ A \circ C^{-1}(\infty) = \infty$ and hence $D = C \circ A \circ C^{-1}$ has the form $D(z) = az + b$ with $a \neq 0$. Since A is parabolic, so is D. Hence $a = 1$. We conclude A is parabolic if and only if A is conjugate to a translation $z \mapsto z + b$ (and thus also conjugate to the translation $z \mapsto z + 1$). Similarly, it is easy to establish that an element A with two fixed points is conjugate to $z \mapsto \lambda z$, $\lambda \neq 0, 1$ and that

$$A \text{ is loxodromic} \Leftrightarrow |\lambda| \neq 1, \qquad \lambda \neq 0,$$
$$A \text{ is hyperbolic} \Leftrightarrow \lambda \in \mathbb{R}, \qquad \lambda > 0, \lambda \neq 1,$$

and

$$A \text{ is elliptic} \Leftrightarrow |\lambda| = 1, \qquad \lambda \neq 1.$$

The number λ is called the *multiplier* of the motion A. It is a root of unity if and only if A is elliptic of finite order.

IV.6.3. Theorem. *The only Riemann surface M which has as universal covering the sphere, is the sphere itself.*

PROOF. The covering group of M would necessarily have fixed points if $\pi_1(M) \neq \{1\}$. □

IV.6.4. The fixed point free elements in Aut $\mathbb{C}$ are of the form $z \mapsto z + a$, $a \in \mathbb{C}$. Since a covering group of a Riemann surface must be discrete, we see that (consult Ahlfors' book *Complex Analysis*) we have the following

Theorem. *If the (holomorphic) universal covering space of M is $\mathbb{C}$, then M is conformally equivalent to $\mathbb{C}$, $\mathbb{C}^*$, or a torus.*

These correspond to $\pi_1(M)$ being trivial, $\cong \mathbb{Z}$, and $\cong \mathbb{Z} \oplus \mathbb{Z}$. Note that if $\pi_1(M) \cong \mathbb{Z}$, then we can take as generator for the covering group of M, the translation $z \mapsto z + 1$. The covering map

$$\pi: \mathbb{C} \to \mathbb{C}^*$$

is given by $\pi(z) = \exp(2\pi i z)$.

IV.6.5. All surfaces except those listed in Theorems IV.6.3 and IV.6.4 have the unit disc or equivalently the upper half plane U as their holomorphic universal covering space. We have shown in III.6.4, that every torus has

$\mathbb{C}$ as its holomorphic universal covering space (we will reprove this below). Since no surface can have both U and $\mathbb{C}$ as its holomorphic universal covering space, a torus cannot be written as U modulo a fixed point free subgroup of Aut U. More, however, is true (see Theorem IV.6.7).

Since Aut $U \cong PSL(2,\mathbb{R})$, we see that each elliptic element $A \in$ Aut U fixes a point $z \in U$ (and $\bar{z}$ as well). Conversely, every $A \in$ Aut U that fixes an element of U is elliptic. Thus a covering group of a Riemann surface cannot contain elliptic elements. (As an exercise, prove that a discrete subgroup of Aut U cannot contain elliptic elements of infinite order.) Since every element of Aut U of finite order is elliptic, we have established the following topological result.

Proposition. *The fundamental group of a Riemann surface is torsion free.*

Remark. Using Fuchsian groups, one can determine generators and relations for fundamental groups of surfaces.

IV.6.6. If $A \in$ Aut U is loxodromic, we can choose an element $B \in$ Aut U such that

$$C(0) = B \circ A \circ B^{-1}(0) = 0,$$

and

$$C(\infty) = B \circ A \circ B^{-1}(\infty) = \infty.$$

Thus $C(z) = \lambda z$ with $\lambda \in \mathbb{R}$, $\lambda > 0$, $\lambda \neq 1$. Thus Aut U does not contain any non-hyperbolic loxodromic elements, and hence the covering group of a Riemann surface (whose universal covering space is U) consists only of parabolic and hyperbolic elements.

Lemma. *Let A and B be Möbius transformations which commute, with neither A nor B the identity. Then we have:*

a. *If A is parabolic, so is B and both have the same fixed point.*
b. *If one of A and B is not parabolic (then neither is the other by (a)), then either A and B have the same fixed points, or both of them are elliptic of order 2 and one permutes the fixed points of the other.*

Proof. Say $A \circ B = B \circ A$. If A is parabolic, it involves no loss of generality to assume $A(z) = z + 1$ (by conjugating A and B by the same element C). Write $B = \left[\begin{smallmatrix} a & b \\ c & d \end{smallmatrix}\right] \in SL(2,\mathbb{C})$, $A = \left[\begin{smallmatrix} 1 & 1 \\ 0 & 1 \end{smallmatrix}\right]$. Thus the statement A commutes with B gives

$$\begin{bmatrix} a+c & b+d \\ c & d \end{bmatrix} = \begin{bmatrix} a & a+b \\ c & c+d \end{bmatrix} \quad \text{in } PSL(2,\mathbb{C}).$$

From which we conclude that $c = 0$, thus $ad = 1$ and $d = a$; showing that $B(z) = z + \beta$.

If neither A nor B have precisely one fixed point, then by conjugation we may assume $A(z) = \lambda z$ with $\lambda \neq 0, 1$. The commutativity relation now reads

$$\begin{bmatrix} \lambda a & \lambda b \\ c & d \end{bmatrix} = \begin{bmatrix} \lambda a & b \\ \lambda c & d \end{bmatrix} \quad \text{in } PSL(2,\mathbb{C}).$$

There are now two possibilities: $c \neq 0$ or $c = 0$. If $c \neq 0$, then we must have the equality

$$\begin{bmatrix} \lambda^2 a & \lambda^2 b \\ \lambda c & \lambda d \end{bmatrix} = \begin{bmatrix} \lambda a & b \\ \lambda c & d \end{bmatrix}.$$

Thus $a = 0$ and if we assume that $\lambda \neq -1$, also $b = 0$ (hence $ad - bc \neq 1$). Thus this case is impossible and $\lambda = -1$. The matrix identity now reads

$$\begin{bmatrix} 0 & b \\ -c & -d \end{bmatrix} = \begin{bmatrix} 0 & b \\ -c & d \end{bmatrix};$$

from which we conclude $a = 0 = d$ and $bc = -1$; from which we conclude $B(z) = k/z$. The remaining case is $c = 0$. Thus $d \neq 0$ and the matrix equality becomes

$$\begin{bmatrix} \lambda a & \lambda b \\ 0 & d \end{bmatrix} = \begin{bmatrix} \lambda a & b \\ 0 & d \end{bmatrix}.$$

Thus $b = 0$ and $B(z) = kz$. □

Corollary. *Two commuting loxodromic transformations have the same fixed point set.*

IV.6.7. Theorem. *Let M be a Riemann surface with $\pi_1(M) \cong \mathbb{Z} \oplus \mathbb{Z}$, then the holomorphic universal covering space of M is $\mathbb{C}$.*

PROOF. Assume the covering space is U. The covering group of M is an abelian group on two generators. Let A be one of the generators. If A is parabolic, then we may assume $A(z) = z + 1$. Let B be another free generator. Since B commutes with A, B must also be parabolic. Thus $B(z) = z + \beta$, $\beta \in \mathbb{R}\backslash\mathbb{Q}$. The group generated by A and B is not discrete.

So assume A is hyperbolic (it cannot be anything else if it is not parabolic). We may assume $A(z) = \lambda z$, $\lambda > 1$. The other generator B must be of the form $B(z) = \mu z$ with $\mu > 1$ and $\lambda^n \neq \mu^m$ for all $(n,m) \in \mathbb{Z} \oplus \mathbb{Z}\backslash\{0\}$. Again, such a group cannot be discrete (take logarithms to transform to a problem in discrete modules). □

IV.6.8. To finish the proof of Theorem IV.6.1, we must establish the following

Theorem. *Let M be a Riemann surface with holomorphic universal covering space U. Then $M \cong \Delta$, Δ^*, or Δ_r, provided $\pi_1(M)$ is commutative.*

PROOF. We have seen in the proof of the previous theorem that if M is covered by U and $\pi_1(M)$ is commutative, then the covering group G of M must be cyclic. The two possibilities (other than the trivial one) are the generator A of G is parabolic (without loss of generality $z \mapsto z + 1$) or hyperbolic ($z \mapsto \lambda z$, $\lambda > 1$). In the first case the map

$$U \to \Delta^*$$

is given by

$$z \mapsto \exp(2\pi i z).$$

For $z = re^{i\theta}$, $0 \leq \theta < 2\pi$, we define

$$\log z = \log r + i\theta$$

(the principal branch of the logarithm). The covering map (for the case of hyperbolic generator)

$$U \to \Delta_r$$

is then given by

$$z \mapsto \exp\left(2\pi i \frac{\log z}{\log \lambda}\right).$$

From this we also see that

$$r = \exp(-2\pi^2/\log \lambda). \qquad (6.8.1) \quad \square$$

IV.6.9. As an application we prove the following

Theorem. *Let D be a domain in $\mathbb{C} \cup \{\infty\}$ such that $\mathbb{C} \cup \{\infty\} \backslash D$ consists of two components α, β. Then $D \cong \mathbb{C}^*$, Δ^*, or Δ_r.*

PROOF. The complement of α (in $\mathbb{C} \cup \{\infty\}$) is simply connected ($\neq \mathbb{C} \cup \{\infty\}$) and thus conformally equivalent to $\mathbb{C}$ or Δ. Thus it suffices to assume α is a point or the unit circle. Now ($\mathbb{C} \cup \{\infty\} \backslash \beta$) is also equivalent to $\mathbb{C}$ or Δ. Thus we are reduced to the case where δD consists of points or analytic arcs. In either case, it is now easy to see that $\pi_1(D) \cong \mathbb{Z}$, and the result follows by our classification of Riemann surfaces M with commutative $\pi_1(M)$. $\square$

IV.7. Two Problems on Moduli

The general problem of moduli of Riemann surfaces may be stated as follows: Given two topologically equivalent Riemann surfaces, find necessary and sufficient conditions for them to be conformally equivalent. What does the "space" of conformally inequivalent surfaces of the same topological type look like? The solution to this general problem is beyond the scope of this book; however, two simple cases (the case of the annulus and the case of the torus) will be solved completely in this section.

IV.7.1. Let M_j be a Riemann surface with $\pi_j : \tilde{M}_j \to M_j$, the holomorphic universal covering map ($j = 1,2$). First, recall that the covering group G_j

of M_j is determined up to conjugation in Aut $\tilde{M}_j$. Furthermore, if

$$f: M_1 \to M_2$$

is a topological, holomorphic, conformal, etc., map, then there exists a map

$$\tilde{f}: \tilde{M}_1 \to \tilde{M}_2$$

of the same type so that

$$\pi_2 \circ \tilde{f} = f \circ \pi_1.$$

(Of course G_j is the set of lifts of the identity map $M_j \to M_j$.) The map $\tilde{f}$ is not uniquely determined. It may be replaced by $A_2 \circ \tilde{f} \circ A_1$ with $A_j \in G_j$.

IV.7.2. How many conformally distinct annuli are there? We have seen that every annulus can be written as U/G where G is the group generated by $z \mapsto \lambda z$, $\lambda \in \mathbb{R}$, $\lambda > 1$. If A_1 and A_2 are two such annuli with the corresponding λ_1 and λ_2, when do they determine the same conformal annulus? They determine the same annulus if and only if there is an element $T \in$ Aut U such that

$$T(\lambda_1 z) = \lambda_2 T(z), \qquad z \in U.$$

From this it follows rather easily that $\lambda_1 = \lambda_2$ (by direct examination or using the fact that the trace of a Möbius transformation is invariant under conjugation).

We conclude

Theorem. *The set of conformal equivalence classes of annuli is in one-to-one canonical correspondence with the open real interval* $(1, \infty)$.

Of course, we could substitute for the word "annuli" the words "domains in $\mathbb{C} \cup \{\infty\}$ with two non-degenerate boundary components."

IV.7.3. Let T_1 and T_2 be two conformally equivalent tori. As we have seen, the covering group G_j of T_j may be chosen to be generated by the translations $z \mapsto z + 1$, $z \mapsto z + \tau_j$ with Im $\tau_j > 0$. Thus we are required to find when two distinct points in U determine the same torus. Let f be the conformal equivalence between T_1 and T_2 and let $\tilde{f}$ be its lift to the universal covering space:

$$\begin{array}{ccc} \mathbb{C} & \xrightarrow{\tilde{f}} & \mathbb{C} \\ \downarrow{\scriptstyle \pi_1} & & \downarrow{\scriptstyle \pi_2} \\ T_1 & \xrightarrow{f} & T_2 \end{array}$$

The mapping $\tilde{f}$ is conformal (thus affine; that is, $z \mapsto az + b$), and it induces an isomorphism

$$G_1 \ni g \overset{\theta}{\mapsto} \tilde{f} \circ g \circ \tilde{f}^{-1} \in G_2.$$

We abbreviate the Möbius transformation $z \mapsto z + c$ by c. Thus

$$\begin{aligned} a &= \theta(1) = \alpha 1 + \beta \tau_2 \quad \text{with } \alpha, \beta \in \mathbb{Z} \\ a\tau_1 &= \theta(\tau_1) = \gamma 1 + \delta \tau_2 \quad \text{with } \gamma, \delta \in \mathbb{Z}. \end{aligned} \tag{7.3.1}$$

Further, since θ is an isomorphism the matrix $\left[\begin{smallmatrix}\alpha & \beta \\ \gamma & \delta\end{smallmatrix}\right]$ is invertible. Thus $\alpha\delta - \beta\gamma = \pm 1$. Considering the symbols in (7.3.1) once again as numbers, we see that

$$\tau_1 = \frac{\gamma + \delta\tau_2}{\alpha + \beta\tau_2},$$

and since both τ_1 and $\tau_2 \in U$ we see that $\beta\gamma - \alpha\delta = -1$. Conversely, two τ's related as above correspond to conformally equivalent tori. Let Γ denote the unimodular group of Möbius transformations; that is, $PSL(2,\mathbb{Z})$. It is easy to see that Γ is a discrete group (hence discontinuous), and with some work that $U/\Gamma \cong \mathbb{C}$. Thus we have

Theorem. *The set of equivalence classes of tori is in one-to-one canonical correspondence with the points in* $\mathbb{C}$.

IV.8. Riemannian Metrics

In this section we show that on every Riemann surface we can introduce a complete Riemannian metric of constant (usually negative) curvature. We also develop some of the basic facts of non-Euclidean geometry that will be needed in IV.9.

IV.8.1. We introduce Riemannian metrics on the three simply connected Riemann surfaces. The metric on M will be of the form

$$\lambda(z)|dz|, \qquad z \in M. \tag{8.1.1}$$

We set

$$\lambda(z) = \frac{2}{1 + |z|^2}, \quad \text{for } M = \mathbb{C} \cup \{\infty\},$$

$$\lambda(z) = 1, \quad \text{for } M = \mathbb{C},$$

$$\lambda(z) = \frac{2}{1 - |z|^2}, \quad \text{for } M = \Delta = \{z \in \mathbb{C}; |z| < 1\}.$$

The definition of λ for $M = \mathbb{C} \cup \{\infty\}$ is, of course, only valid for $z \neq \infty$. At infinity, invariance leads to the form of λ in terms of the local coordinate $\zeta = 1/z$.

We will now explain each of the above metrics.

IV.8.2. The compact surface of genus 0, $\mathbb{C} \cup \{\infty\}$, has been referred to (many times) as the Riemann sphere; but up to now no "sphere" has appeared. Consider hence the unit sphere S^2,

$$\xi^2 + \eta^2 + \zeta^2 = 1$$

in $\mathbb{R}^3$. We map $S^2\backslash\{(0,0,1)\}$ onto $\mathbb{C}$ by (*stereographic projection*)

$$(\xi,\eta,\zeta) \mapsto \frac{\xi + i\eta}{1 - \zeta} = z.$$

It is a trivial exercise to find an inverse and to show that the above map establishes a diffeomorphism between $S^2\backslash\{(0,0,1)\}$ and $\mathbb{C}$ that can be extended to a diffeomorphism between S^2 and $\mathbb{C} \cup \{\infty\}$. The inverse is

$$z \mapsto \left(\frac{2 \operatorname{Re} z}{|z|^2 + 1}, \frac{2 \operatorname{Im} z}{|z|^2 + 1}, \frac{|z|^2 - 1}{|z|^2 + 1}\right). \tag{8.2.1}$$

(Write $x = \operatorname{Re} z$, $y = \operatorname{Im} z$.) The Euclidean metric on $\mathbb{R}^3$

$$ds^2 = d\xi^2 + d\eta^2 + d\zeta^2$$

induces a metric on S^2, which in turn induces a metric on $\mathbb{C} \cup \{\infty\}$. Equation (8.2.1) shows that

$$d\xi = \frac{2(1 - x^2 + y^2)\,dx - 4xy\,dy}{(1 + |z|^2)^2}$$

$$d\eta = \frac{2(1 + x^2 - y^2)\,dy - 4xy\,dx}{(1 + |z|^2)^2}$$

$$d\zeta = \frac{4x\,dx + 4y\,dy}{(1 + |z|^2)^2}.$$

A lengthy, but routine, calculation now shows that

$$ds^2 = \frac{4(dx^2 + dy^2)}{(1 + |z|^2)^2}.$$

The curvature K of a metric (8.1.1) is given by

$$K = -\frac{\Delta \log \lambda}{\lambda^2}$$

(here Δ is the Laplacian, not the unit disc). A calculation shows that

$$K = +1,$$

and that the area of $\mathbb{C} \cup \{\infty\}$ in the metric is

$$\operatorname{Area}(\mathbb{C} \cup \{\infty\}) = \iint_{\mathbb{C}} \lambda(z)^2\,dx\,dy = \int_0^{2\pi} \int_0^{\infty} \frac{4}{(1 + r^2)^2}\,r\,dr\,d\theta = 4\pi.$$

Proposition. *An element* $T \in \operatorname{Aut}(\mathbb{C} \cup \{\infty\})$ *is an isometry in the metric* $(2/(1 + |z|^2))|dz|$ *if and only if as a matrix*

$$T = \begin{bmatrix} a & -\bar{c} \\ c & \bar{a} \end{bmatrix}, \qquad |a|^2 + |c|^2 = 1. \tag{8.2.2}$$

PROOF. Write

$$T = \begin{bmatrix} a & b \\ c & d \end{bmatrix}, \qquad ad - bc = 1.$$

Of course, T is an isometry if and only if

$$\lambda(Tz)|T'(z)| = \lambda(z), \qquad z \in \mathbb{C} \cup \{\infty\},$$

or equivalently if and only if

$$\frac{1}{|cz + d|^2 + |az + b|^2} = \frac{1}{1 + |z|^2}. \tag{8.2.3}$$

If (8.2.2) holds, then so does (8.2.3), as can be shown by the calculation

$$\frac{1}{|cz + \bar{a}|^2 + |az - \bar{c}|^2} = \frac{1}{(cz + \bar{a})(\overline{cz} + a) + (az - \bar{c})(\overline{az} - c)} = \frac{1}{1 + |z|^2}.$$

Conversely, if (8.2.3) holds, we have

$$1 + |z|^2 = |cz + d|^2 + |az + b|^2.$$

Expanding the right-hand side gives

$$1 + |z|^2 = (|a|^2 + |c|^2)|z|^2 + (c\bar{d} + a\bar{b})z + (d\bar{c} + b\bar{a})\bar{z} + |b|^2 + |d|^2,$$

and the only way this identity can hold is if $|a|^2 + |c|^2 = 1$, $|b|^2 + |d|^2 = 1$, $c\bar{d} + a\bar{b} = 0$. We now add to this the determinant condition $-cb + ad = 1$. The last four equations yield $c = -\bar{b}$, $a = \bar{d}$. □

Remark. Note that for T not the identity

$$\text{Trace } T = a + \bar{a} = 2 \text{ Re } a,$$

and (since Re $a = \pm 1$ implies that T is the identity)

$$-2 < \text{Trace } T < 2.$$

Hence T is elliptic. As a matter of fact, T is an arbitrary elliptic element fixing an arbitrary point z *and* $-\bar{z}^{-1}$.

EXERCISE

Determine the geodesics of the metric of constant positive curvature on the sphere.

IV.8.3. The metric $|dz|$ on $\mathbb{C}$ needs little explanation. It is, of course, the Euclidean metric. It has constant curvature zero. Geodesics are the straight lines; and automorphisms of $\mathbb{C}$ of the form

$$z \mapsto e^{i\theta}z + b, \qquad \theta \in \mathbb{R}, b \in \mathbb{C},$$

are isometries in the metric.

IV.8.4. We turn now to the most interesting case: the unit disk Δ. Let us try to define a metric invariant under the full automorphism group Aut Δ. The metric should be conformally equivalent to the Euclidean metric; hence of the form $\lambda(z)|dz|$. Set $\lambda(0) = 2$ and note that $\lambda(A0)|A'(0)| = \lambda(0)$ for all $A \in \text{Aut}\,\Delta$ that fix 0 (since $A(z) = e^{i\theta}z$). Let $z_0 \in \Delta$ be arbitrary and choose $A \in \text{Aut}\,\Delta$ such that $A(0) = z_0$. Define (to have an invariant metric)

$$\lambda(z_0) = 2|A'(0)|^{-1}.$$

Note that $A(z) = (z + z_0)/(1 + \bar{z}_0 z)$ (it suffices to consider only motions of this form). From which we conclude that

$$\lambda(z_0) = \frac{2}{1 - |z_0|^2}.$$

A simple calculation shows that this metric is indeed invariant under Aut Δ. (In particular, the elements of Aut Δ are isometries in this metric.)

Before proceeding let us compute the formula for the metric on the upper half plane U. Since we want the metric $\tilde{\lambda}(z)|dz|$ to be invariant under Aut U, we must require

$$\tilde{\lambda}(z_0) = 2|A'(0)|^{-1},$$

where A is a conformal map of Δ onto U taking 0 onto z_0. A calculation similar to our previous one shows that

$$\tilde{\lambda}(z) = \frac{1}{\operatorname{Im} z}.$$

Computations now show that the metric we have defined has constant curvature -1.

Let $c: I \to \Delta$ be a smooth path. Then the length of this path $l(c)$ is defined by

$$l(c) = \int_0^1 \lambda(c(t))|c'(t)|\,dt. \tag{8.4.1}$$

We can now introduce a distance function on Δ by defining

$$d(a,b) = \inf\{l(c);\ c \text{ is a piecewise smooth path joining } a \text{ to } b \text{ in } \Delta\}. \tag{8.4.2}$$

Take $a = 0$ and $b = x$, real, $0 \leq x < 1$, and compute the length $l(c)$ of an arbitrary piecewise smooth path c joining 0 to x,

$$\begin{aligned} l(c) &= \int_0^1 \frac{2}{1 - |c(t)|^2}\,|c'(t)|\,dt \\ &\geq \left|\int_0^1 \frac{2}{1 - c(t)^2}\,c'(t)\,dt\right| = \left|\int_0^x \frac{2}{1 - u^2}\,du\right| \\ &= \left|\log\frac{1 + x}{1 - x}\right| \geq \log\frac{1 + x}{1 - x}. \end{aligned}$$

We conclude that

$$d(0,x) = \log \frac{1+x}{1-x},$$

and that $c(t) = tx$, $0 \leq t \leq 1$, is the unique geodesic joining 0 to x. We now let $0 \neq b = z \in \Delta$ be arbitrary (and keep $a = 0$). Since

$$\zeta \mapsto e^{-i\theta}\zeta$$

($\theta = \arg z$) is an isometry in the non-Euclidean metric $\lambda(z)|dz|$, we see that

$$d(0,z) = \log \frac{1+|z|}{1-|z|}, \tag{8.4.3}$$

and $c(t) = tz$, $0 \leq t \leq 1$, is the unique geodesic joining 0 to z.

Proposition

a. *The distance induced by the metric $2/(1 - |z|^2)|dz|$ is complete.*
b. *The topology induced by this metric is equivalent to the usual topology on Δ.*
c. *The geodesics in the metric are the arcs of circles orthogonal to the unit circle $\{z \in \mathbb{C};\ |z| = 1\}$.*
d. *Given any two points $a, b \in \Delta$, $a \neq b$, there is a unique geodesic between them.*

PROOF. Note first that (8.4.3) shows that every Euclidean circle with center at the origin is also a non-Euclidean circle and vice-versa. Also note that

$$\lim_{z \to 0} \frac{d(0,z)}{|z|} = 2 = \lambda(0)$$

showing that at the origin the topologies agree. Since there is an isometry taking an arbitrary point $z \in \Delta$ to the origin, the topologies agree everywhere. This establishes (b).

We have already shown that the unique geodesic between 0 and b is the segement of the straight line joining these two points. Now let z_1 and z_2 be arbitrary. Choose $A \in \operatorname{Aut} \Delta$ such that $A(z_1) = 0$. Let $b = A(z_2)$. The geodesic between z_1 and z_2 is $A^{-1}(c)$, where c is the portion of the diameter joining 0 to b. Since A^{-1} preserves circles and angles, (c) and (d) are verified.

Now let $\{z_n\}$ be a Cauchy sequence in the d-metric. Then $\{d(0,z_n)\}$ is bounded. Hence $\{z_n\}$ is also Cauchy in the Euclidean metric and $\{|z_n|\}$ is bounded away from 1. This completes the proof of the proposition.

IV.8.5. If D is any Lebesgue-measurable subset of Δ, then

$$\text{Area } D = \frac{1}{2} \iint_D \lambda(z)^2 |dz \wedge d\bar{z}|$$

defines the non-Euclidean area of D.

The following technical proposition will be needed in the next section.

Proposition. *The area of a non-Euclidean triangle (that is, a triangle whose sides are geodesics) with angles α, β, γ is*

$$\pi - (\alpha + \beta + \gamma).$$

PROOF. It is convenient to work with the upper half plane U instead of the unit disk. We will allow "infinite triangles"—that is, triangles one or more of whose sides are geodesics of infinite length (with a vertex on $\delta U = \mathbb{R} \cup \{\infty\}$). If a triangle has a vertex on δU, it must make a zero degree angle (because the geodesics in U are semi-circles (including straight lines = semi-circle through ∞) perpendicular to $\mathbb{R}$). We call such a vertex a *cusp*.

Case I. The triangle has one cusp at ∞ (see Figure IV.1), and one right angle. The angles of the triangle are hence: $0, \pi/2, \alpha$. (Note we are *not* excluding the possibility that $\alpha = 0$; that is, a cusp at B.) The area of the triangle is then

$$\int_0^a dx \int_{\sqrt{\rho^2 - x^2}}^{\infty} \frac{dy}{y^2} = \int_0^a \frac{dx}{\sqrt{\rho^2 - x^2}} = \operatorname{arc\,sin} \frac{a}{\rho} = \frac{\pi}{2} - \alpha$$
$$= \pi - \left(0 + \frac{\pi}{2} + \alpha\right).$$

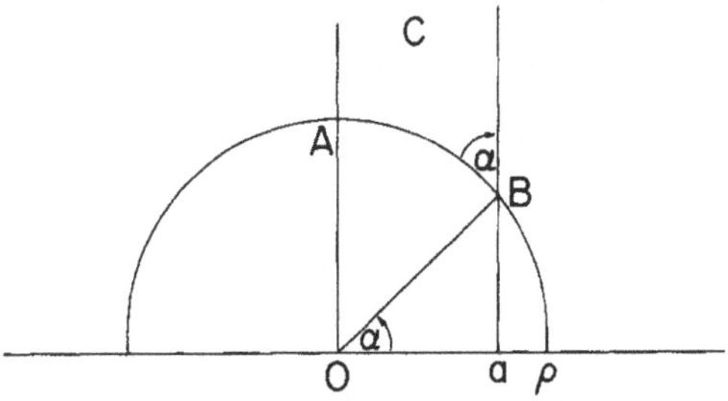

Figure IV.1

Case II. The triangle has a cusp at ∞. (See Figure IV.2.) It has angles $0, \alpha, \beta$. We are not excluding the possibility that either $\alpha = 0$ or $\beta = 0$ or both $\alpha = \beta = 0$. Choose any point D on the arc AB and draw a geodesic "through

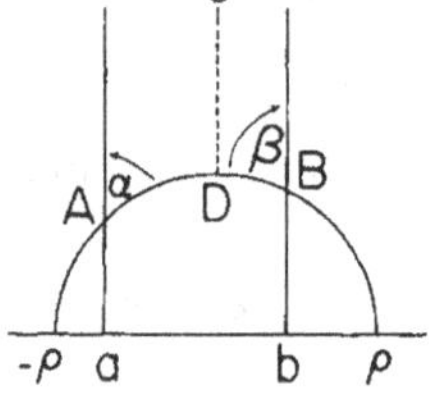

Figure IV.2

D and ∞." The area of the triangle is the sum of the areas of two triangles:

$$\pi - \left(\alpha + \frac{\pi}{2}\right) + \pi - \left(\frac{\pi}{2} + \beta\right) = \pi - \alpha - \beta.$$

The case in which A and B lie on the same side of D is treated similarly.

Case III. If the triangle has a cusp, then it is equivalent under Aut U to the one treated in Case II.

Case IV. The triangle has no cusps. (See Figure IV.3.) Extend the geodesic AB until it intersects $\mathbb{R}$ at D. Join C to D by a non-Euclidean straight line. The area in question is the difference between the areas of two triangles and thus equals

$$[\pi - (\alpha + \gamma + \varepsilon)] - [\pi - (\varepsilon + \eta)] = \eta - (\alpha + \gamma) = \pi - (\alpha + \beta + \gamma). \quad \square$$

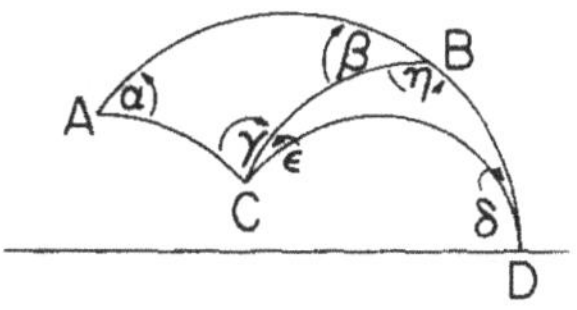

Figure IV.3

IV.8.6. Theorem. *Let M be an arbitrary Riemann surface. We can introduce on M a complete Riemannian metric of constant curvature: the curvature is positive for $M = \mathbb{C} \cup \{\infty\}$, zero if $\mathbb{C}$ is the holomorphic universal covering space of M, and negative otherwise.*

PROOF. Let $\tilde{M}$ be the holomorphic universal covering space of M and G the corresponding group of cover transformations. We have seen that G is a group of isometries in the metric of constant curvature we have introduced. Since the metric is invariant under G, it projects to a metric on M. Since curvature is locally defined, it is again constant on M. (The reader should verify that the curvature is well-defined on M; that is, that it transforms correctly under change of local coordinates.) The (projected) Riemannian metric on M (denoted in terms of the local coordinate z by $\lambda(z)|dz|$) allows us to define length of curves on M by formula (8.4.1) and hence a distance function (again denoted by d) by (8.4.2). Let

$$\pi \colon \tilde{M} \to M$$

be the canonical holomorphic projection. Let $x, y \in M$. We claim that

$$d(x,y) = \inf\{d(\xi,\eta);\ \pi(\xi) = x,\ \pi(\eta) = y\}. \tag{8.6.1}$$

Note that we have shown that in $\tilde{M}$ the infimum in (8.4.2) is always assumed for some smooth curve c. Let d be the infimum in (8.6.1); that is, the right-hand side. Let c be any curve joining ξ to η in $\tilde{M}$ with $\pi(\xi) = x$, $\pi(\eta) = y$. Then

$\pi \circ c$ is a curve in M joining x to y. Clearly, $l(\pi \circ c) = l(c)$. Thus

$$d(x, y) \leq l(\pi \circ c) = l(c),$$

and since c is arbitrary,

$$d(x,y) \leq d(\xi,\eta).$$

Finally, since ξ, η are arbitrary in the preimage of x, y,

$$d(x,y) \leq d.$$

Next, if c is any curve in M joining x to y, it can be lifted to a curve $\tilde{c}$ joining ξ to η ($\pi(\xi) = x$, $\pi(\eta) = y$) of the same length (because π is a local isometry). Thus,

$$l(c) = l(\tilde{c}) \geq d(\xi,\eta) \geq d.$$

Since c is arbitrary, we conclude

$$d(x,y) \geq d.$$

It is now easy to show that the metric d on M is complete. We leave it as an exercise to the reader. □

Remarks

1. If a Riemann surface M carries a Riemannian metric, then the metric can be used to define an element of area dA and a curvature K. The Gauss–Bonnet formula gives

$$\iint_M K\, dA = 2\pi(2 - 2g),$$

if M is compact of genus g. Thus, in particular, the sign of the metric of constant curvature is uniquely determined by the topology of the surface. We will not prove the Gauss–Bonnet theorem in this book. The next section will, however, contain similar results.
2. Let M be an arbitrary Riemann surface with holomorphic universal covering space $\tilde{M}$. On $\tilde{M}$

$$\lambda(z)^2\, dz \wedge d\bar{z}$$

is a nowhere vanishing 2-form that is invariant under the covering group of $\tilde{M} \to M$. Thus the form projects to a nowhere vanishing smooth 2-form on M. (Recall that in I.3.7, we promised to produce such a form.)

IV.9. Discontinuous Groups and Branched Coverings

In this section we discuss the elementary Kleinian groups and establish a general uniformization theorem. We show that a Riemann surface with ramification points can be uniformized by a unique (up to conjugation)

Fuchsian or elementary group. This establishes an equivalence between the theory of Fuchsian (and elementary) groups and the theory of Riemann surfaces with ramification points. We have not always included the complete arguments in this section. The reader should fill in the details where only the outline has been supplied.

IV.9.1. Let M be a Riemann surface. We shall say that M is of *finite type* if there exists a compact Riemann surface $\check{M}$ such that $\check{M}\backslash M$ consists of finitely many points. The *genus* of M is defined as the genus of $\check{M}$ (it is well defined). By a *puncture* on a Riemann surface M, we mean a domain $D_0 \subset M$ with D_0 conformally equivalent to $\{z \in \mathbb{C}; 0 < |z| < 1\}$ and such that every sequence $\{z_n\}$ with $z_n \to 0$ is discrete on M. We shall identify $z = 0$ with the puncture D_0.

Let D be a connected, open, G-invariant subset of $\Omega(G)$, where G is a Kleinian group. We must use a little care in defining a puncture on D/G because we want to view this space as a Riemann surface with ramification points. Regard for the moment D/G as an abstract Riemann surface. It is possible that a puncture on this surface is an accumulation point of branch values of the canonical projection. We will ignore such punctures. Thus we require that a deleted neighborhood of a puncture on D/G consist only of images of ordinary points of D (points with trivial stabilizers). Let $\{x_1, x_2, \ldots\}$ be the set of points that are either punctures on D/G or points $x \in D/G$ with the canonical projection $\pi: D \to D/G$ ramified at $\pi^{-1}(x)$. Let v_j = ramification index of $\pi^{-1}(x_j)$ for such an x_j, and let $v_j = \infty$ for punctures. The set $\{x_j\}$ is at most countable, since the punctures and ramified points each form discrete sets.

Let us now assume that D/G is of finite type. If in addition, π is ramified over only finitely many points, then we shall say that *G is of finite type over D*. In this latter case, we let p be the genus of D/G. We may arrange the sequence x_j so that

$$2 \le v_1 \le v_2 \le \cdots \le v_n \le \infty$$

(here n = cardinality of $\{x_1, x_2, \ldots\} < \infty$), and we call

$$(p; v_1, v_2, \ldots, v_n)$$

the *signature of* G (*with respect to* D).

We also introduce the rational number, called the *characteristic of* G (*with respect to* D),

$$\chi = 2p - 2 + \sum_{j=1}^{n} \left(1 - \frac{1}{v_j}\right),$$

where, as usual, $1/\infty = 0$, whenever G is of finite type over D and $\chi = \infty$, otherwise.

IV.9.2. Definition. Let G be a Kleinian group. Let $D \subset \Omega(G)$ be a G-invariant open set. By a *fundamental domain for G with respect to D* we mean an open subset ω of D such that

i. no two points of ω are equivalent under G,
ii. every point of D is G-equivalent to at least one point of $\mathrm{Cl}\,\omega$,
iii. the relative boundary of ω in D, $\delta\omega$, consists of piecewise analytic arcs, and
iv. for every arc $c \subset \delta\omega$, there is an arc $c' \subset \delta\omega$ and an element $g \in G$ such that $gc = c'$.

By a *fundamental domain for* G we mean a fundamental domain for G with respect to $\Omega(G)$.

The next proposition will only be needed to discuss examples. We will hence only outline its proof.

Proposition. *Every Kleinian group G has a fundamental domain.*

PROOF. Let $\pi:\Omega \to \Omega/G = \bigcup_j S_j$ be the canonical projection. Let Ω_j be a component of $\pi^{-1}(S_j)$, and let G_j be the stabilizer of Ω_j. If ω_j is a fundamental domain for G_j with respect to Ω_j, then $\bigcup_j \omega_j$ is a fundamental domain for G. Thus it suffices to consider only the case where Ω is connected.

We can, therefore, assume that we are given a discontinuous group G of automorphisms of a domain $D \subset \mathbb{C} \cup \{\infty\}$. We now introduce a construction that is very useful in studying *function groups*. (These are precisely the groups that have an invariant component.) Let

$$\rho : \tilde{D} \to D$$

be a holomorphic universal covering map of D and define

$$\Gamma = \{\gamma \in \operatorname{Aut} \tilde{D};\ \rho \circ \gamma = g \circ \rho \text{ for some } g \in G\}.$$

It is now easy to check that $\tilde{D}/\Gamma \cong D/G$ (thus G is discontinuous if and only if Γ is). For a given $\gamma \in \Gamma$ there is just one $g \in G$ with $\rho \circ \gamma = g \circ \rho$. Hence we can define a homomorphism

$$\rho^* : \Gamma \to G$$

that satisfies

$$\rho \circ \gamma = \rho^*(\gamma) \circ \rho, \qquad \gamma \in \Gamma.$$

We let $H = \operatorname{Kernel} \rho^*$. Then

$$\{1\} \to H \hookrightarrow \Gamma \xrightarrow{\rho^*} G \to \{1\}$$

is an exact sequence of groups and group homomorphisms. Now let $\tilde{\omega}$ be a fundamental domain for Γ with respect to $\tilde{D}$. It is quite easy to show that $\omega = \rho(\tilde{\omega})$ is a fundamental domain for G in D. Thus we need only establish fundamental domains for groups acting discontinuously on $\mathbb{C} \cup \{\infty\}$, $\mathbb{C}$, or U. The first case ($\mathbb{C} \cup \{\infty\}$) is trivial (because the group is finite). The second case ($\mathbb{C}$) will be exhausted by listing all the examples. If $\tilde{D} = U$, then we choose a point $z_0 \in U$ with trivial stabilizer and set

$$\omega = \{z \in U;\ d(z,z_0) < d(gz,z_0) \text{ all } g \in G,\ g \neq 1\}.$$

It must now be shown that ω is indeed a fundamental domain. We omit (the non-trivial) details. □

Definition. The group Γ constructed above will be called the *Fuchsian equivalent of G (with respect to D).*

We now take a slight detour and on the way encounter some interesting special cases.

IV.9.3. Let us now classify the discontinuous groups G with $\Omega(G) = \mathbb{C} \cup \{\infty\}$. Since $\mathbb{C} \cup \{\infty\}$ is compact, G is finite (let $N = |G|$) and $M = (\mathbb{C} \cup \{\infty\})/G$ is compact. Let

$$\pi : \mathbb{C} \cup \{\infty\} \to M$$

be the canonical projection. Let $(\gamma; \nu_1, \ldots, \nu_n)$ be the signature of G. Clearly $2 \le \nu_j \le N$ and each ν_j divides N. The Riemann–Hurwitz relation (Theorem I.2.7; see also V.1.3) reads

$$-2 = N(2\gamma - 2) + N \sum_{j=1}^{n} \left(1 - \frac{1}{\nu_j}\right). \tag{9.3.1}$$

In particular, the characteristic of G,

$$\chi = -\frac{2}{N} < 0,$$

is negative. Since $(1 - 1/\nu_j) > 0$ all j, we conclude that $\gamma = 0$. Thus (9.3.1) becomes

$$2 - \frac{2}{N} = \sum_{j=1}^{n} \left(1 - \frac{1}{\nu_j}\right).$$

Assume $N > 1$ (discarding the trivial case). If $N = 2$, then $\nu_j = 2$, all j and $n = 2$. Assume now $N > 2$. Then

$$1 < 2 - \frac{2}{N} < 2$$

and ($n > 0$) since

$$\frac{1}{2} \le 1 - \frac{1}{\nu_j} < 1,$$

we conclude that

$$n > 1, \qquad n < 4.$$

Thus $n = 2$ or $n = 3$. If $n = 2$, then

$$\frac{2}{N} = \frac{1}{\nu_1} + \frac{1}{\nu_2}.$$

Now since $1/\nu_j \ge 1/N$ it must be that $\nu_1 = \nu_2 = N$. If $n = 3$, then

$$1 + \frac{2}{N} = \frac{1}{\nu_1} + \frac{1}{\nu_2} + \frac{1}{\nu_3}. \tag{9.3.2}$$

Observe (again) that

$$1 < 1 + \frac{2}{N} \leq \frac{5}{3}.$$

Thus $\nu_1 = 2$ (recall that $\nu_1 \leq \nu_2 \leq \nu_3$). We replace (9.3.2) by

$$\frac{1}{2} + \frac{2}{N} = \frac{1}{\nu_2} + \frac{1}{\nu_3}.$$

Now, N is even since $\nu_1 = 2$ and ν_1 divides N (thus $N \geq 4$),

$$\frac{1}{2} < \frac{1}{2} + \frac{2}{N} \leq 1.$$

If $\nu_2 = 2$, then ν_3 is (apparently) arbitrary ≥ 2. If $\nu_2 > 2$, then $\nu_2 = 3$ and hence

$$\frac{1}{6} + \frac{2}{N} = \frac{1}{\nu_3}.$$

Thus

$$\frac{1}{6} < \frac{1}{\nu_3} \quad \text{or} \quad 6 > \nu_3.$$

Thus $\nu_3 = 3$, 4 or 5.

We summarize below the signatures and cardinalities of the groups G that could possibly act on $\mathbb{C} \cup \{\infty\}$.

Signature of G	$\lvert G \rvert$
$(0;—)$	1
$(0;\nu,\nu)$	$\nu \ (2 \leq \nu < \infty)$
$(0;2,2,\nu)$	$2\nu \ (2 \leq \nu < \infty)$
$(0;2,3,3)$	12
$(0;2,3,4)$	24
$(0;2,3,5)$	60

Remark. Existence and uniqueness of groups with $\chi < 0$ will be established in IV.9.12. We have also determined all possible negative characteristics of Kleinian groups.

IV.9.4. Before proceeding to the next special case, we must establish the following

Lemma. *Let A and B be two non-parabolic transformations with exactly one fixed point in common, then $C = A^{-1} \circ B \circ A \circ B^{-1}$ is parabolic.*

PROOF. Without loss of generality (by conjugation) we may assume that A fixes 0 and ∞ and B fixes 1 and ∞. Thus $A(z) = K_1 z$ and $B(z) = K_2(z - 1) + 1$,

$K_1 \neq 1 \neq K_2$. A computation shows that

$$C(z) = z + \frac{(K_1 - 1)(K_2 - 1)}{K_1},$$

which is parabolic. □

Corollary. *Let G be a group of Möbius transformations and $z \in \mathbb{C} \cup \{\infty\}$. If G_z, the stabilizer of z in G, is finite, it must be cyclic.*

Exercise

Let M be a Riemann surface whose universal covering space is the unit disk and let $P \in M$. Let G_P be the stabilizer of P in Aut M. For most Riemann surfaces, G_P is finite. Find the exceptional Riemann surfaces; that is, find those M for which G_P is infinite for at least one $P \in M$. Prove that your list is complete. Discuss the same problem for Riemann surfaces M whose universal covering space is the complex plane. (*Hint*: See also V.4.)

IV.9.5. Next we determine the groups G that can possibly act discontinuously on $\mathbb{C}$. Since G leaves $\mathbb{C}$ invariant, $G \subset \text{Aut}\, \mathbb{C}$, and the elements g of G are of the form

$$g : z \mapsto az + b, \qquad a \neq 0. \tag{9.5.1}$$

Observe that a, if $\neq 1$, must be a root of unity, since a is the multiplier of g and g must in this case be elliptic of finite order. Map the element $g \in G$ of (9.5.1) onto $a \in S^1$. This map is a homomorphism since

$$\begin{bmatrix} a & b \\ 0 & 1 \end{bmatrix} \begin{bmatrix} \alpha & \beta \\ 0 & 1 \end{bmatrix} = \begin{bmatrix} a\alpha & a\beta + b \\ 0 & 1 \end{bmatrix}.$$

Let G_0 be the kernel of this homomorphism. Clearly G_0 is the unique maximal fixed point free (normal) subgroup of G.

If G_0 is trivial, then by the preceding lemma, all elements of G must fix a common point $z_0 \in \mathbb{C}$. We conclude that G is a finite cyclic group acting discontinuously on $\mathbb{C} \cup \{\infty\}$. We will not consider such groups to be acting on $\mathbb{C}$.

We consider now the case with G_0 non-trivial. It is then easily seen that G_0 is a free abelian group on 1 or 2 generators. In studying groups of Möbius transformations, we are always permitted to conjugate the group. Conjugation does not change the Riemann surfaces represented by the group. In the case under consideration, we may conjugate G by elements of Aut $\mathbb{C}$ (rather than elements of $\text{Aut}(\mathbb{C} \cup \{\infty\}) = PSL(2,\mathbb{C})$).

Assume that G_0 is cyclic. Then (without loss of generality G_0 is generated by $z \mapsto z + 1$) $\mathbb{C}/G_0$ is equivalent to the punctured plane $\mathbb{C}^*$ (via the map $z \mapsto e^{2\pi i z}$), and G/G_0 acts as a group of conformal automorphisms of $\mathbb{C}/G_0$.

The automorphisms of $\mathbb{C}^*$ are of the form $z \mapsto kz$ and $z \mapsto k/z$ (see V.4.3). The former lift to automorphisms of $\mathbb{C}$ of the form $(z \mapsto z + (\log k)/2\pi i)$ and must therefore be in G_0; the latter, to automorphisms of the form $z \mapsto -z + (\log k)/2\pi i$ whose squares must be in G_0. Thus G/G_0 is trivial or isomorphic to $\mathbb{Z}_2$. In the second case, G consists of mappings of the form $z \mapsto \pm z + n$, $n \in \mathbb{Z}$.

We "draw" fundamental domains for the groups encountered above.

G generated by $z \mapsto z + 1$ (Figure IV.4)

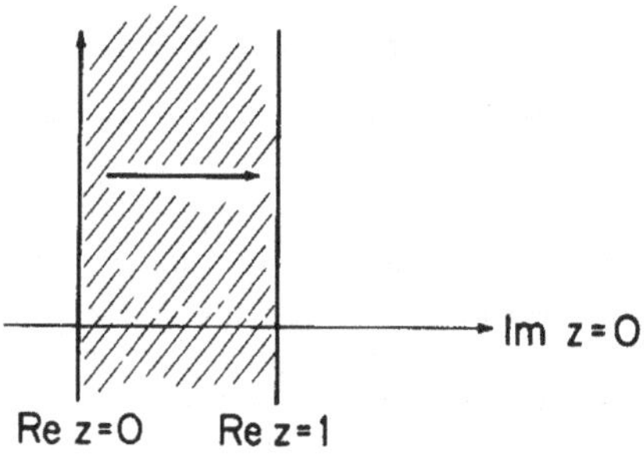

Figure IV.4. Signature $(0;\infty,\infty)$.

G generated by $z \mapsto -z$, $z \mapsto z + 1$ (Figure IV.5)

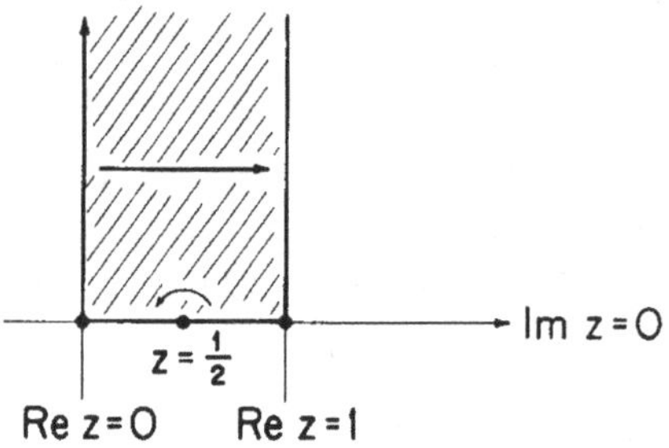

Figure IV.5. Signature $(0;2,2,\infty)$ where the first two twos correspond to the points $z = 0$ and $z = \frac{1}{2}$.

We now consider the case where G_0 has rank 2. Then we may assume that G_0 is generated by the two translations $z \mapsto z + 1$, $z \mapsto z + \tau$, $\operatorname{Im} \tau > 0$, and $\mathbb{C}/G_0$ is a torus. Without loss of generality, we may assume $|\tau| \geq 1$. Again G/G_0 acts as a group of automorphisms of $\mathbb{C}/G_0$.

Assume now that G/G_0 is non-trivial. Let $g: z \mapsto z + 1$ and $h: z \mapsto az + b$ be parabolic and elliptic elements of G. We conclude that

$$h \circ g \circ h^{-1}: z \mapsto z + a$$

is an element of G_0, or that the multipliers of elements of G are periods of G_0. Since there are only finitely many periods on the unit circle, we can find a "primitive" multiplier $K = e^{2\pi i/\mu}$ with $\mu \in \mathbb{Z}$, $\mu > 1$ and μ maximal. By conjugating G once again (by an element $z \mapsto z + c$) by a conjugation that does not destroy the previous normalization, we may assume that G contains the element $z \mapsto Kz$. Hence, we see that G/G_0 is a finite cyclic group $\cong \mathbb{Z}_\mu$. The group G hence consists of motions of the form

$$z \mapsto K^\nu z + n + m\tau, \qquad \nu = 0, \ldots, \mu - 1, n \in \mathbb{Z}, m \in \mathbb{Z}.$$

Now we let γ be the genus of $\mathbb{C}/G(=(\mathbb{C}/G_0)/(G/G_0))$. Let $x_1 = 0, x_2, \ldots, x_r$ be the fixed points of G/G_0 on $\mathbb{C}/G_0$ and let $\nu_1, \ldots, \nu_r$ be the orders of the respective stability subgroups. We use Riemann–Hurwitz (see V(1.3.1))

$$0 = 2\gamma - 2 + \sum_{j=1}^{r} \left(1 - \frac{1}{\nu_j}\right), \tag{9.5.2}$$

and conclude that if $r > 0$ (G/G_0 non-trivial) then $\gamma = 0$. Thus (9.5.2) becomes

$$2 = \sum_{j=1}^{r} \left(1 - \frac{1}{\nu_j}\right).$$

Let N be the order of G/G_0. Then

$$2 \le \nu_j \le N.$$

Since $1 > (1 - 1/\nu_j) \ge \frac{1}{2}$, $r = 3$ or 4. If $r = 4$, then $\nu_1 = \nu_2 = \nu_3 = \nu_4 = 2$, and the ramification is completely accounted for by the element $z \mapsto -z$, and $G/G_0 \cong \mathbb{Z}_2$.

Fundamental domain for G consisting of $\{z \mapsto z + n + m\tau\}$ (Figure IV.6)

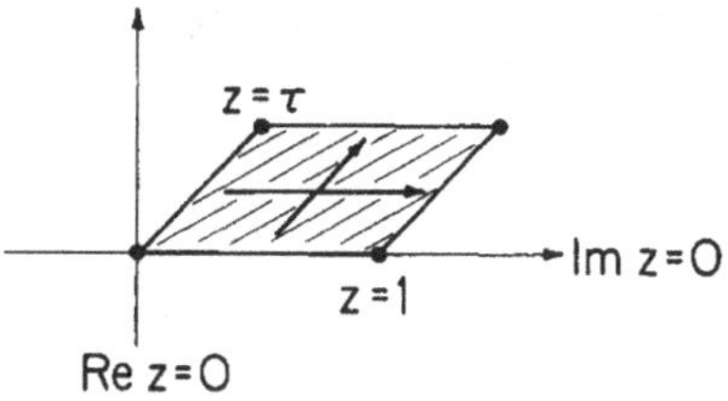

Figure IV.6. Signature (1;–).

Fundamental domain for G consisting of $\{z \mapsto \pm z + n + m\tau\}$ (Figure IV.7)

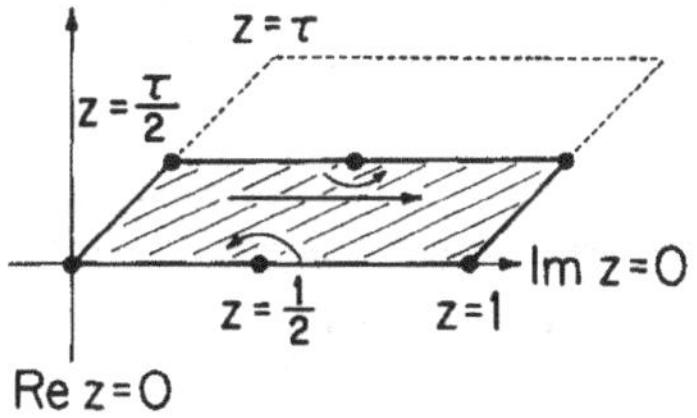

Figure IV.7. Signature (0;2,2,2,2) where the twos correspond to the points $z = 0$, $z = \frac{1}{2}$, $z = \tau/2$, $z = (\tau + 1)/2$.

If $r = 3$, then the only possibilities are as follows (because $1 = 1/\nu_1 + 1/\nu_2 + 1/\nu_3$):

ν_1	ν_2	ν_3	G/G_0
2	3	6	$\mathbb{Z}_6$
2	4	4	$\mathbb{Z}_4$
3	3	3	$\mathbb{Z}_3$

The computation of the ν_j appearing in the above table is routine. The description of G/G_0 requires some explanation (that is, the determination of μ). Let $z \mapsto Kz$ be the generator of G/G_0, which is also the generator of the stability subgroup of 0. Since K is a period, there are integers n, m such that

$$K = n + m\tau.$$

Now $|K| = 1$, $|\tau| \geq 1$, implies either $n = 0$ or $m = 0$. If $m = 0$, then $n = -1 = K$. Thus G is the group of signature (0;2,2,2,2) which we previously discussed. If $n = 0$, then $m = \pm 1$. Since Im $\tau > 0$ and Im $K \geq 0$, we conclude that $m = 1$ and $K = \tau$ (also $\mu > 2$). Since $z \mapsto Kz$ generates the stabilizer of 0, $\mu = \nu_3$.

Fundamental domain for $\{z \mapsto K^\nu z + n + mK\}$, $K = e^{2\pi i/6}$ (Figure IV.8)

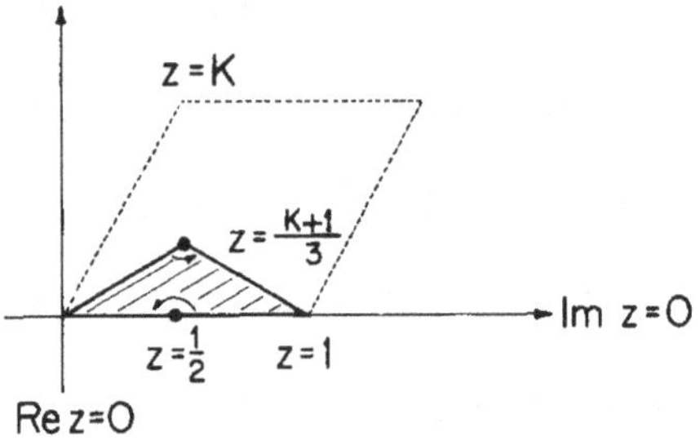

Figure IV.8. Signature (0;2,3,6) where the two corresponds to the point $z = \frac{1}{2}$, the three corresponds to the point $z = (K + 1)/3$ and the six corresponds to the point $z = 0$.

Fundamental domain for $\{z \mapsto K^\nu z + n + mK\}$ $K = e^{2\pi i/4} = i$ (Figure IV.9)

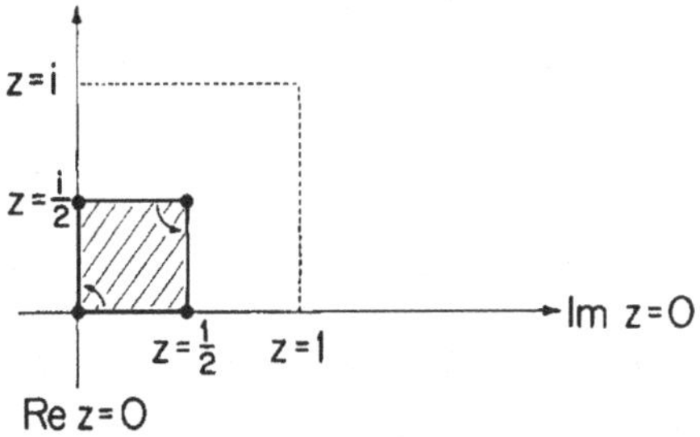

Figure IV.9. Signature (0;2,4,4) where the two corresponds to $z = \frac{1}{2}$, one four corresponds to $z = (1 + i)/2$ and the last four corresponds to $z = 0$.

Fundamental domain for $\{z \mapsto K^\nu z + n + mK\}$, $K = e^{2\pi i/3}$ (Figure IV.10)

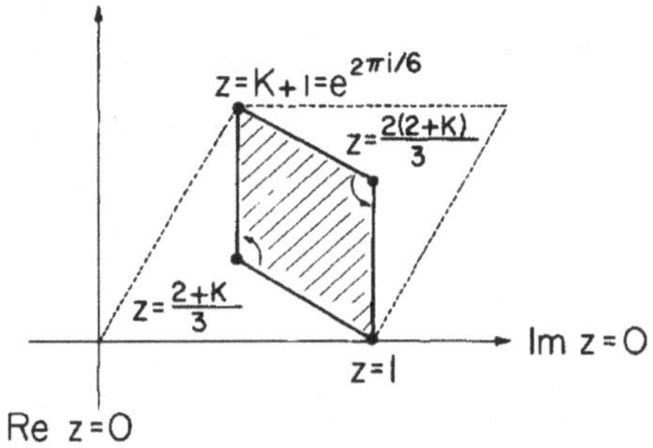

Figure IV.10. Signature (0;3,3,3) where the threes correspond to the points $z = 0$, $z = (K + 2)/3$, $z = (2K + 1)/3$. (Each of these three points is equivalent to a corresponding point in the figure.)

Remark. We have established above the existence and uniqueness of all the elementary groups with one limit point. We have also determined all possible signatures that yield zero characteristic.

IV.9.6. To classify the remaining elementary groups, we must study the groups with two limit points; without loss of generality the limit points are 0 and ∞. The elements of such a group G are of the form

$$z \mapsto kz \quad \text{and} \quad z \mapsto k/z.$$

By passing to the Fuchsian equivalent of the group G, it is easy to classify these groups. The details are left to the reader.

IV.9.7. Let G be a Kleinian group. We have seen that $\chi < 0$ (respectively, $\chi = 0$) whenever $\Omega(G) = \mathbb{C} \cup \{\infty\}$ (respectively, $\Omega(G) = \mathbb{C}$). We want to show now that for Fuchsian groups $\chi > 0$. In addition, we want to interpret the characteristic in geometric terms. We need some preliminaries.

Lemma. *Let G be a Fuchsian group operating on the upper half plane U. Assume that P is a puncture on U/G (and that the natural projection π is unramified over a deleted neighborhood of P). Then there exists a deleted neighborhood D of P in U/G, a disc $\tilde{D}$ in U, and a parabolic element $T \in G$ such that $D \cong \tilde{D}/G_0$, where G_0 is the cyclic group generated by T, and two points in $\tilde{D}$ are equivalent under G if and only if they are equivalent under G_0.*

Proof. Let $D \cup \{P\}$ be a simply connected neighborhood of the puncture P such that π is unramified over D. Let $\tilde{D}$ be a connected component of $\pi^{-1}(D)$ and let G_0 be the stabilizer of $\tilde{D}$ in G;

$$G_0 = \{g \in G; g\tilde{D} = \tilde{D}\}.$$

We must, of course, have that $\tilde{D}/G_0 = D$. Let $\tilde{\tilde{D}}$ be the holomorphic universal covering space of $\tilde{D}$ and let $\tilde{G}_0$ be the Fuchsian model of G_0 with respect to $\tilde{D}$. Then $\tilde{G}_0$ is isomorphic to the fundamental group of D and thus infinite cyclic. Hence G_0 (being a homomorphic image of $\tilde{G}_0$) is also cyclic. It cannot be generated by an elliptic element. If G_0 were generated by an elliptic element, the fixed point of its generator would have to be on the extended real axis. But an elliptic element with a fixed point on $\mathbb{R} \cup \{\infty\}$ cannot fix U. Thus G_0 is generated by a parabolic or hyperbolic element T. Further, $\tilde{D}$ must be simply connected since only a disk can cover non-trivially a punctured disk. We clearly have an inclusion

$$\tilde{D}/G_0 \hookrightarrow U/G_0.$$

If T were hyperbolic, then U/G_0 would be an annulus that contains the punctured disc $\tilde{D}/G_0$. Thus the puncture P on $\tilde{D}/G_0$ would be a point in U/G_0 and hence would be the image of a point $z \in U$ under the map $\pi_0: U \to U/G_0$. We conclude that $\tilde{D}$ must contain many punctured discs. But G_0 is isomorphic to $\pi_1(D)$ which is mapped isomorphically into $\pi_1(U/G_0)$ by the inclusion of $\tilde{D}/G_0$ into U/G_0. We have obtained an obvious contradiction. Thus T is parabolic. By conjugating G in the group of real Möbius transformations, we may assume that $T(z) = z + 1$. By shrinking D, we may assume that $\tilde{D} = \{z \in \mathbb{C}; \operatorname{Im} z > b\}$ for some $b > 0$. □

Definition. *We shall call $\tilde{D}$, a disc (half plane) corresponding to the puncture P.*

IV.9.8. To establish the converse of the above lemma, and to show that D can always be chosen to be the half plane $\{z \in \mathbb{C}; \operatorname{Im} z > 1\}$ we prove the following

Lemma. *Let $A(z) = (az + b)(cz + d)^{-1}$ with $a, b, c, d \in \mathbb{C}$ and $ad - bc = 1$. Let $T(z) = z + 1$. Then the group generated by A and T is not discrete if $0 < |c| < 1$.*

PROOF. Let $A_0 = A$ and for $n \in \mathbb{Z}$, $n \geq 0$, set

$$A_{n+1} = A_n \circ T \circ A_n^{-1}.$$

We show that if $0 < |c| < 1$, then

$$\lim_{n\to\infty} A_n = T.$$

We write as matrices in $SL(2,\mathbb{C})$,

$$A_n = \begin{bmatrix} a_n & b_n \\ c_n & d_n \end{bmatrix},$$

and compute

$$A_{n+1} = \begin{bmatrix} a_n & b_n \\ c_n & d_n \end{bmatrix}\begin{bmatrix} 1 & 1 \\ 0 & 1 \end{bmatrix}\begin{bmatrix} d_n & -b_n \\ -c_n & a_n \end{bmatrix}$$
$$= \begin{bmatrix} 1 - a_n c_n & a_n^2 \\ -c_n^2 & 1 + a_n c_n \end{bmatrix} = \begin{bmatrix} a_{n+1} & b_{n+1} \\ c_{n+1} & d_{n+1} \end{bmatrix}.$$

We conclude that

$$c_n = -c^{2^n} \quad \text{and} \quad \lim_{n\to\infty} c_n = 0 \tag{9.8.1}$$

(in particular, the sequence $\{A_n\}$ is distinct). Now we choose γ so that $0 < (1 - |c|)^{-1} < \gamma$ and $|a| < \gamma$. By choice, $|a_0| = |a| < \gamma$. Suppose $|a_n| < \gamma$ for some n, then also

$$|a_{n+1}| = |1 - a_n c_n| \leq 1 + |a_n|\,|c_n| = 1 + |a_n|\,|c|^{2^n}$$
$$< 1 + \gamma|c|^{2^n} < 1 + \gamma|c| < \gamma.$$

Hence, by induction

$$|a_n| < \gamma \quad \text{for all } n.$$

From the relation

$$a_{n+1} = 1 - a_n c_n$$

and (9.8.1), we conclude that

$$\lim_{n\to\infty} a_n = 1.$$

Thus also

$$\lim_{n\to\infty} b_n = \lim_{n\to\infty} a_{n-1}^2 = 1,$$

and

$$\lim_{n\to\infty} d_n = \lim_{n\to\infty} (1 + a_{n-1}c_{n-1}) = 1. \qquad \square$$

IV.9.9. Lemma. *Let G be a Fuchsian group operating on the upper half plane U. Assume that G contains a parabolic element T with fixed point $x \in \mathbb{R} \cup \{\infty\}$. Then G_x, the stabilizer of x in G is infinite cyclic.*

Proof. We claim that every element $A \in G_x$ is parabolic. Without loss of generality $x = \infty$. Assume that G_∞ contains an element A with another fixed point which may be taken to be the point 1 (since it must be on the real axis because A cannot be elliptic). Thus we have $T(z) = z + a$, $a \in \mathbb{R}$, and $A(z) = \alpha(z - 1) + 1$, $\alpha \in \mathbb{R}$, $\alpha > 0$. Without loss of generality we may take $0 < \alpha < 1$ (replace A by A^{-1}, if necessary). Now $A^n(z) = \alpha^n(z - 1) + 1$, and hence

$$A_n \circ T \circ A_n^{-1}(z) = z + a\alpha^n.$$

Thus we have constructed a distinct sequence in G approaching the identity. We have shown that every element of G_∞ is of the form $z \mapsto z + a$. Taking, the minimum of all such positive a's, we obtain a generator for G_∞. $\square$

IV.9.10. Theorem. *Let G be a Fuchsian group operating on the upper half plane U. Let $T(z) = z + 1$ be an element of G. Assume that T is the generator of the stabilizer of ∞, G_∞. If $A \in G \backslash G_\infty$, then*

$$AU_1 \cap U_1 = \varnothing,$$

where $U_1 = \{z \in \mathbb{C};\ \operatorname{Im} z > 1\}$.

Proof. Let $A(z) = (az + b)/(cz + d)$ with $a, b, c, d \in \mathbb{R}$ such that $ad - bc = 1$. Lemma IV.9.9 shows that $c \neq 0$ and we may assume $c > 0$. Lemma IV.9.8 yields that $c \geq 1$. Now AU_1 is a disc whose boundary is a circle tangent to $\mathbb{R}$ at $A(\infty) = a/c \neq \infty$. It is clear that the diameter of this circle is the maximum of

$$\left|\frac{a(x + i) + b}{c(x + i) + d} - \frac{a}{c}\right|, \qquad x \in \mathbb{R},$$

which equals

$$\left|\frac{-1}{c(cx + d + ci)}\right| = \frac{1}{c}\frac{1}{\sqrt{(cx + d)^2 + c^2}}.$$

Thus the maximum occurs at $x = -d/c$, and the diameter of the circle is $1/c^2 \leq 1$. Hence AU_1 is contained in a disc bounded by a circle tangent to $\mathbb{R}$ of diameter ≤ 1. $\square$

Corollary 1. *Let G be a Fuchsian group operating on U. If G contains a parabolic element, then U/G contains a puncture. Furthermore, the punctures on*

U/G are in one-to-one correspondence with the conjugacy classes of parabolic elements in G.

Corollary 2. *Let G be a Fuchsian group. If G is the covering group of a compact Riemann surface of genus $p \geq 2$, then G contains only hyperbolic elements.*

Remark. For the second part of Corollary 1, we must recall the discussion of IV.9.1. The punctures on U/G, viewed as a "Riemann surface with ramification points", are in one-to-one correspondence with the conjugacy classes of parabolic elements of G. By a puncture on such a surface, we mean a puncture D on the abstract surface U/G such that $\pi: U \to U/G$ is unramified over D (that is, there are no elliptic fixed points in $\pi^{-1}(D)$). We shall see (Theorem IV.9.12) that without this assumption we can have punctures on the abstract surface U/G that are limits of points over which π is ramified.

IV.9.11. We consider now a Fuchsian group G operating on the upper half plane U (or the unit disc Δ). In IV.8, we introduced the Poincaré metric

$$\lambda_U(z)|dz| = \frac{2}{|z - \bar{z}|}|dz|$$

for U which is clearly invariant under G (since $G \subset \operatorname{Aut} U$). Thus we may project $\lambda_U(z)|dz|$ to U/G and obtain a metric with singularities at the images of the elliptic fixed points. We want to show that these singularities are not too bad in the sense that the Poincare metric on U/G is locally square integrable at these points and that the same is true even in a deleted neighborhood of a puncture. Let

$$\pi: U \to U/G$$

be the canonical projection. Denote the Poincaré metric on U/G by $\tilde{\lambda}(Z)|dZ|$ in terms of the local coordinate Z. If Z can be expressed as a function of $z \in U$, then of course, the relation

$$\tilde{\lambda}(Z)|dZ| = \lambda_U(z)|dz|$$

allows us to solve for $\tilde{\lambda}(Z)$.

Let us assume that $z_0 \in U$ is an elliptic fixed point. By conjugation we may replace U by Δ and assume that $z_0 = 0$. The stabilizer G_0 is then generated by an element of the form $z \mapsto e^{2\pi i/\nu}z$, for some $\nu \in \mathbb{Z}$, $\nu \geq 2$. Thus $Z = z^\nu$ is a local coordinate at $\pi(0)$. Since

$$\lambda_\Delta(z) = \frac{2}{(1 - |z|^2)},$$

we see that

$$\tilde{\lambda}(Z) = \frac{2|Z|^{(1/\nu)-1}}{\nu(1 - |Z|^{2/\nu})}.$$

Since $(1/\nu) - 1 > -1$, $\tilde{\lambda}(Z)^2|dZ \wedge d\bar{Z}|$ is integrable in a sufficiently small neighborhood of $Z = 0$.

It remains to investigate a neighborhood of a puncture. Here it is more convenient to assume that G acts on U, and that the puncture corresponds to the cyclic subgroup of G generated by $z \mapsto z + 1$. Thus $Z = e^{2\pi i z}$ is a local coordinate on U/G vanishing at the puncture, and

$$\tilde{\lambda}(Z) = \frac{1}{|Z| \log\left|\frac{1}{Z}\right|},$$

which is readily seen to be square integrable.

We are thus able to define

$$\text{Area}(U/G) = \iint_{U/G} \tilde{\lambda}^2(Z)\left|\frac{dZ \wedge d\bar{Z}}{2}\right|.$$

Theorem. *Let G be a Fuchsian group operating on U. Assume that G is of finite type over U with characteristic χ, then*

$$\text{Area}(U/G) = 2\pi\chi.$$

PROOF. Consider the natural projection $\pi: U \to U/G$. Triangulate the compact surface $\overline{U/G}$ so that

i. each ramified point is a vertex of the triangulation (however, there may (and as a result of (ii) there must) be other vertices),
ii. every triangle of the triangulation contains at most one ramified point,
iii. if a triangle Δ contains no ramified points, then there exists a connected neighborhood $\tilde{\Delta}$ in U so that $\pi(\tilde{\Delta}) \supset \Delta$ and $\pi|\tilde{\Delta}$ is one-to-one,
iv. if a triangle Δ contains a finite ramified point with ramification number ν, then there exists a connected neighborhood $\tilde{\Delta}$ in U so that $\pi(\tilde{\Delta}) \supset \Delta$ and $\pi|\tilde{\Delta}$ is ν-to-one, and
v. if a triangle Δ contains an infinite ramification point, then each component of $\pi^{-1}(\Delta)$ is contained in a half plane corresponding to the puncture.

Having constructed a triangulation as above, it is easy to replace it by another triangulation in which each triangle is the image under π of a non-Euclidean triangle in U. Let

$$(p; \nu_1, \ldots, \nu_n)$$

be the signature of G with respect to U. Let us assume that in the above triangulation we have c_j j-simplices ($j = 0,1,2$). Let α_k, β_k, γ_k be the three angles of the kth triangle, $k = 1, \ldots, c_2$. Then

$$\begin{aligned}\text{Area}(U/G) &= \sum_{k=1}^{c_2} (\pi - \alpha_k - \beta_k - \gamma_k)\\ &= \pi c_2 - \sum_{k=1}^{c_0} \delta_k,\end{aligned}$$

where δ_k = sum of the angles at the kth vertex.

Now $\delta_k = 2\pi$, if the kth vertex is an unramified point,

$= 2\pi/\nu$, if the kth vertex is a point of finite ramification number ν, and

$= 0$, if the kth vertex is an infinite ramification point.

Hence

$$\sum_{k=1}^{c_0} \delta_k = 2\pi(c_0 - n) + \sum_{k=1}^{n} 2\pi/\nu_k.$$

Thus we see that

$$\text{Area}(U/G) = 2\pi\left(\frac{1}{2}c_2 - c_0\right) + 2\pi \sum_{k=1}^{n}\left(1 - \frac{1}{\nu_k}\right).$$

Since each edge in the triangulation appears in exactly two triangles, we have

$$3c_2 = 2c_1,$$

and hence (using the fact that for a compact surface of genus p, $c_2 - c_1 + c_0 = 2 - 2p$)

$$\text{Area}(U/G) = 2\pi\left(2p - 2 + \sum_{k=1}^{n}\left(1 - \frac{1}{\nu_k}\right)\right). \qquad \square$$

Corollary. *Let G be a group of conformal automorphisms of a Riemann surface D. If D is simply connected and G operates discontinuously on D, it is possible to define the characteristic χ of G with respect to D. Furthermore:*

i. $\chi < 0$ *if and only if* $D \cong \mathbb{C} \cup \{\infty\}$,
ii. $\chi = 0$ *if and only if* $D \cong \mathbb{C}$, *and*
iii. $\chi > 0$ *if and only if* $D \cong \Delta$.

IV.9.12. Theorem. *Let M be a Riemann surface and $\{x_1, x_2, \ldots\}$ a discrete sequence on M. To each point x_k we assign the symbol ν_k which is an integer ≥ 2 or ∞. If $M = \mathbb{C} \cup \{\infty\}$ we exclude two cases:*

i. $\{x_1, x_2, \ldots\}$ *consists of one point and* $\nu_1 \neq \infty$, *and*
ii. $\{x_1, x_2, \ldots\}$ *consists of two points and* $\nu_1 \neq \nu_2$.

Let $M' = M - \bigcup_{\nu_k = \infty}\{x_k\}$, $M'' = M \backslash \bigcup_k \{x_k\}$. Then there exists a simply connected Riemann surface $\tilde{M}$, a Kleinian group G of self mappings of $\tilde{M}$ such that

a. $\tilde{M}/G \cong M'$, $\tilde{M}_G/G \cong M''$, *where $\tilde{M}_G$ is $\tilde{M}$ with the fixed points of the elliptic elements of G deleted, and*
b. *the natural projection $\pi: \tilde{M} \to M'$ is unramified except over the points x_k with $\nu_k < \infty$ where $b_\pi(\tilde{x}) = \nu_k - 1$ for all $\tilde{x} \in \pi^{-1}(\{x_k\})$.*

Further, G is uniquely determined up to conjugation in the full group of automorphisms of $\tilde{M}$. The conformality type of $\tilde{M}$ is uniquely determined by the characteristic of the data: that is, by the genus p of M and the sequence *of integers* $\{\nu_1,\nu_2,\ldots\}$. *(If we set* $\chi = 2p - 2 + \sum_j(1 - 1/\nu_j)$, *then*

$$\chi < 0 \quad \text{if and only if} \quad \tilde{M} \cong \mathbb{C} \cup \{\infty\},$$
$$\chi = 0 \quad \text{if and only if} \quad \tilde{M} \cong \mathbb{C}, \text{ and}$$
$$\chi > 0 \quad \text{if and only if} \quad \tilde{M} \cong U.)$$

PROOF. Let $\pi_1(M'') = \pi_1(M'',b)$ be the fundamental group of M'' with base point $b \in M''$. Let $[\alpha]$ denote the homotopy equivalence class of the closed loop α on M'' beginning at b. Let α_k^ν denote a closed curve beginning at b extending to a point x near x_k, (by x being near to x_k we mean that a closed disc D_k around x_k that contains x and does not contain any x_l with $l \neq k$) winding around x_k exactly ν times in D_k, and then returning from x to b along the original path.

Let Γ_0 be the smallest subgroup of $\pi_1(M'')$ containing $\{[\alpha_k^{\nu_k}]; \nu_k < \infty\}$ and let Γ be the smallest normal subgroup of $\pi_1(M'')$ containing Γ_0. Observe that for any k, $[\alpha_k^l] \notin \Gamma$ for $l \in \mathbb{Z}$ with $0 < l < \nu_k$. The normal subgroup Γ of $\pi_1(M'')$ defines, of course, an unbounded, unramified, regular (Galois) covering $(\tilde{M}'',\pi)$ of M''. We recall the (familiar) topological construction of $(\tilde{M}'',\pi)$. Consider the set $\mathscr{C}$ of curves on M'' beginning at b. Two such curves α, β are called equivalent $(\sim)$ provided both have the same end point and $[\alpha\beta^{-1}] \in \Gamma$. The surface $\tilde{M}''$ is defined as $\mathscr{C}/\sim$, and the mapping π sends a curve α into its end point. Choose $\tilde{b} \in \tilde{M}''$ such that $\pi(\tilde{b}) = b$. Then $\Gamma \cong \pi_1(\tilde{M}'',\tilde{b})$. A closed curve α on M'' beginning at b lifts to a closed curve $\tilde{\alpha}$ on $\tilde{M}''$ beginning at $\tilde{b}$ if and only if $\alpha \in \Gamma$. The topology and complex structure of $\mathscr{C}/\sim$ are defined so that the natural projection $\pi:\mathscr{C}/\sim \;\to M''$ is holomorphic

Choose a point x_k with $\nu_k < \infty$. Let D_k be a small deleted disc around x_k. Consider a connected component $\tilde{D}_k$ of $\pi^{-1}(D_k)$. Choose a point $\tilde{x} \in \tilde{D}_k$. We may assume that $\tilde{x}$ corresponds to a curve c from b to $x \in D_k$. Let d be a closed loop around x_k beginning and ending at x. Then $\pi(cd^\nu) = x$, and $[cd^\nu]$, $\nu = 0,\ldots,\nu_k - 1$, represent ν_k distinct points in the same component of $\pi^{-1}(D_k)$. Thus $\pi|\tilde{D}_k$ is at least ν_k-to-one. Since any closed path in D_k is homotopic to d^ν for some ν, it is not too hard to see that by choosing D_k sufficiently small $\pi|\tilde{D}_k$ is precisely ν_k-to-one. In particular, the lift $\tilde{\alpha}_k$ of α_k consists of a path from $\tilde{b}$ to a point $\tilde{x}$ with $\pi(\tilde{x})$ near x_k and then a homotopically nontrivial loop beginning and ending at $\tilde{x}$, and finally a path from $\tilde{x}$ back to $\tilde{b}$ (the original path in reverse direction). We want to show next that we can fill in a puncture in $\tilde{D}_k$ over x_k.

Let G be the covering group of π; that is, G is the group of conformal automorphisms g of $\tilde{M}''$ such that $\pi \circ g = \pi$. Since Γ is a normal subgroup of $\pi_1(M'')$, $G \cong \pi_1(M'')/\Gamma$, and $\tilde{M}''/G \cong M''$ (and G is transitive and acts fixed point freely on $\tilde{M}''$). Let $\tilde{G}$ be the Fuchsian equivalent of G with respect to $\tilde{M}''$. We assume that $\tilde{G}$ acts on the upper half plane U (the cases where $\tilde{G}$ acts on the plane or the sphere are left to the reader). Let

$$\rho: U \to \tilde{M}'' \quad \text{and} \quad \rho^*: \tilde{G} \to G$$

be the holomorphic covering map and the associated homomorphism of IV.9.3. Then $U/(\text{Kernel } \rho^*) \cong \tilde{M}''$ and $U/\tilde{G} \cong M''$. Every "puncture" x_k on M'' with $\nu_k < \infty$ is determined by a parabolic element of $\tilde{G}$ by Lemma IV.9.6.

Recall the exact sequence

$$\{1\} \hookrightarrow (\text{Kernel } \rho^*) \hookrightarrow \tilde{G} \xrightarrow{\rho^*} G \to \{1\}.$$

We may assume that the half plane $U_1 = \{z \in \mathbb{C};\ \operatorname{Im} z > 1\}$ gets mapped by $\pi \circ \rho$ onto the deleted neighborhood D_k of the puncture x_k, and that the corresponding parabolic subgroup of $\tilde{G}$ is generated by $z \mapsto z + 1$. The intersection of the subgroup with the kernel of ρ^* is a cyclic subgroup. We claim that it must be the cyclic subgroup generated by $z \mapsto z + \nu_k$. It is clear that $\rho(U_1)$ is a component $\tilde{D}_k$ of $\pi^{-1}(D_k)$. Since $\rho(U_1)$ is conformally equivalent to U_1 modulo a parabolic subgroup, $\tilde{D}_k$ is indeed a punctured disc.

Let $\tilde{M}$ be $\tilde{M}''$ together with all the punctures on the punctured discs $\tilde{D}_k$ corresponding to the punctured discs D_k corresponding to the punctures x_k with $\nu_k < \infty$. We show now that $\tilde{M}$ is simply connected. Let $\tilde{\alpha}$ be a closed path on $\tilde{M}$. We must show that $\tilde{\alpha}$ is homotopic to a point. It clearly involves no loss of generality to assume that $\tilde{\alpha}$ begins at $\tilde{b}$ and that $\tilde{\alpha}$ is actually a closed path in $\tilde{M}''$. Now $\alpha = \pi(\tilde{\alpha})$ is a closed path on M'' beginning at b, and $[\alpha] \in \Gamma$. Thus it suffices to show that for every closed path α on M'' that begins at b with $[\alpha] \in \Gamma$, the lift $\tilde{\alpha}$ of α beginning at $\tilde{b}$ can be contracted to a point on $\tilde{M}$. Now Γ is the subgroup of $\pi_1(M'')$ generated by $\{\gamma \Gamma_0 \gamma^{-1}\}$ with $\gamma \in \pi_1(M'')$. Thus it suffices to show that every generator α_k of Γ_0 lifts to a curve $\tilde{\alpha}_k$ in $\tilde{M}''$ that is contractible on $\tilde{M}$. But this is entirely obvious.

The elements of G trivially extend to be conformal self maps of the simply connected Riemann surface $\tilde{M}$. From now on we may assume that $\tilde{M}$ is either the sphere, plane, or upper half plane and that G is a Kleinian group. Thus the stabilizer of each puncture is a cyclic subgroup of finite order.

It remains to verify uniqueness. Let G and G_1 be two groups of conformal mappings of the same simply connected Riemann surface U with canonical projections π and π_1 onto the same Riemann surface M such that for each $x \in M$, $|G_{\pi^{-1}(x)}| = |(G_1)_{\pi_1^{-1}(x)}|$, then there exists a conformal mapping $A: U \to U$ such that

$$\begin{array}{ccc} U & \xrightarrow{A} & U \\ & {\scriptstyle\pi}\searrow \quad \swarrow{\scriptstyle\pi_1} & \\ & M & \end{array}$$

commutes. Let

$$\begin{aligned} M' &= \{x \in M;\ |G_{\pi^{-1}(x)}| = 1\} \\ &= \{x \in M;\ |(G_1)_{\pi_1^{-1}(x)}| = 1\}. \end{aligned}$$

Let $U' = \pi^{-1}(M')$ and $U_1' = \pi_1^{-1}(M')$. Choose $\tilde{b} \in U'$, $\tilde{b}_1 \in U_1'$ such that $\pi(\tilde{b}) = \pi_1(\tilde{b}_1) = b \in M'$. Let $\Gamma = \pi^*(\pi_1(U', \tilde{b}))$ and $\Gamma_1 = \pi_1^*(\pi_1(U_1', \tilde{b}_1))$. It is quite easy to see that $\Gamma_1 \supset \Gamma$ (and hence $\Gamma_1 = \Gamma$). From this it follows that π and π_1 are equivalent coverings. □

Remark. The above theorems also establish the existence of finite Kleinian groups.

EXERCISE

Prove that the theorem is not true for the two excluded cases.

IV.10. Riemann–Roch: An Alternate Approach

The existence theorems of Chapters II and III were established under the assumption that every Riemann surface is triangulable. Also, we used differentials as a basis for constructing functions. The development in this chapter did not require *a priori* knowledge that Riemann surfaces are triangulable. In this section we show that the Riemann–Roch theorem can be obtained in a different manner, using only results of this chapter. This shows, among other things, that Chapter II can be completely dispensed with and the proof of the Riemann–Roch theorem needs a lot less machinery than was used to establish it in Chapter III. However, the proof in this section is much more *ad-hoc* than the one in III.4.

IV.10.1. Let M be a compact Riemann surface of genus g. We have seen (Remark II.5.5) that the existence of meromorphic functions on M implies the triangulability of M and thus Proposition II.5.4. Clearly the Riemann–Hurwitz relation (Theorem I.2.7) required only the triangulability of compact surfaces. (The knowledge how to compute the Euler characteristic was, of course, also required. Note that the fact that the zeroth and second Betti numbers for a compact surface agree follows, at once, from any of the usual duality theorems in algebraic topology.) Furthermore, the second proof of Riemann–Hurwitz (III.4.12) can be turned around to establish the following

Proposition. *Let D be a q-canonical divisor on M, then*

$$\deg D = q(2g - 2).$$

PROOF. For $q = 0$, the result follows by Proposition I.1.6. If $\omega_1 \neq 0$ and $\omega_2 \neq 0$ are two q-differentials, then ω_1/ω_2 is a meromorphic function and thus

$$0 = \deg(\omega_1/\omega_2) = \deg(\omega_1) - \deg(\omega_2).$$

Thus it suffices to establish the proposition for a single non-zero q-differential. Let f be a non-constant meromorphic function on M. Then $(df)^q$ is a meromorphic q-differential, and

$$\deg((df)^q) = q \deg(df).$$

Hence, we must only show that

$$\deg(df) = 2g - 2.$$

Recall that for $f(P) \neq \infty$

$$\operatorname{ord}_P df = b_f(P), \tag{10.1.1}$$

and if $f(P) = \infty$, then

$$\operatorname{ord}_P df = \operatorname{ord}_P f - 1. \tag{10.1.2}$$

By following f by a Möbius transformation, we may assume f is unramified over ∞. (Assume $\deg f = n$. Choose $w_1 \in \mathbb{C}$ such that $\operatorname{card}\{f^{-1}(w_1)\} = n$. Replace f by

$$\frac{1}{f - w_1}.)$$

For such an f we have from (10.1.1) and (10.1.2)

$$\deg(df) = B - 2n.$$

Using Riemann–Hurwitz with $\gamma = 0$ (Theorem I.2.7), the result follows.

IV.10.2. We now prove one-half of the Riemann–Roch theorem in the following

Lemma. *Let $D \in \operatorname{Div}(M)$, then*

$$r(D^{-1}) \geq \deg D - g + 1.$$

PROOF. Write

$$D = \frac{P_1^{\alpha_1} P_2^{\alpha_2} \cdots P_r^{\alpha_r}}{Q_1^{\beta_1} Q_2^{\beta_2} \cdots Q_s^{\beta_s}} \quad \text{with } \alpha_j > 0,\ \beta_k > 0.$$

The dimension (over $\mathbb{R}$) of the space H of harmonic functions on M with "poles" of order $\leq \alpha_j$ at P_j is precisely

$$2\left(\sum_{j=1}^{r} \alpha_j + 1\right)$$

by Theorem IV.3.11. Not all the functions $u \in H$ have single valued harmonic conjugates. Let c_j be a small circle around P_j. Notice that

$$\int_{c_j} {*du} = 0, \qquad j = 1, \ldots, r, \text{ all } u \in H.$$

(To see this last equality, recall that the singularity of u at P_j is of the form $\operatorname{Re} z^{-m}$ or $\operatorname{Im} z^{-m}$, with $1 \leq m \leq \alpha_j$. Thus the meromorphic differential $du + i{*du}$ has no residues.) A necessary and sufficient condition for u to have a harmonic conjugate is that

$$\int_{\alpha} {*du} = 0$$

for all closed curves α on M. Since $H_1(M)$ is generated by $2g$ elements, the dimension over $\mathbb{R}$ of the space of harmonic functions on M with "poles" of orders $\leq \alpha_j$ at P_j which have harmonic conjugates is

$$\geq 2\left[\sum_{j=1}^{r} \alpha_j + 1 - g\right].$$

Thus the dimension over $\mathbb{C}$ of the space of meromorphic functions on M with poles of order $\leq \alpha_j$ at P_j is

$$\geq \sum_{j=1}^{r} \alpha_j + 1 - g.$$

For such a function to belong to $L(D^{-1})$, it must vanish at Q_j of order $\geq \beta_j$. This imposes at most $\sum_{j=1}^{s} \beta_j$ linear conditions. Hence

$$\begin{aligned} r(D^{-1}) &\geq \sum_{j=1}^{r} \alpha_j - \sum_{j=1}^{s} \beta_j + 1 - g \\ &= \deg D + 1 - g. \end{aligned}$$ □

IV.10.3. We now establish the Riemman–Roch theorem in the form of the following

Theorem. *For $D \in \operatorname{Div}(M)$,*

$$r(D^{-1}) = \deg D - g + 1 + r(DZ^{-1}),$$

where Z is any canonical divisor on M.

PROOF. We break up the argument into a series of steps.

(1) If $\deg D < 0$, then $r(D^{-1}) = 0$. This is Proposition III.4.5.

(2) For all divisors D, $r(D^{-1}) \geq \deg D + 1 - g$. This is the content of Lemma IV.10.2.

(3) For $P_0 \in M$, $r(D^{-1}P_0^{-1}) \leq r(D^{-1}) + 1$. Write

$$D = \prod_{P \in M} P^{\alpha_D(P)}.$$

Say f_1 and $f_2 \in L(D^{-1}P_0^{-1}) \backslash L(D^{-1})$, then

$$\operatorname{ord}_{P_0} f_j = -\alpha_D(P_0) - 1 = \mu \qquad (j = 1,2).$$

Let z be a local coordinate vanishing at P_0, then

$$f_j(z) = \sum_{n=\mu}^{\infty} b_{j,n} z^n,\ b_{j,\mu} \neq 0 \qquad (j = 1,2).$$

Thus $f_1 - (b_{1,\mu}/b_{2,\mu}) f_2 \in L(D^{-1})$, and (3) has been established.

(4) Let $P_0 \in M$, and assume $r(D^{-1}P_0^{-1}) = r(D^{-1}) + \varepsilon$ and $r(DZ^{-1}) = r(DZ^{-1}P_0) + \varepsilon'$. Then $0 \leq \varepsilon + \varepsilon' \leq 1$.

Assume for contradiction $\varepsilon = 1 = \varepsilon'$ (by (3) no other possibilities exist). Let $Z = (\alpha)$, with α an abelian differential on M. We have assumed the existence of functions

$$f_1 \in L(D^{-1}P_0^{-1}) \backslash L(D^{-1}),$$
$$f_2 \in L(DZ^{-1}) \backslash L(DZ^{-1}P_0).$$

Thus

$$(f_1)DP_0 \geq 1 \quad \text{and} \quad (f_1)D \ngeq 1,$$

and

$$(f_2)ZD^{-1} \geq 1 \quad \text{and} \quad (f_2)ZD^{-1}P_0^{-1} \ngeq 1.$$

Now

$$(f_1)D = B/P_0 \quad \text{with } B \geq 1 \text{ and } \alpha_B(P_0) = 0,$$

and

$$(f_2\alpha)D^{-1} = C \quad \text{with } C \geq 1 \text{ and } \alpha_C(P_0) = 0.$$

Combining the last two equalities, we get

$$(f_1 f_2 \alpha) = BC/P_0,$$

or $f_1 f_2 \alpha$ is an abelian differential holomorphic except at P_0 ($BC \geq 1$) with a simple pole at P_0 ($\alpha_{B+C}(P_0) = 0$). This contradiction (of Proposition II.5.4) establishes (4).

Definition. $\varphi(D) = r(D^{-1}) - \deg D - r(DZ^{-1})$.

(5) For all $P_0 \in M$, $\varphi(DP_0) \leq \varphi(D)$.
We expand

$$\begin{aligned}\varphi(DP_0) &= r(D^{-1}P_0^{-1}) - \deg(DP_0) - r(DP_0Z^{-1}) \\ &= r(D^{-1}) + \varepsilon - \deg D - 1 - r(DZ^{-1}) + \varepsilon' \\ &= \varphi(D) + \varepsilon + \varepsilon' - 1 \leq \varphi(D).\end{aligned}$$

(6) $\varphi(D) \geq 1 - g$.
Choose $k \in \mathbb{Z}$ large so that

$$\deg(ZD^{-1}P_0^{-k}) < 0$$

where $P_0 \in M$ is arbitrary. By (5) and (1)

$$\varphi(D) \geq \varphi(DP_0^k) = r(D^{-1}P_0^{-k}) - \deg(DP_0^k).$$

Hence, from (2) we get (6).
(7) $\varphi(D) \leq 1 - g$.
Choose k large so that $\deg(D/P_0^k) < 0$. Then by the argument which is established in (5) and (1),

$$\varphi(D) \leq \varphi(D/P_0^k) = -\deg(D/P_0^k) - r(D/P_0^k Z).$$

We shall now use (2) and Proposition IV.10.1 to obtain

$$\begin{aligned} r(D/P_0^kZ) &\geq \deg(P_0^kZ/D) + 1 - g \\ &= \deg(P_0^k/D) + (2g - 2) + 1 - g \\ &= \deg(P_0^k/D) - 1 + g, \end{aligned}$$

establishing (7).

The Riemann–Roch theorem follows from (6) and (7). □

Remark. The above is certainly a shorter and more elegant proof of the Riemann–Roch theorem than the one in III.4. The earlier proof was, in our opinion, more transparent than this one.

The reader should now re-establish (using the Riemann–Roch theorem) the following:

(1) The dimension of the space of holomorphic abelian differentials is g.

(2) The dimension of the space of harmonic differentials is $2g$.

(3) Let $x_1, \ldots, x_n$ be n distinct points of M and $(c_1, \ldots, c_n) \in \mathbb{C}^n$. A necessary and sufficient condition that there exist an abelian differential ω, with ω regular on $M \backslash \{x_1, \ldots, x_n\}$,

$$\operatorname{ord}_{x_j} \omega = -1, \qquad \operatorname{Res}_{x_j} \omega = c_j$$

is that

$$\sum_{j=1}^{n} c_j = 0.$$

(4) Let $x_1, \ldots, x_n$ be $n \geq 1$ distinct points on M. Let z_j be a local coordinate vanishing at x_j, and let

$$\{a_{j,k_j}, \ldots, a_{j,-1}\}$$

be complex numbers with $k_j \leq -1$, $k_j \in \mathbb{Z}$, $j = 1, \ldots, n$. If $\sum_{j=1}^{n} a_{j,-1} = 0$, then there exists a meromorphic differential ω regular on $M \backslash \{x_1, \ldots, x_n\}$ such that

$$\omega(z_j) = \left(\sum_{\nu = k_j}^{\infty} a_{j,\nu} z_j^{\nu} \right) dz_j, \qquad j = 1, \ldots, n.$$

Furthermore, ω is unique modulo differentials of the first kind.

IV.11. Algebraic Function Fields in One Variable

In this section, we explore the algebraic nature of compact Riemann surfaces. We show that there is a one-to-one correspondence between the set of conformal equivalence classes of compact Riemann surfaces and the set of birational equivalence classes of algebraic function fields in one variable. We determine the structure of the field of meromorphic functions on compact surfaces and describe all valuations on these fields.

IV.11.1. In III.9 and IV.4, we briefly considered the space of germs of holomorphic functions. We gave a topology and complex structure to the set $\mathcal{M}(M)$ of germs of meromorphic functions over a Riemann surface M. Let us change our point of view slightly. The elements of $\mathcal{M}(M)$ will now merely represent convergent Laurent series with finitely many "negative terms". Let us assume $M = \mathbb{C} \cup \{\infty\}$. Thus we are dealing only with convergent series f of two forms (here N is an arbitrary integer)

$$\sum_{n=N}^{\infty} a_n(z - z_0)^n, \qquad z_0 \in \mathbb{C}, a_N \neq 0, \tag{11.1.1}$$

or

$$\sum_{n=N}^{\infty} a_n\left(\frac{1}{z}\right)^n, \qquad a_N \neq 0. \tag{11.1.2}$$

Recall the projection and evaluation functions introduced earlier. For f given by (11.1.1), $\operatorname{proj} f = z_0$; while for f defined by (11.1.2), $\operatorname{proj} f = \infty$. Also, $\operatorname{eval} f = \infty$ if $N < 0$, $\operatorname{eval} f = a_0$ if $N = 0$, and $\operatorname{eval} f = 0$ if $N > 0$.

IV.11.2. Let us take a component $\mathscr{F}$ of $\mathcal{M}(\mathbb{C} \cup \{\infty\})$ which is "spread" over a simply connected domain $D \subset \mathbb{C} \cup \{\infty\}$; that is,

$$\operatorname{proj}: \mathscr{F} \to D$$

is surjective and every curve c in D can be lifted to $\mathscr{F}$. Thus every germ in $\mathscr{F}$ can be continued analytically over D and the Monodromy theorem shows that $\mathscr{F}$ is determined by a meromorphic function on D; that is, proj is conformal and there exists an $F \in \mathscr{K}(D)$ such that for all $z \in D$, $\operatorname{proj}^{-1} z$ is the germ of the function F at z.

Let us now assume that $D = \Delta^*$ is the punctured disc $\{0 < |z| < 1\}$. We keep all the other assumptions on $\mathscr{F}$ (surjectivity of proj and path lifting property). Let us take an element $f \in \mathscr{F}$. It is represented by a convergent series of the form (11.1.1) with $0 < |z_0| < 1$. (Without loss of generality we assume $|z_0| = z_0$—this can be achieved by a rotation.) Since analytic continuation is possible over all paths, we may continue f along a generator c of $\pi_1(\Delta^*)$ which we take to be the curve $z_0 e^{i\theta}$, $0 \leq \theta \leq 2\pi$. There are now two possibilities: continuation of f along c^k $(k \in \mathbb{Z})$ never returns to the original f (in which case the singularity of $\mathscr{F}$ at 0 is *algebraically essential*) or there is a smallest positive k such that continuation of f around c^k leads back to the original function element f.

In this latter case, we consider a k-sheeted unramified covering of Δ^* by Δ^* given by the map $\rho: \zeta \mapsto \zeta^k$. We define a function F on Δ^* as follows: If $\zeta_0 \in \Delta^*$, we join $z_0^{1/k}$ to ζ_0 by a smooth curve γ in Δ^* and set $F(\zeta_0)$ to be the evaluation of the germ obtained by continuing f along $\rho(\gamma)$. It is clear that F is a locally well defined meromorphic function. We must show it is globally well defined. Let γ_1 be another path joining $z_0^{1/k}$ to ζ_0. Then $\rho(\gamma\gamma_1^{-1})$ is homotopic to a power of c^k and hence continuation of f along this path leads back to f. Let us assume that f is regular over a deleted neighborhood of zero; that is, eval $f \in \mathbb{C}$ whenever $|\text{proj} f|$ is sufficiently small. The function F is now represented by a convergent Laurent series

$$F(\zeta) = \sum_{n=-\infty}^{\infty} a_n \zeta^n, \qquad 0 < |\zeta| < \varepsilon, \varepsilon \text{ small.}$$

If there are infinitely many negative integers n with $a_n \neq 0$, then $\mathscr{F}$ has an *essential analytic* singularity at 0. Let us assume that the resulting F is meromorphic.

$$F(\zeta) = \sum_{n=N}^{\infty} a_n \zeta^n, \qquad 0 < |\zeta| < \varepsilon, a_N \neq 0. \tag{11.2.1}$$

We can now represent $\mathscr{F}$ by a single series (known as a *Puiseaux* series) of the form

$$f(z) = \sum_{n=N}^{\infty} a_n z^{n/k}. \tag{11.2.2}$$

Without loss of generality we assume that the ε appearing in (11.2.1) is 1. How does f of (11.2.2) determine $\mathscr{F}$? In a neighborhood of each point $z_1 \in \Delta^*$, $z^{1/k}$ yields k distinct analytic functions. Substituting these into f gives k different function elements lying over z_1. These function elements determine all the germs in $\text{proj}^{-1}\{z_1\}$. These germs also define a neighborhood of the Puiseaux series (11.2.2) in the expanded space $\mathscr{M}^*$ (consisting of Laurent *and* Puiseaux series).

Clearly we can extend the domains of the functions proj and eval to include such Puiseaux series. We must be a little bit more careful than in dealing with Laurent series only. We must introduce an equivalence relation among Puiseaux series. The same germs in a neighborhood of f would be produced by the series

$$\sum_{n=N}^{\infty} a_n \varepsilon^n z^{n/k}$$

with $\varepsilon \in \mathbb{C}$ and $\varepsilon^k = 1$. Two such series will henceforth be identified.

Note that proj and eval are still meromorphic functions on $\mathscr{M}^*$ and that a deleted neighborhood of a Puiseaux series consists only of Laurent series. We will now formally review all that we have done.

IV.11.3. Let $\mathscr{M}^*$ be the collection of convergent Puiseaux series ω of the form

$$\sum_{n=N}^{\infty} a_n (z - z_0)^{n/k}, \qquad a_N \neq 0, z_0 \in \mathbb{C}, \tag{11.3.1}$$

$$\sum_{n=N}^{\infty} a_n \left(\frac{1}{z}\right)^{n/k}, \qquad a_N \neq 0. \tag{11.3.2}$$

We exclude the constant series ($\omega = a_0$) from $\mathscr{M}^*$. In the above, $k \in \mathbb{Z}$, $k \geq 1$. If $k > 1$ we assume that there exists an $n > 0$ with $a_n \neq 0$ and $n/k \notin \mathbb{Z}$ (this can always be achieved), and that k has been chosen as small as possible with this property. We call such a k the *ramification index of* ω, ram ω.

The space $\mathcal{M}^*$ comes equipped with two functions (into $\mathbb{C} \cup \{\infty\}$):

$$\begin{array}{ccc} \mathcal{M}^* & \xrightarrow{\text{proj}} & \mathbb{C} \cup \{\infty\} \\ \Big\downarrow{\scriptstyle \text{eval}} & & \\ \mathbb{C} \cup \{\infty\} & & \end{array}$$

$$\text{proj}\ \omega = \begin{cases} z_0 & \text{if } \omega \text{ is given by (11.3.1)}, \\ \infty & \text{if } \omega \text{ is given by (11.3.2)}, \end{cases}$$

$$\text{eval}\ \omega = \begin{cases} \infty & \text{if } N < 0, \\ a_0 & \text{if } N = 0, \\ 0 & \text{if } N > 0. \end{cases}$$

Before proceeding, we must introduce an equivalence relation on $\mathcal{M}^*$. We shall say that $\omega \in \mathcal{M}^*$ is *equivalent* to $\tilde{\omega} \in \mathcal{M}^*$ (given by (11.3.1) or (11.3.2) with coefficients $\tilde{a}_n$, z_0, k, N) provided

$$\text{proj}\ \omega = \text{proj}\ \tilde{\omega}, \qquad \text{ram}\ \omega = \text{ram}\ \tilde{\omega} = k,$$

and there exists an $\varepsilon \in \mathbb{C}$ with $\varepsilon^k = 1$ such that

$$\tilde{a}_n = \varepsilon^n a_n, \qquad n = N, N + 1, \ldots.$$

From now on $\mathcal{M}^*$ will stand for the previously defined object with the same symbol modulo this equivalence relation. Note that ram, proj, and even eval are well defined on $\mathcal{M}^*$ (the new $\mathcal{M}^*$!).

We topologize $\mathcal{M}^*$ as follows. Let ω_0 be given by (11.3.1). Assume that the series coverges for $|z - z_0| < \varepsilon$. Let $0 < r < \varepsilon$. We define a neighborhood $U(\omega_0,r)$ of ω_0. Let $z_1 \in \mathbb{C}$ with $0 < |z_1 - z_0| < r$. The "multivalued" function $(z - z_0)^{1/k}$ determines k single valued analytic functions converging in $\{|z - z_1| < \varepsilon - r\}$—thus k distinct Taylor series in $(z - z_1)$. Substituting these Taylor series into (11.3.1) we obtain k convergent Laurent series. The collection of all such Laurent series (for all z_1 with $0 < |z_1 - z_0| < r$) plus the original ω_0 forms the neighborhood $U(\omega_0,r)$ of ω_0. Next assume that ω_0 is given by (11.3.2) where the series converges for $|z| > 1/\varepsilon$. Choose, again, $0 < r < \varepsilon$. For $|z_1| > 1/r$, the "multivalued" function $z^{-1/k}$ determines k convergent Taylor series in $(z - z_1)$ with radius of convergence $\geq 1/r - 1/\varepsilon$. Substitute these functions into (11.3.2) and proceed as before.

We see that with this "topology" (we will show next that the sets we have introduced do form a *basis* for a topology), a deleted neighborhood of any point $\omega_0 \in \mathcal{M}^*$ can be chosen to consist only of $\omega \in \mathcal{M}^*$ with proj $\omega \neq \infty$, eval $\omega \neq \infty$, and ram $\omega = 1$.

Let $\omega \in U(\omega_1,r_1) \cap U(\omega_2,r_2)$. Assume that proj $\omega = z_0$, proj $\omega_1 = z_1$, and proj $\omega_2 = z_2$ all belong to $\mathbb{C}$, and that ω is given by (11.3.1). We must find an $r > 0$ such that $U(\omega,r) \subset U(\omega_1,r_1) \cap U(\omega_2,r_2)$. Let us first assume that

$\omega \neq \omega_1$. Thus ram $\omega = 1$. Let $\varepsilon, \varepsilon_j$ be the radii of convergence of ω, ω_j $(j = 1,2)$. Then $0 \le |z_j - z_0| < r_j$ and $\varepsilon \ge \varepsilon_j - r_j$ $(j = 1,2)$. Choose $r > 0$ such that $\{|z - z_0| < r\} \subset \{|z - z_1| < r_1\} \cap \{|z - z_2| < r_2\}$. Then $U(\omega,r) \subset U(\omega_1,r_1) \cap U(\omega_2,r_2)$.

We are then left with the case that $\omega = \omega_1 = \omega_2$. Here we merely choose $r = \min(r_1,r_2)$. If proj ω or proj $\omega_j = \infty$ $(j = 1,2)$, analogous arguments apply. Hence, the sets $U(\omega,r)$ are a basis for the topology.

The topology is actually Hausdorff. Let $\omega_j \in \mathcal{M}^*$ with $\omega_1 \neq \omega_2$. If proj $\omega_1 \neq$ proj ω_2, it is trivial to produce neighborhoods $U(\omega_1,r_1)$ and $U(\omega_2,r_2)$ such that $U(\omega_1,r_1) \cap U(\omega_2,r_2)$ is empty. So we assume $z_0 =$ proj $\omega_1 =$ proj $\omega_2 \in \mathbb{C}$. If ram $\omega_1 = 1 =$ ram ω_2, then the ω_j determine meromorphic functions f_j in some disk of radius $r > 0$ about z_0. Since $f_1 \neq f_2$ (otherwise $\omega_1 = \omega_2$), $U(\omega_1,r) \cap U(\omega_2,r) = \emptyset$. (Otherwise, choose ω in the intersection and obtain that in some open subset of $|z - z_0| < r$ both f_1 and f_2 agree with the function defined by ω.) We are left to consider the case ram $\omega_1 > 1$. Let ω_1 be given by (11.3.1), and ω_2 similarly with coefficients $\tilde{a}_n$, $\tilde{z}_0 = z_0$, $\tilde{k}$. We have seen that ω_j determines a single valued function F_j on a punctured disc which covers the given punctured disk $k(\tilde{k})$ times. If $F_1 \neq F_2$, then clearly we can choose non-intersecting neighborhoods of ω_1 and ω_2 by the above method. So let us assume that $F_1 = F_2$. Since $\omega_1 \neq \omega_2$ we must have ram $\omega_1 \neq$ ram ω_2. It is now easy once again to choose non-intersecting neighborhoods of ω_1 and ω_2. (The case with proj $\omega_1 \in \mathbb{C} \cup \{\infty\}$ is treated similarly.)

We have shown that $\mathcal{M}^*$ is a Hausdorff space. Now we introduce local coordinates in $\mathcal{M}^*$. Let ω be given by (11.3.1) we introduce a local coordinate t vanishing at ω by

$$(z - z_0) = t^{\text{ram}\,\omega}.$$

If ω is given by (11.3.2), t is defined by

$$z^{-1} = t^{\text{ram}\,\omega}.$$

It is clear that we have introduced a conformal structure on $\mathcal{M}^*$. Proj and eval are meromorphic functions in $\mathcal{M}^*$ (note that proj is no longer locally univalent.) Each component of $\mathcal{M}^*$ is a Riemann surface known as an *analytic configuration.*

An analytic configuration S is determined by a single $\omega_0 \in S$. Look at all $\omega \in S$ with ram $\omega = 1$. Two points in $\mathcal{M}^*$ belong to the same component S if and only if they can be joined by a curve in $\mathcal{M}^*$. If the initial and terminal points of the curve are unramified, we may assume that the entire curve consists of unramified points. We thus see that an analytic configuration S can be described as follows:

1. Take a fixed germ ω_0 of a meromorphic function (= Laurent series).
2. Continue this germ ω_0 in all possible ways to obtain a Riemann surface $S_0 \subset \mathcal{M}^*$.
3. Take the closure S of S_0 in $\mathcal{M}^*$.

EXERCISE

Let f be a "multivalued meromorphic function". Then f^{-1} (the *inverse* function) is again a "multivalued meromorphic function". Show that the map $f \mapsto f^{-1}$ defines a conformal involution on $\mathscr{M}^*$. Use this fact to prove that the analytic configuration of a function and its inverse are conformally equivalent.

IV.11.4. Theorem. *Let S be an analytic configuration. The Riemann surface S is compact if and only if there exists an irreducible polynomial P in two variables such that* $P(\text{proj } \omega, \text{eval } \omega) = 0$, *all* $\omega \in S$.

PROOF. Assume S is a compact analytic configuration. Let $\tilde{\omega}_1, \ldots, \tilde{\omega}_r$ be the collection of points with either $\text{proj } \tilde{\omega}_j = \infty$ or $\text{ram } \tilde{\omega}_j > 1$. (Since $f =$ proj is a meromorphic function on the compact Riemann surface S, f is a finite sheeted covering. The points $\omega \in S$ with $\text{ram } \omega > 1$ correspond to points ω with $b_f(\omega) > 0$.) Let n be the degree of the function proj. Let $z_j = \text{proj } \tilde{\omega}_j$, and call $z_1, \ldots, z_r$ (∞ is one of these points) the *excluded* points. If $z_0 \in \mathbb{C}$ is a non-excluded point, then there exist n distinct function elements $\omega_1, \ldots, \omega_n \in S$ with $\text{proj } \omega_j = z_0$. Form the elementary symmetric functions $a_0, \ldots, a_n$ of $\omega_1, \ldots, \omega_n$. Note that ω_j is a Laurent series in $(z - z_0)$. Thus

$$a_0(z) = 1$$
$$a_1(z) = -\omega_1(z) - \cdots - \omega_n(z)$$
$$a_2(z) = \omega_1(z)\omega_2(z) + \omega_1(z)\omega_3(z) + \cdots + \omega_{n-1}(z)\omega_n(z)$$
$$= \sum_{\substack{i,j=1 \\ i<j}}^{n} \omega_i(z)\omega_j(z)$$
$$\vdots$$
$$a_\nu(z) = (-1)^\nu \sum_{\substack{n_1, \ldots, n_\nu = 1 \\ n_1 < n_2 < \cdots < n_\nu}}^{n} \omega_{n_1}(z)\omega_{n_2}(z) \cdots \omega_{n_\nu}(z)$$
$$\vdots$$
$$a_n(z) = (-1)^n \omega_1(z) \cdots \omega_n(z).$$

The functions $a_1, \ldots, a_n$ are meromorphic functions on $\mathbb{C} \cup \{\infty\} \backslash \{z_1, \ldots, z_r\}$. These functions are independent of the point z_0. Near the excluded point z_j, they grow like a power of $|z - z_j|$. Thus the functions a_k are rational functions of z ($\in \mathbb{C}(z) = \mathscr{K}(\mathbb{C} \cup \{\infty\})$). Clearly at the non-excluded points the functions elements ω satisfy the equation

$$P(z,w) = w^n + a_1(z)w^{n-1} + \cdots + a_n(z) = 0. \tag{11.4.1}$$

Equation (11.4.1) may be viewed as an equation over $\mathbb{C}(z)$ for the elements $\omega \in S$, or as an equation satisfied by two meromorphic functions $z =$ proj, $w =$ eval on S. Thus (11.4.1) also holds at the excluded points. We claim that every Puiseaux series ω that satisfies (11.4.1) is a point of S. Over every non-excluded point, we have exhibited n solutions of (11.4.1). These are all the solutions over such points, since the equation (11.4.1) has only n formal (not necessarily convergent) power or Puiseaux series solutions at every point. Let z_m be an excluded point. Over z_m we have certain series, say $\tilde{\omega}_1, \ldots, \tilde{\omega}_k$. Let $n_j = \operatorname{ram} \tilde{\omega}_j$. Then $\sum_{j=1}^{k} n_j = n$ and recall that such equivalence classes of Puiseaux series represent n_j distinct formal series.

We postpone the proof of the irreducibility of P until after we have established the converse. To establish the converse, let $P(z,w)$ be an irreducible polynomial in two variables. Assume that

$$P(z,w) = a_0(z)w^n + a_1(z)w^{n-1} + \cdots + a_n(z)$$

with $a_i \in \mathbb{C}[z]$. Let R be the discriminant of P. Then R is a polynomial in z of lowest degree with (here p and q are polynomials) such that

$$R = pP + q\frac{\partial P}{\partial w}.$$

Furthermore, for $z_0 \in \mathbb{C}$,

$$R(z_0) = 0 \Leftrightarrow P(z_0,w) = 0 \text{ has multiple roots.}$$

We again exclude the point ∞ and the zeros of R and the zeros of a_0. By the implicit function theorem, for any non-excluded z_0, there are n distinct function elements (Taylor series)

$$\omega_1, \ldots, \omega_n$$

such that

$$P(z,\omega_j(z)) = 0. \tag{11.4.2}$$

If we continue one of these function elements ω_j to another non-excluded point z_1, we get another function element (over z_1) that satisfies (11.4.2).

Let us start with a fixed function element, say ω_1 (that satisfies (11.4.2)) over a non-excluded point $z_0 \in \mathbb{C}$. It is clear that ω_1 can be continued along any path that avoids excluded points. Let z_1 be an excluded point. Without loss of generality we may assume that $z_1 = 0$ and that the unit disc Δ contains no other excluded points. It is also obvious that continuing any function elements ω (with proj $\omega \in \Delta^*$) around the origin will eventually lead back to ω. Thus we are in the situation discussed in IV.11.2. The Laurent series ω about zero obtained in this way clearly satisfies (11.4.2). By including all the functions elements over the excluded points obtained in the above manner, we get a compact Riemann surface S (since it is a finite sheeted covering of $\mathbb{C} \cup \{\infty\}$). The elements of S satisfy a polynomial equation $P_1(z,w) = 0$ by the first part of the theorem. Since they also satisfy the irreducible equation $P(z,w) = 0$, P must divide P_1 and we have shown that all solutions of (11.4.2) are in S. The above also shows that the equation (11.4.1) is irreducible. Write (11.4.1) as

$$P = P_1^{\alpha_1} \cdots P_r^{\alpha_r}, \tag{11.4.3}$$

with P_j irreducible and $P_k \neq P_j$ for $k \neq j$. Note that P is not a power of an irreducible polynomial, since the n function elements lying over most points are distinct. Furthermore, each P_j in (11.4.3) clearly determines a compact Riemann surface $S_j \subset S$. Thus, $S_j = S$ and $j = 1$. □

Definition. We shall say that S is the analytic configuration *corresponding* to the irreducible polynomial P.

Corollary (of Proof). *For every irreducible polynomial P, there is an analytic configuration corresponding to P.*

IV.11.5. Theorem. *Let M be any Riemann surface. Let $w, z \in \mathscr{K}(M)\backslash\mathbb{C}$. There exists a canonical holomorphic mapping*

$$\varphi: M \to \mathscr{M}^*$$

such that for all $P \in M$,

$$w(P) = \operatorname{eval}(\varphi(P)),$$
$$z(P) = \operatorname{proj}(\varphi(P)).$$

PROOF. Let t be a local coordinate vanishing at P. Without loss of generality we may assume that for some $n \in \mathbb{Z}$, $n \geq 1$,

$$z(t) = t^{-n} \quad \text{or} \quad z(t) = z(P) + t^n, \quad \text{near } P. \tag{11.5.1}$$

If

$$w(t) = \sum_{j \geq j_0} \alpha_j t^j,$$

then we set $\varphi(P)$ to be either

$$\sum_{j \geq j_0} \alpha_j z^{-j/n} \quad \text{or} \quad \sum_{j \geq j_0} \alpha_j (z - z(P))^{j/n},$$

depending on which case applied in (11.5.1). □

IV.11.6. Proposition. *If M is compact and $w, z \in \mathscr{K}(M)\backslash\mathbb{C}$, then there exists a polynomial P such that $P(z,w) \equiv 0$.*

PROOF. Let φ be the map of the previous theorem. Since the image of φ is a compact component of $\mathscr{M}^*$, z and w satisfy an algebraic equation by Theorem IV.11.4. □

Remark. The mapping φ of Theorem IV.11.5 is one-to-one if, for example, the pair z, w separate points on M. The converse is not true. We begin with a

Definition. We say that $z, w \in \mathscr{K}(M)\backslash\mathbb{C}$ form a *primitive pair* provided the map φ of Theorem IV.11.5 is one-to-one.

IV.11.7. Proposition. *If M is a compact Riemann surface and $z \in \mathscr{K}(M)\backslash\mathbb{C}$, then there exists $w \in \mathscr{K}(M)\backslash\mathbb{C}$ such that z, w form a primitive pair.*

PROOF. Let $n = \deg z$. Choose $a \in \mathbb{C}$ such that $z^{-1}(a)$ consists of n distinct points $x_1, \ldots, x_n$. By the Riemann inequality, there exists a non-constant meromorphic function w_j on M such that

w_j has a pole of order $\nu_j \geq 1$ at x_j, and

w_j is holomorphic on $M\backslash\{x_j\}$.

Choose positive integers $k_1, \ldots, k_n$ such that

$$k_1 v_1 < k_2 v_2 < \cdots < k_n v_n,$$

and set

$$w = w_1^{k_1} + \cdots + w_n^{k_n}.$$

Then $w \in \mathscr{K}(M)$ and the polar divisor of w is $x_1^{k_1 v_1} \cdots x_n^{k_n v_n}$. Let φ be the mapping of Theorem IV.11.5 determined by z and w. We obtain this way n function elements $\omega_1, \ldots, \omega_n$ with proj $\omega_j = a$. These elements are distinct since the Laurent series for the different ω_j have different order poles at the point $z = a$. Let $S = \text{Image } \varphi$. Then

$$M \xrightarrow{\varphi} S \xrightarrow{\text{proj}} \mathbb{C} \cup \{\infty\}.$$

Since deg proj $\circ\, \varphi = n = \deg z =$ deg proj, and

$$\deg \text{proj} \circ \varphi = (\deg \text{proj})(\deg \varphi),$$

we conclude that $\deg \varphi = 1$. □

IV.11.8. Corollary. *If M is a compact Riemann surface and $z \in \mathscr{K}(M)\backslash\mathbb{C}$, then there exists an irreducible polynomial P in two variables z and w such that M is conformally equivalent to the analytic configuration corresponding to P with z being the proj function.*

IV.11.9. Proposition. *Let M be a compact Riemann surface and z, w a primitive pair. Assume $n = \deg z$, $m = \deg w$. Let P be an irreducible polynomial satisfied by the pair z, w. Then $n = \deg_w P$, and $m = \deg_z P$.*

PROOF. We have

$$P(z,w) = 0.$$

Let $x_0 \in M$ be such that $z^{-1}(w(x_0))$ consists of precisely n points. Thus the degree of P in w is precisely n. □

IV.11.10. Proposition. *Let M be a compact Riemann surface and z, w a primitive pair. Assume $\deg z = n$. Let $f \in \mathscr{K}(M)$. Then there exist rational functions a_j, $j = 0, \ldots, n-1$, such that*

$$f = \sum_{j=0}^{n-1} a_j(z) w^j. \tag{11.10.1}$$

PROOF. Let $\zeta \in \mathbb{C}$ be such that $z^{-1}(\zeta)$ consists of n distinct points $x_1, \ldots, x_n$ and such that $w(x_1), \ldots, w(x_n)$ consists of n distinct complex values ($\neq \infty$). Let $a_0(\zeta), \ldots, a_{n-1}(\zeta)$ be the unique solutions of the linear system

$$f(x_k) = \sum_{j=0}^{n-1} a_j(\zeta) w(x_k)^j, \qquad k = 1, \ldots, n.$$

Note that

$$a_j(\zeta) = \frac{\det W_j}{\det W}, \qquad j = 0, \ldots, n-1,$$

where

$$W = \begin{bmatrix} 1 & w(x_1) & \cdots & w(x_1)^{n-1} \\ 1 & w(x_2) & \cdots & w(x_2)^{n-1} \\ \vdots & \vdots & & \vdots \\ 1 & w(x_n) & \cdots & w(x_n)^{n-1} \end{bmatrix}$$

is the Vandermonde determinant (and thus $\det W = \prod_{k<j} (w(x_j) - w(x_k)) \neq 0$), and W_j is the matrix obtained by substituting the column

$$^t[f(x_1), f(x_2), \ldots, f(x_n)]$$

for the $(j+1)$-st column of W. The functions $a_j(\zeta)$ are independent of the ordering of the x_i. They are defined except for finitely many $\zeta \in \mathbb{C}$. It is clear that they extend to meromorphic functions of ζ—thus rational functions. Equation (11.10.1) holds now for all but a finite number of points. Hence, everywhere by continuity. □

Corollary. *If M is a compact surface and z, w is a primitive pair as above, satisfying the irreducible equation P, then*

$$\mathscr{K}(M) \cong \mathbb{C}(z)[w]/P(z,w).$$

PROOF. The last equality means that $\mathscr{K}(M)$ is isomorphic (as a field) to an algebraic extension of the field of rational functions. The isomorphism is the obvious one given by the previous proposition. If $f \in \mathbb{C}(z)[w]$, then clearly

$$f = \sum_{j=0}^{N} a_j(z) w^j.$$

Thus f defines an element of $\mathscr{K}(M)$. Since $P(z,w)$ goes to the zero element, of $\mathscr{K}(M)$, this mapping factors to a mapping from $\mathbb{C}(z)[w]$ into $\mathscr{K}(M)$. This mapping is surjective by the proposition, and injective because the domain is a field and the mapping is non-trivial. □

IV.11.11. Example. Consider a hyperelliptic surface M of genus $g > 0$. Let $z \in \mathscr{K}(M)$ with $\deg z = 2$. We have seen (III.7) that z has $2g + 2$ branch points. Now Proposition IV.11.7 shows that there exists a w such that the pair z and w are primitive. By Proposition IV.11.6, z and w satisfy a polynomial equation

$$w^2 - 2aw + c = 0,$$

where $a, c \in \mathbb{C}(z)$. Completing the square in the usual manner, we rewrite the above as

$$(w - a)^2 + c - a^2 = 0.$$

We make now a birational transformation

$$(w - a) = w_1, \qquad z = z_1,$$

and obtain

$$w_1^2 = a^2 - c.$$

Now $a^2 - c \in \mathbb{C}(z)$ and thus can be written as

$$a^2 - c = b^2 \prod_{j=1}^{r} (z - e_j),$$

with distinct $e_j \in \mathbb{C}$ and $b \in \mathbb{C}(z)$. We now define one more birational transformation

$$w_2 = \frac{w_1}{b}, \qquad z_2 = z_1,$$

and obtain the equation (dropping the subscript 2)

$$w^2 = \prod_{j=1}^{r} (z - e_j).$$

Since all our birational transformations kept z fixed, the complex numbers e_j can, of course, be identified. They are the finite branch values of z. Thus $r = 2g + 1$ if ∞ is a branch value, and $r = 2g + 2$ if ∞ is not a branch value.

We have now reproven Proposition III.7.4 without the use of "paste and scissors".

EXERCISE

Return to the situation studied in III.6.7. Let us take, for the function z on our surface M, the Weierstrass $\wp$-function.

(a) Observe that $\wp$ is an even function (that is,

$$\wp(\zeta) = \wp(-\zeta) \quad \text{for every complex number } \zeta)$$

and hence its derivative $\wp'$ is an odd function.

(b) Show that the only branch points for $\wp$ are at the half-periods (the half-periods are the four points $0, \frac{1}{2}, \tau/2, \frac{1}{2} + \tau/2$ and their images under the group G).

(c) Let $e_1 = \wp(\frac{1}{2})$, $e_2 = \wp(\tau/2)$, and $e_3 = \wp(\frac{1}{2} + \tau/2)$. Show that there exists a doubly periodic function f on the plane such that

$$f^2 = 4(\wp - e_1)(\wp - e_2)(\wp - e_3).$$

(d) Prove that $f = \wp'$. *Hint*: Compare the divisors (f) and $(\wp')$ on M.

IV.11.12. Definition. Let K be a field. By a *(discrete) valuation (of rank* 1) v on K, we mean a surjective homomorphism

$$v: K^* \to \mathbb{Z}$$

(here K^* is the multiplicative group of non-zero elements of K, and $\mathbb{Z}$ is the additive group of integers), such that

$$v(f + g) \geq \min\{v(f), v(g)\}, \qquad f, g \in K^*. \tag{11.12.1}$$

Remark. By setting $v(0) = +\infty$, we see that for all $f, g \in K$

$$v(fg) = v(f) + v(g),$$

and that (11.12.1) holds provided we use the usual conventions regarding $+\infty$.

IV.11.13. Lemma. *If v is a valuation on the field K, then for f and $g \in K$*

$$v(f + g) = \min\{v(f), v(g)\}$$

provided $v(f) \neq v(g)$.

Proof. We may, without loss of generality, assume $f \neq 0 \neq g$ and $v(f) < v(g)$. Note that $v(1) = 0 = v(-1)$. Since for every odd integer n, we have

$$v(1) = v(1^n) = nv(1),$$

and

$$v(-1) = v((-1)^n) = nv(-1).$$

Also, $v(f) = v(-f)$ for all $f \in K$ since

$$v(-f) = v(-1) + v(f).$$

Now we compute

$$\begin{aligned} v(f) = v(f + g - g) &\geq \min\{v(f + g), v(-g)\} \\ &\geq \min\{\min\{v(f), v(g)\}, v(g)\} = \min\{v(f), v(g)\} = v(f). \end{aligned}$$

Hence all the inequalities must be equalities, and in particular

$$v(f) = \min\{v(f + g), v(g)\} = v(f + g).$$

□

IV.11.14. To classify all valuations on $\mathscr{K}(M)$ we need the following

Lemma. *Let $\{P_0, \ldots, P_n\}$ be $n + 1$ distinct points on a compact Riemann surface M. Then there exists an $f \in \mathscr{K}(M)$ that separates these points ($f(P_j) \neq f(P_k)$, $j \neq k$) with $(df)(P_j) \neq 0, \infty$, $j = 0, \ldots, n$.*

Proof. It suffices to show that for $k = 0, \ldots, n$, there exists an $f_k \in \mathscr{K}(M)$, such that

$$f_k(P_k) = 1, \qquad df_k(P_k) \neq 0, \infty,$$

and

$$f_k(P_j) = 0 = df_k(P_j), \qquad j \neq k.$$

Having established the existence of the f_k, we may set $f = \sum_{j=0}^{n} (j + 1) f_j$. It suffices to show existence of f_0. Let $P \in M$ with $P \neq P_j$, all j. Set

$$D = \frac{P^m}{P_1^2 \cdots P_n^2},$$

with $m \in \mathbb{Z}$ so large that $\deg D > \max\{g, 2g - 1\}$. Then

$$r\left(\frac{1}{D}\right) = \deg D + 1 - g > 1,$$

and

$$r\left(\frac{P_0}{D}\right) = \deg D - g > 0.$$

We choose $h_0 \in L(D^{-1})\backslash L(P_0 D^{-1})$; that is,

$$(h_0) \geq \frac{P_1^2 \cdots P_n^2}{P^m}, \qquad (h_0) \ngeq \frac{P_1^2 \cdots P_n^2 P_0}{P^m}.$$

Thus $\operatorname{ord}_{P_j} h_0 \geq 2$, $j = 1, \ldots, n$, and $\operatorname{ord}_{P_0} h_0 = 0$. If $\operatorname{ord}_{P_0} dh_0 = 0$, we are done ($f_0 = h_0/h_0(P_0)$). Otherwise, construct h_1 such that

$$(h_1) \geq \frac{P_0 P_1^2 \cdots P_n^2}{P^{m+1}}, \qquad (h_1) \ngeq \frac{P_0^2 P_1^2 \cdots P_n^2}{P^{m+1}},$$

and set

$$f_0 = \frac{h_0}{h_0(P_0)} + h_1. \qquad \square$$

IV.11.15. Theorem. *Let M be a compact Riemann surface. Then v is a valuation on $\mathscr{K}(M)$ if and only if there exists (a unique) $x \in M$ such that*

$$v(f) = \operatorname{ord}_x f, \quad \textit{all } f \in \mathscr{K}(M). \tag{11.15.1}$$

Proof. It is clear that (11.15.1) defines a valuation on $\mathscr{K}(M)$, so we need only prove the converse. Let v be given. Choose $f \in \mathscr{K}(M)$ such that $v(f) = 1$. Clearly $f \notin \mathbb{C}$. (If $\lambda \in \mathbb{C}$, then $v(\lambda) = nv(\lambda^{1/n})$, all $n \in \mathbb{Z}$. Thus $v(\lambda) = 0$.) Let r be a rational function, then

$$v(r(f)) = \operatorname{ord}_0 r. \tag{11.15.2}$$

To establish (11.15.2), recall (that up to a non-zero constant multiple)

$$r \circ f = \frac{\prod_{j=0}^{n} f - \alpha_j}{\prod_{j=0}^{m} f - \beta_j}, \qquad \alpha_j, \beta_j \in \mathbb{C},$$

and thus,

$$v(r \circ f) = \sum_{j=0}^{n} v(f - \alpha_j) - \sum_{j=0}^{m} v(f - \beta_j).$$

By Lemma IV.11.13, $v(f - \lambda) = 0$ unless $\lambda = 0$, in which case $v(f - \lambda) = v(f) = 1$. This establishes (11.15.2). Let $P_1, \ldots, P_n$ be the zeros of f listed according to their multiplicities. Let $F \in \mathscr{K}(M)$ be arbitrary. By Propositions IV.11.6 and IV.11.9, F satisfies an equation

$$F^n + \sum_{j=0}^{n-1} r_j(f) F^j = 0, \quad \text{with } r_j \text{ rational functions.}$$

Thus, by (11.15.2)

$$nv(F) \geq \min\{\text{ord}_0 r_j + jv(F), j = 0, \ldots, n-1\}.$$

If $v(F) < 0$, then $\text{ord}_0 r_j < 0$ for some $j = 0, \ldots, n-1$. Since $r_j(0)$ is a symmetric function of the values of F at the points $P_1, \ldots, P_n$, we see that F must have a pole at one of these points. Similarly, if $v(F) > 0$, then $v(F^{-1}) < 0$ and F must vanish at one of these points. We conclude that $v(F) = 0$ whenever $0 \neq F(P_j) \neq \infty, j = 1, \ldots, n$. From now on we assume (without loss of generality) that $\{P_1, \ldots, P_n\}$ are the distinct zeros of f. (We are changing the meaning of the symbol n.) Let h be a function holomorphic and non-zero at $P_1, \ldots, P_n$. Further, choose an h that separates these points with dh not zero at these points. We have $v(h) = 0$. Consider the function

$$\frac{f}{\prod_{j=1}^{n} (h - h(P_j))^{\text{ord}_{P_j} f}}$$

that is holomorphic and non-zero at $P_1, \ldots, P_n$. Thus

$$1 = v(f) = \sum_{j=1}^{n} (\text{ord}_{P_j} f) v(h - h(P_j)).$$

Since $v(h - h(P_j)) \geq 0$, there is a unique k such that

$$v(h - h(P_k)) = 1 = \text{ord}_{P_k} f.$$

(and $v(h - h(P_j)) = 0, j \neq k$). Now take an arbitrary $F \in \mathscr{K}(M)$, and consider the function

$$\frac{F}{\prod_{j=1}^{n} (h - h(P_j))^{\text{ord}_{P_j} F}}$$

whose value under v is zero. Hence

$$v(F) = \sum_{j=1}^{n} (\text{ord}_{P_j} F) v(h - h(P_j)) = \text{ord}_{P_k} F. \qquad \square$$

IV.11.16. Theorem. *Let M and N be compact Riemann surfaces. Let*

$$F : \mathscr{K}(N) \to \mathscr{K}(M) \tag{11.16.1}$$

be a $\mathbb{C}$-algebra homomorphism (thus injective). Then there exists a unique holomorphic map

$$F^* : M \to N \tag{11.16.2}$$

such that

$$(Ff)(x) = f(F^*(x)), \quad \textit{all } f \in \mathscr{K}(N), \textit{ all } x \in M. \tag{11.16.3}$$

Proof. Let $x \in M$. Define a valuation v_x on $\mathscr{K}(N)$ by

$$v_x(f) = \text{ord}_x Ff, \qquad f \in \mathscr{K}(N).$$

By Theorem IV.11.15, there exists a unique point $F^*x \in N$ such that

$$\text{ord}_x Ff = v_x(f) = \text{ord}_{F^*x} f, \; f \in \mathscr{K}(N). \tag{11.16.4}$$

This defines our mapping F^* of (11.16.2). We claim that F^* has the property (11.16.3). This is clear for those x with $f(F^*(x)) = 0$ or ∞, and is clear in general from the following argument. Let $\lambda \in \mathbb{C}$. Then by (11.16.4)

$$\begin{aligned} f(F^*(x)) = \lambda &\Leftrightarrow \operatorname{ord}_{F^*(x)}(f - \lambda) > 0 \\ &\Leftrightarrow \operatorname{ord}_x F(f - \lambda) = \operatorname{ord}_x(Ff - \lambda) > 0 \\ &\Leftrightarrow (Ff)(x) = \lambda. \end{aligned}$$

We claim that F^* is a continuous mapping. Let $\{x_n\}$ be a sequence on M, with $\lim_n x_n = x$. If F^*x_n does not converge to F^*x, then (since N is compact) there must be a subsequence that converges to $y \neq F^*x$. We may assume that the entire sequence $\{F^*x_n\}$ converges to y. Now there is an $f \in \mathcal{K}(N)$ with $f(y) = 0$ and $f(F^*x) = 1$. Thus

$$f(F^*x_n) = (Ff)(x_n) \to (Ff)(x) = f(F^*x) = 1,$$

and

$$f(F^*x_n) \to f(y) = 0.$$

This contradiction establishes the continuity of F^*.

We show next that F^* is holomorphic. Let $x \in M$ be arbitrary. Choose an $f \in \mathcal{K}(N)$ such that f is univalent in a disc U about $F^*(x)$. Clearly $Ff \notin \mathbb{C}$. There is thus a disc V about x such that

$$F^*(V) \subset U \quad \text{and} \quad (Ff)(V) \subset f(U).$$

Thus $F^* = f^{-1} \circ Ff$ in V.

Uniqueness of F^* follows from the fact that $\mathcal{K}(N)$ separates points. □

IV.11.17. The map $F \mapsto F^*$ is a functor, since we have

Corollary 1

a. *Let $F: \mathcal{K}(M) \to \mathcal{K}(M)$ be the identity, then $F^*: M \to M$ is also the identity.*
b. *If $\mathcal{K}(M_1) \xrightarrow{F_1} \mathcal{K}(M_2) \xrightarrow{F_2} \mathcal{K}(M_3)$ are $\mathbb{C}$-algebra homomorphisms, then*

$$(F_2 \circ F_1)^* = F_1^* \circ F_2^*.$$

Corollary 2. *If F is an isomorphism (surjective), then F^* is a homeomorphism.*

Since the holomorphic mapping (11.16.2) between compact Riemann surfaces induces the homomorphism (11.16.1) between their function fields that is defined by (11.16.3), the functor under discussion establishes an equivalence between two categories.

Remark. Theorems similar to III.11.16 can be established for the ring of holomorphic functions and the field of meromorphic functions on open Riemann surfaces. If M is an open Riemann surface, $\{x_n\}$ is a discrete sequence of points of M and $\{v_n\}$ is a sequence of positive integers, then there exists a holomorphic function f ($\in \mathscr{H}(M)$) such that f is non-zero on $M - \{x_n\}$ and $\text{ord}_{x_n} f = \nu_n$. This result easily implies that every maximal principal ideal J in M is the ideal of functions vanishing at a point $x \in M$. This classification of (some of) the ideals of $\mathscr{H}(M)$ can be used to prove that $\mathbb{C}$-algebra isomorphisms between rings (fields) of holomorphic (meromorphic) functions on open surfaces are induced by conformal maps between the corresponding surfaces. For the study of the field of meromorphic functions one needs to show that Theorem III.11.15 is valid on open surfaces (thus an algebra isomorphism between fields of meromorphic functions maps holomorphic functions to holomorphic functions).

CHAPTER V

Automorphisms of Compact Surfaces—Elementary Theory

In this chapter we develop the basic results on the automorphism group of a compact Riemann surface, continuing the study began in III.7. Some of the deeper results will have to await the creation of more powerful machinery.

Using quite elementary methods, we study the action of the group of automorphisms on various spaces of differentials.

V.1. Hurwitz's Theorem

Throughout this section, M is a compact Riemann surface of genus g usually (but not always) ≥ 2, and Aut M denotes the group (under composition) of conformal automorphisms of M.

The most important result of this section is the theorem referred to in the title (Theorem V.1.3). We also obtain various bounds on the order of an automorphism of a compact Riemann surface of genus ≥ 2 in terms of the number of fixed points of the automorphism. We also generalize the concept of hyperellipticity. The methods of proof in this section are mostly combinatorial and involve the examining of many special cases.

V.1.1. The next result will be strengthened considerably in V.1.5.

Proposition. *If* $1 \neq T \in \text{Aut } M$, *then* T *has at most* $2g + 2$ *fixed points.*

PROOF. Since M is compact and the fixed point set of T is discrete, the fixed point set is finite. Choose a $P \in M$ that is not a fixed point of T. There is a meromorphic function f on M whose polar divisor is P^r with $1 \leq r \leq g + 1$ ($r = g + 1$ if $g \geq 2$ and P is not a Weierstrass point or $g \leq 1, r \leq g$, otherwise).

Consider the function $h = f - f \circ T$. Its polar divisor is $P^r(T^{-1}P)^r$. The function h thus has $2r \leq 2g + 2$ zeros. Each fixed point of T is a zero of h. Thus T has at most $2g + 2$ fixed points. □

V.1.2. We assume now that $g \geq 2$. Let $W(M)$ be the (finite) set of Weierstrass points on M. Recall that $W(M)$ consists of at least $2g + 2$ points and precisely $2g + 2$ points if and only if M is hyperelliptic.

Proposition. *If* $T \in \operatorname{Aut} M$, *then* $T(W(M)) = W(M)$.

Proof. As a matter of fact the gap sequences (with respect to q-differentials, $q \geq 1$) at $P \in M$ and at TP are the same. □

We define $\operatorname{Perm}(W(M))$ as the permutation group of the Weierstrass points on M.

Corollary 1. *There is a homomorphism*

$$\lambda : \operatorname{Aut} M \to \operatorname{Perm}(W(M)).$$

Furthermore, λ is injective unless M is hyperelliptic, in which case Kernel $\lambda = \langle J \rangle$, where J is the hyperelliptic involution.

Proof. The existence of λ follows from the proposition. If M is not hyperelliptic, then there are more than $2g + 2$ Weierstrass points. By Proposition V.1.1, only the identity fixes all the Weierstrass points and thus λ is injective.

If M is hyperelliptic, then by Proposition III.7.11, an element $T \notin \langle J \rangle$ has at most $4 (< 2g + 2)$ fixed points. Of course, J fixes all the Weierstrass points. □

Corollary 2 (Schwarz). *If M is a surface of genus $g \geq 2$, then* Aut M *is a finite group.*

Proof. We have produced a homomorphism of Aut M into a finite group (the permutation group of a finite set), and the homomorphism has a finite kernel. □

V.1.3. Having seen that Aut M is a finite group, we naturally want to get a bound on its order.

Theorem (Hurwitz). *Let N be the order of* Aut M, *where M is compact Riemann surface of genus $g \geq 2$, then*

$$N \leq 84(g - 1).$$

Proof. Consider (recall the discussion in III.7.8) the holomorphic projection (abbreviate Aut M by G)

$$\pi : M \to M/G.$$

We know that π is of degree N, and M/G is a compact Riemann surface of genus γ. The mapping π is branched only at the fixed points of G and

$$b_\pi(P) = \text{ord } G_P - 1, \qquad \text{all } P \in M.$$

Let $P_1, \ldots, P_r$ be a maximal set of inequivalent (that is, $P_j \neq h(P_k)$, all $h \in G$, all $j \neq k$) fixed points of elements of $G \backslash \{1\}$.

Let $\nu_j = \text{ord } G_{P_j}$. Then (using either group theory or covering space theory) there are N/ν_j distinct points on M equivalent under G to P_j—each with a stability subgroup of order ν_j (if h takes P to Q, then $G_Q = hG_Ph^{-1}$). Thus the total branch number of π is given by

$$B = \sum_{j=1}^{r} \frac{N}{\nu_j}(\nu_j - 1) = N \sum_{j=1}^{r} \left(1 - \frac{1}{\nu_j}\right).$$

The Riemann-Hurwtiz relation now reads

$$2g - 2 = N(2\gamma - 2) + N \sum_{j=1}^{r} \left(1 - \frac{1}{\nu_j}\right). \tag{1.3.1}$$

Note that $\nu_j \geq 2$ and thus $\frac{1}{2} \leq 1 - 1/\nu_j < 1$. The rest of the proof consists of an analysis of (1.3.1). It is clear (since we may assume that $N > 1$) that $g > \gamma$. We consider possibilities:

Case I: $\gamma \geq 2$.

In this case we obtain from (1.3.1) that

$$2g - 2 \geq 2N \quad \text{or} \quad N \leq g - 1.$$

Case II: $\gamma = 1$.

In this case (1.3.1) becomes

$$2g - 2 = N \sum_{j=1}^{r} \left(1 - \frac{1}{\nu_j}\right). \tag{1.3.2}$$

If $r = 0$, then also $g = 1$ (we assumed $g > 1$). This is the basic fact (the left hand side of (1.3.1) is ≥ 2) that will be used repeatedly. Thus (1.3.2) implies

$$2g - 2 \geq \tfrac{1}{2}N \quad \text{or} \quad N \leq 4(g - 1).$$

Case III: $\gamma = 0$.

We rewrite (1.3.1) as

$$2(g - 1) = N\left(\sum_{j=1}^{r} \left(1 - \frac{1}{\nu_j}\right) - 2\right), \tag{1.3.3}$$

and conclude that

$$r \geq 3$$

(since $2(g - 1) > 0$, $N > 1$, and $(1 - 1/\nu_j) < 1$ for each j).

If $r \geq 5$, then (1.3.3) gives

$$2(g-1) \geq \tfrac{1}{2}N \quad \text{or} \quad N \leq 4(g-1).$$

If $r = 4$, then it cannot be that all the v_j are equal to 2. Thus, at least one is ≥ 3, and (1.3.3) gives

$$2(g-1) \geq N(\tfrac{3}{2} + \tfrac{2}{3} - 2) \quad \text{or} \quad N \leq 12(g-1).$$

It remains to consider the case $r = 3$. Without loss of generality we assume

$$2 \leq v_1 \leq v_2 \leq v_3. \tag{1.3.4}$$

Clearly $v_3 > 3$ (otherwise the right hand side of (1.3.3) is negative). Furthermore, $v_2 \geq 3$. If $v_3 \geq 7$, then (1.3.3) yields

$$2(g-1) \geq N(\tfrac{1}{2} + \tfrac{2}{3} + \tfrac{6}{7} - 2) \quad \text{or} \quad N \leq 84(g-1).$$

If $v_3 = 6$ and $v_1 = 2$, then $v_2 \geq 4$ and

$$N \leq 24(g-1).$$

If $v_3 = 6$ and $v_1 \geq 3$ (recall (1.3.4)), then

$$N \leq 12(g-1).$$

If $v_3 = 5$ and $v_1 = 2$, then $v_2 \geq 4$ and

$$N \leq 40(g-1).$$

If $v_3 = 5$ and $v_1 \geq 3$, then

$$N \leq 15(g-1).$$

If $v_3 = 4$, then $v_1 \geq 3$ and

$$N \leq 24(g-1).$$

This exhaustion (of cases) completes the proof.

Exercise

We outline below an alternate proof of Hurwitz's theorem.

(1) Represent M as U/Γ where U is the upper half plane and Γ is a fixed point free Fuchsian group.

(2) Show that $N(\Gamma)$ = normalizer of Γ in Aut U is Fuchsian and that Aut $M \cong N(\Gamma)/\Gamma$.

(3) Use the fact that $\pi: U/\Gamma \to U/N(\Gamma)$ is holomorphic to conclude that $N(\Gamma)$ is of finite type over U.

(4) Observe that

$$\deg \pi = [N(\Gamma):\Gamma],$$

where $\deg \pi$ = degree of π, and $[N(\Gamma):\Gamma]$ = the index of Γ in $N(\Gamma)$.

(5) Show that

$$\frac{\text{Area}(U/\Gamma)}{\text{Area}(U/N(\Gamma))} = [N(\Gamma):\Gamma].$$

(6) Prove that for any Fuchsian group F of finite type over U, we have

$$\text{Area}(U/F) \geq \frac{\pi}{21}.$$

(7) Since $\text{Area}(U/\Gamma) = 4\pi(g-1)$, conclude Hurwitz's theorem.

V.1.4. To simplify the statement of results, we introduce the following notation. For f a non-constant meromorphic function on a compact Riemann surface $M(f \in \mathscr{K}(M)\backslash\mathbb{C})$, let $\deg f$ be the degree of f, then (of course)

$$\deg f = \deg D,$$

where $D = f^{-1}(\infty)$. For $T \in \text{Aut}\, M$, ord T will denote the order of T and $\nu(T)$ the number of fixed points of T. Our next result is similar to Proposition V.1.1.

Proposition. *Let* $f \in \mathscr{K}(M)\backslash\mathbb{C}$ *and* $1 \neq T \in \text{Aut}\, M$. *Assume that* $\nu(T) > 2 \deg f$. *Then* $f = f \circ T$ *and* $\deg f$ *is a multiple of* ord T. *If in addition,* $\deg f$ *is prime, then* ord T *is prime and* $M/\langle T\rangle \cong \mathbb{C} \cup \{\infty\}$.

PROOF. Consider $h = f - f \circ T$. If $h \notin \mathbb{C}$, then $\deg h \leq 2 \deg f - r$, where $0 \leq r \leq \deg f$ and r is the number of poles of f fixed by T. Each fixed point of T that is not a pole of f is a zero of h. Thus, h has $\nu(T) - r > 2 \deg f - r$ zeros. This contradiction shows that $h \in \mathbb{C}$, and since T fixes points which are not poles of f, $h = 0$. Thus, f projects to a well-defined function $\tilde{f}$ on $M/\langle T\rangle$. In particular,

$$\deg f = (\deg \tilde{f})(\text{ord}\, T).$$

If $\deg f$ is prime, then $\deg \tilde{f} = 1$ and thus $M/\langle T\rangle \cong \mathbb{C} \cup \{\infty\}$. □

Remark. The fact that f is invariant under T shows that T has finite order.

V.1.5. In this section, M is a compact surface of genus $g \geq 2$.

Proposition. *For* $1 \neq T \in \text{Aut}\, M$,

$$\nu(T) \leq 2 + \frac{2g}{\text{ord}\, T - 1}. \tag{1.5.1}$$

PROOF. Apply the Riemann–Hurwitz relation to the natural projection $M \to M/\langle T\rangle$. If the range has genus γ, then

$$(2g - 2) = (\text{ord}\, T)(2\gamma - 2) + \sum_{j=1}^{\text{ord}\, T - 1} \nu(T^j). \tag{1.5.2}$$

We must explain the evaluation of the total branch number B of the projection appearing in the above formula. Clearly B is the weighted sum of the fixed points of $\langle T\rangle$; each fixed point appearing one less time than the order of its stability subgroup. This is exactly the contribution for B in

(1.5.2). We use now the obvious inequality $\nu(T) \leq \nu(T^j), j = 1, \ldots, \text{ord } T - 1$, and conclude

$$(2g - 2) \geq (\text{ord } T)(2\gamma - 2) + \nu(T)(\text{ord } T - 1),$$

or

$$\nu(T) \leq 2 + \frac{2g}{\text{ord } T - 1} - \frac{(2\gamma)(\text{ord } T)}{\text{ord } T - 1}$$

from which (1.5.1) follows. □

Corollary 1. *If* ord T *is prime, then*

$$\nu(T) = 2 + \frac{2g - 2\gamma(\text{ord } T)}{\text{ord } T - 1},$$

where γ is the genus of $M/\langle T \rangle$. In this case, equality holds in (1.5.1) *if and only if $\gamma = 0$.*

Proof. If ord T is prime, then $\nu(T) = \nu(T^j), j = 1, \ldots, \text{ord } T - 1$. □

Remark. In general, (even if ord T is *not* prime) if equality holds in (1.5.1), then $\gamma = 0$.

Corollary 2. *In general, $\nu(T) \leq 2g - 1$ if M is not hyperelliptic.*

Remark. For hyperelliptic surfaces we have, in general, better bounds (provided $g \geq 3$). See Proposition III.7.11.

Proof of Corollary. From Corollary 1, an automorphism of order 2 has precisely $2g + 2 - 4\gamma$ fixed points. Since M is not hyperelliptic, $\gamma \geq 1$. Hence for such an automorphism $\nu(T) \leq 2g - 2$. If ord $T \geq 3$, then $\nu(T) \leq 2 + g$. Now $2g - 1 \geq 2 + g$ unless $g = 2$. But in genus 2, every surface is hyperelliptic. □

V.1.6. We now give two examples to show that our results are sharp.

Example 1. Consider the hyperelliptic Riemann surface of genus $g \geq 2$,

$$w^2 = (z - e_1) \cdots (z - e_{2g+2}),$$

here $e_1, \ldots, e_{2g+2}$ are $2g + 2$ distinct complex numbers. The hyperelliptic involution is described by

$$(z,w) \mapsto (z, -w).$$

It clearly has $2g + 2$ fixed points: $z^{-1}(e_j), j = 1, \ldots, 2g + 2$; and the points over infinity (that is, the two points in $z^{-1}(\infty)$) cannot be fixed. We introduce some symmetry now by setting $e_{g+1+j} = -e_j, j = 1, \ldots, g + 1$. (Hence we are assuming also $e_j \neq 0$, for every j.) Thus the hyperelliptic surface is represented by

$$w^2 = (z^2 - e_1^2) \cdots (z^2 - e_{g+1}^2).$$

Here we have the additional automorphism T of period 2 given by

$$(z,w) \mapsto (-z,w).$$

How many fixed points does this automorphism T have? Clearly the quotient surface is

$$w^2 = (z - e_1^2) \cdots (z - e_{g+1}^2),$$

which is of genus $(g-1)/2$ if g is odd, and of genus $g/2$ if g is even. In the former case, $\nu(T) = 4$ and in the latter $\nu(T) = 2$. What are these fixed points? They clearly must be the two points over 0 in the later case, and the four points lying over 0 and ∞ in the former case.

We now introduce hyperelliptic surfaces with another symmetry: Let $\varepsilon = \exp(2\pi i/(2g+2))$ and let $e_j = \varepsilon^j$, $j = 1, \ldots, 2g+2$ and define an automorphism T of order $2g+2$ by

$$(z,w) \mapsto (\varepsilon z,w).$$

The only possible fixed points are over 0 and ∞. We will now examine the Taylor series of w at the origin and at ∞ to show that T has 2 fixed points (lying over zero). Note that the Riemann surface is represented by the equation

$$w^2 = z^{2g+2} - 1.$$

The two function elements lying over zero are then

$$w(z) = \pm i\left(\sum_{j=0}^{\infty} a_j z^{(2g+2)j}\right) \quad \text{with } a_0 = 1.$$

Clearly these two are fixed under T. At infinity $\zeta = 1/z$ is a good local coordinate, and the function elements over ∞ are

$$w(\zeta) = \pm\zeta^{-1-g}\left(\sum_{j=0}^{\infty} b_j \zeta^{(2g+2)j}\right) \quad \text{with } b_0 = 1.$$

Thus T interchanges these two function elements. Similarly, T^k fixes the two points lying over 0, and T^k sends a function element lying over ∞ onto the element times $\varepsilon^{k(g+1)}$. Thus T^k has 4 fixed points if and only if $k \equiv 0 \bmod 2$, and T^k has only 2 fixed points otherwise.

EXAMPLE 2. Consider the Riemann surface defined by the equation

$$w^3 = (z - e_1)^2(z - e_2) \cdots (z - e_r),$$

with $r \geq 2$ and $e_1, \ldots, e_r$ distinct complex numbers. We first compute its genus. The surface is a 3-sheeted covering of the sphere. It is branched (with each branch point of order 2) over the points $e_1, \ldots, e_r$. It is branched over ∞ if and only if $r + 1 \not\equiv 0 \bmod 3$. Thus, the genus of the surface is

$$r - 2, \quad \text{if } r \equiv 2 \bmod 3,$$

and

$$r - 1, \quad \text{if } r \not\equiv 2 \bmod 3.$$

An automorphism T of this surface given by

$$(z,w) \mapsto (z,\varepsilon w), \quad \text{with } \varepsilon = \exp\left(\frac{2\pi i}{3}\right).$$

For $r = 5$ (thus $g = 3$) it has $5(=2g - 1)$ fixed points. More generally, for $r \equiv 2 \bmod 3$, the surface has genus $r - 2$ and the automorphism has r fixed points $(=g + 2)$. Similarly, if $r \not\equiv 2 \bmod 3$ the surface has genus $r - 1$ and the automorphism has $r + 1$ fixed points $(=g + 2)$.

Exercise

Let α be an arbitrary complex number and let M_α be the Riemann surface of the algebraic curve

$$\left(1 + z + \frac{\alpha}{z}\right)\left(1 + w + \frac{1}{w}\right) = 1.$$

(a) Compute the genus of M_α.
(b) Consider the case $\alpha = 1$. Show that M_1 is conformally equivalent to the Riemann sphere and find explicitly a conformal map of M_1 onto $\mathbb{C} \cup \{\infty\}$.
(c) The surface M_1 has several automorphisms; consider the group of automorphisms generated by

$$z \mapsto 1/z \quad \text{and} \quad w \mapsto w,$$

$$z \mapsto z \quad \text{and} \quad w \mapsto 1/w,$$

and

$$z \mapsto w \quad \text{and} \quad w \mapsto z.$$

Under the isomorphism from part (b), identify this group with a subgroup of $PSL(2,\mathbb{C})$.

V.1.7. Theorem. *Let M be a compact surface of genus $g \geq 2$. Let $1 \neq T \in \operatorname{Aut} M$. If $\nu(T) > 4$, then every fixed point of T is a Weierstrass point.*

Proof. We may assume that $n = \operatorname{ord} T$ is prime (if not, there is a j such that T^j is of prime order, and clearly every fixed point of T is a fixed point of T^j). Let

$$\pi: M \to M/\langle T \rangle = \tilde{M}$$

be the canonical projection, and let γ be the genus of $\tilde{M}$. The Riemann–Hurwitz formula yields

$$2g - 2 = n(2\gamma - 2) + (n - 1)\nu(T). \tag{1.7.1}$$

Let $P \in M$ be a fixed point of T and $\tilde{P} = \pi(P) \in \tilde{M}$. There is a non-constant function $\tilde{f} \in \mathscr{K}(\tilde{M})$ such that $\tilde{f}$ is holomorphic on $\tilde{M}\backslash\{\tilde{P}\}$ and $\tilde{f}$ has a pole at $\tilde{P}$ of order $\leq \gamma + 1$. Let $f = \tilde{f} \circ \pi$ (f is the lift of $\tilde{f}$ to M). Then $f \in \mathscr{K}(M)$, f is holomorphic on $M\backslash\{P\}$, and $\operatorname{ord}_P f \leq n(\gamma + 1)$. We now use (1.7.1) and the hypothesis $\nu(T) > 4$ to conclude

$$2g - 2 = n(2\gamma - 2) + (n - 1)\nu(T) > n(2\gamma - 2) + 4(n - 1),$$

or

$$g + 1 > n(\gamma + 1).$$

It follows that P is a Weierstrass point on M. □

V.1.8. Theorem. *Let M be a compact surface of genus $g \geq 2$. Let $1 \neq \tilde{T} \in \operatorname{Aut} M$ with $n = \operatorname{ord} \tilde{T}$, a prime. Let $\tilde{g}$ be the genus of $\tilde{M} = M/\langle \tilde{T} \rangle$ (thus, $\nu(\tilde{T}) = 2 + 2(g - n\tilde{g})/(n - 1)$). Suppose that $g > n^2\tilde{g} + (n - 1)^2$, and that there is a $1 \neq T \in \operatorname{Aut} M$ with $\nu(T) > 2n(\tilde{g} + 1)$. Then:*

Each fixed point of $\tilde{T}$ is a fixed point of T and $\operatorname{ord} T \leq \operatorname{ord} \tilde{T} = n$. (1.8.1)

If $\operatorname{ord} T = n$, then $T \in \langle \tilde{T} \rangle$. (1.8.2)

$\langle \tilde{T} \rangle$ is a normal subgroup of $\operatorname{Aut} M$. (1.8.3)

If $n = 2$, $\tilde{T}$ is in the center of $\operatorname{Aut} M$. (1.8.4)

PROOF. Let $P_1, \ldots, P_{\nu(\tilde{T})}$ be the fixed points of $\tilde{T}$. As in the proof of the last theorem, for each j, there is a non-constant f_j holomorphic on $M \backslash \{P_j\}$ with a pole at P_j of order $\leq n(\tilde{g} + 1)$. It therefore follows from Proposition V.1.4 that $f_j = f_j \circ T$ for each j and thus $T(P_j) = P_j$. We have thus verified the first part of (1.8.1). In particular, $\nu(T) \geq \nu(\tilde{T})$. We rewrite

$$g > n^2\tilde{g} + (n - 1)^2$$

in the form (add $(n - 1)g$ to both sides and simplify)

$$\frac{g - n\tilde{g}}{n - 1} > \frac{g + n - 1}{n}. \tag{1.8.5}$$

We continue with

$$\nu(T) \geq \nu(\tilde{T}) = 2 + \frac{2(g - n\tilde{g})}{n - 1} > 2 + 2\frac{g + n - 1}{n}$$

$$= 2 + \frac{2g}{n} + 2\frac{n - 1}{n} > 2 + \frac{2g}{n}.$$

(The first strict inequality being a consequence of (1.8.5).) By Proposition V.1.5,

$$\nu(T) \leq 2 + \frac{2g}{\operatorname{ord} T - 1},$$

and hence $\operatorname{ord} T \leq n$, finishing the proof of (1.8.1).

Assume $\operatorname{ord} T = \operatorname{ord} \tilde{T}$. Both T and $\tilde{T}$ are in the stability subgroup (of $\operatorname{Aut} M$) of P_1. Since this stability subgroup is a cyclic rotation group (Corollary to Proposition III.7.7), it is clear that $\langle T \rangle = \langle \tilde{T} \rangle$.

Next, let $B \in \operatorname{Aut} M$ and set $C = B^{-1} \circ \tilde{T} \circ B$. Then $\operatorname{ord} C = \operatorname{ord} \tilde{T}$; and, by the inequality on g, $\nu(C) = \nu(\tilde{T}) = 2 + 2(g - n\tilde{g})/(n - 1) > 2n(\tilde{g} + 1)$. Thus by (1.8.2) $C \in \langle \tilde{T} \rangle$ or $\langle \tilde{T} \rangle$ is normal in Aut M. We have established (1.8.3) and hence also (1.8.4). □

Remark. The above generalizes the fact that on a hyperelliptic surface M (of genus ≥ 2) any automorphism with more than $4(=2 \cdot 2(0 + 1))$ fixed points must (be the hyperelliptic involution and hence must) commute with every element of Aut M.

V.1.9. We wish to generalize slightly the concept of hyperellipticity. A compact Riemann surface M will be called *γ-hyperelliptic* ($\gamma \in \mathbb{Z}$, $\gamma \geq 0$) provided there is a compact Riemann surface $\tilde{M}$ of genus γ and a holomorphic mapping of degree 2

$$\pi: M \to \tilde{M}.$$

Thus, a γ-hyperelliptic surface is a two sheeted covering of a compact surface of genus γ, with "0-hyperelliptic" corresponding to our previous notion of "hyperelliptic".

It is immediately obvious that on every γ-hyperelliptic surface M, there is a γ-hyperelliptic involution; that is, a $J_\gamma \in$ Aut M with

$$\text{ord } J_\gamma = 2, \qquad \nu(J_\gamma) = 2g + 2 - 4\gamma.$$

Theorem. *Let M be a γ-hyperelliptic surface of genus $g > 4\gamma + 1$. We have:*

$$\text{If } 1 \neq T \in \text{Aut } M \text{ with } \nu(T) > 4(\gamma + 1), \text{ then } T = J_\gamma. \tag{1.9.1}$$

$$J_\gamma \text{ is in the center of Aut } M. \tag{1.9.2}$$

$$\text{If } f \in \mathcal{K}(M)\backslash\mathbb{C}, \text{ and } \deg f < g + 1 - 2\gamma, \text{ then } \deg f \text{ is even.} \tag{1.9.3}$$

Proof. Statements (1.9.1) and (1.9.2) are immediate consequences of the preceding theorem. Note that $2 \deg f < 2g + 2 - 4\gamma = \nu(J_\gamma)$, thus by Proposition V.1.4, $f = f \circ J_\gamma$ and $\deg f$ is a multiple of ord $J_\gamma = 2$. □

Corollary 1. *The γ-hyperelliptic involution on a surface of genus g is unique (if it exists) provided $g > 4\gamma + 1$.*

Proof. The γ-hyperelliptic involution has $2g + 2 - 4\gamma$ fixed points, and $2g + 2 - 4\gamma > 4\gamma + 4$ if and only if $g > 4\gamma + 1$. The uniqueness now follows from (1.9.1). □

Corollary 2. *If γ, r are non-negative integers with $r > 0$ and $g > 4\gamma + 1 + 2r$, then a surface of genus g cannot be both γ-hyperelliptic and $(\gamma + r)$-hyperelliptic. In particular for $g > 3$, a surface of genus g cannot be both hyperelliptic and 1-hyperelliptic.*

Remark. In V.1.6 we have seen examples of surfaces of genus 2 and 3 that are both hyperelliptic and 1-hyperelliptic

Proof of Corollary 2. Suppose M is both $(\gamma + r)$- and γ-hyperelliptic, then $J_{\gamma+r}$ has

$$2g + 2 - 4(\gamma + r) > 4(\gamma + 1)$$

fixed points. Thus $J_{\gamma+r} = J_\gamma$, which is only possible if $r = 0$. □

V.1.10. Proposition (Accola). *Let M be a compact surface of genus $g \geq 2$. Let G, G_j be subgroups of* Aut M *such that $G = \bigcup_{j=1}^k G_j$ and $G_i \cap G_j = \{1\}$ for $i \neq j$. Assume* ord $G_j = n_j$, ord $G = n$, *and let γ = genus of M/G, γ_j = genus of M/G_j. Then*

$$(k-1)g + n\gamma = \sum_{j=1}^k n_j\gamma_j. \tag{1.10.1}$$

Proof. Let us denote by B and B_j the total branch number of the canonical projections

$$M \to M/G, \qquad M \to M/G_j.$$

Obviously

$$B = \sum_{P \in M} (\text{ord } G_P - 1)$$

(with G_P, the stability subgroup of G at $P \in M$), with a similar formula for B_j. Now if G_P is nontrivial, it is cyclic and generated by some $T \in G$. Clearly $T \in (G_j)_P$ for some j. It follows easily from these observations that

$$B = \sum_{j=1}^k B_j.$$

We now use Riemann–Hurwitz

$$2g - 2 = n(2\gamma - 2) + B,$$
$$2g - 2 = n_j(2\gamma_j - 2) + B_j.$$

Summing the last expressions over j, subtracting the first, and using the fact that $n = \sum_{j=1}^k n_j - (k-1)$, we get (1.10.1). □

Corollary. *Let M be a surface of genus 3. Assume there is a subgroup G of* Aut M *with $G \cong Z_2 \oplus Z_2$, and M/G of genus zero. If one element of $G \backslash \{1\}$ operates fixed point freely, then M is hyperelliptic.*

Proof. The group G can be decomposed as in the theorem with $k = 3$ and $n_1 = n_2 = n_3 = 2$. Hence

$$3 = \gamma_1 + \gamma_2 + \gamma_3.$$

Since one of the elements of G operates fixed point freely, one of the γ_j must be equal to 2. The other two cannot both be positive. □

A consequence of the corollary is the

Theorem. *A smooth two-sheeted holomorphic cover of a surface of genus 2 is hyperelliptic.*

First Proof. By Riemann–Hurwitz, the cover M has genus 3. The fact that M is a two-sheeted cover shows that it carries an involution E (the sheet interchange); the involution is fixed point free because the cover is unramified. To apply the corollary, we must produce another involution on M. The second involution J is the lift of the hyperelliptic involution from the surface of genus 2. We need to explain why the involution on the surface of genus 2 can be lifted to M. The general theory tells us that an automorphism can be lifted to a smooth cover provided the automorphism preserves the defining subgroup of the cover. In the case of abelian covers, those for which the covering group is commutative, there is a simplification. In this case, the defining subgroup contains the commutator subgroup of the fundamental group. Therefore, we can in effect consider the action on the hyperelliptic involution on the image of the defining subgroup in the first homology group of the surface which is isomorphic to the fundamental group factored by its commutator subgroup (see also V.3.3). The hyperelliptic involution acts as minus the identity on the homology group; so all subgroups are preserved.

The lift J of the hyperelliptic involution can always be chosen so that it has fixed points (because we are lifting a map with fixed points). The map J can have four or eight fixed points. In general, by the corollary we need only show that E and J generate the Klein four group. This follows from the observation that $E \circ J$ and $J \circ E$ are also lifts to M of the hyperelliptic involution on the surface of genus 2 and must hence coincide.

Second (more elegant) Proof. As we shall see in V.2.4, the fixed point free involution E on M acts on the space of abelian differentials of the first kind (on M). The eigenvalues for this action are ± 1. The eigenspace for the eigenvalue 1 is 2-dimensional (this eigenspace consists of the lifts to M of the holomorphic differentials on the surface of genus 2 and these differentials are, of course, invariant under the action of E); let θ_1 and θ_2 be two linearly independent differentials in this space. The eigenspace for the eigenvalue -1 is 1-dimensional (this eigenspace consists of anti-invariant differentials); let ω be a non-trivial element of this space. Each of the three differentials ω, θ_1, and θ_2 have the property that they vanish at a point $Q \in M$ if and only if they vanish at $E(Q)$. Let P be a zero of ω. Clearly we can find a non-trivial linear combination θ of θ_1 and θ_2 that vanishes at P. Now ω and θ are linearly independent and vanish at P and $E(P) \neq P$. Thus $i(PE(P)) = 2$ and by Riemann–Roch $r(1/PE(P)) = 2$. □

V.1.11. We record for future use the following simple

Proposition. *Let M be compact Riemann surface of genus $g \leq 2$. If $T \in$ Aut M is of prime order n, then $n \leq 2g + 1$.*

Proof. *Let* $\nu = \nu(T)$ and use Riemann–Hurwitz with respect to the projection $\pi: M \to \tilde{M} = M/\langle T \rangle$ onto a surface of genus $\tilde{g}$.

$$2g - 2 = n(2\tilde{g} - 2) + \nu(n - 1). \tag{1.11.1}$$

Assume $n \geq 2g$.

If $\tilde{g} \geq 2$, then $n(2\tilde{g} - 2) \geq 2n \geq 4g$; a contradiction.
If $\tilde{g} = 1$, then $v \neq 0$ and $v(n-1) \geq 2g - 1$; again a contradiction.
If $\tilde{g} = 0$, then $v \geq 3$. Assume first that $v \geq 4$, then $-2n + v(n-1) \geq 2n - 4 \geq 4g - 4$; a contradiction.
If $v = 3$, then (1.11.1) reads $2g - 2 = n - 3$ or $n = 2g + 1$.

Remark. Proposition V.1.11 has shown that the maximal prime order of an automorphism is $2g + 1$ and that in this case the automorphism T has 3 fixed points and $M/\langle T \rangle$ is the Riemann sphere. Examples of such Riemann surfaces are those defined by the equations

$$w^{2g+1} = (z - e_1)^{\alpha_1}(z - e_2)^{\alpha_2}(z - e_3)^{\alpha_3},$$

where e_1, e_2, e_3 are distinct complex numbers and $\alpha_1, \alpha_2, \alpha_3$ are positive integers such that $\alpha_1 + \alpha_2 + \alpha_3 \equiv 0 \operatorname{mod}(2g + 1)$, and $2g + 1$ is a prime number. The automorphism T sends w into $\exp(2\pi i/(2g + 1))w$ and fixes z. A natural question to ask is what are some other possible prime orders. We shall see in the next section that the only prime order $>g$ other than $2g + 1$ is $g + 1$. In this case T will turn out to have 4 fixed points and once again $M/\langle T \rangle$ will be the Riemann sphere.

V.2. Representations of the Automorphism Group on Spaces of Differentials

Let M be a compact Riemann surface of genus $g \geq 2$ and Aut M the group of conformal automorphisms of M. In this section we study the representation of Aut M on the space of holomorphic q-differentials on M, $\mathscr{H}^q(M)$, $q \geq 0$. With the aid of the representations, we will improve many results of the previous section.

V.2.1. Let us begin by observing that a product of a holomorphic q_1-differential with a holomorphic q_2-differential is a holomorphic $(q_1 + q_2)$-differential. Thus the direct sum

$$\mathscr{H}(M) = \bigoplus_{q=0}^{\infty} \mathscr{H}^q(M)$$

is a commutative graded algebra.

Let $T \in \operatorname{Aut} M$ and $\varphi \in \mathscr{H}^q(M)$, then T acts on φ to produce

$$T\varphi = \varphi \circ T^{-1}.$$

(Say that φ in terms of the local coordinate z vanishing at $P \in M$ is given by $\mu(z)\,dz^q$. Choose a local coordinate ζ vanishing at TP. Assume that in terms of these local coordinates T^{-1} is given by $z = f(\zeta)$. Then $T\varphi$ in terms of the local coordinate ζ is given by $\mu(f(\zeta))f'(\zeta)^q\,d\zeta^q$.) We note several properties of this action of T on differentials:

a. For $T_1, T_2 \in \operatorname{Aut} M$, $\varphi \in \mathscr{H}^q(M)$,

$$(T_1 \circ T_2)\varphi = T_1(T_2\varphi).$$

b. For each $T \in \operatorname{Aut} M$,

$$T: \mathscr{H}^q(M) \to \mathscr{H}^q(M)$$

is a $\mathbb{C}$-linear isomorphism.

c. For $T \in \operatorname{Aut} M$, $\varphi_j \in \mathscr{H}^{q_j}(M)$, $j = 1, 2$,

$$T(\varphi_1\varphi_2) = (T\varphi_1)(T\varphi_2).$$

d. For $0 \neq \varphi \in \mathscr{H}^q(M)$, $T \in \operatorname{Aut} M$, $P \in M$,

$$\operatorname{ord}_{TP} T\varphi = \operatorname{ord}_P \varphi.$$

Proposition. *We have defined a representation of* Aut M on $\mathscr{H}^q(M)$ *(that is, a homomorphism of* Aut M *into* Aut$(\mathscr{H}^q(M))$*) as well as a graded representation of* Aut M *on* $\mathscr{H}(M)$. *For* $q \geq 1$, *the representation on* $\mathscr{H}^q(M)$ *is faithful, except for* $g = 2 = q$ *in which case the kernel of the representation is precisely* $\langle J \rangle$, J = *hyperelliptic involution.*

Proof. Assume $T \neq 1$, $T \in \operatorname{Aut} M$. Exclude for a moment the case where $T = J$ (in particular M is hyperelliptic in the excluded cases). There is thus a Weierstrass point $P \in M$ with $TP \neq P$. Therefore, there is a $0 \neq \varphi \in \mathscr{H}^1(M)$ such that $\operatorname{ord}_P \varphi \geq g$. If $T\varphi^q = \varphi^q$, then φ^q has at least gq zeros at P and gq zeros at TP. Hence φ^q has $2gq > q(2g - 2)$ zeros and we have that $\varphi = 0$. Hence $T \neq 1$ cannot induce the identity in Aut$(\mathscr{H}^q(M))$, and the representation is faithful.

We examine now the action of J on $\mathscr{H}^q(M)$. In this case M is hyperelliptic and we may assume that M is represented by

$$w^2 = (z - e_1) \cdots (z - e_{2g+2}),$$

with $\{e_1, \ldots, e_{2g+2}\}$ a set of $2g + 2$ distinct complex numbers. Recall the basis for $\mathscr{H}^1(M)$ introduced in III.7.5:

$$\varphi_j = \frac{z^{j-1}\,dz}{w}, \qquad j = 1, \ldots, g.$$

The automorphism J can be represented by

$$(z,w) \mapsto (z, -w).$$

It is thus clear that J does not act as the identity on $\mathscr{H}^q(M)$ for odd q. (Note that $J\,dz = dz$, and thus $J\varphi_j = -\varphi_j$ and $J\varphi_j^q = (-1)^q\varphi_j^q$.) For $g = 2$, a basis for $\mathscr{H}^2(M)$ is given by

$$\frac{dz^2}{w^2}, \quad \frac{z\,dz^2}{w^2}, \quad \frac{z^2\,dz^2}{w^2}$$

from which it follows clearly that $J\varphi = \varphi$ for all $\varphi \in \mathscr{H}^2(M)$. For $g \geq 2$ and $q \geq 2$ even, we consider

$$J\left(\frac{dz^q}{w^{q-1}}\right) = -\frac{dz^q}{w^{q-1}}.$$

We must only verify that dz^q/w^{q-1} is holomorphic. Use the notation of III.7.4 and III.7.5,

$$\left(\frac{dz^q}{w^{q-1}}\right) = P_1 \cdots P_{2g+2} Q_1^{(g+1)(q-1)-2q} Q_2^{(g+1)(q-1)-2q}.$$

Thus the differential is not holomorphic if and only if $(g+1)(q-1) < 2q$ if and only if $q = 2$ and $g = 2$ (as was to be expected from the exclusion we already made). □

V.2.2. Let G be a subgroup of Aut M. Let $\varphi \in \mathscr{H}^q(M)$, we shall say that φ is *G-invariant* provided $T\varphi = \varphi$ for all $T \in G$. The vector space of G-invariant holomorphic q-differentials on M will be denoted by $\mathscr{H}^q_G(M)$. For the moment let us ignore the fact that $q \geq 1$ (assume q is arbitrary) and the fact that φ is holomorphic (merely assume that φ is meromorphic). Let π denote the natural projection

$$\pi: M \to \tilde{M} = M/G.$$

Recall that we have shown in III.4.12 how to lift an arbitrary q-differential $\tilde{\varphi}$ on $\tilde{M}$ to a q-differential φ on M. It is clear that φ is, in this case, G-invariant. Conversely, every G-invariant q-differential φ on M projects to a q-differential on $\tilde{M}$.

We let for $P \in M$, $\nu(P) = \operatorname{ord} G_P$, then as we saw (III(4.12.2))

$$\operatorname{ord}_P \varphi = \nu(P) \operatorname{ord}_{\pi(P)} \tilde{\varphi} + q(\nu(P) - 1).$$

Thus, we see that

$$\operatorname{ord}_P \varphi \geq 0 \Leftrightarrow \operatorname{ord}_{\pi(P)} \tilde{\varphi} \geq -q\left(1 - \frac{1}{\nu(P)}\right).$$

Since $\operatorname{ord}_{\pi(P)} \tilde{\varphi} \in \mathbb{Z}$, we see that

$$\operatorname{ord}_P \varphi \geq 0 \Leftrightarrow \operatorname{ord}_{\pi(P)} \tilde{\varphi} \geq -\left[q\left(1 - \frac{1}{\nu(P)}\right)\right],$$

where $[x]$ is the greatest integer $\leq x$.

Since the projection π is defined by a group, $\nu(P)$ depends only on the orbit of P under G. Thus, for $\tilde{P} \in \tilde{M}$ we can define $\nu(\tilde{P}) = \nu(\pi^{-1}(\tilde{P}))$, and it is thus convenient to introduce a *q-canonical ramification divisor* of π by

$$Z^{(q)} = \prod_{\tilde{P} \in \tilde{M}} \tilde{P}^{-[q(1-1/\nu(\tilde{P}))]}. \tag{2.2.1}$$

If, for an arbitrary divisor D on M, we define

$$\mathscr{H}^q(M; D) = \{\varphi;\ \varphi \text{ is a meromorphic } q\text{-differential and } (\varphi)/D \text{ is integral}\},$$

then we have established the following

Proposition. *For* $q \geq 1$

$$\mathscr{H}^q_G(M) \cong \mathscr{H}^q(\tilde{M}, Z^{(q)}). \tag{2.2.2}$$

Corollary. *Let* $\tilde{g}$ *be the genus of* $\tilde{M} = M/G$, *then*

$$\dim \mathscr{H}^1_G(M) = \tilde{g}, \tag{2.2.3}$$

and for $q \geq 2$

$$\dim \mathscr{H}^q_G(M) = (2q-1)(\tilde{g}-1) + \sum_{\tilde{P} \in \tilde{M}} \left[q\left(1 - \frac{1}{\nu(\tilde{P})}\right)\right]. \tag{2.2.4}$$

PROOF. The proof is analogous to the one given in III.5.2. We use the isomorphism of the proposition and the Riemann-Roch theorem. Let Z be a canonical divisor on $\tilde{M}$. It is easy to see that

$$L\left(\frac{Z^{(q)}}{Z^q}\right) \cong \mathscr{H}^q(\tilde{M}, Z^{(q)}).$$

The isomorphism is given by

$$L\left(\frac{Z^{(q)}}{Z^q}\right) \ni f \mapsto f\varphi \in \mathscr{H}^q(\tilde{M}, Z^{(q)}),$$

where φ is a q-differential and $(\varphi) = Z^q$. Now for $q = 1$, $Z^{(1)} = 1$ and thus $\mathscr{H}^1(\tilde{M}, Z^{(1)}) = \mathscr{H}^1(\tilde{M}) =$ the $\tilde{g}$-dimensional space of abelian differentials of the first kind on $\tilde{M}$. For $q \geq 2$, we compute using Riemann-Roch

$$\begin{aligned} r\left(\frac{Z^{(q)}}{Z^q}\right) &= \deg\left(\frac{Z^q}{Z^{(q)}}\right) - \tilde{g} + 1 + i\left(\frac{Z^q}{Z^{(q)}}\right) \\ &= q(2\tilde{g}-2) + \sum_{\tilde{P}\in\tilde{M}} \left[q\left(1 - \frac{1}{\nu(\tilde{P})}\right)\right] - (\tilde{g}-1) + i\left(\frac{Z^q}{Z^{(q)}}\right). \end{aligned}$$

Thus to complete the proof of the corollary, it suffices to show that $i(Z^q/Z^{(q)}) = 0$ which follows from the inequality $\deg(Z^q/Z^{(q)}) > 2\tilde{g} - 2$. We have already used the fact that

$$\deg\left(\frac{Z^q}{Z^{(q)}}\right) = q(2\tilde{g}-2) + \sum_{\tilde{P}\in\tilde{M}} \left[q\left(1 - \frac{1}{\nu(\tilde{P})}\right)\right].$$

Thus the desired inequality clearly holds for $\tilde{g} \geq 2$. For $\tilde{g} \leq 1$, we must examine

$$\sum_{\tilde{P}\in\tilde{M}} \left[q\left(1 - \frac{1}{\nu(\tilde{P})}\right)\right]. \tag{2.2.5}$$

If $\tilde{g} = 1$, then we saw in the proof of Hurwitz's theorem that for some $\tilde{P} \in \tilde{M}$, $\nu(\tilde{P}) \geq 2$, thus the sum in (2.2.5) is greater than or equal to $[q/2] \geq 1$. For

$\tilde{g} = 0$, we must show that the sum in (2.2.5) exceeds $2(q-1)$. Now

$$\sum_{\tilde{P} \in \tilde{M}} \left[q\left(1 - \frac{1}{v(\tilde{P})}\right)\right] \geq \sum_{\tilde{P} \in \tilde{M}} \left\{ q - \frac{q}{v(\tilde{P})} - \frac{v(\tilde{P}) - 1}{v(\tilde{P})} \right\}$$
$$= (q-1) \sum_{\tilde{P} \in \tilde{M}} \left(1 - \frac{1}{v(\tilde{P})}\right).$$

But we saw in the proof of Hurwitz's theorem (equation (1.3.3)) that $\sum_{\tilde{P} \in \tilde{M}} (1 - (1/v(\tilde{P}))) > 2$. □

Remark. Let $q = 2$ and let $n =$ the number of points $\tilde{P} \in \tilde{M}$ with $v(\tilde{P}) > 1$, then

$$\dim \mathscr{H}^2_G(M) = 3\tilde{g} - 3 + n \qquad (\geq 0).$$

V.2.3. We start now a more detailed investigation of the action of $T \in \operatorname{Aut} M$ on $\mathscr{H}^q(M)$, $q \geq 1$.

Proposition. *There exists a basis for $\mathscr{H}^q(M)$ such that the action of T on $\mathscr{H}^q(M)$ is represented by a diagonal matrix.*

PROOF. Since Aut M is a finite group, there is an integer n such that $T^n = 1$. Thus (because Aut M is represented on $\mathscr{H}^q(M)$) $T^n = 1$ as a matrix with respect to any basis for $\mathscr{H}^q(M)$. In particular, $\lambda^n - 1$ is an annihilating polynomial for T and the minimal polynomial (a factor of $\lambda^n - 1$) has distinct roots. Thus, by a well known result of linear algebra, T can be diagonalized.

We can say something more. Each entry on the diagonal must be an n-th root of unity. We will continue with this analysis in the next section. □

Remark. We shall actually be able (in most cases) to exhibit a basis for $\mathscr{H}^q(M)$ with respect to which T is diagonal.

V.2.4. We continue with the problem introduced in the last section. We wish to determine the eigenvalues (and their multiplicities) of T.

Assume that $T \in \operatorname{Aut} M$, ord $T = n > 1$, and $\langle T \rangle$ operates fixed point freely on M. (If n is prime, then $\langle T \rangle$ operates fixed point freely if and only if T is fixed point free. If n is not prime, the backwards implication need not hold.) Let us recall that the eigenvalues must be of the form ε^j where $\varepsilon = \exp(2\pi i/n)$ with $j = 0, \ldots, n-1$. The eigenspace of 1 is precisely $\mathscr{H}^q_{\langle T \rangle}(M)$ and thus has dimension equal to $(2q-1)(\tilde{g}-1)$ for $q > 1$ and equal to $\tilde{g}$ for $q = 1$, where as usual $\tilde{g} =$ genus of $\tilde{M} = M/\langle T \rangle$. Next assume that λ is any eigenvalue. Then for $\varphi \in \mathscr{H}^q(M)$ with $T\varphi = \lambda\varphi$, we have that (φ) is invariant under T (and hence also under $\langle T \rangle$). Let us denote by E_λ the eigenspace of λ. If we now choose any non-trivial $\varphi_0 \in E_\lambda$ we see that for $\varphi \in E_\lambda$, φ/φ_0 is a $\langle T \rangle$-invariant meromorphic function on M whose divisor is a multiple of $(\varphi_0)^{-1}$. The divisor (φ_0) projects (because it is $\langle T \rangle$-invariant) to an integral divisor D on $\tilde{M}$. It is thus easy to see that $E_\lambda \cong L(1/D)$. Now

$\deg D = 2q(g-1)/n$ and Riemann–Hurwitz (or elementary arithmetic) shows that $(2g-2) = n(2\tilde{g}-2)$, and thus $\deg D = 2q(\tilde{g}-1)$. Hence by Riemann–Roch (recall that $\tilde{g} \geq 2$)

$$r\left(\frac{1}{D}\right) = \begin{cases} \deg D - (\tilde{g}-1) + i(D) = (2q-1)(\tilde{g}-1), & \text{for } q > 1, \\ \tilde{g} - 1 + i(D), & \text{for } q = 1. \end{cases}$$

For $q = 1$, $i(D) = 1$ if D is canonical and $i(D) = 0$ otherwise. Now the divisor D is canonical if and only if φ_0 projects to a holomorphic differential on $\tilde{M}$; that is, if and only if φ_0 is G-invariant (if and only if $\lambda = 1$). We conclude

$$\dim E_\lambda = \begin{cases} (2q-1)(\tilde{g}-1), & q > 1, \lambda^n = 1, \\ \tilde{g}, & q = 1, \lambda = 1, \\ \tilde{g} - 1, & q = 1, \lambda \neq 1, \lambda^n = 1. \end{cases}$$

Since $\sum_{\lambda \in \mathbb{C}} \dim E_\lambda = \dim \mathscr{H}^q(M)$ and $(g-1) = n(\tilde{g}-1)$, we see that all the n-th roots of unity must be eigenvalues and that their multiplicities are given by the above formulae.

V.2.5. We extend next the considerations of the previous section to the general case ($G = \langle T \rangle$ is assumed to have fixed points). The development given below is due to I. Guerrero.

To fix notation (in addition to what was introduced in V.2.4), we let

$$n_j = \dim E_{\varepsilon^j}, \qquad j = 0, \ldots, n-1.$$

Formulae (2.2.3) and (2.2.4) compute n_0. To compute n_j for $1 \leq j \leq n-1$, we partition the branch set X of $\pi: M \to M/G$ into a disjoint union

$$X = \bigcup_{l=1}^{n-1} X_l,$$

where

$$X_l = \{P \in M;\ T^l P = P \text{ and } T^k P \neq P \quad \text{for } 0 < k \leq l-1\}.$$

We claim that if $X_l \neq \varnothing$, then $l | n$ (l divides n). Suppose $P \in X_l$ and $2 \leq l \leq n-1$. Write $n = \mu l + k$, $0 \leq k \leq l-1$. Then $T^k P = P$, and thus $k = 0$. Hence we see $l | n$. Furthermore, the orbit of each point $P \in X_l$ consists of the l points $P, TP, \ldots, T^{l-1}P$ and are also in X_l. It is thus clear that if for non-empty X_l, we set

$$\pi(X_l) = \{Q_{ml};\ 1 \leq m \leq x_l\}, \qquad 1 \leq l \leq n-1,$$

then $\pi(X_l)$ consists of x_l distinct points and X_l consists of $l x_l$ points. We define x_l to be zero for empty X_l. We can choose a local coordinate z for each point in $\pi^{-1}(Q_{ml})$ such that T^{-l} is given by

$$T^{-l}: z \mapsto \eta_{ml} z,$$

where η_{ml} is a primitive n/l-root of unity (note that η_{ml} depends only on Q_{ml} and *not* on the choice of $P \in \pi^{-1}(Q_{ml})$).

Suppose now that $0 \neq \varphi_0 \in E_{\varepsilon^j}$, $1 \leq j \leq n-1$. As before, in this case we also have

$$n_j = r\left(\frac{1}{D_j}\right),$$

where D_j is a certain integral divisor on M/G.

To find what D_j is, let us consider the equation for $0 \neq \varphi \in E_{\varepsilon^j}$,

$$T\varphi = \varepsilon^j \varphi.$$

Thus we also have

$$T^l \varphi = \varepsilon^{jl} \varphi. \tag{2.5.1}$$

Let us look at the Taylor series expansion of φ, at a point in $\pi^{-1}(Q_{jl})$:

$$\varphi = \left(\sum_{k=0}^{\infty} \mathscr{A}_k z^k\right) dz^q.$$

Thus equation (2.5.1) reads

$$\sum_{k=0}^{\infty} \mathscr{A}_k \eta_{ml}^{k+q} z^k = \varepsilon^{jl}\left(\sum_{k=0}^{\infty} \mathscr{A}_k z^k\right).$$

In particular, for each $k = 0, 1, 2, \ldots,$

$$\mathscr{A}_k(\eta_{ml}^{k+q} - \varepsilon^{jl}) = 0.$$

Choose the unique integer λ_{mlj} such that

$$1 \leq \lambda_{mlj} \leq n/l \quad \text{and} \quad \eta_{ml}^{\lambda_{mlj}} = \varepsilon^{jl}.$$

Note that $\lambda_{ml(n/l)} = n/l$ (this defines λ_{mln}) and we set (for further use) $\lambda_{ml0} = 0$. We conclude that

$$\mathscr{A}_k = 0 \quad \text{unless } k + q \equiv \lambda_{mlj}\left(\operatorname{mod} \frac{n}{l}\right).$$

Let α_{ml} be the order of φ_0 at $\pi^{-1}(Q_{ml})$ (this integer only depends on Q_{ml} and not on the point P we choose in its preimage). By our previous remarks there is an integer β_{mlj} such that

$$\alpha_{ml} = \beta_{mlj}\frac{n}{l} + \lambda_{mlj} - q \geq 0$$

(note that $\beta_{mlj} \geq 0$).

Let $Q_1, \ldots, Q_s$ be the projections to M/G of the zeros of φ_0 not in X. Let $\alpha_k = \operatorname{ord}_{\pi^{-1}(Q_k)} \varphi_0$. We note that a function $\tilde{f} \in \mathscr{K}(M/G)$ is such that $f\varphi_0 \in \mathscr{H}^q(M)$, where $f = \tilde{f} \circ \pi$, if and only if

i. $\tilde{f}$ is holomorphic except at the points Q_k, Q_{ml},
ii. $\operatorname{ord}_{Q_k} \tilde{f} \geq -\alpha_k$, and
iii. $\operatorname{ord}_{Q_{ml}} \tilde{f} \geq -\beta_{mlj} + (q - \lambda_{mlj})/(n/l)$.

Now we know how to define the divisor D_j (which depends on φ_0):

$$D_j = \prod_{k=1}^{s} Q_k^{\alpha_k} \prod_{l|n} \prod_{m=1}^{x_l} Q_{ml}^{\beta_{mlj} + [(\lambda_{mlj} - q)/(n/l)]}.$$

Remark. With our convention for $j = 0$, the formula for D_j is valid for all j, $0 \leq j \leq n - 1$.

Note that $\deg(\varphi_0) = 2q(g - 1)$, and hence

$$n \sum_{k=1}^{s} \alpha_k + \sum_{l|n} \sum_{m=1}^{x_l} l\alpha_{ml} = 2q(g - 1). \tag{2.5.2}$$

Now from the definition of D_j,

$$\deg D_j = \sum_{k=1}^{s} \alpha_k + \sum_{l|n} \sum_{m=1}^{x_l} \beta_{ml} + \left[\frac{\lambda_{mlj} - q}{n/l}\right]. \tag{2.5.3}$$

Next we use Riemann–Hurwitz in the form

$$\begin{aligned} 2g - 2 &= n(2\tilde{g} - 2) + \sum_{l|n} \sum_{m=1}^{x_l} l\left(\frac{n}{l} - 1\right) \\ &= n(2\tilde{g} - 2) + \sum_{l|n} x_l(n - l). \end{aligned}$$

We return to (2.5.3) and continue our calculation (using (2.5.2) and Riemann–Hurwitz)

$$\begin{aligned} \deg D_j &= \sum_{k=1}^{s} \alpha_k + \sum_{l|n} \sum_{m=1}^{x_l} \left(\frac{\alpha_{ml} + q - \lambda_{mlj}}{n/l} + \left[\frac{\lambda_{mlj} - q}{n/l}\right]\right) \\ &\geq \frac{1}{n}\left(n \sum_{k=1}^{s} \alpha_k + \sum_{l|n} \sum_{m=1}^{x_l} l\alpha_{ml}\right) - \sum_{l|n} \sum_{m=1}^{x_l} \frac{n/l - 1}{n/l} \\ &= \frac{1}{n} 2q(g - 1) - \frac{1}{n} \sum_{l|n} x_l(n - l) = \frac{1}{n}\left(2g - 2 - \sum_{l|n} x_l(n - l)\right) \\ &\quad + \frac{1}{n}(2g - 2)(q - 1) = 2\tilde{g} - 2 + \frac{1}{n}(2g - 2)(q - 1) \\ &\geq 2\tilde{g} - 2. \end{aligned}$$

We have also shown that for $q > 1$

$$\deg D_j > 2\tilde{g} - 2.$$

We claim that even if $q = 1$ and even if $\deg D_j = 2\tilde{g} - 2$, $i(D_j) = 0$ for $j > 0$. Otherwise D_j is canonical. In this case, there exists an abelian differential $\tilde{\varphi}$ on M/G such that $(\tilde{\varphi}) = D_j$. Note that the only way for $\deg D_j = 2\tilde{g} - 2$ is to have equality all along the way in our previous calculation. In particular, $\{(n/l) - 1\}|(\lambda_{mlj} - 1)$, and since $1 \leq \lambda_{mlj} \leq (n/l)$, we conclude that $\lambda_{mlj} = n/l$. Thus

$$D_j = \prod_{k=1}^{s} Q_k^{\alpha_k} \prod_{l|n} \prod_{m=1}^{x_l} Q_{ml}^{\beta_{mlj}}.$$

The lift φ of $\tilde{\varphi}$ will have a zero of order α_k at each preimage of Q_k, and a zero of order

$$\beta_{mlj}\frac{n}{l}+\left(\frac{n}{l}-1\right)=\alpha_{ml}$$

at each point of $\pi^{-1}(Q_{ml})$. Thus φ will have the same divisor as φ_0. But this means that φ is a constant multiple of φ_0. Since φ is G-invariant, $\varphi \in E_1$. This contradicts the fact that $\varphi_0 \in E_{\varepsilon^j}$, $1 \le j \le n-1$.

We have concluded that if E_{ε^j} is non-trivial, then it has dimension $r(1/D_j) = \deg D_j + 1 - \tilde{g}$, except for the case $q = 1$, $j = 0$, and no fixed points.

Let us turn to the case $q = 1$. Note that for $j > 0$, $[(\lambda_{mlj} - 1)/(n/l)] = 0$. Thus for $j > 0$,

$$\deg D_j = \sum_{k=1}^{s} \alpha_k + \sum_{l|n}\sum_{m=1}^{x_l} \frac{\alpha_{ml} + 1 - \lambda_{mlj}}{n/l},$$

and from (2.5.2)

$$\deg D_j = \frac{2g-2}{n} - \frac{1}{n}\sum_{l|n}\sum_{m=1}^{x_l} l(\lambda_{mlj} - 1).$$

Hence we conclude that (using Riemann–Hurwitz)

$$r\left(\frac{1}{D_j}\right) = \deg D_j + 1 - \tilde{g} = \tilde{g} - 1 + \frac{1}{n}\sum_{l|n}\sum_{m=1}^{x_l}(n - l\lambda_{mlj})$$

$$= \tilde{g} - 1 + \sum_{l|n} x_l - \frac{1}{n}\sum_{l|n}\sum_{m=1}^{x_l} l\lambda_{mlj}. \tag{2.5.4}$$

We compute next

$$\sum_{j=1}^{n-1} \lambda_{mlj}.$$

Recall that η_{ml} is a primitive (n/l)-root of unity and that $\eta_{ml}^{\lambda_{mlj}} = \varepsilon^{jl}$. For $j = 1, 2, \ldots, (n/l)$, $\{\varepsilon^{jl}\}$ is the set of all (n/l)-roots of unity. We thus conclude that

$$j \overset{\sigma}{\mapsto} \lambda_{mlj}$$

is a permutation of $\{1, \ldots, (n/l)\}$ with $\sigma(n/l) = (n/l)$. Further for $j = k(n/l) + r$, $\lambda_{mlj} = \lambda_{mlr}$. Therefore

$$\sum_{j=1}^{n} \lambda_{mlj} = l\left(1 + 2 + \cdots + \frac{n}{l}\right) = \frac{n}{2}\left(1 + \frac{n}{l}\right).$$

Thus

$$\sum_{j=0}^{n-1} n_j = \tilde{g} + \sum_{j=1}^{n-1} r\left(\frac{1}{D_j}\right)$$

$$= n(\tilde{g} - 1) + 1 + (n-1)\sum_{l|n} x_l - \frac{1}{n}\sum_{l|n}\sum_{m=1}^{x_l}\left(\frac{1}{2}nl + \frac{1}{2}n^2 - n\right)$$

$$= n(\tilde{g} - 1) + 1 + \frac{1}{2}\sum_{l|n}(n - l)x_l$$

$$= g.$$

We finally conclude that once again for each j, ε^j is an eigenvalue (and we have computed its multiplicity). Note that some of the multiplicities may be zero (for example, if $\tilde{g} = 0$). It involves no loss of generality to consider these as eigenvalues with zero multiplicity.

V.2.6. Let us now consider the case $q > 1$. We calculate for $0 \leq j \leq n-1$,

$$\begin{aligned} r\left(\frac{1}{D_j}\right) &= \deg D_j + 1 - \tilde{g} \\ &= \sum_{k=1}^{s} \alpha_k + \sum_{l|n} \sum_{m=1}^{x_l} \left(\frac{\alpha_{ml} + q - \lambda_{mlj}}{n/l} + \left[\frac{\lambda_{mlj} - q}{n/l}\right]\right) + 1 - \tilde{g} \\ &= \frac{2q(g-1)}{n} + \sum_{l|n} \sum_{m=1}^{x_l} \left(\frac{q - \lambda_{mlj}}{n/l} + \left[\frac{\lambda_{mlj} - q}{n/l}\right]\right) + 1 - \tilde{g} \\ &= q\left(2\tilde{g} - 2 + \frac{1}{n}\sum_{l|n} x_l(n-l)\right) + \sum_{l|n} \sum_{m=1}^{x_l} \left(\frac{q - \lambda_{mlj}}{n/l} + \left[\frac{\lambda_{mlj} - q}{n/l}\right]\right) + 1 - \tilde{g}. \end{aligned}$$

Next we compute

$$\sum_{j=0}^{n-1} r\left(\frac{1}{D_j}\right) = (2q-1)n(\tilde{g}-1) + q\sum_{l|n} x_l(n-l) - \frac{1}{2}\sum_{l|n} x_l(n-l).$$

To explain the last term, we write

$$q = \mu\frac{n}{l} + v^{(l)}, \quad \mu \in \mathbb{Z},\ v^{(l)} \in \mathbb{Z},\ \mu \geq 0,\ 1 \leq v^{(l)} \leq \frac{n}{l}.$$

(We will write v for $v^{(1)}$.) Thus we have

$$\begin{aligned} &\sum_{l|n} \sum_{m=1}^{x_l} \left(\mu + \frac{v^{(l)} - \lambda_{mlj}}{n/l} + \left[\frac{\lambda_{mlj}}{n/l} - \mu - \frac{v^{(l)}}{n/l}\right]\right) \\ &\quad = \sum_{l|n} \sum_{m=1}^{x_l} \left(\frac{v^{(l)} - \lambda_{mlj}}{n/l} + \left[\frac{\lambda_{mlj} - v^{(l)}}{n/l}\right]\right). \end{aligned}$$

Recall now that $\{\lambda_{mlj};\ j = 1, \ldots, n/l\}$ is a permutation of $\{1, \ldots, n/l\}$. Thus the above summed from $j = 1$ to $j = n/l$ yields

$$\sum_{l|n}\left(v^{(l)} - \frac{1}{2}\left(1 + \frac{n}{l}\right) - (v^{(l)} - 1)\right)x_l = \sum_{l|n}\left(1 - \frac{1}{2}\left(1 + \frac{n}{l}\right)\right)x_l.$$

Thus the sum from $j = 0$ to $n - 1$ yields

$$\sum_{l|n}\left(1 - \frac{1}{2}\left(1 + \frac{n}{l}\right)\right)lx_l = \sum_{l|n}\left(\frac{1}{2} - \frac{1}{2}\frac{n}{l}\right)lx_l = \frac{1}{2}\sum_{l|n} x_l(l-n).$$

In conclusion (using Riemann–Hurwitz once again)

$$\begin{aligned} \sum_{j=0}^{n-1} r\left(\frac{1}{D_j}\right) &= (2q-1)n(\tilde{g}-1) + \left(q - \frac{1}{2}\right)\sum_{l|n} x_l(n-l) \\ &= (2q-1)n(\tilde{g}-1) + \left(q - \frac{1}{2}\right)[(2g-2) - n(2\tilde{g}-2)] \\ &= (2q-1)(g-1). \end{aligned}$$

We have just proved that in all cases

$$r\left(\frac{1}{D_j}\right) = n_j.$$

V.2.7. We now want to compute the trace of $T(=\operatorname{tr} T)$. Note that for $q = 1$

$$\operatorname{tr} T = \tilde{g} + \sum_{j=1}^{n-1} n_j \varepsilon^j.$$

First observe that

$$\sum_{j=1}^{n-1} \varepsilon^j = -1,$$

and thus using (2.5.4)

$$\begin{aligned} \operatorname{tr} T &= \tilde{g} + 1 - \tilde{g} - \sum_{l|n} x_l - \frac{1}{n} \sum_{l|n} \sum_{m=1}^{x_l} \sum_{j=1}^{n-1} l \lambda_{mlj} \varepsilon^j \\ &= 1 - \sum_{l|n} x_l - \frac{1}{n} \sum_{l|n} \sum_{m=1}^{x_l} l \sum_{j=1}^{n-1} \lambda_{mlj} \varepsilon^j. \qquad (2.7.1) \end{aligned}$$

As before, we fix m and l and compute $\sum_{j=1}^{n-1} \lambda_{mlj}\varepsilon^j$. Calculate for $l > 1$

$$\begin{aligned} \sum_{j=1}^{n} \lambda_{mlj}\varepsilon^j &= \lambda_{ml1}\varepsilon + \lambda_{ml2}\varepsilon^2 + \cdots + \lambda_{ml(n/l)}\varepsilon^{n/l} \\ &\quad + \lambda_{ml1}\varepsilon^{(n/l)+1} + \cdots + \lambda_{ml(n/l)}\varepsilon^{2(n/l)} \\ &\quad \vdots \\ &\quad + \lambda_{ml1}\varepsilon^{(l-1)(n/l)+1} + \cdots + \lambda_{ml(n/l)}\varepsilon^{n} \\ &= \sum_{j=1}^{n/l} \lambda_{mlj} \sum_{k=0}^{l-1} \varepsilon^{k(n/l)+j} \\ &= \sum_{j=1}^{n/l} \lambda_{mlj}\varepsilon^j \sum_{k=0}^{l-1} \varepsilon^{k(n/l)} = 0 \end{aligned}$$

(since $\sum_{k=0}^{l-1} \varepsilon^{k(n/l)} = 0$). Hence for $l > 1$

$$\sum_{j=1}^{n-1} \lambda_{mlj}\varepsilon^j = -\lambda_{ml(n/l)} = -\frac{n}{l}.$$

Finally, for $l = 1$,

$$\sum_{j=1}^{n-1} \lambda_{m1j}\varepsilon^j = \sum_{j=1}^{n-1} \lambda_{m1j}\, \eta_{m1}^{\lambda_{m1j}}$$

(and because $\{\lambda_{m1j};\ j = 1, \ldots, n-1\}$ is a permutation of $\{1, \ldots, n-1\}$)

$$= \sum_{j=1}^{n-1} j\eta_{m1}^j = \frac{-n}{1 - \eta_{m1}}.$$

We now return to (2.7.1) and continue with our computation after changing notation: $x_1 = t =$ number of fixed points of T and $\eta_{m1} = \varepsilon^{\nu_m}$,

$$\operatorname{tr} T = 1 - \sum_{l|n} x_l - \frac{1}{n}\sum_{\substack{l|n\\ l\neq 1}} \sum_{m=1}^{x_1} l\left(-\frac{n}{l}\right) - \frac{1}{n}\sum_{m=1}^{x_1} \frac{-n}{1-\eta_{m1}}$$

$$= 1 - \sum_{l|n} x_l + \sum_{\substack{l|n\\ l\neq 1}} x_l + \sum_{m=1}^{t} \frac{1}{1-\varepsilon^{\nu_m}}$$

$$= 1 - t + \sum_{m=1}^{t} \frac{1}{1-\varepsilon^{\nu_m}}$$

$$= 1 + \sum_{m=1}^{t} \frac{\varepsilon^{\nu_m}}{1-\varepsilon^{\nu_m}}.$$

V.2.8. As usual, we proceed to the case $q > 1$. Here we compute

$$\operatorname{tr} T = \sum_{j=0}^{n-1} n_j \varepsilon^j$$

$$= \left((2q-1)(\tilde{g}-1) + \frac{q}{n}\sum_{l|n} x_l(n-l)\right)\sum_{j=0}^{n-1} \varepsilon^j$$

$$+ \sum_{l|n}\sum_{m=1}^{x_l}\sum_{j=0}^{n-1}\left(\frac{q-\lambda_{mlj}}{n/l} + \left[\frac{\lambda_{mlj}-q}{n/l}\right]\right)\varepsilon^j.$$

The first sum, as before, is equal to zero. Thus we conclude (simplying further as in V.2.6)

$$\operatorname{tr} T = -\frac{1}{n}\sum_{l|n} l \sum_{m=1}^{x_l}\sum_{j=0}^{n-1} \lambda_{mlj}\varepsilon^j + \sum_{l|n}\sum_{m=1}^{x_l}\sum_{j=0}^{n-1}\left[\frac{\lambda_{mlj}-\nu^{(l)}}{n/l}\right]\varepsilon^j$$

$$= -\frac{1}{n}\sum_{\substack{l|n\\ l\neq 1}} l\left(-\frac{n}{l}\right)x_l - \frac{1}{n}\sum_{m=1}^{x_1}\frac{-n}{1-\eta_{m1}} + \sum_{l|n}\sum_{m=1}^{x_l}\sum_{j=0}^{n-1}\left[\frac{\lambda_{mlj}-\nu^{(l)}}{n/l}\right]\varepsilon^j.$$

As we remarked before,

$$\gamma_{mlj} = \left[\frac{\lambda_{mlj}-\nu^{(l)}}{n/l}\right] = \begin{cases} 0 & \text{if } \nu^{(l)} \leq \lambda_{mlj} \leq \frac{n}{l}, \\ -1 & \text{if } 0 \leq \lambda_{mlj} < \nu^{(l)}. \end{cases}$$

We proceed essentially as before for $l > 1$

$$\sum_{j=1}^{n} \gamma_{mlj}\varepsilon^j = \gamma_{ml1}\varepsilon + \gamma_{ml2}\varepsilon^2 + \cdots + \gamma_{ml(n/l)}\varepsilon^{n/l}$$

$$+ \gamma_{ml1}\varepsilon^{(n/l)+1} + \cdots + \gamma_{ml(n/l)}\varepsilon^{2(n/l)}$$

$$\vdots$$

$$+ \gamma_{ml1}\varepsilon^{(l-1)(n/l)+1} + \cdots + \gamma_{ml(n/l)}\varepsilon^{n}$$

$$= \gamma(1 + \varepsilon^{n/l} + \varepsilon^{2(n/l)} + \cdots + \varepsilon^{(l-1)(n/l)}) = 0,$$

where $\gamma = \gamma_{ml1}\varepsilon + \gamma_{ml2}\varepsilon^2 + \cdots + \gamma_{ml(n/l)}\varepsilon^{n/l}$. Therefore for $l > 1$,

$$\sum_{j=0}^{n-1} \gamma_{mlj}\varepsilon^j = \gamma_{ml0} - \gamma_{mln} = -1.$$

For $l = 1$

$$\begin{aligned}\sum_{j=1}^{n-1} \gamma_{m1j}\varepsilon^j &= \sum_{j=1}^{n-1} \gamma_{m1j}\eta_{m1}^{\lambda_{m1j}} \\ &= \sum_{j=1}^{n-1}\left[\frac{\lambda_{m1j} - \nu}{n}\right]\eta_{m1}^{\lambda_{m1j}} \\ &= \sum_{j=1}^{n-1}\left[\frac{j-\nu}{n}\right]\eta_{m1}^{j} \\ &= -\sum_{j=1}^{\nu-1}\eta_{m1}^j = -\frac{\eta_{m1} - \eta_{m1}^{\nu}}{1 - \eta_{m1}}.\end{aligned}$$

We can finally complete the calculation:

$$\begin{aligned}\operatorname{tr} T &= \sum_{\substack{l|n \\ l\neq 1}} x_l + \sum_{m=1}^{x_1} \frac{1}{1-\eta_{m1}} - \sum_{\substack{l|n \\ l\neq 1}} x_l + \sum_{m=1}^{x_1} \frac{-1+\eta_{m1}^{\nu}}{1-\eta_{m1}} \\ &= \sum_{m=1}^{t} \frac{\varepsilon^{\nu_m \nu}}{1-\varepsilon^{\nu_m}}.\end{aligned}$$

V.2.9. We summarize our results in the following

Theorem (Eichler Trace Formula). *Let T be an automorphism of order $n > 1$ of a compact Riemann surface M of genus $g > 1$. Represent T by a matrix via its action on $\mathscr{H}^q(M)$. Let t be the number of fixed points of T. Let $\varepsilon = \exp(2\pi i/n)$. Let $P_1, \ldots, P_t$ be the fixed points of T. For each $m = 1, \ldots, t$ choose a local coordinate z at P_m and an integer ν_m such that $1 \le \nu_m \le n-1$ and such that T^{-1} is given near P_m by*

$$T^{-1}: z \mapsto \varepsilon^{\nu_m} z$$

(note that ν_m must be relatively prime to n). Then

$$\operatorname{tr} T = 1 + \sum_{m=1}^{t} \frac{\varepsilon^{\nu_m}}{1-\varepsilon^{\nu_m}} \quad \textit{for } q = 1,$$

and

$$\operatorname{tr} T = \sum_{m=1}^{t} \frac{\varepsilon^{\nu_m \nu}}{1-\varepsilon^{\nu_m}} \quad \textit{for } q > 1,$$

where $0 < \nu \le n$ is chosen as the unique integer such that $q = \mu n + \nu$ with $\mu \in \mathbb{Z}$. In each case the sum is taken to be zero whenever $t = 0$.

Note that only the fixed points of T contribute to tr T (not the points that are fixed by a power of T but not T itself).

Corollary (Lefschetz Fixed Point Formula). *For* $q = 1$,

$$\operatorname{tr} T + \overline{\operatorname{tr} T} = 2 - t.$$

PROOF. We merely note that for $\theta \in \mathbb{C}$, $\theta \neq 1$, $|\theta| = 1$, $2\,\mathrm{Re}(\theta/(1-\theta)) = -1$. □

Remark. For $q = 1$, $\operatorname{tr} T + \overline{\operatorname{tr} T}$ is the trace of the matrix representing T on the space of harmonic differentials—which is, of course, by duality related to the representation of T on $H_1(M)$.

V.2.10. We now turn to the case where T has prime order n and at least one fixed point. Say $P \in M$ is the fixed point of T. Choose a basis $\{\varphi_1, \ldots, \varphi_d\}$ of $\mathscr{H}^q(M)$ adapted to $P(d = \dim \mathscr{H}^q(M))$; that is, in terms of some local coordinate z vanishing at P, we have

$$\varphi_j = (z^{\gamma_j - 1} + O(|z|^{\gamma_j}))\,dz^q, \quad j = 1, \ldots, d,$$

where $1 = \gamma_1 < \gamma_2 < \cdots < \gamma_d < 2q(g-1) + 2$ is the "q-gap" sequence at P (III.5.8). Furthermore, we may assume without loss of generality that the γ_k coefficient of the Taylor series expansion for φ_j in terms of z vanishes for $k \neq j$ (all j) and that the automorphism T^{-1} is given by

$$z \mapsto \varepsilon z, \qquad \varepsilon = e^{2\pi i/n}.$$

It is of course clear that T^{-l} is given in the above form for some l, $1 \leq l \leq n-1$. If we are interested only in fixed points (because n is prime), T can be replaced by T^l. Note also that φ is an eigenvector for the eigenvalue ε^j of T if and only if φ is an eigenvector of the eigenvalue ε^{jl} for T^l. Thus T^l and T have the same eigenvalues and the same traces.

It follows that

$$T\varphi_j = ((\varepsilon z)^{\gamma_j - 1} + O(|z|^{\gamma_j}))\,\varepsilon^q\,dz^q,$$

and in fact

$$T\varphi_j = \varepsilon^{\gamma_j - 1 + q}\varphi_j, \qquad j = 1, \ldots, d.$$

On the basis of the above development, we can strengthen (slightly) a previous result (Theorem V.1.7).

Theorem. *Let* $T \in \operatorname{Aut} M$ *be of prime order* n, *and assume that* T *fixes a non-Weierstrass point* P. *Then* $2 \leq \nu(T) \leq 4$. *Further, the genus* $\tilde{g}$ *of* $M/\langle T \rangle$ *is given by* $\tilde{g} = [g/n]$, *and writing* $g = \tilde{g}n + r$ $(0 \leq r \leq n-1)$, *there are only three possibilities*:

a. $r = 0$, $g = \tilde{g}n$, $\nu(T) = 2$,
b. $r = \frac{1}{2}(n-1)$, $g = (\tilde{g} + \frac{1}{2})n - \frac{1}{2}$, $\nu(T) = 3$, *or*
c. $r = n - 1$, $g = (\tilde{g} + 1)n - 1$, $\nu(T) = 4$.

PROOF. Since T fixes a non Weierstrass point, we have $\gamma_j = j$, $j = 1, \ldots, g$ and the action of T on $\mathscr{H}^1(M)$ is given by the diagonal matrix

$$\operatorname{diag}(\varepsilon, \varepsilon^2, \ldots, \varepsilon^g)$$

with $\varepsilon = \exp(2\pi i/n)$. The genus $\tilde{g}$ of $M/\langle T\rangle$ is the multiplicity of the eigenvalue 1, and thus $\tilde{g} = [g/n]$. Now we write

$$g = \tilde{g}n + r, \qquad r \in \mathbb{Z}, 0 \le r \le n-1.$$

We use the Riemann–Hurwitz formula (Corollary 1 to Proposition V.1.5)

$$g = n(\tilde{g} - 1) + \tfrac{1}{2}(n-1)v(T) + 1, \tag{2.10.1}$$

and conclude that

$$v(T) = 2\frac{r}{n-1} + 2.$$

We conclude that $v(T) \ge 2$. Further, since $0 < v(T) \in \mathbb{Z}$ we see that there are only three possible cases: $r = 0$, $r = \frac{1}{2}(n-1)$, or $r = n-1$. □

V.2.11. We have seen that if an automorphism of prime order fixes a non-Weierstrass point, it must fix at least two points (and at most 4 points). What happens if all the fixed points are Weierstrass points?

Theorem. *Let $T \in \operatorname{Aut} M$ be of prime order n. If T has a fixed point, it must have at least two.*

PROOF. Assume T has precisely one fixed point P. The Riemann–Hurwitz relation yields

$$2g - 2 = (2\tilde{g} - 2)n + (n-1) \quad \text{or} \quad n = \frac{2g-1}{2\tilde{g}-1}.$$

If $\varphi \in \mathscr{H}^1(M)$ and $T\varphi = \varphi$, then $\operatorname{ord}_P \varphi > 0$. Since it is not the case that for all $\varphi \in \mathscr{H}^1(M)$ $\operatorname{ord}_P \varphi > 0$, we can find a $\varphi \in \mathscr{H}^1(M)$ with $T\varphi = \varepsilon\varphi$, $\varepsilon \ne 1$, $\varepsilon^n = 1$, $\operatorname{ord}_P \varphi = 0$. Since φ does not vanish at P (the unique fixed point of T) and (φ) is invariant under T, $2g - 2 = kn$ for some positive integer k. Thus we see that

$$1 = (2g-1) - (2g-2) = n(2\tilde{g} - 1) - nk = n(2\tilde{g} - 1 - k).$$

In particular, n cannot be > 1, and this contradiction establishes the theorem.

Remark. The Riemann–Hurwitz formula that we applied used the fact that n is prime. This is essential; for the theorem is not true if n is not a prime. See VII.3.11.

V.2.12. We consider now the representation of an automorphism T of prime order n on $\mathscr{H}^q(M)$, $q > 1$. We assume that P is a fixed point of T and that P is not a q-Weierstrass point. Then the representation of T on $\mathscr{H}^q(M)$ is

$$\operatorname{diag}(\varepsilon^q, \ldots, \varepsilon^{(2q-1)(g-1)+q-1}).$$

Let us take $q \equiv 1 \bmod n$, $q = kn + 1$, $k \ge 1$, then the multiplicity of the eigenvalue 1 is

$$\left[\frac{1}{n}(2kn+1)(g-1)\right] = \left[\left(2k + \frac{1}{n}\right)(g-1)\right] = 2k(g-1) + \left[\frac{1}{n}(g-1)\right].$$

Abbreviate $\nu(T)$ by ν and use the Corollary to Proposition V. 2.2,

$$\dim \mathscr{H}^q_{\langle T\rangle}(M) = (2kn+1)(\tilde{g}-1) + \nu\left[(kn+1)\left(1-\frac{1}{n}\right)\right]$$
$$= (2kn+1)(\tilde{g}-1) + \nu k(n-1),$$

where $\tilde{g}$ is the genus of $M/\langle T\rangle$. Combining these results with the Riemann–Hurwitz formula (2.10.1) we see that

$$(2kn+1)\{n(\tilde{g}-1) + \tfrac{1}{2}\nu(n-1)\} = \{(2kn+1)(\tilde{g}-1) + \nu k(n-1)\}n + r,$$

with $0 \le r \le n-1$. Thus

$$\nu = \frac{2r}{n-1}$$

and there is just one possibility (recall $\nu(T) \neq 1$) $r = n-1$ and $\nu = 2$, and we have established the following

Theorem. *Let $T \in \operatorname{Aut} M$ be of prime order n with $\nu(T) > 2$. Then every fixed point of T is a q-Weierstrass point for every $q > 1$, $q \equiv 1 \bmod n$.*

Corollary. *Let $1 \neq T \in \operatorname{Aut} M$. If $\nu(T) > 2$, then every fixed point of T is a q-Weierstrass point for some q.*

An alternate proof of the theorem is obtained by equating $2k(g-1) + [(1/n)(g-1)]$ to $\dim \mathscr{H}^q_{\langle T\rangle}(M)$. This leads to the inequality $\nu < 2 + 2/(n-1)$. If $n \ge 3$, we find $\nu = 2$. If $n = 2$, we find $\nu < 4$. Since an involution has an even number of fixed points $\nu = 2$, also in this case.

V.2.13. We can now reprove and generalize slightly some of the results on automorphisms of hyperelliptic surfaces (compare with III.7.11).

Theorem. *Let M be a hyperelliptic surface of genus $g \ge 2$. A conformal involution on M has either no fixed points, only non-Weierstrass fixed points, or is the hyperelliptic involution J.*

Let T be an automorphism of prime order n with $2 < n < 2g$ that fixes a Weierstrass point. Let $M/\langle T\rangle$ have genus $\tilde{g}$. Then there are two possibilities:

a. *$g = \tilde{g}n$, $\nu(T) = 2$, and both fixed points are Weierstrass points, or*
b. *$n = (2g+1)/(2\tilde{g}+1)$ $(\tilde{g} \ge 1)$, $\nu(T) = 3$, and the other fixed points are P and $J(P)$ with P not a Weierstrass point.*

PROOF. Let T be of order 2. Assume T fixes a Weierstrass point. Represent T on $\mathscr{H}^1(M)$. The action is given by (see V.2.10)

$$\operatorname{diag}(\underbrace{-1,\ldots,-1}_{g\text{-times}}).$$

This is, however, the representation of J on $\mathscr{H}^1(M)$. Since the representation of Aut M on $\mathscr{H}^1(M)$ is faithful $T = J$.

Next we consider an automorphism T of prime order n, $2 < n < 2g$, that fixes a Weierstrass point. Its action on $\mathscr{H}^1(M)$ is given by ($\varepsilon = \exp(2\pi i/n)$)

$$\operatorname{diag}(\varepsilon, \varepsilon^3, \ldots, \varepsilon^{2g-1}).$$

Write $2g - 1 = ln + r$, $0 \le r \le n - 1$. Then (since $\tilde{g}$ is the multiplicity of the eigenvalue 1) l even implies $\tilde{g} = l/2$ and l odd implies that $\tilde{g} = \frac{1}{2}(l + 1)$. Using Riemann–Hurwitz (in a by now familiar way) we see that

$$l \text{ even} \Rightarrow \nu(T) = 2 + \frac{r+1}{n-1} = 3 \qquad (r = n - 2) \Rightarrow n = \frac{2g+1}{2\tilde{g}+1},$$

$$l \text{ odd} \Rightarrow \nu(T) = 1 + \frac{r}{n-1} = 2 \qquad (r = n - 1) \Rightarrow n = \frac{g}{\tilde{g}}.$$

Assume T fixes three points. We view the action of T on the $2g + 2$ Weierstrass points. Clearly T is a permutation of prime order n, and can be written as a product of r_1 cycles of length n and r_2 singletons (cycles of length one). Thus

$$r_1 n + r_2 = 2g + 2.$$

From the above and $2g + 2 = (2\tilde{g} + 1)n + 1$, we see that

$$(2\tilde{g} + 1 - r_1)n = r_2 - 1.$$

Since $n > 2$, the last equation implies $r_2 = 1$ or $r_2 > 3$. Since $r_2 \le \nu(T)$, we see that $r_2 = 1$. Thus T has a fixed point P that is not a Weierstrass point. Since, J is central in Aut M,

$$T(J(P)) = J(T(P)) = J(P),$$

and $J(P)$ ($\neq P$) is also a fixed point.

If T fixes two points, both must be Weierstrass points by an argument similar to the one given above. □

Corollary. *If* $1 \neq T \in$ Aut M *fixes a Weierstrass point on the hyperelliptic surface* M *of genus* $g \ge 2$ *and* ord T *is prime and* $< 2g$, *then* ord $T \le g$.

Remark. It is possible for ord $T = 2g + 1$ (recall Proposition V.1.11 and its corollary).

V.2.14. Proposition. *If* $T \in$ Aut M *is of prime order* $n > g$, *then* $n = g + 1$ *or* $n = 2g + 1$. *In each of these cases* $M/\langle T\rangle$ *is of genus* 0. *In the first case,* $\nu(T) = 4$ *and in the second case,* $\nu(T) = 3$.

Proof. We start with (1.11.1), and assume that $\tilde{g} \ge 2$. Then $n(2\tilde{g} - 2) \ge 2n > 2g$. Thus the only possibilities are $\tilde{g} = 1$ or $\tilde{g} = 0$. If $\tilde{g} = 1$, then

$2g - 2 = \nu(n-1)$, and $\nu > 0$. Thus $2 \le \nu$ and

$$n = \frac{2g-2}{\nu} + 1.$$

Now (with $\nu \ge 2$, $g \ge 2$) $(2g-2)/\nu + 1 \le g$, and is thus impossible.

It remains to consider $\tilde{g} = 0$. Here

$$2g - 2 = -2n + \nu(n-1) = (\nu - 2)n - \nu,$$

or

$$n = \frac{2g - 2 + \nu}{\nu - 2} = \frac{2g}{\nu - 2} + 1.$$

We must have $\nu \ge 3$. If $\nu \ge 6$, then

$$n \le \tfrac{1}{2}g + 1 \le g.$$

If $\nu = 5$, then

$$n = \tfrac{2}{3}g + 1 \le g \quad \text{for } g \ge 3,$$

and is impossible for $g = 2$ ($n = \frac{7}{3}$). If $\nu = 4$, then $n = g + 1$; and if $\nu = 3$, then $n = 2g + 1$. □

V.3. Representation of Aut M on $H_1(M)$

In this section we study the action of Aut M on $H_1(M) = H_1(M,\mathbb{Z})$, the first homology group (with integral coefficients), where M, as usual, is a compact Riemann surface of genus $g \ge 1$. Let ${}^t\{\kappa\} = \{\kappa_1, \ldots, \kappa_{2g}\} = \{a_1, \ldots, a_g, b_1, \ldots, b_g\} = {}^t\{a,b\}$ be a canonical homology basis for M.

The main result is that the action of Aut M on $H_1(M)$ is faithful. This result is generalized in two directions. If $T \in$ Aut M fixes enough homology classes, then $T = 1$. Also, in most cases, the action of Aut M on the finite group of homology classes mod n is already faithful.

V.3.1. If $T \in$ Aut M, then $\kappa' = T\kappa$ is again a canonical homology basis for M. If we denote this basis by ${}^t\{a',b'\}$, we immediately see that there exists a $2g \times 2g$ matrix X with integer entries x_{mn} whose inverse is also an integer matrix (thus det $X = \pm 1$) such that

$$\kappa' = X\kappa; \tag{3.1.1}$$

that is

$$\kappa_j' = \sum_{l=1}^{2g} x_{jl}\kappa_l, \qquad j = 1, \ldots, 2g. \tag{3.1.1$'$}$$

Since T is orientation preserving, det $X = 1$.

We now write $X = \left[\begin{smallmatrix} A & B \\ C & D \end{smallmatrix}\right]$ in $g \times g$ blocks, and hence

$$\begin{bmatrix} a' \\ b' \end{bmatrix} = \begin{bmatrix} A & B \\ C & D \end{bmatrix} \begin{bmatrix} a \\ b \end{bmatrix}. \tag{3.1.2}$$

Recall (III.1.2) that the statement κ is a canonical homology basis means that the intersection matrix for κ is $J_0 = \begin{bmatrix} 0 & I \\ -I & 0 \end{bmatrix} = (\kappa_j \cdot \kappa_k)$. We can now ask: what are the conditions on X so that κ' also be a canonical basis? The condition is simply that $(\kappa_j' \cdot \kappa_k') = XJ_0{}^tX = J_0$. Hence we see that in addition to det $X = \pm 1$, it is also necessary that X satisfy the equation $XJ_0{}^tX = J_0$. Matrices X with integer entries satisfying det $X = 1$ are called *unimodular matrices*. The set of $n \times n$ unimodular matrices forms the group $SL(n,\mathbb{Z})$. The remarks we have made lead us to make the following

Definition. The set of $2g \times 2g$ unimodular matrices X which satisfy

$$XJ_0{}^tX = J_0 \tag{3.1.3}$$

is called the *symplectic group of genus g* and is denoted by $Sp(g,\mathbb{Z})$.

Remark. The definition is meaningful and the set is indeed a group. All that we need verify is that if $X \in Sp(g,\mathbb{Z})$ so is X^{-1}. It is easy to check that if $X = \begin{bmatrix} A & B \\ C & D \end{bmatrix}$ and $X \in Sp(g,\mathbb{Z})$ then $X^{-1} = \begin{bmatrix} {}^tD & -{}^tB \\ -{}^tC & {}^tA \end{bmatrix}$. This last statement is equivalent to (3.1.3). Hence $Sp(g,\mathbb{Z})$ is a subgroup of $SL(2g,\mathbb{Z})$.

Theorem. *There is a natural homomorphism h:*Aut $M \to Sp(g,\mathbb{Z})$. *For $g \geq 2$, h is a monomorphism.*

PROOF. The homomorphism h has already been described in the remarks preceding the statement of the theorem. We shall therefore assume that $g \geq 2$ and $T \in$ Aut M. Suppose now that $h(T) = I$. Let $\{\zeta_1, \ldots, \zeta_g\}$ be a basis of $\mathscr{H}^1(M)$ dual to ${}^t\{a,b\}$; that is, $\int_{a_k} \zeta_j = \delta_{jk}$. Since Ta_j is assumed to be homologous to a_j, we have $\int_{a_j} T\zeta_k = \int_{Ta_j} T\zeta_k$. Now for any smooth differential ω it is clear that

$$\int_{Tc} T\omega = \int_c \omega. \tag{3.1.4}$$

Hence, in particular, we have $\int_{Ta_j} T\zeta_k = \int_{a_j} \zeta_k = \delta_{kj}$. It follows that $\int_{a_j} T\zeta_k = \delta_{kj}$. We therefore conclude that $\{T\zeta_1, \ldots, T\zeta_g\}$ is also a basis of $\mathscr{H}^1(M)$ dual to ${}^t\{a,b\}$. Recall however (Proposition III.2.8) that the dual basis is unique; so that it necessarily is the case that $T\zeta_j = \zeta_j$, $j = 1, \ldots, g$.

We have already seen that the representation of Aut M on $\mathscr{H}^1(M)$ is faithful. Hence we conclude that $T = 1$. □

Corollary. *If $T(a_j) = a_j$ in $H_1(M)$ for $j = 1, \ldots, g$, then $T = 1$.*

V.3.2. There is a very simple relation between the action of an automorphism on $H_1(M)$ and the representation of the automorphism on the space of harmonic differentials on M. (Recall II.3.6.)

Let ${}^t\{\kappa\}$ be a canonical homology basis for M and let ${}^t\{\alpha\} = \{\alpha_1, \ldots, \alpha_{2g}\}$ be the basis of the complex valued harmonic differentials on M which is dual to ${}^t\{\kappa\}$; that is, $\int_{\kappa_j} \alpha_k = \delta_{kj}$, $k, j = 1, \ldots, 2g$. It is clear that this basis is unique and we have, as before, $\int_{T\kappa_j} T\alpha_l = \int_{\kappa_j} \alpha_l = \delta_{lj}$. Let the $2g \times 2g$ matrix $X = (x_{jk})$ represent the action of T on ${}^t\{\kappa\}$. We then have

$$I = (\delta_{lj}) = \left(\int_{T\kappa_j} T\alpha_l\right) = \left(\int_{\sum_{k=1}^{2g} x_{jk}\kappa_k} T\alpha_l\right) = \left(\sum_{k=1}^{2g} x_{jk} \int_{\kappa_k} T\alpha_l\right).$$

Let us denote by T the matrix whose (l,k) entry is $\int_{\kappa_k} T\alpha_l$. Thus the above equation reads

$$I = T{}^tX,$$

or

$$T = {}^tX^{-1}.$$

Since $T\alpha_l$ is determined by its periods we have established the following

Theorem. *Let X be an element of $Sp(g,\mathbb{Z})$ which represents the automorphism T on $H_1(M,\mathbb{Z})$ with respect to the canonical homology basis ${}^t\{\kappa\}$. Let ${}^t\{\alpha\} = \{\alpha_1, \ldots, \alpha_{2g}\}$ be a basis of the harmonic differentials on M dual to $\{\kappa\}$, in the sense that $\int_{\kappa_j} \alpha_k = \delta_{kj}$, $k, j = 1, \ldots, 2g$. Then the matrix which represents T on the vector space of harmonic differentials with respect to the basis ${}^t\{\alpha\}$ is ${}^t(X^{-1})$.*

Remark. If T is represented in $g \times g$ blocks by $\begin{bmatrix} A & B \\ C & D \end{bmatrix}$, then ${}^t(X^{-1})$ is represented by $\begin{bmatrix} D & -C \\ -B & A \end{bmatrix}$.

V.3.3. In this section we wish to strengthen Theorem V.3.1 and its corollary. We prove the following

Theorem. *Let M be of genus $g \geq 2$ and assume $T \in$ Aut M satisfies $T(a_1) = a_1$, $T(a_2) = a_2$, $T(b_1) = b_1$ and $T(b_2) = b_2$ in $H_1(M)$. Then $T = 1$.*

In order to prove this theorem it is convenient to first derive some preliminary results concerning $SL(n,\mathbb{Z})$ and $Sp(g,\mathbb{Z})$.

Lemma. *Let a be a vector in $\mathbb{Z}^n$. A necessary and sufficient condition for ${}^ta$ to be the first row of an element of $SL(n,\mathbb{Z})$ is that the components of a be relatively prime.*

Proof. Let, as usual, $e^{(j)}$ be the jth column of the $n \times n$ identity matrix. It suffices to prove that the orbit of the vector $e^{(1)}$ under $SL(n,\mathbb{Z})$ consists of all vectors a with relatively prime components or equivalently that for any a with relatively prime components there is an $X \in SL(n,\mathbb{Z})$ such that $Xa = e^{(1)}$.

We consider the following problem: Find

$$\min_{X \in SL(n,\mathbb{Z})} |Xa|,$$

where a is a fixed vector in $\mathbb{Z}^n$ with relatively prime components, and $|-|$ is the usual (Euclidean) norm. We wish to show that the minimum is precisely 1, and therefore that there is an $X \in SL(n,\mathbb{Z})$ such that $Xa = e^{(j)}$ for some j. Without loss of generality we can then assume $j = 1$.

Note first that the lemma is obviously true for $n = 2$, since the condition

$$\det\begin{bmatrix} a_1 & b_1 \\ a_2 & b_2 \end{bmatrix} = 1$$

is equivalent to the condition that a_1 and a_2 be relatively prime. Furthermore if (in general) the components of $a \in \mathbb{Z}^n$ have a common divisor $v > 1$, then

$$\operatorname*{Min}_{X \in SL(n,\mathbb{Z})} |Xa| \geq v.$$

Thus in particular for $n = 2$, the above minimum is 1 if and only if a has relatively prime components. We have also established necessity in general.

Now assume that $n > 2$, and that a has relatively prime components. Suppose X is a solution to the minimum problem. Let $b = Xa$. Clearly b has relatively prime components. (Otherwise $a = X^{-1}b$ would not have relatively prime components.) Assume that b has two non-zero components. We may assume that these are the first two components b_1, b_2. Let $v \geq 1$ be the greatest common divisor of b_1, b_2. Choose $x \in SL(2,\mathbb{Z})$ such that

$$x^t(b_1,b_2) = v^t(1,0).$$

Let $X' = \left[\begin{smallmatrix} x & 0 \\ 0 & I_{n-2} \end{smallmatrix}\right]$. Then $X' \in SL(n,\mathbb{Z})$ and $|X'Xa|$

$$= \left|\begin{bmatrix} x & 0 \\ 0 & I_{n-2} \end{bmatrix}{}^t(b_1,b_2,\ldots,b_n)\right| = \sqrt{v^2 + \sum_{j=3}^{n} b_j^2} < |b| = \sqrt{\sum_{j=1}^{n} b_j^2}.$$

Thus we see that b must be an integral multiple of some $e^{(j)}$. Since the components of b are relatively prime, $b = \pm e^{(j)}$, for some j. □

Lemma. *Let a and b be elements of $\mathbb{Z}^n$. A necessary and sufficient condition for $({}^ta,{}^tb)$ to be the first row of an element of $Sp(n,\mathbb{Z})$ is that its components be relatively prime.*

Proof. Just as in the case of the previous lemma, it suffices to show that there is an element X in $Sp(n,\mathbb{Z})$ such that $X\left[\begin{smallmatrix} a \\ b \end{smallmatrix}\right] = \left[\begin{smallmatrix} e^{(1)} \\ 0 \end{smallmatrix}\right]$.

We fix $\left[\begin{smallmatrix} a \\ b \end{smallmatrix}\right]$ with relatively prime components and for $X \in Sp(n,\mathbb{Z})$, let $X\left[\begin{smallmatrix} a \\ b \end{smallmatrix}\right] = \left[\begin{smallmatrix} \delta \\ \delta' \end{smallmatrix}\right]$. Let $u \geq 0$ be the greatest common divisor of the components of δ and $v \geq 0$ the greatest common divisor of the components of δ' ($u = 0$ if and only if $\delta = 0$; $v = 0$ if and only if $\delta' = 0$). We now consider the problem of finding

$$\operatorname*{Min}_{X \in Sp(n,\mathbb{Z})} u + v.$$

We can assume without loss of generality that $u \geq v$. (Let $X' = J_0X$. Then $X' \in Sp(n,\mathbb{Z})$ and $X'\left[\begin{smallmatrix} a \\ b \end{smallmatrix}\right] = \left[\begin{smallmatrix} \delta' \\ -\delta \end{smallmatrix}\right]$.) We shall now show that $v = 0$.

Assume $v \neq 0$. Then δ'/v has relatively prime components and thus by the previous lemma, is the first row of an element T of $SL(n,\mathbb{Z})$. Choose now an integral vector x such that each component of the vector $-T\delta + vx$ has absolute value less than v and thus less than u. Consider now

$$X' = \begin{bmatrix} 0 & {}^tT^{-1} \\ -T & -\tau{}^tT^{-1} \end{bmatrix},$$

where τ is any $n \times n$ symmetric matrix, with first column $-x$, and observe that X' is an element of $Sp(n,\mathbb{Z})$.

Let X be a solution to the minimization problem, and suppose $X\left[\begin{smallmatrix} a \\ b \end{smallmatrix}\right] = \left[\begin{smallmatrix} \delta \\ \delta' \end{smallmatrix}\right]$. Then $X'X\left[\begin{smallmatrix} a \\ b \end{smallmatrix}\right] = \left[\begin{smallmatrix} ve^{(1)} \\ -T\delta + vx \end{smallmatrix}\right]$, and this contradicts the minimality of X. Hence $v = 0$. It thus follows that $u = 1$. We can therefore conclude that any solution X to the minimization problem satisfies $X\left[\begin{smallmatrix} a \\ b \end{smallmatrix}\right] = \left[\begin{smallmatrix} \delta \\ 0 \end{smallmatrix}\right]$ with the components of δ relatively prime. By the previous lemma we can always find a $U \in SL(n,\mathbb{Z})$ such that $U\delta = e^{(1)}$, thus we have $\left[\begin{smallmatrix} U & 0 \\ 0 & {}^tU^{-1} \end{smallmatrix}\right]\left[\begin{smallmatrix} \delta \\ 0 \end{smallmatrix}\right] = \left[\begin{smallmatrix} e^{(1)} \\ 0 \end{smallmatrix}\right]$. □

We now proceed with the proof of the theorem. Let $\{\alpha_1, \ldots, \alpha_{2g}\}$ be the basis of the harmonic differentials dual to the given homology basis. It follows from Theorem V.3.2 (and the remark immediately following) that $T(\alpha_1) = \alpha_1$, $T(\alpha_2) = \alpha_2$, $T(\alpha_{g+1}) = \alpha_{g+1}$, and $T(\alpha_{g+2}) = \alpha_{g+2}$.

Furthermore, for any harmonic differential ω on M (see III(2.3.1)),

$$(\omega, {}^*\alpha_j) = \iint_M \alpha_j \wedge \omega = \begin{cases} \int_{b_j} \omega, & j = 1, \ldots, g, \\ -\int_{a_{j-g}} \omega, & j = g+1, \ldots, 2g. \end{cases} \tag{3.3.1}$$

Let us for the moment assume that T has a fixed point say P. Let $Q \in M$ be arbitrary, then we conclude from (3.1.4) that

$$\int_P^Q \alpha_j = \int_P^{T(Q)} \alpha_j \pmod{\mathbb{Z}}, \qquad j = 1, 2, g+1, g+2.$$

It therefore follows that

$$\int_Q^{T(Q)} \alpha_j = \int_Q^P \alpha_j + \int_P^{T(Q)} \alpha_j \in \mathbb{Z}, \quad \text{for } j = 1, 2, g+1, g+2. \tag{3.3.2}$$

We consider the surface $\tilde{M} = M/\langle T \rangle$ and the natural projection $\pi: M \to \tilde{M}$. Let $j = 1, 2, g+1$, or $g+2$. Because of $T(\alpha_j) = \alpha_j$, the differentials α_j project to harmonic differentials $\pi\alpha_j$ on $\tilde{M}$. (This assertion is not quite obvious. Its verification is left to the reader.) If $\tilde{c}$ is a closed curve on $\tilde{M}$, then we lift it to a curve c on M beginning at some $Q \in M$ and ending at T^kQ. By (3.3.2) we see that

$$\int_c \pi\alpha_j \in \mathbb{Z}, \quad j = 1, 2, g+1, g+2.$$

(Note that since we are only interested in the homology classes of the curves, we may assume that c does not pass through any fixed points of T.) Let us construct a canonical homology basis $\{\tilde{a}_1, \ldots, \tilde{a}_{\tilde{g}}, \tilde{b}_1, \ldots, \tilde{b}_{\tilde{g}}\}$ on $\tilde{M}$ and a basis $\{\tilde{\alpha}_1, \ldots, \tilde{\alpha}_{2\tilde{g}}\}$ of harmonic differentials on $\tilde{M}$ dual to the given homology basis. Since $\pi\alpha_j$ has integral periods over every cycle on $\tilde{M}$, there are integers n_{jk} such that

$$\pi\alpha_j = \sum_{k=1}^{2\tilde{g}} n_{jk}\tilde{\alpha}_k, \qquad j = 1, 2, g+1, g+2.$$

It follows from (3.3.1) (viewed on $\tilde{M}$) that

$$(\pi\alpha_j, {}^*\pi\alpha_{g+j}) \in \mathbb{Z}, \qquad j = 1, 2.$$

It is also quite obvious that (for $j = 1,2$)

$$(\text{ord } T)(\pi\alpha_j, {}^*\pi\alpha_{g+j}) = (\alpha_j, {}^*\alpha_{g+j}) = -1.$$

Note that the second equality follows from (3.3.1). Thus we conclude that for the hypothesis of the theorem to hold, T must be fixed point free (if it is not the identity). For this assertion we need only one of the above equations (not two).

To finish the proof, we must study fixed point free automorphisms. By our previous considerations no power of $T(\neq 1)$ can have fixed points. Hence by Riemann-Hurwitz

$$2g - 2 = n(2\tilde{g} - 2),$$

with $g \geq 2$ and $\tilde{g} \geq 2$. Thus we are studying smooth n-sheeted cyclic coverings of a surface $\tilde{M}$ of genus $\tilde{g} \geq 2$. A smooth n-sheeted covering is described by the selection of a normal subgroup G of $\pi_1(\tilde{M})$ such that $\pi_1(\tilde{M})/G \cong \mathbb{Z}_n$ ($=$ integers mod n). Thus G appears in a short exact sequence of groups and group homomorphisms

$$\{1\} \to G \to \pi_1(\tilde{M}) \xrightarrow{\varphi} \mathbb{Z}_n \to \{0\}.$$

Without loss of generality we assume that the canonical homology basis on $\tilde{M}$ comes from a set of generators for $\pi_1(\tilde{M})$ denoted by the same symbols. The homomorphism φ is described by its action on the generators. Setting $\varphi(\tilde{a}_j) = \varepsilon_j$, and $\varphi(\tilde{b}_j) = \varepsilon'_j$, we see that φ is described by the $2 \times \tilde{g}$ matrix

$$\chi = \begin{bmatrix} \varepsilon \\ \varepsilon' \end{bmatrix} = \begin{bmatrix} \varepsilon_1, \ldots, \varepsilon_{\tilde{g}} \\ \varepsilon'_1, \ldots, \varepsilon'_{\tilde{g}} \end{bmatrix}$$

of integers mod n.

Since φ is surjective the vector $(\varepsilon, \varepsilon')$ has relatively prime components.

We review the above situation in the language of multiplicative functions. Construct on $\tilde{M}$ a multiplicative function f belonging to the n-characteristic χ. This function lifts to a single valued function on the cover corresponding to the homomorphism φ.

By a change of homology basis (since $\mathbb{Z}_n$ is commutative we may work with $H_1(\tilde{M})$ instead of $\pi_1(\tilde{M})$) we may assume that χ is of the form $\left[\begin{smallmatrix} 1 & 0 & \cdots & 0 \\ 0 & 0 & \cdots & 0 \end{smallmatrix}\right]$. This follows immediately from our second lemma. More precisely, if χ is

$\left[\begin{smallmatrix}\varepsilon\\ \varepsilon'\end{smallmatrix}\right]$ then we can consider the change of homology basis effected by a matrix ${}^tX \in Sp(\tilde{g},\mathbb{Z})$ such that

$$X\begin{bmatrix}{}^t\varepsilon\\ {}^t\varepsilon'\end{bmatrix} = \begin{bmatrix}e^{(1)}\\ 0\end{bmatrix}.$$

Such an X exists by the lemma and the χ for this new basis is $\left[\begin{smallmatrix}1 & 0 & \cdots & 0\\ 0 & 0 & \cdots & 0\end{smallmatrix}\right]$. The last assertion follows from the fact that X has integral entries. If $c \in H_1(\tilde{M})$ is written as $\sum_{j=1}^{\tilde{g}} n_j\tilde{a}_j + \sum_{j=1}^{\tilde{g}} m_j\tilde{b}_j$, then

$$\chi(c) = (\varepsilon,\varepsilon')\begin{bmatrix}n\\ m\end{bmatrix}.$$

(As usual $n = {}^t(n_1, \ldots, n_{\tilde{g}})$, $m = {}^t(m_1, \ldots, m_{\tilde{g}})$.) Thus

$$\chi({}^tXc) = (\varepsilon,\varepsilon')({}^tXc),$$

and in particular for $\tilde{\kappa}'_j = {}^tX\tilde{\kappa}_j$

$$\begin{aligned}\chi(\tilde{\kappa}_j) &= (\varepsilon,\varepsilon')(j\text{th column of } {}^tX)\\ &= j\text{th entry of } X\begin{bmatrix}{}^t\varepsilon\\ {}^t\varepsilon'\end{bmatrix}\\ &= j\text{th entry of} \begin{bmatrix}e^{(1)}\\ 0\end{bmatrix}.\end{aligned}$$

It is now a simple matter to complete the proof. We examine the cover corresponding to $X = \left[\begin{smallmatrix}1 & 0 & \cdots & 0\\ 0 & 0 & \cdots & 0\end{smallmatrix}\right]$. It is clear that if $\tilde{a}_1, \ldots, \tilde{a}_{\tilde{g}}, \tilde{b}_1, \ldots, \tilde{b}_{\tilde{g}}$ is the homology basis for $\tilde{M}$ such that $X = \left[\begin{smallmatrix}1 & 0 & \cdots & 0\\ 0 & 0 & \cdots & 0\end{smallmatrix}\right]$, then $\tilde{a}_1, 2\tilde{a}_1, \ldots, (n-1)\tilde{a}_1$ lift to open curves on M and $n\tilde{a}_1$ lifts to a closed curve on M. Further $\tilde{b}_1$ has n disjoint lifts all of which are homologous on M. Each of the other cycles also have n disjoint lifts on M; however, these are all homologously independent. The action of T on M fixes the lift of $n\tilde{a}_1$ and permutes the lifts of $\tilde{b}_1$; however, since the lifts are all homologous we can say that (as far as the induced action on homology) T fixes the lift of $\tilde{b}_1$. No other cycles are fixed. Hence we finally conclude that T must be the identity and this concludes the proof of the theorem.

V.3.4. In this paragraph we strengthen (in another direction) Theorem V.3.1. The group Aut M acts not only on $H_1(M)$ but also on $H_1(M,\mathbb{Z}_n)$, the first homology group on M with coefficients in the integers mod n. Choose $T \in$ Aut M. We have seen that T acts on $H_1(M)$ by a square integral matrix $A \in Sp(g,\mathbb{Z})$ and A has finite order m. We first prove the following

Lemma (Serre). *Let $A \in SL(k,\mathbb{Z})$ have finite order $m > 1$. If $A \equiv I$ mod n, then $m = n = 2$.*

PROOF. (Due to C. J. Earle). Let p be a prime factor of n. Then $A \equiv I$ mod p^l, where $l \geq 1$. If we show that $p = m = 2$ and $l = 1$, then the lemma will follow easily. We break the argument down into a series of steps.

a. If $A \equiv I \bmod p^l$, then $l = 1$.

PROOF. Let q be a prime factor of m. So $m = qr$. Then $B = A^r$ has order q, and $B \equiv I \bmod p^l$. So $B = (I + p^s X)$, where X is integral, with some element of X prime to p, and $s \geq l \geq 1$. Now

$$I = (I + p^s X)^q = I + qp^s X + p^{2s} Y.$$

Thus

$$qX = -p^s Y,$$

and so

$$q = p \quad \text{and} \quad X = -p^{s-1} Y.$$

Again, since p does not divide X, $s = 1$ and since $s \geq l \geq 1$, also $l = 1$.

b. A has order p.

PROOF. Since p is prime, we need only show $A^p = 1$. Clearly A^p has finite order, and by (a)

$$A^p = (I + pX)^p = I + p^2 X + p^2 Y = I + p^2 Y',$$

so $A^p \equiv I \bmod p^2$. This contradicts (a) unless $A^p = I$.

c. $p = 2$.

PROOF. Otherwise p is an odd prime, and $p \geq 3$, and

$$I = (I + pX)^p = I + ppX + \frac{p(p-1)}{2} p^2 X^2 + p^3 Y.$$

Thus

$$-X = p\left(\frac{p-1}{2}\right) X^2 + pY.$$

But p is odd, and so p divides the right hand side. This is the final contradiction. □

The above has as an immediate consequence the following

Theorem. *There is a natural homomorphism*

$$h: \mathrm{Aut}\, M \to Sp(g, \mathbb{Z}_n).$$

For $g \geq 2$, $n \geq 3$, h is injective. For $g \geq 2$ and $n = 2$ only automorphisms of order 2 are in the kernel of h.

V.4. The Exceptional Riemann Surfaces

We have seen in IV.6, that there are a few Riemann surfaces with commutative fundamental groups—the so called "exceptional surfaces". We shall now encounter the same surfaces in a slightly different context.

V.4.1. We want to describe all Riemann surfaces M for which Aut M is not discrete. We shall not bother defining what we mean by Aut M *not* being discrete. We will establish a result which will hold with any reasonable concept of discreteness for Aut M. We state it as a

Theorem. Aut M *is not a discrete group if and only if* M *is conformally equivalent to one of the following Riemann surfaces*:

1. $\mathbb{C} \cup \{\infty\}$,
2. $\mathbb{C}$,
3. $\Delta = \{z \in \mathbb{C}, |z| < 1\}$,
4. $\mathbb{C}^* = \mathbb{C}\backslash\{0\}$,
5. $\Delta^* = \Delta\backslash\{0\}$,
6. $\Delta_r = \{z \in \mathbb{C}, r < |z| < 1\}, 0 < r < 1$,
7. *a torus*, $\mathbb{C}/G$, *where* G *is the free group generated by the two translations* $z \mapsto z + 1, z \mapsto z + \tau, \operatorname{Im} \tau > 0$.

V.4.2. In any standard book on complex analysis, the reader will find a description of the automorphism groups of the three simply connected Riemann surfaces. We have used this information already in IV.5.2. We repeat it here:

$$\operatorname{Aut}(\mathbb{C} \cup \{\infty\}) \cong PSL(2,\mathbb{C})$$

$$\operatorname{Aut} \mathbb{C} \cong P\Delta(2,\mathbb{C})$$

$$\operatorname{Aut} \Delta \cong \operatorname{Aut} U \cong PSL(2,\mathbb{R}).$$

The above are clearly not discrete subgroups of $PSL(2,\mathbb{C})$. As a matter of fact they are Lie groups. The first two are complex Lie groups (of complex dimensions 3 and 2, respectively). The last is a real Lie group (of real dimension 3).

V.4.3. It is clear that every automorphism A of $\mathbb{C}^*$ extends to one of $\mathbb{C} \cup \{\infty\}$. Furthermore, A must either fix 0 and ∞ or interchange these points. Thus the automorphism group of $\mathbb{C}^*$ consists of Möbius transformations of the form

$$z \mapsto kz \quad \text{or} \quad z \mapsto \frac{k}{z}, \quad \text{with } k \in \mathbb{C}^*.$$

Hence Aut $\mathbb{C}^*$ is isomorphic to a $\mathbb{Z}_2$-extension of $\mathbb{C}^*$.

V.4.4. Every automorphism of Δ^* extends to an automorphism of Δ that fixes 0. Thus Aut Δ^* is the rotation group

$$z \mapsto e^{i\theta}z, \qquad \theta \in \mathbb{R},$$

which is isomorphic to the circle group S^1.

V.4.5. Let M be a Riemann surface with holomorphic universal covering map $\pi:\tilde{M} \to M$. Let G be the covering group of π. We claim that

$$\text{Aut } M \cong N(G)/G,$$

where $N(G)$ is the normalizer of G in Aut $\tilde{M}$. The verification is easy and straightforward—and left to the reader. We shall show that in the non-exceptional cases $N(G)$ is already a discrete subgroup of Aut $\tilde{M}$.

V.4.6. Let us consider the surface Δ_r. Its holomorphic universal covering space is U and its covering group G is generated by $C:z \mapsto \lambda z$ for some $\lambda > 1$ (the relation between λ and r was given by IV(6.8.1)). Let $A \in N(G)$. Thus

$$A \circ C \circ A^{-1} = C^{\pm 1},$$

since conjugation by A is an automorphism of G. Note that $A \circ C \circ A^{-1} = C$ implies (by Lemma IV.6.6) that A is hyperbolic with 0 and ∞ as fixed points; that is, a mapping of the form

$$z \mapsto kz, \qquad k > 0. \tag{4.6.1}$$

We are left with the case $A \circ C \circ A^{-1} = C^{-1}$. Since $A \circ C^{-1} \circ A^{-1} = C$, we conclude that A^2 commutes with C and is of the form (4.6.1). Thus $N(G)$ is trivial or a $\mathbb{Z}_2$-extension of the hyperbolic subgroup of Aut U fixing 0 and ∞ (this group is isomorphic to the multiplicative positive numbers). To show we have a $\mathbb{Z}_2$-extension, we must find just one Möbius transformation that conjugates C into C^{-1}. Clearly, $z \mapsto -(1/z)$ will do this. We wish to see what $N(G)/G$ looks like as a group of motions of Δ_r. First let us consider the mapping (4.6.1). Recall the holomorphic universal covering

$$\rho: U \to \Delta_r$$

is given by

$$\rho(z) = \exp\left(2\pi i \frac{\log z}{\log \lambda}\right).$$

Thus a (multivalued) inverse of ρ is given by

$$z \mapsto \exp\left(\frac{1}{2\pi i} \log \lambda \log z\right).$$

From this it follows that $z \mapsto kz$ induces the automorphism (a rotation) $z \mapsto e^{ik_0}z$ of Δ_r, where $k_0 = (2\pi \log k)/\log \lambda$. Note that these automorphisms are extendable to $\delta\Delta_r$ and fix each boundary component (as a set). The automorphism corresponding to $z \mapsto -(1/z)$ is $z \mapsto r/z$. This map also has a continuous extension to $\delta\Delta_r$ and switches the boundary components. We see that Aut Δ_r is isomorphic to a $\mathbb{Z}_2$-extension of the circle group.

V.4.7. We examine next the torus. Let the covering group G of the torus T be generated by $z \mapsto z + 1$ and $z \mapsto z + \tau$, Im $\tau > 0$. It is quite easy to see that $A(z) = az + b$ ($A \in \text{Aut}\ \mathbb{C}$) is in $N(G)$ if and only if a and $a\tau$ are generators for the lattice G. In particular, all the translations ($a = 1$) are in $N(G)$. Thus, Aut T contains $T(\cong \mathbb{C}/G)$ as a commutative subgroup. (We have thus shown that Aut T is never discrete.)

When is Aut T a proper extension of T? First, since a and $a\tau$ are two linearly independent (over $\mathbb{Z}$) periods that generate the group G we have

$$a = \alpha + \beta\tau, \qquad a\tau = \alpha\tau + \beta\tau^2 = \gamma + \delta\tau, \tag{4.7.1}$$

with $X = \left[\begin{smallmatrix} \alpha & \beta \\ \gamma & \delta \end{smallmatrix}\right] \in SL\ (2,\mathbb{Z})$. In particular, we see that for (Aut $T)/T$ to be non-trivial, τ must satisfy the quadratic algebraic equation with integral coefficients

$$\beta\tau^2 + (\alpha - \delta)\tau - \gamma = 0. \tag{4.7.2}$$

If $\beta = 0$, then $\gamma = 0$ and $\alpha = \delta = \pm 1$ and the equation (4.7.2) reduces to the trivial equation. Since $\alpha = 1$ corresponds to a translation, the only new automorphism ($a = -1$) in this case is the one corresponding to $z \mapsto -z$. Thus Aut T always contains a $\mathbb{Z}_2$-extension of T. If $\beta \neq 0$, the question becomes more complicated. We have of course shown that for most τ (in particular for all transcendental τ), Aut T is a $\mathbb{Z}_2$-extension of T.

Assume that $\beta \neq 0$. Equation (4.7.2) asserts that the Möbius transformation corresponding to $X \in SL(2,\mathbb{Z})$ has a fixed point τ in the upper half plane. Thus X must be an elliptic element.

If $A \in \text{Aut}\ T$, then we can write A uniquely as

$$A = A_1 \circ A_2,$$

where A_1 fixes 0 and $A_2 \in T$. To show that (Aut $T)/T$ is finite we have to show that the stability subgroup of the origin is finite. We may assume that $A \in (\text{Aut}\ T)_0$ is given by $z \mapsto az$. The number $a \in \mathbb{C}^*$ is, of course, subject to the restriction (4.7.1). Simple calculations show that

$$a = \frac{1}{2}\frac{\chi \pm \sqrt{\chi^2 - 4}}{2},$$

with χ the trace of the elliptic element $X \in SL(2,\mathbb{Z})/\pm I$.

Since $SL(2,\mathbb{Z})$ has only finitely many conjugacy classes of elliptic elements, the number of a's is finite. (Since $0 \leq \chi^2 < 4$, we must have $|a| = 1$. As an exercise prove this another way.)

V.4.8. It remains to show that we have exhausted all M with Aut M not discrete. We may restrict our attention to those surfaces whose universal covering space is the unit disc (or the upper half plane U).

Let G be the covering group of M. Suppose there exists a distinct sequence $f_n \in N(G)$ such that $\lim_n f_n = 1$. Choose $1 \neq A \in G$ and observe that $\lim_n f_n \circ A \circ f_n^{-1} \circ A^{-1} = 1$. Since $f_n \circ A \circ f_n^{-1} \circ A^{-1} \in G$, there exists an N such that $f_n \circ A \circ f_n^{-1} \circ A^{-1} = 1$ for $n \geq N$. Thus f_n commutes with A for large n. If A is parabolic, then by Lemma IV.6.6 so is f_n for large n; and A

and f_n have a common fixed point. Since $1 \neq A \in G$ was arbitrary, G is a commutative parabolic group and by Theorem IV.6.1, M is the punctured disc.

Similarly, if A is hyperbolic, every element of G is hyperbolic and commutes with A. Appealing to the same earlier theorem, we see that M is an annulus.

CHAPTER VI

Theta Functions

We have seen in Chapters III and IV how to construct meromorphic functions on Riemann surfaces. In this chapter, we construct holomorphic functions on the Jacobian variety of a compact surface, and via the embedding of the Riemann surface into its Jacobian variety, multivalued holomorphic functions on the surface. The high point of our present development is the Riemann vanishing theorem (Theorem VI.3.5). Along the way, we will reprove the Jacobi inversion theorem.

VI.1. The Riemann Theta Function

In this section we develop the basic properties of Riemann's theta function.

VI.1.1. We fix an integer $g \geq 1$. Let $\mathfrak{S}_g$ denote the space of complex symmetric $g \times g$ matrices with positive definite imaginary part. Clearly, $\mathfrak{S}_g$ is a subset of the $1/2g(g+1)$-dimensional manifold X of symmetric $g \times g$ matrices. To show that $\mathfrak{S}_g$ is a manifold, it suffices to show that it is an open subset of X. But a real symmetric matrix is positive if and only if all its eigenvalues are positive. Thus if $\tau_0 \in \mathfrak{S}_g$, Im τ_0 has positive eigenvalues. Since this also holds for all $\tau \in X$ sufficiently close to τ_0, $\mathfrak{S}_g$ is an open subset of X. The space $\mathfrak{S}_g$ is known as the *Siegel upper half space of genus g*. (Note that $\mathfrak{S}_1$ is the upper half plane, U.) We define Riemann's *theta function* by

$$\theta(z,\tau) = \sum_{N \in \mathbb{Z}^g} \exp 2\pi i\left(\frac{1}{2}{}^tN\tau N + {}^tNz\right), \tag{1.1.1}$$

where $z \in \mathbb{C}^g$ (viewed as a column vector) and $\tau \in \mathfrak{S}_g$, and the sum extends over all integer vectors in $\mathbb{C}^g$. To show that θ converges absolutely and uniformly on compact subsets of $\mathbb{C}^g \times \mathfrak{S}_g$, we review some algebraic and analytic concepts.

First of all, as we already observed, every (real) symmetric matrix has real eigenvalues. Next, every real symmetric positive definite matrix has positive eigenvalues. Furthermore, for a real symmetric matrix T, the smallest eigenvalue is given as the minimum of the quadratic form (here $(\, , \,)$ is the usual inner product in $\mathbb{R}^g$)

$$Q(x) = (Tx,x), \qquad x \in \mathbb{R}^g, \ \|x\| = 1$$

(we use $\|\cdot\|$ to denote the Euclidean norm).

For $\tau \in \mathfrak{S}_g$, we let $\lambda(\tau) =$ minimum eigenvalue of $\operatorname{Im} \tau$. Then $\lambda(\tau) > 0$, and if K is a compact subset of $\mathfrak{S}_g$ we also have

$$\min_{\tau \in K} \{\lambda(\tau)\} \geq \lambda_0 > 0,$$

because

$$\mathfrak{S}_g \times \mathbb{R}^g \ni (\tau,x) \mapsto ((\operatorname{Im} \tau)x,x) \in \mathbb{R}$$

is a continuous function.

To show convergence of θ on compact subsets of $\mathbb{C}^g \times \mathfrak{S}_g$, it suffices to show convergence in a region

$$R = \{(z,\tau) \in \mathbb{C}^g \times \mathfrak{S}_g; \ \|z\| \leq M \text{ and } \lambda(\tau) \geq \lambda_0 > 0\}.$$

We note that

$$\left|\exp 2\pi i(\tfrac{1}{2}{}^t N\tau N + {}^t N z)\right| = \exp \operatorname{Im}(-\pi {}^t N\tau N - 2\pi {}^t N z).$$

Now by the characterization of minimum eigenvalues:

$$\begin{aligned} \operatorname{Im}(-\pi {}^t N\tau N) = -\pi((\operatorname{Im} \tau)N,N) &= -\pi\|N\|^2\left((\operatorname{Im} \tau)\frac{N}{\|N\|}, \frac{N}{\|N\|}\right) \\ &\leq -\pi\|N\|^2\lambda(\tau) \leq -\pi\|N\|^2\lambda_0, \end{aligned}$$

and by the Cauchy-Schwarz inequality

$$\operatorname{Im}(-2\pi {}^t N z) \leq 2\pi|{}^t N z| = 2\pi|(z,N)| \leq 2\pi\|z\| \, \|N\| \leq 2\pi M\|N\|.$$

Thus an upper estimate for the absolute value of the general term of (1.1.1) is

$$\exp(-\pi\|N\|^2\lambda_0 + 2\pi M\|N\|) = \exp\left(-\pi\|N\|^2\lambda_0\left(1 - \frac{2M}{\|N\|\lambda_0}\right)\right).$$

Now only finitely many terms $N \in \mathbb{Z}^g$ satisfy the equation

$$\frac{2M}{\|N\|\lambda_0} \geq \frac{1}{2}$$

(because N is an integer vector). Thus for all but finitely many terms, the general term of (1.1.1) is bounded in absolute value by

$$\exp(-\tfrac{1}{2}\pi\lambda_0\|N\|^2),$$

and our series is majorized by

$$\sum_{N\in\mathbb{Z}^g}\exp\left(-\frac{1}{2}\pi\lambda_0\|N\|^2\right)=\sum_{N\in\mathbb{Z}^g}\exp\left(-\frac{1}{2}\pi\lambda_0\sum_{j=1}^{g}n_j^2\right)$$
$$=\prod_{j=1}^{g}\sum_{n_j=-\infty}^{\infty}\exp\left(-\frac{1}{2}\pi\lambda_0 n_j^2\right),$$

which converges since each factor converges (by the Cauchy root test, for example).

We have shown that θ is a holomorphic function on $\mathbb{C}^g\times\mathfrak{S}_g$. We shall be mostly interested in this function with fixed $\tau\in\mathfrak{S}_g$. In this case, we will abbreviate $\theta(z,\tau)$ by $\theta(z)$.

Remark. The one variable theta function

$$\theta(z,\tau)=\sum_{n=-\infty}^{\infty}\exp 2\pi i\left[\frac{1}{2}n^2\tau+nz\right],\qquad z\in\mathbb{C},\ \tau\in U, \tag{1.1.2}$$

already has a non-trivial theory.

VI.1.2. We now proceed to study the periodicity of the theta function. Let $I_g = I$ denote the $g\times g$ identity matrix.

Proposition. *Let μ, $\mu'\in\mathbb{Z}^g$. Then*

$$\theta(z+I\mu'+\tau\mu,\tau)=\exp 2\pi i[-{}^t\mu z-\tfrac{1}{2}{}^t\mu\tau\mu]\theta(z,\tau), \tag{1.2.1}$$

for all $z\in\mathbb{C}^g$, all $\tau\in\mathfrak{S}_g$.

Proof. We start with the definition (1.1.1)

$$\theta(z+I\mu'+\tau\mu,\tau)$$
$$=\sum_{N\in\mathbb{Z}^g}\exp 2\pi i\left[\frac{1}{2}{}^tN\tau N+{}^tN(z+I\mu'+\tau\mu)\right]$$
$$=\sum_{N\in\mathbb{Z}^g}\exp 2\pi i\left[\frac{1}{2}{}^t(N+\mu)\tau(N+\mu)+{}^t(N+\mu)z-\frac{1}{2}{}^t\mu\tau\mu+{}^tN\mu'-{}^t\mu z\right]$$

(to obtain the above identity use the fact that ${}^t\tau=\tau$)

$$=\sum_{N\in\mathbb{Z}^g}\exp 2\pi i\left[-{}^t\mu z-\frac{1}{2}{}^t\mu\tau\mu\right]\exp 2\pi i\left[\frac{1}{2}{}^t(N+\mu)\tau(N+\mu)+{}^t(N+\mu)z\right]$$

(because ${}^tN\mu'\in\mathbb{Z}$)

$$=\exp 2\pi i[-{}^t\mu z-\tfrac{1}{2}{}^t\mu\tau\mu]\theta(z,\tau)$$

(because $(N+\mu)$ is just as good a summation index as N). ☐

Corollary 1. *Let $e^{(j)}$ be the j-th column of I_g. Then*

$$\theta(z + e^{(j)}, \tau) = \theta(z,\tau), \text{ all } z \in \mathbb{C}^g, \text{ all } \tau \in \mathfrak{S}_g. \tag{1.2.2}$$

PROOF. Take $\mu' = e^{(j)}$, $\mu = 0$. □

Corollary 2. *Let $\tau^{(k)}$ be the k-th column of τ, and τ_{kk} the (k,k)-entry of τ. Then*

$$\theta(z + \tau^{(k)}, \tau) = \exp 2\pi i\left[-z_k - \frac{\tau_{kk}}{2}\right]\theta(z,\tau), \quad \text{all } z \in \mathbb{C}^g, \text{ all } \tau \in \mathfrak{S}_g, \tag{1.2.3}$$

where z_k is the k-th component of z.

PROOF. Take $\mu' = 0$, $\mu = e^{(k)}$. □

Remark. It is also easy to verify that θ is an even function of z. Thus in addition to our basic formula (1.2.1), we also have

$$\theta(-z,\tau) = \theta(z,\tau), \quad \text{all } z \in \mathbb{C}^g, \text{all } \tau \in \mathfrak{S}_g. \tag{1.2.4}$$

VI.1.3. Formula (1.2.1) suggests that we should regard the $2g$ vectors $e^{(j)}$, $j = 1, \ldots, g$, $\tau^{(j)}$, $j = 1, \ldots, g$, as periods of $\theta(z,\tau)$. Of course, only the first g vectors are actually periods; however, if we consider the $g \times 2g$ matrix, $\Omega = (I_g,\tau)$, we can rewrite (1.2.1) as

$$\theta\left(z + \Omega\begin{bmatrix}\mu' \\ \mu\end{bmatrix}, \tau\right) = \exp 2\pi i\left[{}^t\left(\lambda\begin{bmatrix}\mu' \\ \mu\end{bmatrix}\right)z - \frac{1}{2}{}^t\mu\tau\mu\right]\theta(z,\tau), \tag{1.3.1}$$

where λ is the $g \times 2g$ matrix $(0, -I_g)$. If we now denote by $e^{(k)}$, $k = 1, \ldots, 2g$, the $2g$ columns of the $2g \times 2g$ identity matrix we can rewrite (1.2.2) and (1.2.3) as

$$\theta(z + \Omega e^{(k)}, \tau) = \exp 2\pi i[{}^t(\lambda e^{(k)})z + \gamma_k]\theta(z,\tau), \tag{1.3.2}$$

where γ_k is the kth component of the vector γ in $\mathbb{C}^{2g}$ defined by $-\frac{1}{2}{}^t(0, \ldots, 0, \tau_{11}, \ldots, \tau_{gg})$.

A holomorphic function, f, on $\mathbb{C}^g$ which satisfies the conditions

$$f(z + \Omega e^{(k)}) = \exp 2\pi i[{}^t(\lambda e^{(k)})z + \gamma_k]f(z), \qquad k = 1, \ldots, 2g,$$

all $z \in \mathbb{C}^g$, with Ω and λ $g \times 2g$ matrices and $\gamma \in \mathbb{C}^{2g}$, is called a *multiplicative function of type* (Ω,λ,γ). What we have seen here is that $\theta(z,\tau)$ is a multiplicative function of type (Ω,λ,γ) on $\mathbb{C}^g$ with $\Omega = (I,\tau)$, $\lambda = (0,-I)$, $\gamma = -\frac{1}{2}{}^t(0, \ldots, 0, \tau_{11}, \ldots, \tau_{gg})$. Given Ω, λ, and γ, we can ask whether there exist multiplicative functions of this type. We shall, however, not pursue this question any further. There is, of course, an intimate connection between these questions and function theory on complex tori (recall III.11).

VI.1.4. More generally, we shall consider in place of $\theta(z)$ the function obtained by taking a translate of z namely $\theta(z + e)$ for some $e \in \mathbb{C}^g$. It is clear that

$$\theta(z + e + e^{(k)}) = \theta(z + e), \tag{1.4.1}$$

and that

$$\theta(z + e + \tau^{(k)}) = \exp 2\pi i\left[-z_k - e_k - \frac{\tau_{kk}}{2}\right]\theta(z + e). \qquad (1.4.2)$$

The connection between theta functions and the theory of multiply periodic functions on $\mathbb{C}^g$ is seen rather clearly when one considers two distinct points, d and $e \in \mathbb{C}^g$, and sets

$$\frac{\theta(z + e)\theta(z - e)}{\theta(z + d)\theta(z - d)} = f(z). \qquad (1.4.3)$$

It is clear from (1.4.1) and (1.4.2) that $f(z)$ in (1.4.3) is a periodic meromorphic function on $\mathbb{C}^g$ with periods $e^{(j)}$, $\tau^{(j)}$, $j = 1, \ldots, g$.

Recall that the columns of the $g \times (2g)$ period matrix (I,τ) are linearly independent over the reals so that every $e \in \mathbb{C}^g$ can be expressed as $e = I(\varepsilon'/2) + \tau(\varepsilon/2)$ where the number 2 appears in the denominator for technical reasons. Hence

$$\theta(z + e) = \theta\left(z + I\frac{\varepsilon'}{2} + \tau\frac{\varepsilon}{2}\right).$$

This suggests that we should define a function on $\mathbb{C}^g \times \mathscr{S}_g$ by

$$\theta\begin{bmatrix}\varepsilon\\ \varepsilon'\end{bmatrix}(z,\tau) = \sum_{N \in \mathbb{Z}_g} \exp\left\{2\pi i[\tfrac{1}{2}\,{}^t\!\left(N + \frac{\varepsilon}{2}\right)\tau\left(N + \frac{\varepsilon}{2}\right) + {}^t\!\left(N + \frac{\varepsilon}{2}\right)\left(z + \frac{\varepsilon'}{2}\right)\right\}.$$

We observe at once that

$$\theta\begin{bmatrix}0\\ 0\end{bmatrix}(z,\tau) = \theta(z,\tau)$$

and that, in general,

$$\theta\begin{bmatrix}\varepsilon\\ \varepsilon'\end{bmatrix}(z,\tau) = \exp\left\{2\pi i[\tfrac{1}{8}\,{}^t\varepsilon\tau\varepsilon + \tfrac{1}{2}\,{}^t\varepsilon z + \tfrac{1}{4}\,{}^t\varepsilon\varepsilon']\right\}\theta\left(z + I\frac{\varepsilon'}{2} + \tau\frac{\varepsilon}{2}, \tau\right). \qquad (1.4.4)$$

VI.1.5. For our work, the most important case is when ε, ε' are integer vectors. In this case, $\theta\left[\begin{smallmatrix}\varepsilon\\ \varepsilon'\end{smallmatrix}\right](z,\tau)$ is called the *first order theta function with integer characteristic* $\left[\begin{smallmatrix}\varepsilon\\ \varepsilon'\end{smallmatrix}\right]$.

Proposition. *The first order theta function with integer characteristic* $\left[\begin{smallmatrix}\varepsilon\\ \varepsilon'\end{smallmatrix}\right]$ *has the following properties*:

$$\theta\begin{bmatrix}\varepsilon\\ \varepsilon'\end{bmatrix}(z + e^{(k)}, \tau) = \exp 2\pi i\left[\frac{\varepsilon_k}{2}\right]\theta\begin{bmatrix}\varepsilon\\ \varepsilon'\end{bmatrix}(z,\tau), \qquad (1.5.1)$$

$$\theta\begin{bmatrix}\varepsilon\\ \varepsilon'\end{bmatrix}(z + \tau^{(k)}, \tau) = \exp 2\pi i\left[-z_k - \frac{\tau_{kk}}{2} - \frac{\varepsilon'_k}{2}\right]\theta\begin{bmatrix}\varepsilon\\ \varepsilon'\end{bmatrix}(z,\tau), \qquad (1.5.2)$$

$$\theta\begin{bmatrix}\varepsilon\\ \varepsilon'\end{bmatrix}(-z,\tau) = \exp 2\pi i\left[\frac{{}^t\varepsilon\varepsilon'}{2}\right]\theta\begin{bmatrix}\varepsilon\\ \varepsilon'\end{bmatrix}(z,\tau), \qquad (1.5.3)$$

and

$$\theta\begin{bmatrix}\varepsilon + 2\nu \\ \varepsilon' + 2\nu'\end{bmatrix}(z,\tau) = \exp 2\pi i\left[\frac{{}^t\varepsilon\nu'}{2}\right]\theta\begin{bmatrix}\varepsilon \\ \varepsilon'\end{bmatrix}(z,\tau). \tag{1.5.4}$$

(Here ν and ν' are integer vectors in $\mathbb{Z}^g$.)

PROOF. Equations (1.5.1) and (1.5.2) follow from (1.2.1) by specialization. In other words,

$$\begin{aligned}
&\theta(z + I\mu' + \tau\mu, \tau) \\
&\quad = \exp\left\{2\pi i[\tfrac{1}{8}\,{}^t\varepsilon\tau\varepsilon + \tfrac{1}{2}\,{}^t\varepsilon(z + I\mu' + \tau\mu)\right. \\
&\qquad \left. + \tfrac{1}{4}\,{}^t\varepsilon\varepsilon']\right\}\theta\left(z + I\frac{\varepsilon'}{2} + \tau\frac{\varepsilon}{2} + I\mu' + \tau\mu, \tau\right) \\
&\quad = \exp\{2\pi i[\tfrac{1}{8}\,{}^t\varepsilon\tau\varepsilon + \tfrac{1}{2}\,{}^t\varepsilon(z + I\mu' + \tau\mu) + \tfrac{1}{4}\,{}^t\varepsilon\varepsilon']\} \\
&\qquad \times \exp\left\{2\pi i\left[-{}^t\mu\left(z + I\frac{\varepsilon'}{2} + \tau\frac{\varepsilon}{2}\right) - \tfrac{1}{2}\,{}^t\mu\tau\mu\right]\right\}\theta z\left(z + I\frac{\varepsilon'}{2} + \tau\frac{\varepsilon}{2}, \tau\right) \\
&\quad = \exp\left\{2\pi i[-\tfrac{1}{2}\,{}^t\mu\tau\mu - {}^t\mu z + \tfrac{1}{2}({}^t\mu'\varepsilon - {}^t\mu\varepsilon')]\right\}\theta\begin{bmatrix}\varepsilon \\ \varepsilon'\end{bmatrix}(z,\tau).
\end{aligned}$$

Equations (1.5.3) and (1.5.4) are derived in essentially the same manner as (1.2.1) and (1.2.4); the proofs are omitted. □

The importance of the statements (1.5.3) and (1.5.4) is that we immediately have the following

Corollary. *Up to sign there are exactly 2^{2g} different first order theta functions with integer characteristics. Of these $2^{g-1}(2^g + 1)$ are even functions, while $2^{g-1}(2^g - 1)$ are odd. These 2^{2g} functions can be thought to correspond to the 2^{2g} points of order 2 in $\mathbb{C}^g/\langle I,\tau\rangle$, where $\langle I,\tau\rangle$ means the group (lattice) of translations of $\mathbb{C}^g$ generated by the columns of (I,τ).*

PROOF. The only part needing some comment is the number of even and odd functions. This is established by an easy induction argument. □

EXAMPLE. When $g = 1$ the three even functions are: $\theta[{}^0_0](z,\tau)$, $\theta[{}^0_1](z,\tau)$, and $\theta[{}^1_0](z,\tau)$. The odd function is $\theta[{}^1_1](z,\tau)$. When $g = 2$ the six odd functions are:

$$\theta\begin{bmatrix}1 & 0 \\ 1 & 0\end{bmatrix}(z,\tau),\ \theta\begin{bmatrix}1 & 0 \\ 1 & 1\end{bmatrix}(z,\tau),\ \theta\begin{bmatrix}1 & 1 \\ 1 & 0\end{bmatrix}(z,\tau),$$

$$\theta\begin{bmatrix}0 & 1 \\ 0 & 1\end{bmatrix}(z,\tau),\ \theta\begin{bmatrix}0 & 1 \\ 1 & 1\end{bmatrix}(z,\tau),\ \theta\begin{bmatrix}1 & 1 \\ 0 & 1\end{bmatrix}(z,\tau).$$

Remarks

1. For the first order theta functions with integer characteristic, $\theta\left[\begin{smallmatrix}\varepsilon\\ \varepsilon'\end{smallmatrix}\right](z,\tau)$, the even ones are precisely those for which ${}^t\varepsilon\varepsilon' \equiv 0 \pmod 2$ and the odd ones those for which ${}^t\varepsilon\varepsilon' \equiv 1 \pmod 2$.
2. We have seen in the proof of Proposition VI.1.5 that
$$\theta\begin{bmatrix}\varepsilon\\ \varepsilon'\end{bmatrix}(z,\tau) = \exp 2\pi i\left[\frac{1}{2}\frac{{}^t\varepsilon}{2}\tau\frac{\varepsilon}{2} + \frac{{}^t\varepsilon}{2}z + \frac{{}^t\varepsilon}{2}\frac{\varepsilon'}{2}\right]\theta\left(z + I\frac{\varepsilon'}{2} + \tau\frac{\varepsilon}{2}, \tau\right).$$
In particular, if $\theta\left[\begin{smallmatrix}\varepsilon\\ \varepsilon'\end{smallmatrix}\right](z,\tau)$ is an odd function, we have $0 = \theta\left[\begin{smallmatrix}\varepsilon\\ \varepsilon'\end{smallmatrix}\right](0,\tau) = \exp 2\pi i[\frac{1}{2}({}^t\varepsilon/2)\tau(\varepsilon/2) + ({}^t\varepsilon/2)(\varepsilon'/2)]\theta(I(\varepsilon'/2) + \tau(\varepsilon/2), \tau)$. The points $I(\varepsilon'/2) + \tau(\varepsilon/2)$ are points of order 2 in $\mathbb{C}^g/\langle I,\tau\rangle$ and will be called *even* or *odd* depending on whether $\theta\left[\begin{smallmatrix}\varepsilon\\ \varepsilon'\end{smallmatrix}\right](z,\tau)$ is an even or odd function. Hence we immediately see that the Riemann theta function $\theta(z,\tau) = \theta\left[\begin{smallmatrix}0\\ 0\end{smallmatrix}\right](z,\tau)$ always vanishes at the odd points of order two or at what we shall call *odd half-periods.*
3. There is an additional property of first-order theta functions which we shall not need till Chapter VII. We leave the proof as an exercise for the reader.

   For μ, μ' integer vectors in $\mathbb{R}^g$,
$$\theta\begin{bmatrix}\varepsilon\\ \varepsilon'\end{bmatrix}\left(z + I\frac{\mu'}{2} + \tau\frac{\mu}{2}, \tau\right)$$
$$= \exp\left\{2\pi i[-\tfrac{1}{8}\,{}^t\mu\tau\mu - \tfrac{1}{4}\,{}^t\mu(\varepsilon' + \mu') - \tfrac{1}{2}\,{}^t\mu z]\right\}\theta\begin{bmatrix}\varepsilon + \mu\\ \varepsilon' + \mu'\end{bmatrix}(z,\tau).$$

VI.2. The Theta Functions Associated with a Riemann Surface

In the preceding section we have defined and derived some of the basic properties of first order theta functions with characteristics. In this section we show how to associate with a compact Riemann surface a collection of first order theta functions and study the behavior of these functions as multivalued "functions" on the Riemann surface. The basic result of this section is the following: A first order theta function either vanishes identically on the Riemann surface or else has precisely g zeros (g = genus of the surface). In the latter case, we can evaluate the zero divisor of the theta function (more precisely, we can determine the image of the divisor in the Jacobian variety of the surface).

VI.2.1. Let M be a compact Riemann surface of genus $g \geq 1$. Let ${}^t\{a,b\} = \{a_1, \ldots, a_g, b_1, \ldots, b_g\}$ be a canonical homology basis on M, and let ${}^t\{\zeta_1, \ldots, \zeta_g\} = \zeta$ be a basis for $\mathscr{H}^1(M)$ dual to the canonical homology basis (recall III.6.1).

We thus obtain a $g \times 2g$ matrix (I,Π) with $e^{(j)} = \int_{a_j} \zeta$, $\pi^{(j)} = \int_{b_j} \zeta$, where as usual $e^{(j)}$ and $\pi^{(j)}$ are the jth columns of the matrices I and Π respectively.

We have already seen that the matrix Π is a complex symmetric matrix with positive definite imaginary part. It follows therefore, from VI.1.1, that we can define first order theta functions with characteristics using the matrix Π. We therefore obtain, as in VI.1.5, 2^{2g} theta functions $\theta\left[\begin{smallmatrix}\varepsilon\\ \varepsilon'\end{smallmatrix}\right](z,\Pi)$.

VI.2.2. There are, of course, many ways to choose a canonical homology basis on M. If $\{a'_1, \ldots, a'_g, b'_1, \ldots, b'_g\} = {}^t\{a',b'\}$ is another canonical homology basis, and ${}^t\{\zeta'_1, \ldots, \zeta'_g\} = \zeta'$ is the basis for $\mathscr{H}^1(M)$ dual to this new basis, we obtain a new $g \times 2g$ matrix (I,Π'), and can define first order theta functions with characteristics using the matrix Π'. We have seen in Chapter V that

$$\begin{bmatrix} a' \\ b' \end{bmatrix} = \begin{bmatrix} A & B \\ C & D \end{bmatrix} \begin{bmatrix} a \\ b \end{bmatrix},$$

with $\left[\begin{smallmatrix} A & B \\ C & D \end{smallmatrix}\right] \in Sp(g,\mathbb{Z})$. Thus using obvious vector notation

$$a' = Aa + Bb,$$

$$\int_{a'} \zeta = \int_{Aa+Bb} \zeta = A + B\Pi,$$

and hence

$$\zeta' = (A + B\Pi)^{-1}\zeta,$$

and

$$\Pi' = \int_{b'} \zeta' = \int_{Ca+Db} (A + B\Pi)^{-1}\zeta = (C + D\Pi)(A + B\Pi)^{-1}.$$

Thus Π and Π' are related by an element of the symplectic modular group of degree g. The relation among the corresponding theta functions involves then the study of modular forms for the symplectic modular group (which will not be pursued here).

VI.2.3. We shall for the remainder of this chapter assume that the Riemann surface M has on it a fixed canonical homology basis ${}^t\{a,b\} = \{a_1, \ldots, a_g, b_1, \ldots, b_g\}$, and we will study the theta functions associated with M and ${}^t\{a,b\}$ as functions on M. We do this by recalling that we have introduced in III.6 a map $\varphi: M \to J(M) = \mathbb{C}^g/L(M)$, where $L(M)$ is the lattice (over $\mathbb{Z}$) generated by the $2g$ columns of the matrix (I,Π). Recall that φ was defined by choosing a point $P_0 \in M$ and setting $\varphi(P) = \int_{P_0}^{P} \zeta$, where $\zeta = {}^t(\zeta_1, \ldots, \zeta_g)$ is the basis of $\mathscr{H}^1(M)$ dual to ${}^t\{a,b\}$. We now consider $\theta \circ \varphi$ and in this way view θ or $\theta\left[\begin{smallmatrix}\varepsilon\\ \varepsilon'\end{smallmatrix}\right]$, where $\left[\begin{smallmatrix}\varepsilon\\ \varepsilon'\end{smallmatrix}\right]$ is an integer characteristic, as a function on M. The reader is, of course, aware that $\theta \circ \varphi$ is not single valued on M, since φ is not single valued (as a function into $\mathbb{C}^g$) on M, but depends on the path of integration. We have seen (Proposition III.6.1) that the map φ is well defined into $J(M)$, and therefore the function $\theta \circ \varphi$ has a very simple multiplicative behavior. The behavior of this function is not quite as simple as the behavior of the functions considered in III.9, since here the multiplier

χ acquired by continuation over a cycle depends on the variable z as well as the cycle. At any rate, it is immediate from Proposition VI.1.5, that continuation of $\theta[{}^{\varepsilon}_{\varepsilon'}] \circ \varphi$ along the closed curve a_l (respectively, b_l) beginning at a point $P \in M$, multiplies it by

$$\chi(a_l) = \exp 2\pi i[\varepsilon_l/2], \tag{2.3.1}$$

$$\chi(b_l) = \exp 2\pi i[-\varepsilon'_l/2 - \pi_{ll}/2 - \varphi_l(P)], \tag{2.3.2}$$

or we may say more simply that analytic continuation of $\theta[{}^{\varepsilon}_{\varepsilon'}] \circ \varphi$ over the curve a_l (respectively, b_l) beginning at P carries $\theta[{}^{\varepsilon}_{\varepsilon'}] \circ \varphi$ into

$$\exp 2\pi i[\varepsilon_l/2]\theta[{}^{\varepsilon}_{\varepsilon'}] \circ \varphi$$

(respectively, $\exp 2\pi i[-\varepsilon'_l/2 - \pi_{ll}/2 - \varphi_l(P)]\theta[{}^{\varepsilon}_{\varepsilon'}] \circ \varphi$), where φ_l is the lth component of the map φ. It therefore follows that $\theta[{}^{\varepsilon}_{\varepsilon'}] \circ \varphi$ is a holomorphic multiplicative function on M with multipliers given by (2.3.1) and (2.3.2).

VI.2.4. We now wish to study the zeros of $\theta[{}^{\varepsilon}_{\varepsilon'}] \circ \varphi$ on M. We observe that even though $\theta[{}^{\varepsilon}_{\varepsilon'}](z,\Pi)$ is not a single valued function on $J(M) = \mathbb{C}^g/L(M)$, the set of zeros of $\theta[{}^{\varepsilon}_{\varepsilon'}](z,\Pi)$ is a well defined set on $J(M)$ (Proposition VI.1.5). The set of zeros of $\theta[{}^{\varepsilon}_{\varepsilon'}](z,\tau)$, for any symmetric τ with $\operatorname{Im} \tau > 0$, is an analytic set in $\mathbb{C}^g$ of codimension one and in particular for $\tau = \Pi$, the zeros of $\theta[{}^{\varepsilon}_{\varepsilon'}](z,\Pi)$ form an analytic set of codimension one in $J(M)$. The map φ, by Proposition III.6.1, is a holomorphic mapping of maximal rank of M into $J(M)$. The study of the zeros of $\theta[{}^{\varepsilon}_{\varepsilon'}] \circ \varphi$ on M is thus the study of the intersection of the image of M in $J(M)$ under φ and the analytic set consisting of the zeros of $\theta[{}^{\varepsilon}_{\varepsilon'}](z,\Pi)$. Since M is compact there are only two possibilities. Either $\theta[{}^{\varepsilon}_{\varepsilon'}] \circ \varphi$ vanishes identically on M or else $\theta[{}^{\varepsilon}_{\varepsilon'}] \circ \varphi$ has only a finite number of zeros on M. Neglecting for the moment the former possibility, can we determine how many zeros does $\theta[{}^{\varepsilon}_{\varepsilon'}] \circ \varphi$ have on M?

It involves no loss of generality to assume that no zero of $\theta[{}^{\varepsilon}_{\varepsilon'}] \circ \varphi$ lies on any curve chosen as a representative for the canonical homology basis. In order to count the number of zeros, we need to evaluate

$$\frac{1}{2\pi i}\int_{\delta\mathscr{M}} d \log \theta[{}^{\varepsilon}_{\varepsilon'}] \circ \varphi,$$

where $\mathscr{M}$ is the polygon associated with M whose boundary consists of representatives for the canonical homology basis ($\delta\mathscr{M} = \prod_{j=1}^{g} a_j b_j a_j^{-1} b_j^{-1}$). In order to further simplify notation we shall denote $\theta[{}^{\varepsilon}_{\varepsilon'}] \circ \varphi$ by f.

We now compute

$$\begin{aligned}\frac{1}{2\pi i}\int_{\delta\mathscr{M}} \frac{df}{f} &= \frac{1}{2\pi i}\sum_{k=1}^{g} \int_{a_k + b_k + a_k^{-1} + b_k^{-1}} \frac{df}{f} \\ &= \frac{1}{2\pi i}\sum_{k=1}^{g}\left[\int_{a_k}\left(\frac{df}{f} - \frac{df^-}{f^-}\right) + \int_{b_k}\left(\frac{df}{f} - \frac{df^-}{f^-}\right)\right],\end{aligned}$$

where f^- denotes the value of f on the cycle a_k^{-1} or b_k^{-1}. It follows immediately from (2.3.1) and (2.3.2) that if P is a point on a_k

$$f^-(P) = \exp 2\pi i\left[\frac{-\varepsilon_k'}{2} - \frac{\pi_{kk}}{2} - \varphi_k(P)\right] f(P),$$

while if P is a point on b_k,

$$f(P) = \exp 2\pi i\left[\frac{\varepsilon_k}{2}\right] f^-(P).$$

Thus for P a point of b_k, $(df/f - df^-/f^-)(P) = 0$, while for P a point of a_k, $(df/f - df^-/f^-)(P) = 2\pi i\varphi_k'(P)$. It therefore follows that $(1/2\pi i)\int_{\delta\mathcal{M}}(df/f) = (1/2\pi i)\sum_{k=1}^{g}\int_{a_k} 2\pi i\varphi_k'(z)\,dz = g$. We are here exploiting the fact that $\varphi_k'(z)\,dz$ is the kth element of the basis of $\mathscr{H}^1(M)$ dual to the homology basis $\{a_1,\ldots,a_g,b_1,\ldots,b_g\}$. At any rate, we have shown that if $\theta[{}^{\varepsilon}_{\varepsilon'}]\circ\varphi$ is not identically zero it has precisely g zeros on M (counting multiplicities).

The next natural question which arises in regard to the zeros is: can we find the divisor of zeros? What we shall do is find the image $\mathscr{X}$ of the divisor of zeros of $\theta[{}^{\varepsilon}_{\varepsilon'}]\circ\varphi$ under the induced map on integral divisors of degree g, $\varphi: M_g \to J(M)$. To this end we consider $(1/2\pi i)\int_{\delta\mathcal{M}}\varphi(df/f)$, where φ is the column vector ${}^t(\varphi_1,\ldots,\varphi_g)$ and $f = \theta[{}^{\varepsilon}_{\varepsilon'}]\circ\varphi$. We immediately find that

$$\mathscr{X} = \frac{1}{2\pi i}\int_{\delta\mathcal{M}}\varphi\frac{df}{f} = \frac{1}{2\pi i}\sum_{k=1}^{g}\int_{a_k+b_k+a_k^{-1}+b_k^{-1}}\varphi\frac{df}{f}$$
$$= \frac{1}{2\pi i}\sum_{k=1}^{g}\left[\int_{a_k}\left(\varphi\frac{df}{f} - \varphi^-\frac{df^-}{f^-}\right) + \int_{b_k}\left(\varphi\frac{df}{f} - \varphi^-\frac{df^-}{f^-}\right)\right],$$

where φ^- plays the same role for φ as f^- served for f. We once again use the relation between f^- and f and the fact that for P a point on a_k, $\varphi^- = \varphi + \pi^{(k)}$, while for P a point on b_k, $\varphi = \varphi^- + e^{(k)}$. Thus

$$\mathscr{X} = \frac{1}{2\pi i}\sum_{k=1}^{g}\left[\int_{a_k}\varphi\frac{df}{f} - (\varphi + \pi^{(k)})\left(\frac{df}{f} - 2\pi i\varphi_k'(z)\,dz\right)\right.$$
$$\left. + \int_{b_k}(\varphi^- + e^{(k)})\frac{df^-}{f^-} - \varphi^-\frac{df^-}{f^-}\right]$$
$$= \frac{1}{2\pi i}\sum_{k=1}^{g}\int_{a_k}\left[-\pi^{(k)}\frac{df}{f} + 2\pi i\pi^{(k)}\varphi_k'(z)\,dz + 2\pi i\varphi\varphi_k'(z)\,dz\right]$$
$$+ \frac{1}{2\pi i}\sum_{k=1}^{g}\int_{b_k}e^{(k)}\frac{df^-}{f^-}.$$

We now need to evaluate integrals of the form $\int_{a_k}(df/f)$ and $\int_{b_k}(df/f)$. Since $df/f = d\log f$, the integral in question is simply $\log f(P_1) - \log f(P_2)$, where $P_1 = P_2$ are the initial and terminal points of the cycle a_k (or b_k). (The difference is not zero, since we must use different "branches" of f.) We now once again use (2.3.1) and (2.3.2) to obtain for P_1 and P_2 initial and terminal points of a_k, $f(P_2) = \exp[2\pi i(\varepsilon_k/2)]f(P_1)$, and for P_1 and P_2 initial and terminal points of b_k, $f(P_2) = \exp 2\pi i[-(\varepsilon_k'/2) - (\pi_{kk}/2) - \varphi_k(P_1)]f(P_1)$. This shows that

$$\mathscr{L} = \frac{1}{2\pi i}\sum_{k=1}^{g}\left[-\pi^{(k)}\left(2\pi i\frac{\varepsilon_k}{2} + 2\pi i n_k\right) + \pi^{(k)}2\pi i + 2\pi i\int_{a_k}\varphi\varphi_k'(z)\,dz\right]$$
$$+\frac{1}{2\pi i}\sum_{k=1}^{g} e^{(k)}\left[2\pi i\left(-\frac{\varepsilon_k'}{2} - \frac{\pi_{kk}}{2} - \varphi_k\right) + 2\pi i m_k\right],$$

where m_k and n_k are integers. It therefore follows that

$$\mathscr{L} = \frac{1}{2\pi i}\int_{\delta\mathscr{M}}\varphi\frac{df}{f}$$
$$= \sum_{k=1}^{g}\left[-\pi^{(k)}\frac{\varepsilon_k}{2} - e^{(k)}\frac{\varepsilon_k'}{2} + \pi^{(k)}(1 - n_k)\right.$$
$$\left. + \int_{a_k}\varphi\varphi_k'\,dz - e^{(k)}\left(\frac{\pi_{kk}}{2} + \varphi_k(P_1)\right) + e^{(k)}m_k\right],$$

or finally

$$\mathscr{L} = -\Pi\frac{\varepsilon}{2} - I\frac{\varepsilon'}{2} + \Pi n + \mathrm{Im} - \mathscr{K},$$

where

$$\mathscr{K} = -\sum_{k=1}^{g}\left[\int_{a_k}\varphi\varphi_k'\,dz - e^{(k)}\left(\frac{\pi_{kk}}{2} + \varphi_k(P_1)\right)\right]. \tag{2.4.1}$$

Remark and Definition. The vector $\mathscr{K}$ depends, of course, on the choice of the canonical homology basis and the choice of the base point P_0 of the map φ. (There is an illusory dependence on the base point P_1 of $\pi_1(M)$ which can be dispensed with by choosing $P_1 = P_0$ in the above.) The vector $\mathscr{K}$ is known as the *vector of Riemann constants*. It will be denoted by $\mathscr{K}_{P_0}$, when its dependence on the base point is to be emphasized.

Let us consider as an example, the case $g = 1$ and $\varepsilon = 0 = \varepsilon'$. We represent the surface M as the complex plane factored by the lattice generated by 1 and τ, where τ is a complex number with positive imaginary part. (In this case, of course, $M = J(M)$.) Direct calculations yield

$$\mathscr{K} = -\tfrac{1}{2} + \frac{\tau}{2}.$$

This once again shows that $\theta(\cdot,\tau)$ vanishes at the unique odd half-period and nowhere else.

We have proved the following

Theorem. *Let M be a compact Riemann surface of genus $g \geq 1$ with canonical homology basis ${}^t\{a,b\}$. Let $\theta\left[\begin{smallmatrix}\varepsilon\\ \varepsilon'\end{smallmatrix}\right](z,\Pi)$ be the first order theta function associated* *with $(M, {}^t\{a,b\})$ and let φ be the Abel–Jacobi map $M \to J(M)$. Then $\theta[{}^{\varepsilon}_{\varepsilon'}] \circ \varphi$ is either identically zero as a function on M or else has precisely g zeros on M. In this case let $P_1 \cdots P_g$ be the divisor of zeros. We then have $\varphi(P_1 \cdots P_g) = -\Pi\varepsilon/2 - I\varepsilon'/2 + \Pi n + \mathrm{I}m - \mathscr{K}$, where n and m are integer vectors and $\mathscr{K}$ is a vector which depends on the canonical homology basis ${}^t\{a,b\}$, and the base point of the map $\varphi : M \to J(M)$. In particular, since $\Pi n + Im$ is the zero point of $J(M)$, we have $\varphi(P_1 \cdots P_g) + \mathscr{K} = -\Pi\varepsilon/2 - I\varepsilon'/2$.*

Remark. Consider the "function" on M

$$P \mapsto \theta(\varphi(P) - e),$$

with $e \in \mathbb{C}^g$. If this "function" does not vanish identically on M, it has g zeros $P_1, \ldots, P_g$. The theorem asserts that

$$\varphi(P_1 \cdots P_g) + \mathscr{K} = e$$

(where in the last equation we understand by e not the point in $\mathbb{C}^g$, but its projection in $\mathbb{C}^g/L(M)$). Our remark is valid because in the proof of the theorem ε and ε' were arbitrary points in $\mathbb{R}^g$ (we never used the fact that the characteristics were integral).

VI.2.5. We wish next to characterize the zero set of $\theta[{}^{\varepsilon}_{\varepsilon'}](z,\Pi)$ as a subset of $J(M)$. In view of (1.4.4) it suffices to consider only the functions of the form

$$z \mapsto \theta(z - e),$$

with fixed $e \in \mathbb{C}^g$. It is now convenient to review some concepts that we studied in III.11. Recall that for every integer $n \geq 1$, M_n denotes the integral divisors of degree n on M (or equivalently the n-fold symmetric product of M). Using the mapping $\varphi : M_n \to J(M)$, we can view θ as a locally defined holomorphic function on M_n. Also, W_n denotes the image in $J(M)$ of M_n under the mapping φ. Recall that for $D \in M_n$ (Proposition III.11.11(a)), the Jacobian of the mapping φ has rank $n + 1 - r(D^{-1})$, which by Riemann–Roch equals $g - i(D)$. Taking $n = g$ and a non-special divisor $D \in M_g$, we conclude that

$$\varphi : M_g \to J(M)$$

is a local homeomorphism at D. Thus since $\theta \not\equiv 0$, θ does not vanish identically on any open subset of M_g.

VI.3. The Theta Divisor

In this section we begin a study of the divisor of zeros of the theta function, culminating in the Riemann vanishing theorem (Theorem VI.3.5), which prescribes in a rather detailed manner the zero set of the θ-function on $\mathbb{C}^g$.

We also obtain necessary and sufficient conditions for the θ-function to vanish identically on the Riemann surface (Theorem VI.3.3) and as a by-product, an alternate proof of the Jacobi inversion theorem.

In the previous section we saw that the "function" on M,

$$P \mapsto \theta(\varphi(P) - e),$$

if not identically zero, has g zeros on $P_1, \ldots, P_g$ and that $e = \varphi(P_1 \cdots P_g) + \kappa$. If in place of the above function we consider for any integer α, a closely related function,

$$P \mapsto \theta(\alpha\varphi(P) - e),$$

we would find that this "function", if not identically zero on M, has $\alpha^2 g$ zeros, $P_1, \ldots, P_{\alpha^2 g}$ and that

$$\alpha e = \varphi(P_1 \cdots P_{\alpha^2 g}) + \alpha^2 \mathscr{K}.$$

VI.3.1. Theorem. *Let $e \in \mathbb{C}^g$. Then $\theta(e) = 0$ if and only if $e \in W_{g-1} + \mathscr{K}$, where $\mathscr{K}$ is the vector of constants of Theorem* VI.2.4.

Proof. We first show that if $e \in W_{g-1} + \mathscr{K}$, then $\theta(e) = 0$. To this end, choose a point $\zeta = P_1 \cdots P_g$ in M_g with $P_k \neq P_j$ for $k \neq j$ such that $i(\zeta) = 0$. If ζ' is any other point in M_g sufficiently close to ζ, then since $i(\zeta) = i(P_1 \cdots P_g) = g - \operatorname{rank}(\zeta_k(P_j))$, $i(\zeta') = 0$ as well. Set $e = \varphi(\zeta) + \mathscr{K}$ and consider $\psi(P) = \theta(\varphi(P) - e)$ as a function of $P \in M$.

There are now two possibilities to consider. Either ψ is identically zero or not. In the former case, we have for each $k = 1, \ldots, g$,

$$\begin{aligned}\theta(\varphi(P_k) - (\varphi(P_1 \cdots P_g) + \mathscr{K})) &= \theta(-\varphi(P_1 \cdots \hat{P}_k \cdots P_g) - \mathscr{K}) \\ &= \theta(\varphi(P_1 \cdots \hat{P}_k \cdots P_g) + \mathscr{K}) = 0\end{aligned}$$

(where we have used symbol $P_1 \cdots \hat{P}_k \cdots P_g$ to mean P_k does not appear in the divisor) because (by (1.2.4)) θ is an even function. In the latter case, we have from Theorem VI.2.4 that ψ has precisely g zeros on M, say, $Q_1, \ldots, Q_g$, and that $\varphi(Q_1 \cdots Q_g) + \mathscr{K} = e$. Since $\zeta = P_1 \cdots P_g$ was chosen so that $i(P_1 \cdots P_g) = 0$, it follows from the Riemann–Roch theorem and Abel's theorem that $Q_1 \cdots Q_g = P_1 \cdots P_g$. Therefore, even in this case we have $\theta(\varphi(P_1 \cdots \hat{P}_k \cdots P_g) + \mathscr{K}) = 0$ for each $k = 1, \ldots, g$. The remark at the beginning of this proof showed that the divisors ζ of the form under consideration, $\zeta = P_1 \cdots P_g$, with $P_j \neq P_k$ for $j \neq k$ and $i(\zeta) = 0$, form a dense subset of M_g. Thus divisors of the form $P_1 \cdots \hat{P}_k \cdots P_g$ form a dense subset of M_{g-1}. Hence θ vanishes identically on $W_{g-1} + \mathscr{K}$.

Conversely, suppose $\theta(e) = 0$. Let s be the least integer such that $\theta(W_{s-1} - W_{s-1} - e) \equiv 0$ but $\theta(W_s - W_s - e) \not\equiv 0$. Here $W_r - W_r$ has the obvious meaning ($e \in W_r - W_r \Leftrightarrow e = f - h$ with $f, h \in W_r$). Our remarks at the end of VI.2.5 immediately give $1 \leq s \leq g$. The hypothesis we have thus far made assures us of the existence of two points in M_s say $\zeta = P_1 \cdots P_s$, $\omega = Q_1 \cdots Q_s$ such that $\theta(\varphi(P_1 \cdots P_s) - \varphi(Q_1 \cdots Q_s) - e) \neq 0$. We can assume without loss of generality that $P_k \neq P_j$ for $k \neq j$, $Q_k \neq Q_j$ for $k \neq j$, and $P_k \neq Q_j$ for all k, j.

Consider now the function $P \mapsto \theta(\varphi(P) + \varphi(P_2 \cdots P_s) - \varphi(Q_1 \cdots Q_s) - e)$. This function does not vanish identically as a function on M, since $P = P_1$ is not a zero. On the other hand, it is immediately clear that $P = Q_j$, $j = 1, \ldots, s$, is a zero of this function. It therefore follows from Theorem VI.2.4 that for some $T_1 \cdots T_{g-s} \in M_{g-s}$

$$\varphi(Q_1 \cdots Q_s) - \varphi(P_2 \cdots P_s) + e = \varphi(Q_1 \cdots Q_s T_1 \cdots T_{g-s}) + \mathscr{K} \quad (3.1.1)$$

or that

$$e = \varphi(T_1 \cdots T_{g-s} P_2 \cdots P_s) + \mathscr{K}, \quad (3.1.2)$$

which is obviously a point of $W_{g-1} + \mathscr{K}$. (Note that the above arguments work also for $s = 1$.) □

VI.3.2. We now observe that we can really prove a bit more than we have claimed in VI.3.1. The points $\zeta = P_1 \cdots P_s$ and $\omega = Q_1 \cdots Q_s$ utilized in the above proof, were fairly arbitrary in the sense that any other points $\zeta', \omega' \in M_s$ which are sufficiently close to ζ and ω would have worked as well. We also found that $e = \varphi(T_1 \cdots T_{g-s} P_2 \cdots P_s) \in M_{g-1}$ has at least $s - 1$ "arbitrary" points $P_2, \ldots, P_s$ in it. Let $D' = P_2' \cdots P_s'$ be arbitrary but close to $P_2 \cdots P_s$. The argument of VI.3.1 shows

$$\begin{aligned} e &= \varphi(T_1 \cdots T_{g-s} P_2 \cdots P_s) + \mathscr{K} \\ &= \varphi(T_1' \cdots T_{g-s}' P_2' \cdots P_s') + \mathscr{K}. \end{aligned}$$

Thus $T_1 \cdots T_{g-s} P_2 \cdots P_s \sim T_1' \cdots T_{g-s}' P_2' \cdots P_s'$ by Abel's theorem. Lemma III.8.15 now implies that $r(1/T_1 \cdots T_{g-s} P_2 \cdots P_s) \geq s$ and thus (by Riemann–Roch) that $i(T_1 \cdots T_{g-s} P_2 \cdots P_s) \geq s$.

On the other hand, suppose that $e = \varphi(\Delta) + \mathscr{K}$ with $\Delta \in M_{g-1}$ and $i(\Delta) = s$. Then Δ has precisely $s - 1$ "arbitrary" points. Any point X of $W_{s-1} - W_{s-1} - e$ is of the form $\varphi(P_1 \cdots P_{s-1}) - \varphi(Q_1 \cdots Q_{s-1}) - e$, with $e = \varphi(\Delta) + K = \varphi(P_1 \cdots P_{s-1}\delta) + \mathscr{K}$ and $\delta \in M_{g-s}$. Therefore, we have

$$\begin{aligned} X &= \varphi(P_1 \cdots P_{s-1}) - \varphi(Q_1 \cdots Q_{s-1}) - \varphi(P_1 \cdots P_{s-1}\delta) - \mathscr{K} \\ &= -\varphi(Q_1 \cdots Q_{s-1}\delta) - \mathscr{K}; \end{aligned}$$

and $X \in -(W_{g-1} + \mathscr{K})$; and $\theta(X) = 0$.

We can therefore now strengthen slightly Theorem VI.3.1 to the following

Theorem. *For $e \in \mathbb{C}^g$, $\theta(e) = 0$ if and only if $e \in W_{g-1} + \mathscr{K}$. If $e \in W_{g-1} + \mathscr{K}$, and $e = \varphi(\zeta) + \mathscr{K}$ with $\zeta \in M_{g-1}$ and $i(\zeta) = s\ (\geq 1)$, then $\theta(W_{s-1} - W_{s-1} - e) \equiv 0$. Conversely, if s is the least integer such that $\theta(W_{s-1} - W_{s-1} - e) \equiv 0$ but $\theta(W_s - W_s - e) \not\equiv 0$, then $e = \varphi(\zeta) + \mathscr{K}$ with $\zeta \in M_{g-1}$ and $i(\zeta) = s$.*

PROOF. All but the last remark is contained in the discussion preceding the statement of the theorem. In the discussion we only showed $i(\zeta) \geq s$; however, the statement $i(\zeta) = s$ is an immediate consequence of the one preceding it. If $i(\zeta) > s$, we would have $\theta(W_s - W_s - e) \equiv 0$ contrary to the hypothesis. □

VI.3.3. In the above analysis we have several times encountered the multi-valued holomorphic function on M

$$\psi : P \mapsto \theta(\alpha\varphi(P) - e)$$

with fixed $\alpha \in \mathbb{Z}$ and fixed $e \in J(M)$. (Recall that this function depends on a choice of base point for φ.) We have observed that either ψ vanishes identically or else has precisely $\alpha^2 g$ zeros. When does each of these possibilities occur? The answer is given by the following:

Theorem. *Let P_0 be the base point of the map φ. Then ψ vanishes identically on M if and only if $e = \varphi_{P_0}(Q_1 \cdots Q_{g-1}) + \mathscr{K}_{P_0}$ and $i(P_0^\alpha Q_1 \cdots Q_{g-1}) > 0$. In particular, for $\alpha = 1$ we have that $e = \varphi_{P_0}(D) + \mathscr{K}_{P_0}$ with $D \in M_g$, then ψ is not identically zero if and only if $i(D) = 0$ and D is the divisor of zeros of ψ.*

PROOF. If ψ vanishes identically (so that, in particular, $\theta(e) = 0$), then for each point $P \in M$, we have

$$\varphi_{P_0}(P^\alpha) - \phi_{P_0}(Q_1 \cdots Q_{g-1}) - \mathscr{K}_{P_0} = -\varphi_{P_0}(R_1 \cdots R_{g-1}) - \mathscr{K}_{P_0}$$

or that

$$\varphi_{P_0}(P^\alpha R_1 \cdots R_{g-1}) = \varphi_{P_0}(P_0^\alpha Q_1 \cdots Q_{g-1}).$$

As a consequence of Abel's theorem, we find that for each $P \in M$ there is a meromorphic function in the vector space $L(1/P_0^\alpha Q_1 \cdots Q_{g-1})$ which has a zero of order at least α at P. This implies, as a consequence of the Riemann–Roch theorem, that $i(P_0^\alpha Q_1 \cdots Q_{g-1}) > 0$.

Conversely, if $i(P_0^\alpha Q_1 \cdots Q_{g-1}) > 0$, then by the Riemann–Roch theorem $r(1/P_0^\alpha Q_1 \cdots Q_{g-1}) \geq \alpha + 1$ so that there is a function in the space that vanishes to order at least α at any point $P \in M$. Therefore,

$$\begin{aligned}\theta(\varphi_{P_0}(P^\alpha - \varphi_{P_0}(P_0^\alpha Q_1 \cdots Q_{g-1}) - \mathscr{K}_{P_0}) &= \theta(\varphi_{P_0}(P^\alpha - \varphi_{P_0}(P^\alpha R_1 \cdots R_{g-1}) - \mathscr{K}_{P_0}) \\ &= \theta(-\varphi_{P_0}(R_1 \cdots R_{g-1}) - \mathscr{K}_{P_0}) = 0.\end{aligned}$$

□

As a consequence of the above theorem we can immediately conclude the following

Corollary. *If $\alpha \geq g$, then for no base point P_0 can ψ vanish identically on M.*

PROOF. Identical vanishing would give by the theorem that $i(P_0^\alpha Q_1 \cdots Q_{g-1}) > 0$ which is clearly impossible. □

Remark. Another immediate consequence of the theorem is that if

$$e = \varphi_{P_0}(Q_1 \cdots Q_{g-1}) + \mathscr{K}_{P_0}$$

and $i(Q_1 \cdots Q_{g-1}) = 1$, then there are only finitely many points P_0 for which we have identical vanishing of ψ. Furthermore, if $i(Q_1 \cdots Q_{g-1}) > 1$, then for every point P_0 we have identical vanishing of ψ for $\alpha = 1$. As a matter of fact, for $\alpha = 1$, we can strengthen the above result. For the convenience of the reader, the next theorem is established without reference to the previous theorem.

Theorem. *Assume* $\alpha = 1$.

a. *If* $e \in J(M)$ *and* $\psi \equiv 0$, *then* $e = \varphi(D) + \mathscr{K}$ *for some* $D \in M_g$ *with* $i(D) = s \geq 1$, *and* s *is the least integer such that*

$$\theta(W_{s+1} - W_s - e) \not\equiv 0.$$

b. *If* $e = \varphi(D) + \mathscr{K}$, $D \in M_g$, *then* $\psi \not\equiv 0$ *if and only if* $i(D) = 0$ *and* D *is the divisor of zeros of* ψ. *In particular,* $\psi \equiv 0$ *if and only if* $i(D) > 0$.

Proof. Let s be the least integer such that $\theta(W_{s+1} - W_s - e) \not\equiv 0$. Then $0 \leq s \leq g - 1$. Choose $P_1, \ldots P_{s+1}, Q_1, \ldots, Q_s$ such that

$$\theta(\varphi(P_1 \cdots P_{s+1}) - \varphi(Q_1 \cdots Q_s) - e) \neq 0.$$

Consider the function

$$P \mapsto \theta(\varphi(P) + \varphi(P_2 \cdots P_{s+1}) - \varphi(Q_1 \cdots Q_s) - e)$$

which does not vanish identically (it is non-zero at P_1). It then has (by the minimality of s) g zeros

$$Q_1, \ldots, Q_s, T_1, \ldots, T_{g-s}.$$

Thus, again by Theorem VI.2.4, we conclude that

$$\varphi(Q_1 \cdots Q_s T_1 \cdots T_{g-s}) + \mathscr{K} = \varphi(Q_1 \cdots Q_s) - \varphi(P_2 \cdots P_{s+1}) + e;$$

or that

$$e = \varphi(T_1 \cdots T_{g-s} P_2 \cdots P_{s+1}) + \mathscr{K} \in W_g + \mathscr{K}.$$

As before, the divisor $D = T_1 \cdots T_{g-s} P_2 \cdots P_{s+1}$ has s free points. Thus $r(D^{-1}) \geq s + 1$ and (by Riemann-Roch) $i(D) \geq s$. We have established part (a), since $\psi \equiv 0$ implies $s \geq 1$.

To establish part (b), note that by (a) if $i(D) = 0$, then $\psi \not\equiv 0$. Conversely, suppose $i(D) > 0$ and $\psi \not\equiv 0$. Choose $Q \in M$ such that $\psi(Q) \neq 0$. Lemma III.8.15, implies that there exists a $D' \in M_{g-1}$ such that $\varphi(D) = \varphi(QD')$. We now use Theorem VI.3.1 to conclude that

$$\begin{aligned}\psi(Q) = \theta(\varphi(Q) - e) &= \theta(\varphi(D) - \varphi(D') - e)\\ &= \theta(-\varphi(D') - \mathscr{K}) = \theta(\varphi(D') + \mathscr{K}) = 0.\end{aligned}$$

This contradiction shows that $\psi \equiv 0$.

To conclude the proof of part (b) we must verify that whenever $e = \varphi(D) + \mathscr{K}$, $D \in M_g$, and $i(D) = 0$, then D is indeed the divisor of zeros of ψ. Let D' be the divisor of zeros of ψ, then $\varphi(D) = \varphi(D')$. Since $i(D) = 0$, this implies that $D = D'$. □

VI.3.4. As a consequence of Theorems VI.3.1 and VI.3.3, we obtain an alternate proof of the Jacobi inversion

Theorem. *Given $a \in J(M)$, there exists a $D \in M_g$ such that $\varphi(D) = a$.*

PROOF. View a as a point in $\mathbb{C}^g$. Let $e = a + \mathscr{K}$. Consider the function $P \mapsto \theta(\varphi(P) - e)$. If this function does not vanish identically, it has as its zero set a divisor $D \in M_g$ with $\varphi(D) + \mathscr{K} = e = a + \mathscr{K}$, by Theorem VI.2.4. If the function does vanish identically, then the last equation follows from Theorem VI.3.3. □

VI.3.5. Lemma. *The condition $\theta(W_r - W_r - e) \equiv 0$ implies that all partial derivatives of θ of order $\leq r$ vanish at e.*

PROOF. Since the lemma clearly holds for $r = 0$, we take $r > 0$. We have $\theta(\varphi(P_1 \cdots P_r) - \varphi(Q_1 \cdots Q_r) - e) = 0$ for all $D = P_1 \cdots P_r \in M_r$, all $D' = Q_1 \cdots Q_r \in M_r$. Thus

$$\theta(\varphi(P) + \varphi(P_2 \cdots P_r) - \varphi(Q_1 \cdots Q_r) - e)$$

vanishes identically on M as a function of P. If we now expand this function in a Taylor series about the point Q_1 and set $P = Q_1$, we find the coefficient of a local coordinate at Q_1 to be

$$\sum_{j=1}^{g} (\partial\theta/\partial z_j)(\varphi(P_2 \cdots P_r) - \varphi(Q_2 \cdots Q_r) - e)\zeta_j(Q_1),$$

which must vanish. Moreover, since Q_1 on M is arbitrary, we find that $(\partial\theta/\partial z_j)(\varphi(P_2 \cdots P_r) - \varphi(Q_2 \cdots Q_r) - e) = 0$ (by the linear independence of the ζ_j's for $j = 1, \ldots, g$), and this holds for all points $P_2 \cdots P_r$, $Q_2 \cdots Q_r \in M_{r-1}$. We conclude that $(\partial\theta/\partial z_j)(W_{r-1} - W_{r-1} - e) \equiv 0$, for $j = 1, \ldots, g$. We now simply repeat the argument for each of these functions and continue until we find all partial derivatives of θ of order $\leq r$ vanish at $-e$, and therefore also at e. □

This lemma together with Theorem VI.3.2 gives us a substantial portion of Riemann's theorem. We have seen that any zero, e, of θ is of the form $e = \varphi(\zeta) + \mathscr{K}$ with $\zeta \in M_{g-1}$. Furthermore, we have seen that if $i(\zeta) = s$, then $\theta(W_{s-1} - W_{s-1} - e) \equiv 0$, and therefore all partial derivatives of θ of order less than s vanish at e. We can thus say that in this case θ vanishes to order at least s at e. The Riemann vanishing theorem asserts that s is the precise order of vanishing at e; that is, at least one partial derivative of order s does not vanish at e.

We now return to the hypothesis of the last part of Theorem VI.3.2 and observe that this hypothesis allows us to conclude the existence of three distinct points $\zeta, \omega, \tau \in M_s$, with the additional property that the points in these divisors are all distinct, and

$$\theta(\varphi(\zeta) - \varphi(\omega) - e) \neq 0, \tag{3.5.1}$$

$$\theta(\varphi(\zeta) - \varphi(\tau) - e) \neq 0, \tag{3.5.2}$$

$$\theta(\varphi(\omega) - \varphi(\tau) - e) \neq 0. \tag{3.5.3}$$

Equation (3.5.1) is a consequence of the hypothesis $\theta(W_s - W_s - e) \not\equiv 0$, while (3.5.2) follows by continuity from (3.5.1) for all τ near ω. Finally, if for all τ' near ω, $\theta(\varphi(\omega) - \varphi(\tau') - e) = 0$, then by varying ω we would conclude once again that $\theta(W_s - W_s - e) \equiv 0$, contrary to hypothesis.

Consider now

$$\frac{\theta(\varphi(P_1 \cdots P_s) - \varphi(\omega) - e)}{\theta(\varphi(P_1 \cdots P_s) - \varphi(\tau) - e)}$$

as a function on M_s in a neighborhood of ζ. By (3.5.1) and (3.5.2) this function is not identically zero on M_s, since neither numerator nor denominator vanish at ζ. We would rather view this for the moment as a function on M, and in order to do so we consider

$$f: P \mapsto \frac{\theta(\varphi(P) + \varphi(P_2 \cdots P_s) - \varphi(\omega) - e)}{\theta(\varphi(P) + \varphi(P_2 \cdots P_s) - \varphi(\tau) - e)},$$

and observe that as a function of P on M it is also not identically zero.

The same arguments which were used in VI.3.1 give that the divisor of zeros of the numerator is $\omega\gamma$ with $\gamma \in M_{g-s}$, and the divisor of zeros of the denominator is $\tau\delta$. Furthermore, Theorem VI.2.4 when applied to numerator and denominator separately gives

$$e = \varphi(P_2 \cdots P_s) + \varphi(\gamma) + \mathscr{K} = \varphi(P_2 \cdots P_s) + \varphi(\delta) + \mathscr{K};$$

from which we conclude $\varphi(\gamma) = \varphi(\delta)$.

We now claim $\gamma = \delta$. If $\gamma \neq \delta$ then Abel's theorem gives us the existence of a non-constant function in $L(\gamma^{-1})$. The Riemann–Roch theorem gives $r(\gamma^{-1}) = g - s - g + 1 + i(\gamma) = i(\gamma) + 1 - s$. Now $r(\gamma^{-1}) \geq 2$ yields $i(\gamma) \geq s + 1$; which then yields $i(\omega\gamma) \geq 1$ (since $i(\omega\gamma) \geq i(\gamma) - \deg \omega$). We will obtain a contradiction by showing that $i(\omega\gamma) > 0$ implies $\theta(\varphi(P) + \varphi(P_2 \cdots P_s) - \varphi(\omega) - e) \equiv 0$ as a function of P on M. We conclude from Theorem VI.2.4 that $e + \varphi(\omega) - \varphi(P_2 \cdots P_s) = \varphi(\omega\gamma) + \mathscr{K}$ so that we may rewrite our function as

$$P \mapsto \theta(\varphi(P) - (\varphi(\omega\gamma) + \mathscr{K})).$$

By Theorem VI.3.3(b), this function vanishes identically, showing that $\gamma = \delta$. Therefore, the function f viewed as a (multiplicative) function on M

vanishes at the points of ω and has poles at the points of τ (and is holomorphic and non-zero elsewhere).

It is now necessary to analyze the multiplicative behavior of the function f on M. It follows immediately from (1.4.1) and (1.4.2) that continuation of this function over a_k leads back to the original function while continuation over the cycle b_k beginning at P leads to $\exp 2\pi i[\varphi_k(\omega) - \varphi_k(\tau)]f(P)$, where φ_k is the kth component of φ.

Recall now the normalized differentials of third kind τ_{PQ} with simple poles at P and Q with residue $+1$ at P and -1 at Q. We found (III(3.6.3)) that

$$\frac{1}{2\pi i}\int_{b_k} \tau_{PQ} = \varphi_k(P) - \varphi_k(Q).$$

Let us now consider the function

$$g(P) = \exp\left(\int_{P_0}^{P} \sum_{j=1}^{s} \tau_{R_jT_j}\right) \prod_{k=2}^{s} \exp\left(\int_{P_0}^{P_k} \sum_{j=1}^{s} \tau_{R_jT_j}\right),$$

where $\omega = R_1 \cdots R_s$, $\tau = T_1 \cdots T_s$. It is clear that $g(P)$ and $f(P)$ have the same zeros and poles and the same multiplicative behavior. Hence it follows immediately that $f(P) = cg(P)$, with c a constant which may depend on $P_2 \cdots P_s$, since both $f(P)$ and $g(P)$ depend on $P_2 \cdots P_s$. If we write $P = P_1$, however, we observe that both f and g are symmetric in $P_1, \ldots, P_s$; which implies that c is independent of the points P_k and thus is an absolute constant. We therefore have

$$\begin{aligned}\theta(\varphi(P_1 \cdots P_s) - \varphi(\omega) - e) \\ = c\theta(\varphi(P_1 \cdots P_s) - \varphi(\tau) - e) \prod_{k=1}^{s} \exp\left[\int_{P_0}^{P_k} \sum_{j=1}^{s} \tau_{R_jT_j}\right].\end{aligned} \tag{3.5.4}$$

We now differentiate (3.5.4) with respect to a local coordinate z at P_1 and set $P_1 = R_1$ to obtain

$$\begin{aligned}\sum_{j=1}^{g} \frac{\partial\theta}{\partial z_j}(\varphi(P_2 \cdots P_s) - \varphi(R_2 \cdots R_s) - e)\zeta_j(R_1) \\ = c\left[E(R_1)\sum_{j=1}^{g} \frac{\partial\theta}{\partial z_j}(\varphi(R_1P_2 \cdots P_s) - \varphi(\tau) - e)\zeta_j(R_1)\right. \\ \left. + \theta(\varphi(R_1P_2 \cdots P_s) - \varphi(\tau) - e)\,dE(R_1)\right],\end{aligned} \tag{3.5.5}$$

where $E = \Pi_{k=1}^{s} \exp(\int_{P_0}^{P_k} \sum_{j=1}^{s} \tau_{R_jT_j})$. Since R_1 is a simple zero of E we have $E(R_1) = 0$ but $dE(R_1) \neq 0$, where we easily compute the derivative

$$dE(R_1) = \prod_{k=2}^{s} \exp\left(\int_{P_0}^{P_k} \sum_{j=1}^{s} \tau_{R_jT_j}\right),$$

by properly choosing the local coordinate z at R_1 with respect to which we differentiate.

We now continue the process by differentiating (3.5.5) with respect to a local coordinate at P_2 and set $P_2 = R_2$ to obtain

$$\sum_{j,k=1}^{g} \frac{\partial^2 \theta}{\partial z_j \partial z_k}(\varphi(P_3 \cdots P_s) - \varphi(R_3 \cdots R_s) - e)\zeta_j(R_1)\zeta_k(R_2)$$
$$= c\theta(\varphi(R_1 R_2 P_3 \cdots P_s) - \varphi(\tau) - e)(d^2E(R_1,R_2)). \tag{3.5.6}$$

We have used once again that R_2 is a simple zero of E, and therefore $(dE(R_1))(R_2) \neq 0$.

We continue this process arriving finally at the sth stage and obtain

$$\sum_{s=j_1+\cdots+j_g} \frac{\partial^s \theta}{\partial z_1^{j_1} \cdots \partial z_g^{j_g}}(-e)\zeta_1^{j_1}(R_1) \cdots \zeta_g^{j_g}(R_s)$$
$$= c\theta(\varphi(\omega) - \varphi(\tau) - e)(d^sE(R_1, \ldots, R_s)). \tag{3.5.7}$$

It follows from (3.5.3) that the right side of (3.5.7) does not vanish. Hence the same is true for the left side and therefore it is surely not the case that all sth order partial derivatives of θ vanish at $-e$ (and therefore also at e).

Theorem (Riemann). *Let s be the least integer such that $\theta(W_{s-1} - W_{s-1} - e) \equiv 0$ but $\theta(W_s - W_s - e) \not\equiv 0$. Then $e = \varphi(\zeta) + \mathscr{K}$ with $\zeta \in M_{g-1}$ and $i(\zeta) = s$. Furthermore, all partial derivatives of θ of order less than s vanish at e while at least one partial of order s does not vanish at e. Conversely, if e is a point such that all partials of order less than s vanish at e but one partial of order s does not vanish at e then we have $e = \varphi(\zeta) + \mathscr{K}$ with $\zeta \in M_{g-1}$ and $i(\zeta) = s$.*

PROOF. The first claim in the theorem is precisely the last statement of Theorem VI.3.2. The Lemma then gives that all partial derivatives of order less than s vanish at e. The statement that at least one partial of order s does not vanish at e is precisely the remark following (3.5.7). We now turn to the second (converse) part of the theorem.

We assume that all partials of θ of order less than $s \geq 1$ vanish at e. Hence by Theorem VI.3.1 we have $e = \varphi(\zeta) + \mathscr{K}$ with $\zeta \in M_{g-1}$, and the only question is: What is $i(\zeta)$ equal to? First, $i(\zeta)$ cannot be less than s. If it were, then by the first part of the theorem it could not be the case that all partial derivatives of order less than s vanish at e. Similarly $i(\zeta)$ cannot be greater than s. If it were, then by the first part of the theorem it would not be the case that there is a non-vanishing partial derivative of order s. Hence we conclude that $i(\zeta) = s$.

VI.3.6. We have been making reference to the vector $\mathscr{K}$ of Riemann constants since Theorem VI.2.4. In this section, we give a new characterization of $\mathscr{K}$ that partly explains its significance. Up to this point $\mathscr{K}$ has been defined simply by (2.4.1). We show here that the vector $-2\mathscr{K}$ is the image of the divisor of a meromorphic differential under (the extension to divisors of the map) φ.

We have already seen that the divisor of a meromorphic differential on a compact surface of genus g has degree $2g - 2$. Hence we now prove the following

Theorem. *Let Δ be a divisor of degree $2g - 2$ on a compact Riemann surface, M, of genus $g \geq 1$. Then Δ is the divisor of a meromorphic differential on M if and only if $\varphi(\Delta) = -2\mathscr{K}$.*

PROOF. We first show that $-2\mathscr{K}$ is indeed the image of the divisor class of meromorphic differentials. It suffices to show, of course, that $-2\mathscr{K}$ is the image of the divisor of a holomorphic differential. To this end let ζ be an arbitrary integral divisor of degree $g - 1$. It therefore follows from Theorem VI.3.2 that $e = \varphi(\zeta) + \mathscr{K}$ is a zero of the theta function. Since the theta function is even, $-e$ is also a zero and (again by Theorem VI.3.2) $-e = \varphi(\omega) + \mathscr{K}$, with ω an integral divisor of degree $g - 1$ on M. It therefore follows that $\varphi(\zeta\omega) = -2\mathscr{K}$.

We now need show that $\zeta\omega$ is the divisor of a holomorphic differential on M. We need only note that the divisor $\zeta\omega$ has $g - 1$ "arbitrary" points. It therefore follows (by Lemma III.8.15 as in VI.3.2) that $r(1/\zeta\omega) \geq g$; which by the Riemann–Roch theorem implies $i(\zeta\omega) \geq 1$, and therefore $i(\zeta\omega) = 1$. Hence $\zeta\omega$ is the divisor of a holomorphic differential.

Conversely, suppose Δ is a divisor of degree $2g - 2$ on M such that $\varphi(\Delta) = -2\mathscr{K}$. It follows by Abel's theorem that there is a function f on M with divisor $\Delta/(\alpha)$, where α is a holomorphic differential on M and (α) denotes the divisor of α. Thus $f\alpha$ is a differential on M with divisor $(f\alpha) = \Delta$. ☐

In the next chapter we will often make use of the following

Corollary. *The vector $\mathscr{K}$ of Riemann constants is a point of order 2 in $J(M)$ if and only if P_0^{2g-2} is canonical, where P_0 is the base point of the map $\varphi: M \to J(M)$.*

PROOF. The point $\mathscr{K}$ is of order 2 if and only if $-2\mathscr{K} = 0 = \varphi(P_0^{2g-2})$. Now use the theorem. ☐

VI.3.7. In the previous section we have identified the zero set of the θ-function as $W_{g-1} + \mathscr{K}$ where both W_{g-1} (recall the definitions in Chapter III) and $\mathscr{K}$ depend on the base point P_0. It is clear that the zero set of the theta function does not depend on this base point; so that one is led to suspect that the mapping

$$\zeta \in M_{g-1}, \qquad \zeta \mapsto \varphi_{P_0}(\zeta) + \mathscr{K}_{P_0} \in J(M)$$

is also independent of the base point $P_0 \in M$. This, in fact, is the content of our next

Theorem. *The map defined on M_{g-1} which takes ζ to $\varphi_{P_0}(\zeta) + \mathscr{K}_{P_0}$ is independent of the base point $P_0 \in M$.*

Proof. We first observe that if $\theta(e) \neq 0$, then the function $\theta(\varphi_{P_0}(P) - e)$ does not vanish identically so that $e = \varphi_{P_0}(P_1 \cdots P_g) + \mathscr{K}_{P_0}$ and that $P_1, \ldots, P_g$ are the zeroes of this function. Since it is always true that $\varphi_{P_1}(P) = \varphi_{P_1}(P_0) + \varphi_{P_0}(P)$ for all $P_0, P_1, P \in M$, we have

$$\theta(\varphi_{P_0}(P) - e) = \theta(\varphi_{P_1}(P) - \varphi_{P_1}(P_0) - e).$$

From the expression on the right of the equality we find that

$$\varphi_{P_1}(P_0) + e = \varphi_{P_1}(P_1 \cdots P_g) + \mathscr{K}_{P_1}.$$

On the other hand, the expression we already have for e tells us that

$$\mathscr{K}_{P_0} = \varphi_{P_1}(P_0^{g-1}) + \mathscr{K}_{P_1}.$$

The above expression relating $\mathscr{K}_{P_0}$ and $\mathscr{K}_{P_1}$ is what gives us the result we are seeking.

Let ζ be any element of M_{g-1}. Then $\varphi_{P_1}(\zeta) + \mathscr{K}_{P_1} = \varphi_{P_1}(P_0^{g-1}) + \varphi_{P_0}(\zeta) + \mathscr{K}_{P_1} = \varphi_{P_0}(\zeta) + \mathscr{K}_{P_0}$ as was to be shown. □

We record for later use the fact exhibited in the proof

Corollary. *For any two points P and Q on M, we have*

$$\mathscr{K}_P = \mathscr{K}_Q + \varphi_Q(P^{g-1}).$$

We have encountered, in this chapter and in III.11, three very important and beautiful subsets of $J(M)$. The first, is the *zero* set of the θ-function

$$\Theta_{\text{zero}} = W_{g-1} + \mathscr{K};$$

the second, is the *singular* set of the θ-function (the points where θ and its first partials vanish)

$$\Theta_{\text{sing}} = W_{g-1}^1 + \mathscr{K}$$

(see also VII.1.6); the third is the set of points $e \in J(M)$ which lead to the *identically vanishing* of the function ψ of VI.3.3

$$\Theta_{\text{super zero}} = W_g^1 + \mathscr{K}.$$

It is obvious that

$$\Theta_{\text{zero}} \supset \Theta_{\text{super zero}} \supset \Theta_{\text{sing}}. \tag{3.7.1}$$

Note that $\Theta_{\text{super zero}}$ is always non-empty for $g \geq 2$ (take a Weierstrass point, for example), while Θ_{sing} is non-empty for $g \geq 4$ by the Corollary to Theorem III.8.13 and Theorem III.8.7 (also for hyperelliptic surfaces of genus 3). We conclude from Theorem III.11.19 that

$$\dim \Theta_{\text{zero}} = g - 1,$$
$$\dim \Theta_{\text{super zero}} = g - 2,$$

and

$$g - 4 \leq \dim \Theta_{\text{sing}} \leq g - 3,$$

and hence the inclusions of (3.7.1) are all proper. The points corresponding to vanishing of the θ-function, but not identically vanishing (on the surface), are in

$$(W_g \backslash W_g^1) + \mathscr{K} = (J(M) \backslash W_g^1) + \mathscr{K}.$$

The sets Θ_{zero} and Θ_{sing} depend only on the period matrix Π. The fact that such a set is non-empty can, of course, be translated by the Riemann vanishing theorem and the Riemann–Roch theorem into a statement about existence of meromorphic functions (of a certain type) on M.

It should also be remarked that Θ_{zero} and Θ_{sing} are independent of the base point of the map $\varphi: M \to J(M)$. Thus Θ_{sing} describes those points $e \in J(M)$ corresponding to the identically vanishing of ψ for all choices of the base point (see also VII.2.1).

CHAPTER VII
Examples

In this chapter we give applications and examples of the theory developed in the preceding chapters. The examples will consist mainly of taking concrete representations of compact Riemann surfaces given by algebraic functions, and computing on these surfaces various objects of interest. The applications will give rise to new characterizations of surfaces with some given property. The format of this chapter will differ considerably from the format of the preceding ones. Many details will be omitted. One of the aims of this chapter is to convince the reader that computations are quite often possible; and these computations yield beautiful results.

VII.1. Hyperelliptic Surfaces (Once Again)

Our first set of examples will continue to deal with the class of surfaces studied extensively in III.7: the hyperelliptic surfaces. Recall that a hyperelliptic surface M of genus $g \geq 2$ has an essentially unique realization as a two sheeted cover of the sphere branched over $2g + 2$ points. It is the Riemann surface of the algebraic curve

$$w^2 = \prod_{k=1}^{2g+2} (z - e_k), \qquad e_k \neq e_j, \text{ for } k \neq j. \tag{1.0.1}$$

VII.1.1. We adopt the following point of view in the above situation. We think of z as a variable point in $\mathbb{C} \cup \{\infty\}$, and then view M as the Riemann surface on which w is a well defined (single valued) meromorphic function. On this surface, which must be a two sheeted branched covering of $\mathbb{C} \cup \{\infty\}$, the projection map (which we shall also denote by z) onto $\mathbb{C} \cup \{\infty\}$ is the function of degree two, and we can think of w as a function of the point $P \in M$ in the sense that w sends $P \in M$ into $w(z(P))$. Note that as a function of z, w is two-valued; it is defined up to sign.

Alternatively, we can think of M as a surface of genus $g \geq 2$ with a meromorphic function z of degree 2 and another function w satisfying (1.0.1).

Theorem III.7.3 established the result that the $2g + 2$ points $z^{-1}(e_k)$, $k = 1, \ldots, 2g + 2$, are the Weierstrass points on M, and that each of these points has weight $g(g - 1)/2$. Proposition III.7.6 showed that if we choose a Weierstrass point (say $z^{-1}(e_1)$) as the base point for the map φ of M into the Jacobian variety $J(M)$, then for $P \in M$, $\varphi(P)$ is of order 2 if and only if P is a Weierstrass point.

We can actually say more. We can compute precisely which half-periods (points of order 2) are in the image of $\varphi: M \to J(M)$. We need for this purpose a model for a canonical homology basis on M. Recall the model discussed in III.7.4, and set $z^{-1}(e_j) = P_j$, $j = 1, 2, \ldots, 2g + 2$. We wish to amplify (for the convenience of the calculations) that model slightly to the one given in Figure VII.1.

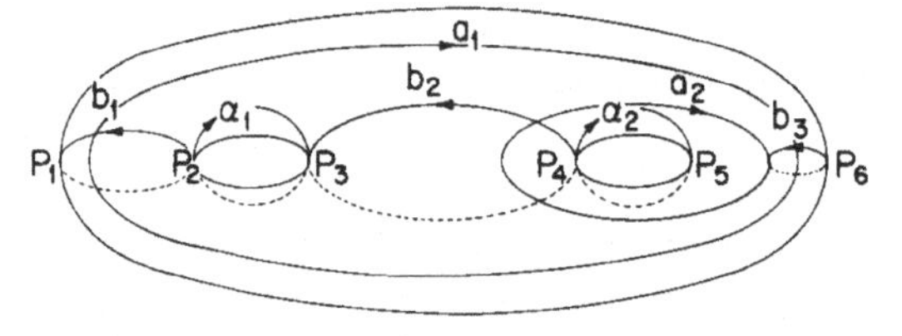

Figure VII.1. Hyperelliptic surface of genus 2.

The hyperelliptic involution J can be viewed as a rotation by π radians about an axis passing through the $2g + 2$ Weierstrass points. For $k = 1, \ldots, g + 1$, let β_k be an oriented curve from P_{2k-1} to P_{2k}. Define b_k as β_k followed by $-J\beta_k$ (that is, followed by $J\beta_k$ is the opposite direction). The curve a_k, $k = 1, \ldots, g$, is all in "one sheet" and it joins a point on b_k to a point on b_{g+1} and then returns to the point on b_k. Whereas b_k is invariant (as a point set) under J, a_k is not. The curves $a_1, \ldots, a_g, b_1, \ldots, b_g$ form a canonical homology basis on M.

In the following calculations, we use standard notation: $\{\zeta\} = {}^t\{\zeta_1, \ldots, \zeta_g\}$ is the normalized basis of $\mathscr{H}^1(M)$ dual to the canonical homology basis, $\pi^{(k)}$ is the k-th column of the period matrix Π, etc. . . . , as well as the fact that J acts as multiplication by -1 on $\mathscr{H}^1(M)$. Now, for $k = 1, \ldots, g$,

$$\pi^{(k)} = \int_{b_k} \zeta = \int_{\beta_k} \zeta - \int_{J\beta_k} \zeta = \int_{\beta_k} \zeta + \int_{J\beta_k} J\zeta$$
$$= 2\int_{\beta_k} \zeta = 2\int_{P_{2k-1}}^{P_{2k}} \zeta.$$

Next, the intersection numbers satisfy for $j = 1, \ldots, g$,

$$a_j \cdot b_{g+1} = -1, \qquad b_j \cdot b_{g+1} = 0.$$

Thus up to homology

$$-b_{g+1} = b_1 + \cdots + b_g,$$

from which we conclude that

$$\int_{P_{2g+1}}^{P_{2g+2}} \zeta = \frac{1}{2}(\pi^{(1)} + \cdots + \pi^{(g)}).$$

To compute some more integrals we introduce the curves $\hat{\alpha}_j$ joining P_{2j} to P_{2j+1} and set $\alpha_j = \hat{\alpha}_j - J\hat{\alpha}_j$. Again we compute intersection numbers ($j = 1, \ldots, g$ and $k = 1, \ldots, g$)

$$\alpha_j \cdot a_k = 0, \qquad \alpha_j \cdot b_k = 0 \quad \text{for } j \neq k, j \neq k+1,$$
$$\alpha_j \cdot b_j = +1, \qquad \alpha_j \cdot b_{j+1} = -1.$$

Thus we conclude that (in $H_1(M)$)

$$\alpha_j = a_j - a_{j+1} \quad \text{for } j = 1, \ldots, g-1, \alpha_g = a_g.$$

It now follows, as above, that

$$\int_{P_{2k}}^{P_{2k+1}} \zeta = \frac{1}{2}(e^{(k)} - e^{(k+1)}) = \frac{1}{2}(e^{(k)} + e^{(k+1)}), \qquad k = 1, \ldots, g-1,$$

$$\int_{P_{2g}}^{P_{2g+1}} \zeta = \frac{1}{2}e^{(g)}.$$

We now combine all the above data in:

$$\varphi(P_1) = 0,$$

$$\varphi(P_2) = \varphi(P_1) + \int_{P_1}^{P_2} \zeta = \frac{1}{2}\pi^{(1)},$$

$$\varphi(P_3) = \varphi(P_2) + \int_{P_2}^{P_3} \zeta = \frac{1}{2}(\pi^{(1)} + e^{(1)} + e^{(2)}),$$

$$\vdots$$

$$\varphi(P_{2k+1}) = \frac{1}{2}(\pi^{(1)} + \cdots + \pi^{(k)} + e^{(1)} + e^{(k+1)}), \qquad k = 1, 2, \ldots, g-1,$$

$$\varphi(P_{2k+2}) = \frac{1}{2}(\pi^{(1)} + \cdots + \pi^{(k+1)} + e^{(1)} + e^{(k+1)}), \qquad k = 1, \ldots, g-1,$$

$$\vdots$$

$$\varphi(P_{2g+1}) = \frac{1}{2}(\pi^{(1)} + \cdots + \pi^{(g)} + e^{(1)}),$$

$$\varphi(P_{2g+2}) = \frac{1}{2}e^{(1)}.$$

Notice that $\sum_{j=1}^{2g+2} \varphi(P_j) = 0$, because $(P_2 \cdots P_{2g+2})/P_1^{2g+1}$ is the divisor of the function w (here we assume that z has a double pole at P_1, and hence in (1.0.1) the product runs over the indices 2 to $2g + 2$).

We compute next

$$\sum_{j=1}^{g} \varphi(P_{2j+1}) = \frac{1}{2}(g\pi^{(1)} + (g-1)\pi^{(2)} + \cdots + \pi^{(g)} + ge^{(1)} + e^{(2)} + \cdots + e^{(g)}). \tag{1.1.1}$$

Consider the map $\varphi: M_g \to J(M)$ at the point $D = P_3P_5 \cdots P_{2g+1}$. Its differential at this point is (the transpose of)

$$\begin{bmatrix} \zeta_1(P_3) & \cdots & \zeta_g(P_3) \\ \zeta_1(P_5) & \cdots & \zeta_g(P_5) \\ \vdots & & \vdots \\ \zeta_1(P_{2g+1}) & \cdots & \zeta_g(P_{2g+1}) \end{bmatrix}. \tag{1.1.2}$$

A change of basis for abelian differentials of the first kind, multiplies the above matrix by a non-singular matrix. Hence to compute the rank of (1.1.2), we may choose a convenient basis for $\mathscr{H}^1(M)$. We shall use the basis given by III(7.5.1):

$$\frac{z^j\,dz}{w}, \qquad j = 0, \ldots, g-1.$$

As before we assume for convenience that z has a double pole at P_1, and vanishes at the point P_2. Note then that

$$\left(\frac{z^j\,dz}{w}\right) = P_1^{2g-2j-2}P_2^{2j}.$$

In particular, dz/w does not vanish at P_{2j+1}, $j = 1, \ldots, g$. Thus we want to compute the rank of

$$\prod_{j=1}^{g} \frac{dz}{w}(P_{2j+1}) \begin{bmatrix} 1 & z(P_3) & \cdots & z^{g-1}(P_3) \\ 1 & z(P_5) & \cdots & z^{g-1}(P_5) \\ & \vdots & & \vdots \\ 1 & z(P_{2g+1}) & \cdots & z^{g-1}(P_{2g+1}) \end{bmatrix}$$

$$= \prod_{j=1}^{g} \frac{dz}{w}(P_{2j+1}) \begin{bmatrix} 1 & e_3 & \cdots & e_3^{g-1} \\ 1 & e_5 & \cdots & e_5^{g-1} \\ & \vdots & & \vdots \\ 1 & e_{2g+1} & \cdots & e_{2g+1}^{g+1} \end{bmatrix}.$$

The last matrix is, of course, the Vandermonde matrix which has already been encountered (in IV.11.10), and is non-singular. We have hence shown (III.11.11) that $i(D) = 0$, and that φ establishes an isomorphism between a neighborhood of D in M_g and its image in $J(M)$.

VII.1.2. In VI.2.4, we have introduced the vector of Riemann constants

$$\mathscr{K} = -\sum_{k=1}^{g} \left[\int_{a_k} \varphi\varphi_k' \, dz - e^{(k)} \left(\frac{\pi_{kk}}{2} \right) \right],$$

where we assumed that the base point for $\pi_1(M)$ is chosen to agree with the base point of the map $\varphi: M \to J(M)$—which in our case was selected to be a Weierstrass, point P_1. It is rather difficult to evaluate $\mathscr{K}$ directly. We know (Theorem VI.3.6) that $-2\mathscr{K}$ is the image in $J(M)$ of the canonical divisor class. Since P_1^{2g-2} is the divisor of an abelian differential of the first kind, the canonical class gets mapped into 0 by φ. Thus $\mathscr{K}$ is a half-period. We want to know which half-period. We shall show that

$$\mathscr{K} = \sum_{j=1}^{g} \varphi(P_{2j+1}). \tag{1.2.1}$$

Assume for the moment that we knew that $\theta(0) \neq 0$, for the θ-function introduced in Chapter VI. In this case $\theta \circ \varphi$ does not vanish identically on M (this is the fact we really need), and hence $\theta \circ \varphi$ has g zeros $Q_1, \ldots, Q_g$ on M. These zeros satisfy

$$\sum_{j=1}^{g} \varphi(Q_j) + \mathscr{K} = 0.$$

(recall Theorem VI.2.4).

We have seen that $\varphi(P_j)$ is a half-period. Such a half-period can be written as $\frac{1}{2}(I\varepsilon' + \Pi\varepsilon)$, where $\left[\begin{smallmatrix}\varepsilon\\ \varepsilon'\end{smallmatrix}\right]$ is an integer characteristic. Thus the half-periods can be classified as odd or even, depending on the parity of the characteristic. Up to a constant non-zero factor,

$$\theta\left(\frac{I\varepsilon' + \Pi\varepsilon}{2}\right) = \theta\begin{bmatrix}\varepsilon\\ \varepsilon'\end{bmatrix}(0).$$

We hence conclude that θ vanishes at odd half-periods. We notice that $\varphi(P_{2j+1})$, $j = 1, \ldots, g$, is an odd half-period. (We also, for the future, notice that $\varphi(P_{2j})$, $j = 1, \ldots, g+1$, is an even half-period.) Thus $\varphi(P_{2j+1})$ is a zero of the function θ on M. Hence

$$\sum_{j=1}^{g} \varphi(P_{2j+1}) = -\mathscr{K} = \mathscr{K}.$$

Note that conversely, the identity (1.2.1) shows that θ does not vanish identically on M. For if θ vanished identically on M, then $\varphi(P_3 P_5 \cdots P_{2g+1}) + \mathscr{K} = 0$ and $i(P_3 \cdots P_{2g+1}) > 0$ by Theorem VI.3.3. This contradicts the previously established fact that $i(P_3 \cdots P_{2g+1}) = 0$. These remarks also show that θ vanishes at 0 if and only if θ vanishes identically on M. The fact that an even theta function vanishes at 0 if and only if it vanishes identically on M is also a consequence of the Riemann vanishing theorem.

Remark. There is an additional property that the $2g + 1$ half-periods $\varphi(P_j)$, $j = 2, \ldots, 2g + 2$ have. We have already mentioned the fact that $\varphi(P_{2j+1})$, $j = 1, \ldots, g$, is odd and $\varphi(P_{2j})$, $j = 1, \ldots, g + 1$, is even. The origin $\varphi(P_1)$, is, of course, also even. If we use the notation $\binom{\varepsilon}{\varepsilon'} = \frac{1}{2}(I\varepsilon' + \Pi\varepsilon)$ for half-periods, and label the half period $\varphi(P_k)$, $k = 2, \ldots, 2g + 2$ by $\binom{\varepsilon(k)}{\varepsilon'(k)}$, then the 2g + 1 half-periods $\binom{\varepsilon(k)}{\varepsilon'(k)}$ have the property that for all $k \neq l$, $\varepsilon(k) \cdot \varepsilon'(l) + \varepsilon'(k) \cdot \varepsilon(l) = 1$. Here $\varepsilon(k) = (\varepsilon_1(k), \ldots, \varepsilon_g(k))$, $\varepsilon'(k) = (\varepsilon'_1(k), \ldots, \varepsilon'_g(k))$, and the operation $(\ ,\cdot, \ )$ is the usual inner product over $\mathbb{Z}_2$.

VII.1.3. We need to establish that θ does not vanish at the origin. We postpone this proof to VII.1.9, since it is more convenient to have some further computations at hand. Note that we never use anything other than the fact that $\mathcal{K}$ is a half-period. We do not anywhere use the exact form of $\mathcal{K}$ (before VII.4).

VII.1.4. It is, of course, true that $\theta(\mathcal{K}) = 0$ by Theorem VI.3.1. The order of vanishing of θ at $\mathcal{K}$ is determined by the Riemann vanishing theorem (VI.3.5). Write

$$\mathcal{K} = \varphi(\zeta) + \mathcal{K}, \qquad \zeta \in M_{g-1}.$$

Clearly we can choose $\zeta = P_1^{g-1}$. The order of vanishing of θ at $\mathcal{K}$ is precisely $i(\zeta)$. Since P_1 is a Weierstrass point, the orders of zeros at P_1 of abelian differentials of the first kind on M are (see III.7.3):

$$0, 2, \ldots, 2g - 2.$$

Thus

$$i(P_1^{g-1}) = \begin{cases} \frac{1}{2}(g + 1), & \text{if } g \text{ is odd,} \\ \frac{1}{2}g, & \text{if } g \text{ is even.} \end{cases}$$

(Using the greatest integer notation, we can write that the order of vanishing is always $[\frac{1}{2}(g + 1)]$.)

For $D \in M_{g-1}$, $r(D^{-1}) = i(D)$, and Clifford's theorem (III.8.4) shows that

$$i(D) \leq \left[\frac{g + 1}{2}\right].$$

We have established that for a hyperelliptic surface there are points at which the θ-function vanishes to the maximum order possible. We have shown that such vanishing occurs at a point of order 2 in $J(M)$.

As a matter of fact, given any non-negative integer $n \leq [\frac{1}{2}(g + 1)]$, there is a point $e \in J(M)$ such that e has order 2 and θ vanishes at e to precisely order n. To see this, we have to construct for $n > 0$, divisors $\zeta \in M_{g-1}$ with ζ containing only Weierstrass points and $i(\zeta) = n$. Then the point $e = \varphi(\zeta) + \mathcal{K}$ will have the desired property (for $n = 0$, we, of course, need a divisor $\zeta \in M_g$ with $i(\zeta) = 0$). The calculation in VII.1.1 established the following fact: If $D \in M_g$ and D consists of *distinct* Weierstrass points, then $i(D) = 0$. From this observation it follows that if (for $0 \leq r \leq g$) $D \in M_{g-r}$ and

D consists of distinct Weierstrass points, then $i(D) = r$. Let

$$1 < j_1 < j_2 < \cdots < j_g \leq 2g + 2$$

be a choice of g distinct integers as indicated. Then

$$\sum_{k=1}^{g} \varphi(P_{j_k}) + \mathscr{K}$$

is a point of order two in $J(M)$ at which the θ-function does not vanish (choosing $j_k = 2k + 1$, we obtain the previously used fact that $\theta(0) \neq 0$). Choosing $g - 1$ distinct integers as above (j_1 is now permitted to be 1), we see that θ vanishes to order 1 at

$$\sum_{k=1}^{g-1} \varphi(P_{j_k}) + \mathscr{K}.$$

More generally, let r and s be non-negative integers such that

$$2s + r = g - 1.$$

Choose s non-negative integers

$$1 \leq j_1 < j_2 < \cdots < j_s \leq 2g + 2$$

and r more non-negative integers

$$1 \leq k_1 < k_2 < \cdots < k_r \leq 2g + 2,$$

such that

$$k_m \neq j_n \quad \text{all } m, n.$$

Recall now (III.7.3) that every $\omega \in \mathscr{H}^1(M)$ that vanishes at a Weierstrass point vanishes to even order, and conclude that

$$i(P_{j_1}^2 P_{j_2}^2 \cdots P_{j_s}^2 P_{k_1} \cdots P_{k_r}) = i(P_{j_1} \cdots P_{j_s} P_{k_1} \cdots P_{k_r}) = g - r - s = s + 1.$$

Thus we see that for $D = P_{j_1}^2 P_{j_2}^2 \cdots P_{j_s}^2 P_{k_1} \cdots P_{k_r}$ the order of vanishing of θ at $\varphi(D) + \mathscr{K}$ is precisely $s + 1$. The possible values of $s + 1$ are, of course, $1, 2, \ldots, [\frac{1}{2}(g + 1)]$.

VII.1.5. We now show how some of the results of III.8 can be restated in terms of vanishing properties of θ-functions. We will obtain this way, among other things, characterizations of hyperelliptic surfaces.

Theorem. *A Riemann surface M of genus 3 is hyperelliptic if and only if θ vanishes at a point of order 2 in $J(M)$ to order 2.*

PROOF. If M is hyperelliptic, then the vector $\mathscr{K}$ of Riemann constants with respect to a Weierstrass point on M is a point of order 2 in $J(M)$, and θ vanishes at $\mathscr{K}$ to order 2. Conversely, if θ vanishes at a point e of order 2 in $J(M)$ to order 2, then $e = \varphi(Q_1 Q_2) + \mathscr{K}$, $i(Q_1 Q_2) = 2$ for some $Q_1 Q_2 \in M_2$ (by the Riemann vanishing theorem). Thus $r(Q_1^{-1} Q_2^{-1}) = 2$ and M is hyperelliptic. □

Note that we have proven somewhat more:

Theorem. *A surface M of genus 3 is hyperelliptic if and only if there is a point $e \in J(M)$ with θ vanishing to order 2 at e.*

VII.1.6. Definition. A *singular point* of the zero set of the θ-function is a point where (θ vanishes and) all the first partial derivatives vanish. We shall denote the *singular set* (that is, the set of singular points) by Θ_{sing}.

It is obvious from the Riemann vanishing theorem that

$$\Theta_{\text{sing}} = W_{g-1}^1 + \mathscr{K}, \tag{1.6.1}$$

and from Theorem III.11.19 that

$$g - 4 \leq \dim \Theta_{\text{sing}} \leq g - 3,$$

with the upper bound attained if and only if the surface is hyperelliptic.

Let us consider $g = 4$. In this case $\Theta_{\text{sing}} \neq \varnothing$ (it is 0 or 1 dimensional). Note that Proposition III.8.6 implies immediately (without Theorem III.11.19) that $W_3^1 \neq \varnothing$. Further Theorem III.8.7 can be reinterpreted as the following

Theorem. *Let M be a compact surface of genus 4. Then one and only one of the following holds:*

a. *Θ_{sing} is 1-dimensional,*
b. *Θ_{sing} consists of precisely one point (a point of order 2 in $J(M)$), or*
c. *Θ_{sing} consists of precisely two points ($a \neq 0$, $a \in J(M)$, and $-a$), neither one of which is of order 2.*

Each case corresponds to the case indexed by the same letter in Theorem III.8.7.

Proof. If M is hyperelliptic then Θ_{sing} is one-dimensional (and conversely). In fact, in this case, Θ_{sing} is analytically equivalent to $W_1 = \varphi(M)$ and hence to M (EXERCISE). Thus case (a) is disjoint from the other two cases. Proposition III.8.6 showed that W_3^1 is non-empty. Assume $a \in W_3^1 + \mathscr{K}$. Assume also $b \in W_3^1 + \mathscr{K}$. The proof of Theorem III.8.7 showed that $a + b = 0$ unless M is hyperelliptic (recall that the image of the canonical class in $J(M)$ is $-2\mathscr{K}$ (Theorem VI.3.6)). Hence the only possibilities are $a = b$ (and hence $2a = 0$) or W_3^1 contains two points and $b = -a \neq 0$. In the latter case a cannot be of order 2. □

Corollary. *A necessary and sufficient condition for a surface M of genus 4 to be hyperelliptic is that θ vanish at two distinct points of order 2 in $J(M)$ to order 2.*

Proof. If M is hyperelliptic, then the vector $\mathscr{K}$ with respect to each of the 10 Weierstrass points is a point of order 2 at which θ vanishes to order (precisely) 2. (The reader should review the dependence of the vector of Riemann constants $\mathscr{K}_{P_0}$ on the base point P_0, and become convinced that we obtain this way ten distinct points of order 2 in $J(M)$.) On the other hand, the theorem implies that whenever θ vanishes at two distinct points of order 2 to order 2, then the surface is hyperelliptic (the points of vanishing to order precisely 2 are the points of $W_3^1 + \mathscr{K}$). □

VII.1.7. To generalize the above corollary, we translate Corollary 2 to Proposition III.8.8.

Theorem. *A necessary and sufficient condition for a surface M of even genus $g \geq 4$ to be hyperelliptic is that θ vanish at two points of order 2 in $J(M)$ to order $\frac{1}{2}g$.*

PROOF. Necessity is established as before. For sufficiency, note that our condition assures the existence of two distinct points of order 2 $a, b \in W_{g-1}^{\frac{1}{2}g-1} + \mathscr{K}$. Write

$$a = \varphi(A) + \mathscr{K}, \qquad A \in M_{g-1}^{\frac{1}{2}g-1},$$
$$b = \varphi(B) + \mathscr{K}, \qquad B \in M_{g-1}^{\frac{1}{2}g-1}.$$

Thus A and B are inequivalent divisors, and a computation yields that $c(A) = 1 = c(B)$. Hence, by the result we are translating, M is hyperelliptic unless A and B are complementary divisors. But in this case $a + b = \varphi(AB) + 2\mathscr{K} = 0$. Thus $a = -b$ (here is the only place that we use the fact that a and b are of order 2), and hence $a = b$ (which contradicts the hypothesis). □

Remark. By Theorem III.8.11, the vanishing of the θ-function at a single point in $J(M)$ to order $\frac{1}{2}g$ implies hyperellipticity except if g is 4 or 6.

VII.1.8. The situation for odd genus is even simpler.

Theorem. *A necessary and sufficient condition for a surface M of odd genus $g \geq 3$ to be hyperelliptic is for θ to vanish at a point of order 2 in $J(M)$ to order $\frac{1}{2}(g + 1)$.*

PROOF. As before, necessity has been established and for sufficiency, we need even less. The vanishing property implies the existence of a divisor $D \in M_{g-1}$ with $r(D^{-1}) = \frac{1}{2}(g + 1)$. Thus by Clifford's theorem (III.8.4), M is hyperelliptic. □

Remark. Again, for sufficiency, we did not need to know that the points at which the θ-function vanished were of order 2. Except for genus 4 or 6, vanishing of the θ-function at any point (to the right order) implies hyperellipticity.

VII.1.9. In this section we complete the proof of (1.2.1). Recall that in VII.1.2 we showed that it suffices to prove that $\theta(0,\Pi) \neq 0$.

We begin by considering the following set of divisors: P_1^{g-1}, the $2g+1$ divisors $P_1^{g-2}P_k$, $k = 2, \ldots, 2g+2$; and the $\binom{2g+1}{2}$ divisors $P_1^{g-3}P_{i_1}P_{i_2}$ such that $i_1 \neq i_2$, $i_1 \neq 1$, and $i_2 \neq 1$. We continue in this fashion ending with the $\binom{2g+1}{g-1}$ divisors $P_{i_1} \cdots P_{i_{g-1}}$ with $i_l \neq 1$, and all indices distinct. Finally we add the $\binom{2g+1}{g}$ divisors $P_{i_1} \cdots P_{i_g}$ with the same assumptions on the indices.

The first observation to be made is that our set consists of 2^{2g} linearly inequivalent divisors. The fact that the divisors are inequivalent follows immediately from Abel's theorem and the Riemann-Roch theorem. The point being, that we would be led to an equivalence of the form $P_{j_1} \cdots P_{j_r} \sim P_1^m P_{i_1} \cdots P_{i_l}$ with $r \leq g$, which is impossible since $i(P_{j_1} \cdots P_{j_r}) = g - r$. Counting the number of divisors we have formed, we have $\sum_{k=0}^{g} \binom{2g+1}{k} = \frac{1}{2} \cdot 2^{2g+1} = 2^{2g}$.

If we now consider D to be one of the divisors in our set, then $\varphi(D) + \mathscr{K}$ is a zero of the theta function of order $i(D)$. Since we know $\varphi(D) + \mathscr{K}$ is a point of order two, and we have here constructed all the points of order two, we have catalogued the exact order of vanishing at each point of order 2. (For example, if g is odd, the arguments of VII.1.4 yield $i(P_1^{g-1}) = \frac{1}{2}(g+1)$, $i(P_1^{g-2}P_k) = \frac{1}{2}(g-1), \ldots, i(P_1 P_{i_1} \cdots P_{i_{g-2}}) = 1$, $i(P_{i_1} \cdots P_{i_{g-1}}) = 1$, and $i(P_{i_1} \cdots P_{i_g}) = 0$.)

The above argument has established that the θ-function does not vanish at the $\binom{2g+1}{g}$ half-periods

$$\varphi(P_{j_1} P_{j_2} \cdots P_{j_g}) + \mathscr{K}, \tag{1.9.1}$$

where $1 < j_1 < j_2 < \cdots < j_g = 2g + 2$. Each of these half-periods must be an even half-period (because θ vanishes at all odd half-periods). The rest of the proof involves showing that if $\theta(0,\Pi) = 0$, then we can construct an odd-half period of the form (1.9.1) and contradict the non-vanishing of the θ-function of this point. (Recall θ vanishes at 0 if and only if $\theta \circ \varphi$ vanishes identically on M.)

Since $\mathscr{K}$ is a half-period, $\mathscr{K}$ can be written as

$$\mathscr{K} = \varphi(P_{i_1} P_{i_2} \cdots P_{i_r}),$$

with $0 \leq r \leq g$, and $1 < i_1 < i_2 < \cdots < i_r \leq 2g + 2$. Assume first that $g = r$, then writing

$$\varphi(P_{i_1} P_{i_2} \cdots P_{i_g}) + \mathscr{K} = 0,$$

and recalling that

$$i(P_{i_1} P_{i_2} \cdots P_{i_g}) = 0,$$

we conclude from Theorem VI.3.3, that $\theta \circ \varphi \not\equiv 0$ on M, and hence also that $\theta(0,\Pi) \neq 0$. We finish by showing that if $r < g$, then it is impossible for (1.9.1) to be an even half-period for all choices of the indices. We show how to construct the set $\{j_1, j_2, \ldots, j_g\}$. First it should contain the set $\{i_1, i_2, \ldots, i_r\}$ and we must add a set of $g - r$ numbers in $\{2, 3, \ldots, 2g+2\}$ that are not in $\{i_1, i_2, \ldots, i_r\}$. Assume this set is $\{k_1, \ldots, k_{g-r}\}$. Then

$$\varphi(P_{j_1} P_{j_2} \cdots P_{j_g}) + \mathscr{K} = \varphi(P_{k_1} P_{k_2} \cdots P_{k_{g-r}}).$$

Recall that the set $\{\varphi(P_2), \ldots, \varphi(P_{2g+2})\}$ contains exactly g odd half-periods and $g + 1$ even half periods. Thus in the set

$$\{2,3,\ldots,2g+2\}\backslash\{i_1,i_2,\ldots,i_r\}$$

there are at least $g - r$ indices corresponding to odd half-periods, and at least $g - r + 1$ corresponding to even half-periods. We choose s odd points and $t = g - r - s$ even points. What is the parity of this divisor? Using the remark at the end of VII.1.2, we see that this divisor has the same parity as

$$s + \binom{s+t}{2} = s + \binom{g-r}{2} = s + \tfrac{1}{2}(g-r)(g-r-1).$$

Thus if $\frac{1}{2}(g-r)(g-r-1)$ is even, we choose s to be odd, and if $\frac{1}{2}(g-r)(g-r-1)$ is odd we choose s to be even. We have enough room in our set to accomplish this.

VII.1.10. We assume that M is a compact Riemann surface of genus $g > 0$. Note that we are *not* requiring that the surface be hyperelliptic. In Chapter VI we have briefly discussed the multivalued function (on the Riemann surface M) $\psi: P \mapsto \theta(\alpha\varphi_{P_0}(P) - e)$ (recall that α is an arbitrary non-zero integer and e an arbitrary point in the Jacobian variety of M) and have observed that for $\alpha \geq g$ this function does not vanish identically for any choice of base point P_0 and that there are $\alpha^2 g$ zeros $P_1, \ldots, P_{\alpha^2 g}$ of the function on the surface which satisfy the relation

$$\alpha e = \varphi_{P_0}(P_1 \cdots P_{\alpha^2 g}) + \alpha^2 \mathscr{K}_{P_0}.$$

In what follows we shall give some applications of this result to Weierstrass points on Riemann surfaces.

Theorem. *Every zero Q of the function*

$$P \mapsto \theta(g\varphi_{P_0}(P) + \mathscr{K}_{P_0}),$$

(we are taking $\alpha = g$ and $e = -\mathscr{K}_{P_0}$) $Q \neq P_0$ is a Weierstrass point on M and conversely every Weierstrass point on M is a zero of this function.

PROOF. Assume $Q \neq P_0$ is a zero of the function. It then follows from the theory of Chapter VI that

$$g\varphi_{P_0}(Q) + \mathscr{K}_{P_0} = \varphi_{P_0}(R_1 \cdots R_{g-1}) + \mathscr{K}_{P_0};$$

so that

$$\varphi_{P_0}(Q^g) = \varphi_{P_0}(R_1 \cdots R_{g-1}) = \varphi_{P_0}(P_0 R_1 \cdots R_{g-1}).$$

By Abel's theorem we therefore have the existence of a non-constant meromorphic function on M in $L(1/Q^g)$; which implies that Q is a Weierstrass point.

Conversely, assume that $Q \neq P_0$ is a Weierstrass point on M. Then $r(1/Q^g) \geq 2$ and therefore there exists an integral divisor $R_1 \cdots R_g$ which we may assume contains P_0 in its support which is equivalent to Q^g. It therefore follows that $g\varphi_{P_0}(Q) + \mathscr{K}_{P_0} = \varphi_{P_0}(R_1 \cdots R_{g-1}) + \mathscr{K}_{P_0}$ and thus Q is a zero of ψ. We note that P_0 is always a zero of this function.

VII.1.11. The number of zeros of the function ψ (of the previous paragraph) is g^3 and these zeros $P_1, \ldots, P_{g^3}$ satisfy

$$\varphi_{P_0}(P_1 \cdots P_{g^3}) = (g(g+1)/2)(-2\mathscr{K}_{P_0}).$$

We have already mentioned in the proof of the previous theorem that P_0 is always a zero of the function in question. It is important for us to observe that the order of vanishing of ψ at P_0 is at least g.

The simplest way of seeing this is to consider the following equalities (valid for arbitrary $P \in M$; of course, P_0 may also be viewed as an arbitrary point on our surface):

$$\theta(g\varphi_{P_0}(P) + \mathscr{K}_{P_0}) = \theta(\varphi_{P_0}(P) + \mathscr{K}_P) = \theta(-\varphi_P(P_0) + \mathscr{K}_P) = \theta(\varphi_P(P_0) - \mathscr{K}_P).$$

In the above we now think of P as being fixed and P_0 as the variable point. Furthermore choose P as a non-Weierstrass point. Then $P_0 \mapsto \theta(\varphi_P(P_0) - \mathscr{K}_P)$ has a zero of order g at $P_0 = P$. If P is a Weierstrass point then the function vanishes identically on the surface. It is thus clear that the order of vanishing of ψ at $P = P_0$ is always at least g.

It thus makes sense to write our relation on the zeros of this function ψ in the form

$$\varphi_{P_0}(P_0^g P_1 \cdots P_{g^3-g}) = (g(g+1)/2)(-2\mathscr{K}_{P_0})$$

or more suggestively

$$\varphi_{P_0}(P_1 \cdots P_{g^3-g}) = (g(g+1)/2)(-2\mathscr{K}_{P_0}).$$

Corollary. *The points $P_1, \ldots, P_{g^3-g}$ are zeros of a holomorphic $g(g+1)/2$-differential on M and Weierstrass points of M.*

If we now recall the definition of Weierstrass points given in Chapter III, we conclude that the holomorphic $g(g+1)/2$-differential referred to above is the Wronskian W of any basis for the holomorphic differentials on M. It is reasonable to suspect (and in fact is true) that the divisor of the Wronskian, (W), equals the divisor of ψ. We have here only proved that the supports of the two divisors agree. We have not shown that the orders of vanishing of the two "functions" agree at all points on the surface. On the "generic" surface all the Weierstrass points are simple and hence the orders in question agree everywhere. The transition from this case to the general case should be possible using the theory of moduli of Riemann surfaces (variation of complex structure on a topological model of the surface). This theory is beyond the scope of this book and the problem under discussion, in a wider context, presents some interesting challenges.

Exercise

For hyperelliptic M

$$(W) = (P_1 \cdots P_{2g+2})^{g(g-1)/2},$$

where $P_1, \ldots, P_{2g+2}$ are the Weierstrass points on the surface. Compute (ψ).

VII.2. Relations Among Quadratic Differentials

VII.2.1. We shall be working with a compact Riemann surface M of genus $g \geq 3$, and using standard notation involving divisors, the Jacobian variety, the θ-function, etc.

We have seen, in III.3.7, that there is a connection between relations among quadratic differentials and integral divisors of degree $g - 1$ with index of specialty ≥ 2. These points in M_{g-1}^1 are related via the map φ to the points in $W_{g-1}^1 \subset J(M)$. This last set is connected to Θ_{sing} by (1.6.1). Our first result elaborates on this connection by strengthening a remark in VI.3.7.

Proposition. *Let $e \in J(M)$. Then*

$$P \mapsto \theta(\varphi(P) \pm e) \tag{2.1.1}$$

vanishes identically on M for every choice of base point P_0 for the map $\varphi: M \to J(M)$ if and only if $e \in \Theta_{\text{sing}}$.

PROOF. The fact is that $e \in \Theta_{\text{sing}}$ is equivalent by (1.6.1) to the existence of points $P_1, \ldots, P_{g-1} \in M$ such that

$$e = \varphi(P_1 \cdots P_{g-1}) + \mathcal{K} \quad \text{and} \quad i(P_i \cdots P_{g-1}) \geq 2.$$

Hence by Theorem VI.3.2, it is equivalent to $\theta(W_1 - W_1 \pm e) \equiv 0$. This is precisely the statement of our proposition. □

VII.2.2. The above proposition has some interesting consequences. For any $e \in J(M)$, (2.1.1) defines (locally) a holomorphic function on M; in particular, it is a function defined in a neighborhood of the base point P_0. Let z be a local coordinate on M vanishing at P_0. The power series expansion of (2.1.1) is then (we are using the function defined by $+e$, and evaluating θ and its partials at e)

$$\begin{aligned}
\theta(\varphi(P) + e) = \theta(e) &+ \left(\sum_{j=1}^{g} \frac{\partial \theta}{\partial u_j}(e) \zeta_j(P_0) \right) z \\
&+ \left(\sum_{j=1}^{g} \frac{\partial \theta}{\partial u_j}(e) \zeta_j'(P_0) + \sum_{j,k=1}^{g} \frac{\partial^2 \theta}{\partial u_j \partial u_k} \zeta_j(P_0) \zeta_k(P_0) \right) \frac{z^2}{2} \\
&+ \left(\sum_j \frac{\partial \theta}{\partial u_j}(e) \zeta_j''(P_0) + 3 \sum_{j,k} \frac{\partial^2 \theta}{\partial u_j u_k}(e) \zeta_j'(P_0) \zeta_k(P_0) \right. \\
&+ \left. \sum_{j,k,l} \frac{\partial^3 \theta}{\partial u_j \partial u_k \partial u_l} \zeta_j(P_0) \zeta_k(P_0) \zeta_l(P_0) \right) \frac{z^3}{3!} + \cdots.
\end{aligned} \tag{2.2.1}$$

(Note that in the above expression, we have identified the normalized differential ζ_j with the holomorphic functions f of z such that $\zeta_j = f(z)\,dz$ near P_0.)

The above expression becomes interesting precisely when $e \in \Theta_{\text{sing}}$ or $-e \in \Theta_{\text{sing}}$. In this case, by Proposition VII.2.1, the coefficient of z^n must vanish for each $n \geq 0$, and each base point. It then follows that $\theta(\pm e) = 0 = (\partial\theta/\partial u_j)(\pm e)$, $j = 1, \ldots, g$. This information is not new. It is a consequence of the Riemann vanishing theorem (VI.3.5). A new result is contained in

Proposition. *For $e \in \Theta_{\text{sing}}$, we have*

$$\sum_{j,k=1}^{g} \frac{\partial^2\theta}{\partial u_j \partial u_k}(\pm e)\zeta_j\zeta_k = 0, \tag{2.2.2}$$

and

$$\sum_{j,k,l=1}^{g} \frac{\partial^3\theta}{\partial u_j \partial u_k \partial u_l}(\pm e)\zeta_j\zeta_k\zeta_l = 0. \tag{2.2.3}$$

Proof. Equation (2.2.2) follows from (2.2.1) and the previous observation that the first partials of θ vanish. The vanishing of the coefficient of z^3 in (2.2.1) yields

$$3\sum_{j,k} \frac{\partial^2\theta}{\partial u_j \partial u_k}(e)\zeta_j'\zeta_k + \sum_{j,k,l} \frac{\partial^3\theta}{\partial u_j \partial u_k \partial u_l}(e)\zeta_j\zeta_k\zeta_l = 0, \tag{2.2.4}$$

$$3\sum_{j,k} \frac{\partial^2\theta}{\partial u_j \partial u_k}(-e)\zeta_j'\zeta_k + \sum_{j,k,l} \frac{\partial^3\theta}{\partial u_j \partial u_k \partial u_l}(-e)\zeta_j\zeta_k\zeta_l = 0. \tag{2.2.5}$$

If we subtract (2.2.5) from (2.2.4) and use the fact that θ is an even function, then we obtain (2.2.3) □

Remark. There are instances when the above proposition is of little value. For example, if $e \in \Theta_{\text{sing}}$, and all the second partials of θ at e vanish ($e \in W_{g-1}^2 + \mathscr{K}$), then (2.2.2) yields no information. It could also happen that all the second order partial derivatives do not vanish at e, but all the third order partial derivatives do vanish. This takes place when $e \in \Theta_{\text{sing}}$ is an even point of order 2. Why? (EXERCISE.)

VII.2.3. We return now to case (c) of Theorem VII.1.6, and consider a surface M of genus 4 where Θ_{sing} consists of precisely two points e and $-e$ (with e not a point of order 2 in $J(M)$). The surface M is not hyperelliptic in this case. We have seen, in III.10, that the abelian differentials of the first kind on M provide an embedding, $M \to \mathbb{P}^3$, of the surface into projective space. Proposition VII.2.2 tells us that the image of M in $\mathbb{P}^3$ is contained in the intersection of the quadric and cubic defined by (2.2.2) and (2.2.3). Note that (2.2.2) is *not* the trivial relation because the highest order of vanishing the θ-function for a surface of genus 4 is 2. It should be observed that the coefficients of the curve (given in this way) in $\mathbb{P}^3$ depend only on Θ_{sing}, which in turn depends only on the period matrix Π of M. Hence the curve in projective space or equivalently the Riemann surface M can be recovered from the period matrix Π, whenever (2.2.2) and (2.2.3) are independent equations.

Remark. (For those familiar with elementary properties of curves.) The intersection of the quadric (2.2.2) and the cubic (2.2.3) certainly contains a curve of degree $2 \cdot 3 = 6$. The degree of the curve M in $\mathbb{P}^3$ is $2 \cdot 4 - 2 = 6$. It is important to observe that while, as stated, the quadric in the above case is non-trivial, we have not really shown that the cubic is non-trivial or that the quadric is not contained in the cubic. In these situations the curve is not recoverable by the above procedure.

Exercise

How much of the above carries over to cases (a) and (b) of Theorem VII.1.6?

VII.2.4. It is not our intention to develop here the theory of moduli compact Riemann surfaces of genus $g > 2$. However, there is one interesting observation that can be made now. We know that the dimension of $\mathscr{H}^2(M)$, the space of holomorphic quadratic differentials on M, is $3g - 3$. We also know that $\zeta_j\zeta_k \in \mathscr{H}^2(M)$, j, $k = 1, \ldots, g$. Hence, if the products of the holomorphic abelian differentials span $\mathscr{H}^2(M)$ (if and only if M is not hyperelliptic, by Noether's theorem (III.11.20)), there must be exactly $\frac{1}{2}(g - 3)(g - 2)$ relations among the products. When $g = 4$, this is precisely the relation (2.2.2). For hyperelliptic surfaces there are more relations. For hyperelliptic surfaces of genus 4, there are $10 - 7 = 3$ linearly independent relations.

We shall now exhibit two *possible* ways of obtaining these additional relations.

One way is quite obvious. Since for hyperelliptic surfaces of genus 4, Θ_{sing} is 1-dimensional, it seems reasonable to expect that one can find three points $e_j \in \Theta_{\text{sing}}$ such that

$$\sum_{k,l=1}^{4} \frac{\partial^2\theta}{\partial u_k\,\partial u_l}(e_j)\zeta_k\zeta_l = 0, \qquad j = 1, 2, 3,$$

are linearly independent. We however, do not know how to effectively prove such independence.

Another possible way is to use more essentially the fact that Θ_{sing} is 1-dimensional. Let P_0 be the base point of $\varphi: M \to J(M)$, and let $J: M \to M$ be the hyperelliptic involution (there should be no confusion with the same letter appearing with two meanings). For $P \in M$, let $P' = J(P)$. Choose $P_1 \in M$ and set

$$e_0 = \varphi(P_1 P_1' P_0) + \mathscr{K} \in \Theta_{\text{sing}}.$$

(Note that since every $\omega \in \mathscr{H}^1(M)$ that vanishes at $P \in M$ also vanishes at P'; $i(PP'Q) = i(PQ) \geq 2$.) The embedding of M into $\Theta_{\text{sing}} \subset J(M)$ is now given by

$$M \ni P \mapsto \varphi(P_1 P_1' P) + \mathscr{K} = e \in \Theta_{\text{sing}}. \tag{2.4.1}$$

Let z be a local coordinate vanishing at P_0. Equation (2.4.1) defines $e(z)$ as a holomorphic function of z. By Proposition VII.2.2,

$$\sum_{j,k=1}^{4} \frac{\partial^2 \theta}{\partial u_j \partial u_k}(e(z))\zeta_j\zeta_k = 0, \tag{2.4.2}$$

for all z in a neighborhood of 0. We now expand $(\partial^2\theta/\partial u_j \partial u_k)(e(z))$ as a power series in z, and obtain

$$\frac{\partial^2 \theta}{\partial u_j \partial u_k}(e(z)) = \frac{\partial^2 \theta}{\partial u_j \partial u_k}(e_0) + \left(\sum_{l=1}^{4} \frac{\partial^3 \theta}{\partial u_j \partial u_k \partial u_l}(e_0)\zeta_l(P_0)\right)z$$
$$+ \left(\sum_{l} \frac{\partial^3 \theta}{\partial u_l \partial u_j \partial u_k}\zeta_l'(P_0) + \sum_{l,m} \frac{\partial^4 \theta}{\partial u_m \partial u_l \partial u_j \partial u_k}(e_0)\zeta_l(P_0)\zeta_m(P_0)\right)\frac{z^2}{2}$$
$$+ \cdots. \tag{2.4.3}$$

Inserting (2.4.3) into (2.4.2), we obtain

$$\sum_{j,k=1}^{4} \frac{\partial^2 \theta}{\partial u_j \partial u_k}(e_0)\zeta_j\zeta_k = 0,$$

$$\sum_{j,k}\left(\sum_{l} \frac{\partial^3 \theta}{\partial u_l \partial u_j \partial u_k}(e_0)\zeta_l(P_0)\right)\zeta_j\zeta_k = 0,$$

and

$$\sum_{j,k}\left(\sum_{l} \frac{\partial^3 \theta}{\partial u_l \partial u_j \partial u_k}(e_0)\zeta_l'(P_0) + \sum_{l,m} \frac{\partial^4 \theta}{\partial u_m \partial u_l \partial u_j \partial u_k}(e_0)\zeta_l(P_0)\zeta_m(P_0)\right)\zeta_j\zeta_k = 0.$$

It seems reasonable to expect that one should be able to extract three linearly independent relations from the huge list obtained above (as one varies the base point P_0). The problem of *specifying* three such relations is apparently still open.

EXERCISE

Is the map $M \to \Theta_{\text{sing}}$, that we have constructed, surjective?

VII.2.5. We shall discuss briefly in this paragraph a question intimately related to the question of relations among the products of the normalized abelian differentials of the first kind. Let $\Pi_0 \in \mathfrak{S}_4$ be a period matrix of a compact Riemann surface of genus 4. (Recall that $\mathfrak{S}_4$ is the Siegel upper half space of genus 4 introduced in VI.1.1.) Theorem VII.1.6 guarantees that Θ_{sing} is non-empty. Assume we are, again, in case (c). By the remark following III.8.7, the rank of the associated relation (2.2.2) among the abelian differentials must be 4. Choose a (small) neighborhood N of $\Pi_0 \in \mathfrak{S}_4$. What is a necessary condition for $\Pi \in N$ to be a period matrix of a compact surface of genus 4? Clearly, the θ-function for Π must have the property that $\Theta_{\text{sing}} \neq \emptyset$. We proceed to write down an equation in $\mathfrak{S}_4$ that expresses this condition. Consider the system of equations on $\mathbb{C}^4 \times \mathfrak{S}_4$

$$\frac{\partial \theta}{\partial z_n}(z,\tau) = 0, \qquad n = 1, 2, 3, 4. \tag{2.5.1}$$

The hypothesis that Π_0 is a period matrix tells us that (z^0,Π_0) solves (2.5.1), whenever $z^0 \in \Theta_{\text{sing}}$ for Π_0. The condition that (2.2.2) have rank four, tells us that the Jacobian matrix of the system (2.5.1) with respect to the z-variables has rank 4, and thus the implicit function theorem asserts that we can solve for $z = {}^t(z_1,z_2,z_3,z_4)$ as holomorphic functions of τ in a neighborhood of (z^0,Π_0) and that

$$\frac{\partial \theta}{\partial z_n}(z(\tau),\tau) = 0 \quad \text{for } n = 1, 2, 3, 4, \tau \in N.$$

A necessary analytic condition for τ to be a period matrix of a compact surface is now easily written down:

$$F(\tau) = \theta(z(\tau),\tau) = 0, \qquad \tau \in N.$$

VII.3. Examples of Non-hyperelliptic Surfaces

We shall study in this section three sheeted covers of the sphere. The calculations will be considerably more involved than in the hyperelliptic case. Other special cases will also be described.

VII.3.1. In Chapter V, we began the study of automorphisms of (compact) Riemann surfaces. At the end of V.1.6, we gave some examples to show that the results of V.1.5 were sharp. We now return to a variation of the second of those examples.

Consider the Riemann surface M:

$$w^3 = z(z-1)(z-\lambda_1)\cdots(z-\lambda_{3k-3}), \qquad k \geq 2. \tag{3.1.1}$$

Here $\lambda_1, \ldots, \lambda_{3k-3}$ are $3k-3$ distinct points in $\mathbb{C}\backslash\{0,1\}$. We view z as a meromorphic function on M of degree 3 (w is of degree $3k-1$). It is branched over $0, 1, \infty, \lambda_1, \ldots, \lambda_{3k-3}$ with branch number 2. For convenience we label:

$$Q_1 = z^{-1}(0), \qquad Q_2 = z^{-1}(1), \qquad Q_3 = z^{-1}(\infty),$$
$$P_j = z^{-1}(\lambda_j), \qquad j = 1, \ldots, 3k-3,$$

and note that the divisor of the function z is

$$(z) = \frac{Q_1^3}{Q_3^3}.$$

The Riemann-Hurwitz formula yields that the genus g of M is $3k-2$ (≥ 4).

Our first task is to compute a basis for $\mathscr{H}^1(M)$—compare III.7.5. Observe that

$$(dz) = \frac{Q_1^2 Q_2^2 P_1^2 \cdots P_{3k-3}^2}{Q_3^4},$$

and

$$(w) = \frac{Q_1 Q_2 P_1 \cdots P_{3k-3}}{Q_3^{3k-1}}.$$

From the above two observations it is easy to conclude that

$$z^j \frac{dz}{w}, \qquad j = 0, \ldots, k-2,$$

and

$$z^l \frac{dz}{w^2}, \qquad l = 0, \ldots, 2k-2,$$

form a basis for $\mathscr{H}^1(M)$; as a matter of fact

$$\left(z^j \frac{dz}{w}\right) = Q_1^{3j+1} Q_2 Q_3^{3(k-j)-5} P_1 \cdots P_{3k-3},$$

and

$$\left(z^l \frac{dz}{w^2}\right) = Q_1^{3l} Q_3^{3(2k-l-2)}.$$

We shall consider the special case $k = 2$ (hence, $g = 4$). Thus the basis for $\mathscr{H}^1(M)$ is:

$$\frac{dz}{w}, \frac{dz}{w^2}, z\frac{dz}{w^2}, z^2\frac{dz}{w^2}. \tag{3.1.2}$$

The above basis is adapted (recall III.5.8) to the point Q_1. As a matter of fact, the Weierstrass "gap" sequence at Q_1 is:

$$1, 2, 4, 7. \tag{3.1.3}$$

Note that the above is also the "gap" sequence at Q_2, Q_3 and P_j, $j = 1, 2, 3$. Thus each of these 6 points is a Weierstrass point of weight 4. We have thus accounted for 24 of the 60 Weierstrass points (here we are counting each Weierstrass point according to its multiplicity). The remaining 36 Weierstrass points will occur in groups of three. Each such group of three will project to the same point in $\mathbb{C} \cup \{\infty\}$ by the map z. This follows from the fact that M has an automorphism of period three that interchanges the sheets of the cover.

VII.3.2. We continue with the case $k = 2$ in (3.1.1). Since M is of genus 4, by Theorem VII.1.6, M is either hyperelliptic (impossible, because M carries a function of degree 3, by Proposition III.7.10) or Θ_{sing} consists of one point of order 2 or Θ_{sing} consists of precisely two points (neither one of order 2). Since Q_1^6 is the divisor of the abelian differential $z^2(dz/w^2)$, the vector $\mathscr{K}$ of Riemann constants with respect to the base point Q_1, is a point of order 2 of $J(M)$ by Theorem VI.3.6. Further, since

$$\mathscr{K} = \varphi(Q_1^3) + \mathscr{K},$$

$\mathscr{K}$ is a zero of the theta function. Also, looking at the "gap" sequence (3.1.3), we see that

$$i(Q_1^3) = 2,$$

which shows that $\mathscr{K} \in \Theta_{\text{sing}}$. We have shown that Θ_{sing} consists of the single point $\mathscr{K}$. It must be the case that there is a relation of rank 3 among the products of the abelian differentials of the first kind. The relation is easily exhibited:

$$\left(z\frac{dz}{w^2}\right)^2 = \left(z^2\frac{dz}{w^2}\right)\left(\frac{dz}{w^2}\right). \tag{3.2.1}$$

Of course, alternatively, (3.2.1) can be used to conclude that Θ_{sing} consists of precisely one point. However, the conclusion (we obtained)

$$\Theta_{\text{sing}} = \{\mathscr{K}\},$$

required a little more analysis.

VII.3.3. We return to the general case (3.1.1) with $k \geq 2$. Using the basis we constructed for $\mathscr{H}^1(M)$, we see that the Weierstrass "gap" sequence at Q_1 (hence also at Q_2, Q_3, P_j, $j = 1, \ldots, 3k - 3$) is:

$$\{3l + 1; l = 0, \ldots, 2k - 2\} \cup \{3j + 2; j = 0, \ldots, k - 2\}.$$

In particular, every function f, holomorphic on $M \backslash \{Q_1\}$, with $\deg f \leq 3k - 4$ must satisfy

$$\deg f \equiv 0 \pmod 3.$$

An easy calculation shows that the weight of each of these $3k$ Weierstrass points is

$$\sum_{l=0}^{2k-2} (3l + 1) + \sum_{j=0}^{k-2} (3j + 2) - \sum_{m=0}^{3k-2} m = (3k - 2)(k - 1).$$

We have accounted for $3k(3k - 2)(k - 1)$ of the $(3k - 3)(3k - 2)(3k - 1)$ Weierstrass points on M. The vector of Riemann constants $\mathscr{K}$ with respect to Q_1 is again a point of order 2 in $J(M)$, since Q_1^{6k-6} is a canonical divisor. The point $\mathscr{K} \in \Theta_{\text{sing}}$, because $i(Q_1^{3k-3}) = k = \frac{1}{3}(g + 2)$.

Finally, we leave it to the reader to explore the question of (linearly independent) relations among the abelian differentials of the first kind. For example, if $k = 3$, and if we label

$$\varphi_1 = \frac{dz}{w^2}, \qquad \varphi_2 = z\frac{dz}{w^2}, \qquad \varphi_3 = z^2\frac{dz}{w^2},$$

$$\varphi_4 = z^3\frac{dz}{w^2}, \qquad \varphi_5 = z^4\frac{dz}{w^2}, \qquad \varphi_6 = \frac{dz}{w}, \qquad \varphi_7 = z\frac{dz}{w},$$

then we can write down the following maximal set of linearly independent relations

$$\varphi_2^2 = \varphi_1\varphi_3, \quad \varphi_3^2 = \varphi_1\varphi_5, \quad \varphi_2\varphi_3 = \varphi_1\varphi_4, \quad \varphi_2\varphi_4 = \varphi_1\varphi_5,$$
$$\varphi_4^2 = \varphi_3\varphi_5, \quad \varphi_2\varphi_5 = \varphi_3\varphi_4, \quad \varphi_1\varphi_7 = \varphi_2\varphi_6, \quad \varphi_2\varphi_7 = \varphi_3\varphi_6,$$
$$\varphi_3\varphi_7 = \varphi_4\varphi_6, \quad \varphi_4\varphi_7 = \varphi_5\varphi_6.$$

VII.3.4. The Riemann surface M of (3.1.1) is an example of a surface with an automorphism T (of period 3, in our case) such that $M/\langle T\rangle \cong \mathbb{C} \cup \{\infty\}$. We have seen that for these surfaces, Θ_{sing} contains a point of order 2. It may seem at first glance that the existence of points of order 2 in Θ_{sing} is caused by the presence of the automorphism. This conclusion is, in general, false. Consider, for example, the surface M defined by

$$w^3 = z(z-1)(z-\lambda_1)^2(z-\lambda_2)^2(z-\lambda_3)^2,$$

where λ_j, $j = 1, 2, 3$, are three distinct points in $\mathbb{C}\backslash\{0,1\}$. The surface M has again an automorphism T of order 3 such that $M/\langle T\rangle$ is conformaly equivalent to $\mathbb{C} \cup \{\infty\}$. Note that M (again) has genus 4. Using notation as in VII.3.1, we conclude that

$$(z) = \frac{Q_1^3}{Q_3^3},$$

$$(w) = \frac{Q_1Q_2P_1^2P_2^2P_3^2}{Q_3^8},$$

and

$$(dz) = \frac{Q_1^2Q_2^2P_1^2P_2^2P_3^2}{Q_3^4}.$$

A basis for $\mathscr{H}^1(M)$ is thus

$$\frac{dz}{w},\ z\frac{dz}{w},\ (z-\lambda_1)(z-\lambda_2)(z-\lambda_3)\frac{dz}{w^2},\ z(z-\lambda_1)(z-\lambda_2)(z-\lambda_3)\frac{dz}{w^2}.$$

From the above we see that the possible orders of zeros of abelian differentials of the first kind at Q_1 are:

$$0, 1, 3, 4,$$

and thus the "gap" sequence at Q_1 is:

$$1, 2, 4, 5.$$

In particular, there is no differential in $\mathscr{H}^1(M)$ all of whose zeros are at Q_1. Now, the point $\mathscr{K}$ belongs to Θ_{sing} (because $i(Q_1^3) = 2$), and if Θ_{sing} contained a point of order 2, $\mathscr{K}$ would have to be that point. Hence $2\mathscr{K} = 0$. By Theorem VI.3.6, we would also have $\varphi(Q_1^6) = 0 = -2\mathscr{K}$ or that Q_1^6 is canonical. We have seen that this does not occur in our example.

VII.3.5. There is, however, one case where the presence of an automorphism T produces points of order 2 in Θ_{sing}: the case when T has period 2.

Suppose now that M is a surface of genus $g \geq 2$, and $T \in \operatorname{Aut} M$ has period 2 and $\nu(T)$ fixed points. The genus $\tilde{g}$ of $M/\langle T\rangle$ is computed (recall V.1.9) by

$$\nu = \nu(T) = 2g + 2 - 4\tilde{g}$$

(and M is called $\tilde{g}$-hyperelliptic).

To show that Θ_{sing} contains a point of order two, it suffices to find an $\omega \in \mathscr{H}^1(M)$ with even order zeros

$$(\omega) = P_1^2 \cdots P_{g-1}^2,$$

such that

$$i(P_1 \cdots P_{g-1}) \geq 2.$$

For then $e = \varphi(P_1 \cdots P_{g-1}) + \mathscr{K} \in \Theta_{\text{sing}}$, and $2e = \varphi((\omega)) + 2\mathscr{K} = 0$.

If $\tilde{g} = 0$, then M is hyperelliptic. Then for $g \geq 3$, there is a point of order 2 in Θ_{sing} and the order of vanishing of the θ-function at this point is $\frac{1}{2}(g+1)$. Hence we now assume $\tilde{g} > 0$.

VII.3.6. Let us generalize by considering an automorphism T of prime order N with $M/\langle T\rangle$ of positive genus. Consider the action of T on $\mathscr{H}^1(M)$. Since $M/\langle T\rangle$ has positive genus $\tilde{g}$, 1 is an eigenvalue. Since M has bigger genus than $\tilde{M} = M/\langle T\rangle$, there is another eigenvalue ε ($\varepsilon^N = 1$, $\varepsilon \neq 1$). In other words, there are holomorphic differentials $\omega, \omega_1 \in \mathscr{H}^1(M)$ such that

$$T\omega = \omega, \qquad T\omega_1 = \varepsilon\omega_1, \qquad \omega \neq 0.$$

The differential ω projects to $\tilde{M}$; while ω_1 projects to a multivalued differential on $\tilde{M}$. The function $f = \omega_1/\omega$ on M projects to an N-valued function on $\tilde{M}$.

To describe the structure of the divisors (ω) and (ω_1), we let $P_1, \ldots, P_\nu$ be the fixed points of T ($\nu = \nu(T)$ could be zero). Calculations similar to the ones performed in V.2 show that

$$\operatorname{ord}_{P_j} \omega = Nr_j + N - 1, \qquad r_j \in \mathbb{Z}, r_j \geq 0.$$

Hence we can write

$$(\omega) = P_1^{N-1} \cdots P_\nu^{N-1} \Delta\Delta^{(1)} \cdots \Delta^{(N-1)}, \tag{3.6.1}$$

where Δ is an integral divisor on M and $\Delta^{(k)} = T^k(\Delta)$. Note that Δ could contain some of the points P_j, $j = 1, \ldots, \nu$.

Similarly,

$$\operatorname{ord}_{P_j} \omega_1 = r_j N + t_j - 1, \qquad r_j \in \mathbb{Z}, t_j \in \mathbb{Z}, 1 \leq t_j \leq N-1, r_j \geq 0.$$

Thus, as before, we can write

$$(\omega_1) = P_1^{t_1 - 1} \cdots P_\nu^{t_\nu - 1} X X^{(1)} \cdots X^{(N-1)}. \tag{3.6.2}$$

From (3.6.1) and (3.6.2) we can compute the divisor of the meromorphic function $f = \omega_1/\omega$ on M, and the projected function $\tilde{f}$ on $\tilde{M}$:

$$(\tilde{f}) = P_1^{(t_1/N) - 1} \cdots P_\nu^{(t_\nu/N) - 1} \frac{X}{\Delta}. \tag{3.6.3}$$

In (3.6.3) we have made an obvious identification of points on M with their images in $\tilde{M}$. Note that $(\tilde{f})$ contains fractional powers because locally $\tilde{f}$ is given by a Puiseaux (not Taylor) series.

The fractional divisors on $\tilde{M}$ can be mapped (locally) into $J(\tilde{M})$. Since $\tilde{f}^N$ is single valued on $\tilde{M}$, the image of the divisor of $\tilde{f}$ in $J(\tilde{M})$ is a point of order N.

In order to study the converse, we begin by remarking that

$$R = \sum_{j=1}^{\nu(T)} \left(1 - \frac{t_j}{N}\right)$$

is an integer (EXERCISE) that satisfies

$$\frac{\nu(T)}{N} \leq R \leq \frac{N-1}{N} \nu(T). \tag{3.6.4}$$

Every multivalued function on $\tilde{M}$ whose divisor has the same fractional exponent at P_j as in (3.6.3) and that has the same image in $J(\tilde{M})$ as $(\tilde{f})$, lifts to a meromorphic function on M.

Consider the divisor D on $\tilde{M}$

$$D = P_1^{N-t_1} \cdots P_\nu^{N-t_\nu} \Delta^N.$$

We now pose the following problem: Does there exist an integral divisor X on $\tilde{M}$ such that

$$\deg X = \tilde{g} - 1 + \left[\frac{1}{2} \sum_{j=1}^{\nu} \left(1 - \frac{t_j}{N}\right)\right]$$

(where, as usual, $[a]$ denotes the greatest integer in a),

$$\varphi(X^{2N}) = \varphi(D), \tag{3.6.5}$$

and

$$\varphi(X^2) - \varphi(P_1^{1-t_1/N} \cdots P_\nu^{1-t_\nu/N} \Delta) \tag{3.6.6}$$

equals the point of order N in $J(\tilde{M})$ determined by the divisor of the function $\tilde{f}$?

The theory of the Jacobi inversion problem asserts that we can always solve the above problem, and in fact do so with $[\frac{1}{2} \sum_{j=1}^{\nu} (1 - t_j/N)] - 1$ free points. The condition that we shall need is

$$R = \sum_{j=1}^{\nu} \left(1 - \frac{t_j}{N}\right) \geq 2,$$

which occurs by (3.6.4) whenever

$$\nu(T) \geq 2N.$$

Furthermore, there are $(2N)^{2\tilde{g}}$ different solutions to (3.6.5) corresponding to the $(2N)^{2\tilde{g}}$ points of order $2N$ in $J(\tilde{M})$. Only $2^{2\tilde{g}}$ of these solutions will satisfy (3.6.6).

There are now two cases to consider:

$$R \text{ is an even integer } (\geq 2), \tag{3.6.7}$$

and

$$R \text{ is an odd integer } (\geq 3). \tag{3.6.8}$$

In case (3.6.7),

$$\frac{X^{2N}}{P_1^{N-t_1} \cdots P_\nu^{N-t_\nu} \Delta^N}$$

is the divisor of a meromorphic function on $\tilde{M}$ whose Nth-root lifts to a meromorphic function on M with divisor

$$\frac{X^2(X^2)^{(1)} \cdots (X^2)^{(N-1)}}{P_1^{N-t_1} \cdots P_\nu^{N-t_\nu} \Delta\Delta^{(1)} \cdots \Delta^{(N-1)}}.$$

Thus we conclude that (multiply the above function by ω)

$$Z_1 = P_1^{t_1-1} \cdots P_\nu^{t_\nu-1} X^2(X^2)^{(1)} \cdots (X^2)^{(N-1)} \tag{3.6.9}$$

is a canonical divisor on M.

In case (3.6.8) we choose $P_1 \in \tilde{M}$ as the base point of the map $\varphi: \tilde{M} \to J(\tilde{M})$. We conclude that

$$\frac{P_1^N X^{2N}}{P_1^{N-t_1} \cdots P_\nu^{N-t_\nu} \Delta^N}$$

is the divisor of a meromorphic function on $\tilde{M}$ whose N-th root lifts to a a meromorphic function on M with divisor

$$\frac{P_1^N X^2(X^2)^{(1)} \cdots (X^2)^{(N-1)}}{P_1^{N-t_1} \cdots P_\nu^{N-t_\nu} \Delta\Delta^{(1)} \cdots \Delta^{(N-1)}}.$$

In this case we conclude that

$$Z_2 = P_1^{N+t_1-1} P_2^{t_2-1} \cdots P_\nu^{t_\nu-1}(X^2)(X^2)^{(1)} \cdots (X^2)^{(N-1)} \tag{3.6.10}$$

is a canonical divisor on M.

The sole purpose of the above series of exercises was to produce a holomorphic differential on M (with enough free points) all of whose zeros are of even order. We have succeeded in this whenever $N = 2$ (because $t_j = 1$ for $j = 1, \ldots, \nu(T)$). In this case

$$R = \sum_{j=1}^{\nu} \left(1 - \frac{t_j}{N}\right) = \frac{\nu(T)}{2},$$

and thus X has precisely

$$\left[\frac{\nu(T)}{4}\right] - 1$$

free points. We conclude from (3.6.9) that

$$i(Z_1) \geq \left[\frac{\nu(T)}{4}\right],$$

with a similar result for (3.6.10).

We have established the following

Theorem. *Let M be a $\tilde{g}$-hyperelliptic surface of genus $g \geq 2\tilde{g} + 3$. Then Θ_{sing} contains $2^{2\tilde{g}}$ points of order 2 with the order of vanishing of the θ-function at these points greater than or equal to $[\frac{1}{4}(2g + 2 - 4\tilde{g})]$.*

Remark. The above analysis for $N = 2$ and $\nu = 0$ does not give points of order two in Θ_{sing}. A slightly different and in some sense simpler analysis, however, does.

VII.3.7. The Riemann surface M of

$$w^{2g+1} = z(z - 1)$$

affords another interesting example of a surface of genus g. It is an immediate consequence of the fact that w is a function of degree 2, that M is hyperelliptic. There is an obvious involution on M:

$$(z,w) \mapsto (1 - z, w).$$

This is the hyperelliptic involution since it fixes the $2g + 1$ points lying over $z = \frac{1}{2}$ and the point $z^{-1}(\infty)$.

VII.3.8. The Riemann surface M of

$$w^4 = z^4 - 1$$

is of genus 3 which obviously has two cyclic groups of order 4 operating on it; the groups generated by

$$(z,w) \mapsto (z, iw), \qquad (z,w) \mapsto (iz, w).$$

Using obvious notational conventions, we record the following facts:

$$(z) = \frac{Q_1 Q_2 Q_3 Q_4}{Q_5 Q_6 Q_7 Q_8},$$

$$(w) = \frac{P_1 P_2 P_3 P_4}{Q_5 Q_6 Q_7 Q_8},$$

and

$$(dz) = \frac{P_1^3 P_2^3 P_3^3 P_4^3}{Q_5^2 Q_6^2 Q_7^2 Q_8^2}.$$

From the above we see that

$$\frac{dz}{w^2}, \frac{dz}{w^3}, z\frac{dz}{w^3}$$

form a basis for $\mathscr{H}^1(M)$. Note that

$$\left(z\frac{dz}{w^3}\right) = Q_1Q_2Q_3Q_4 = z^{-1}(0).$$

More generally for any $c \in \mathbb{C} \cup \{\infty\}$, $z^{-1}(c)$ is a canonical divisor. It follows from these observations that the "gap" sequence at any P_j ($j = 1,2,3,4$) is

$$1, 2, 5.$$

(Note that the P_j are defined by $z(P_j)^4 = 1$.) Thus each P_j is a Weierstrass point of weight 2. We have accounted for 8 of the 24 Weierstrass points. Similarly, the points Q_j ($j = 1,2,3,4$) are Weierstrass points of weight 2, since

$$\frac{dw}{z^2}, \frac{dw}{z^3}, w\frac{dw}{z^3}$$

is also a basis of $\mathscr{H}^1(M)$. Finally, the points Q_j ($j = 5,6,7,8$) are also Weierstrass points of weight 2, since

$$(z,w) \mapsto \left(\frac{z}{w}, \frac{1}{w}\right)$$

is an automorphism (of period 2) of the surface M that interchanges the zeros and poles of w. We have thus accounted for all the Weierstrass points on M. It is a surface of genus 3 with 12 distinct Weierstrass points each of weight 2.

VII.3.9. The next to last example of this section is the most complicated one. It will be used to show that the upper bound on the number of (distinct) Weierstrass points $g^3 - g$ on a surface of genus g (Theorem III.5.11) is attained. Our example will be for $g = 3$. Consider the Riemann surface M of the algebraic curve

$$w^7 = z(z-1)^2. \tag{3.9.1}$$

On the surface, the function z is of degree 7 (and w is of degree 3). The function z is ramified over 0, 1, ∞ (with ramification number 7). A calculation using Riemann-Hurwitz now shows that M has genus 3. In our usual short hand:

$$(z) = \frac{P_1^7}{Q_1^7},$$

$$(dz) = \frac{P_1^6 P_2^6}{Q_1^8},$$

and

$$(w) = \frac{P_1 P_2^2}{Q_1^3}.$$

From the above formulas we see that thc differentials

$$\frac{dz}{w^3}, \ (z-1)\frac{dz}{w^5}, \ (z-1)\frac{dz}{w^6} \tag{3.9.2}$$

have (respectively) divisors

$$P_1^3 Q_1, \ P_1 P_2^3, \ P_2 Q_1^3.$$

It is immediate from this calculation that we have again produced a basis for $\mathscr{H}^1(M)$ and that P_1, P_2, Q_1 are all simple (= weight one) Weierstrass points.

We claim that all the Weierstrass points are simple (hence we must have 24 distinct Weierstrass points). We will compute the Wronskian of our basis of $\mathscr{H}^1(M)$. Since we are no longer interested in points lying over 0, 1, ∞ (via z), the function z is a good local coordinate at such points. Using the notation of III.5.8, we are computing

$$\det\left[\frac{1}{w^3}, \frac{z-1}{w^5}, \frac{z-1}{w^6}\right] = \frac{1}{w^{18}} \det[w^3, w(z-1), z-1]$$

$$= \frac{3!}{z^3} \frac{1}{z^5(z-1)^5}(z^3 - 8z^2 + 5z + 1).$$

(The computation is long and tedious, but routine.) Denoting the cubic polynomial by $p(z)$ we see that

$$p(-1) = -13, \qquad p(0) = +1, \qquad p(1) = -1, \qquad p(8) = 41,$$

and hence p has three distinct real roots (in the open intervals $(-1,0)$, $(0,1)$ and $(1,8)$). Each zero of the Wronskian corresponds to 7 distinct points on M. Thus we have produced a surface of genus 3 with 24 distinct (simple) Weierstrass points.

VII.3.10. The Riemann surface of (3.9.1) actually has $168 = 84(3-1)$ automorphisms, which shows that the maximum number (of Hurwitz's theorem, V.1.3) is achieved. It is easy to exhibit an automorphism of period 7:

$$(z,w) \mapsto (z,\varepsilon w), \qquad \varepsilon = \exp\left(\frac{2\pi i}{7}\right).$$

To exhibit other automorphisms of M we must use some algebraic geometry. The abelian differentials of the first kind provide an embedding of M into $\mathbb{P}^2$ (III.10). Thus setting

$$w = -XY^{-1}, \qquad z - 1 = +X^3Y^{-2},$$

we see from (3.9.1) or (3.9.2) that a projective equation for our curve is

$$X^3Y + Y^3Z + Z^3X = 0.$$

Hence we see that we have another automorphism of M (of order 3) given by the permutation

$$(X,Y,Z) \mapsto (Y,Z,X).$$

We will not proceed with the above line of thought. (The interested reader should consult the work of A. Kuribayashi and K. Komiya for the complete classification of automorphism groups and Weierstrass points for surfaces of genus 3.)

If we are willing to use the fact that M has 168 automorphisms, we can conclude without calculation that each Weierstrass point is simple. We have seen in the proof of Hurwitz's theorem, that the maximum number of automorphisms occur only if $M/\text{Aut } M \cong \mathbb{C} \cup \{\infty\}$ and the canonical projection

$$\pi: M \to \mathbb{C} \cup \{\infty\}$$

is branched over three points with ramification numbers 2, 3, 7. Thus these are the only possible orders of the stability subgroups of points. We conclude that each orbit under Aut M must contain at least $\frac{168}{7} = 24$ points. In particular, the Weierstrass points must be the fixed points of the elements of order 7, and there must be 24 such points. Hence they must all be simple.

VII.3.11. For our last example we consider the Riemann surface of

$$w^2 = z^6 - z,$$

which is a hyperelliptic surface of genus 2. In addition to the hyperelliptic involution this surface permits an automorphism of period 5

$$(z,w) \mapsto (\varepsilon z, \varepsilon^3 w) \quad \text{with } \varepsilon = \exp\left(\frac{2\pi i}{5}\right).$$

Let us denote this automorphism by T and observe that the automorphism JT with J the hyperelliptic involution is of order 10. The automorphism JT is given by

$$(z,w) \mapsto (\varepsilon z, -\varepsilon^3 w),$$

so that the only possible fixed points of JT are $P_1 = z^{-1}(0)$ or Q_1, Q_2 the two points lying over ∞. Since T is of prime order 5, the Riemann-Hurwitz formula gives $2 = 5(2\tilde{g} - 2) + 4\nu(T)$, with $\nu(T)$ as usual the number of fixed points of T. The only way this can be satisfied is with $\tilde{g} = 0$ and $\nu(T) = 3$. Hence T fixes P_1, Q_1, and Q_2. It is therefore obvious that JT fixes P_1, but cannot fix either Q_1 or Q_2, because $JTQ_1 = JQ_1 = Q_2$ and $JTQ_2 = JQ_2 = Q_1$.

The reader should now recall Theorem V.2.11.

VII.4. Branch Points of Hyperelliptic Surfaces as Holomorphic Functions of the Periods

In IV.7, we solved two elementary moduli problems. In this section we will describe one way of obtaining moduli for hyperelliptic surfaces; another elementary case.

VII.4.1. We return once again to the concrete representation of a hyperelliptic surface M of genus $g \geq 2$. We will now assume that our function z of degree two is branched over 0, 1, and ∞, and hence represent M by

$$w^2 = z(z-1)\prod_{k=1}^{2g-1}(z-\lambda_k), \tag{4.1.1}$$

where $\lambda_1, \ldots, \lambda_{2g-1}$ are distinct points in $\mathbb{C}\backslash\{0,1\}$. We are using a slight variation of (1.0.1). To fix notation, we set

$$P_1 = z^{-1}(0), \qquad P_2 = z^{-1}(1),$$
$$P_{j+2} = z^{-1}(\lambda_j), \qquad j = 1, \ldots, 2g-1, \qquad P_{2g+2} = z^{-1}(\infty).$$

We have seen in VII.1.2, that using P_1 as a base point for $\varphi: M \to J(M)$, the vector $\mathscr{K}$ of Riemann constants is given by

$$\mathscr{K} = \frac{1}{2}\left(\Pi\begin{bmatrix} g \\ g-1 \\ \vdots \\ 1\end{bmatrix} + I\begin{bmatrix} g \\ 1 \\ \vdots \\ 1\end{bmatrix}\right),$$

where Π is the period matrix for the canonical homology basis constructed in VII.1.1. We shall show that the λ_j are holomorphic functions of the entries π_{kl} of Π. We will accomplish this by expressing the function z in terms of θ-functions.

The function $z \in \mathscr{K}(M)$ is characterized (up to a constant multiple) by the property that it has a double pole at P_{2g+2}, a double zero at P_1, and is holomorphic and nonzero on $M\backslash\{P_{2g+2}\}$. We will produce such a function in terms of θ-functions. One main tool will be the Riemann vanishing theorem, and our explicit knowledge of the images in $J(M)$ of the Weierstrass points on M.

VII.4.2. We shall see that there are many ways to proceed. We begin with the point of order 2

$$\varphi(P_1P_5P_7\cdots P_{2g+1}) + \mathscr{K} = \varphi(P_1P_3) = \varphi(P_3),$$

by virtue of (1.2.1) and the fact that $\varphi(P^2) = 0$ for every Weierstrass point P on M. We have computed $\varphi(P_3)$ in VII.1.1:

$$\varphi(P_3) = \tfrac{1}{2}(\pi^{(1)} + e^{(1)} + e^{(2)}).$$

Similarly,

$$\varphi(P_{2g+2}P_5P_7\cdots P_{2g+1}) + \mathscr{K} = \varphi(P_{2g+2}P_3) = \tfrac{1}{2}(\pi^{(1)} + e^{(2)}).$$

We also observe that

$$i(P_1P_5P_7\cdots P_{2g+1}) = 0 = i(P_{2g+2}P_5P_7\cdots P_{2g+1})$$

(compare with VII.1.4). We now consider the multiplicative function

$$P \mapsto \frac{\theta\begin{bmatrix}1 & 0 & 0 & \cdots & 0\\ 1 & 1 & 0 & \cdots & 0\end{bmatrix}(\varphi(P),\Pi)}{\theta\begin{bmatrix}1 & 0 & 0 & \cdots & 0\\ 0 & 1 & 0 & \cdots & 0\end{bmatrix}(\varphi(P),\Pi)}. \tag{4.2.1}$$

According to Theorem VI.2.4, the numerator vanishes precisely (it does not vanish identically by Theorem VI.3.3) at $P_1, P_5, P_7, \ldots, P_{2g+1}$ and the denominator at $P_{2g+2}, P_5, P_7, \ldots, P_{2g+1}$. In particular, the function (4.2.1) vanishes to first order at P_1 and has a simple pole at P_{2g+2} and is holomorphic and nonzero elsewhere. Examining the multiplicative behavior of the above function, we see that

$$f(P) = \frac{\theta^2\begin{bmatrix}1 & 0 & 0 & \cdots & 0\\ 1 & 1 & 0 & \cdots & 0\end{bmatrix}(\varphi(P),\Pi)}{\theta^2\begin{bmatrix}1 & 0 & 0 & \cdots & 0\\ 0 & 1 & 0 & \cdots & 0\end{bmatrix}(\varphi(P),\Pi)}$$

is a meromorphic function on M with divisor $P_1^2P_{2g+2}^{-2}$. Hence

$$f = cz, \qquad c \in \mathbb{C}\backslash\{0\}.$$

The constant c is evaluated by $f(P_2) = cz(P_2) = c$. Thus we see that

$$c = f(P_2) = \frac{\theta^2\begin{bmatrix}1 & 1 & 0 & \cdots & 0\\ 1 & 1 & 0 & \cdots & 0\end{bmatrix}(\varphi(P_2),\Pi)}{\theta^2\begin{bmatrix}1 & 0 & 0 & \cdots & 0\\ 0 & 1 & 0 & \cdots & 0\end{bmatrix}(\varphi(P_2),\Pi)},$$

and

$$z(P) = \frac{\theta^2\begin{bmatrix}1 & 0 & 0 & \cdots & 0\\ 0 & 1 & 0 & \cdots & 0\end{bmatrix}\left(\frac{1}{2}\pi^{(1)},\Pi\right)\theta^2\begin{bmatrix}1 & 0 & 0 & \cdots & 0\\ 1 & 1 & 0 & \cdots & 0\end{bmatrix}(\varphi(P),\Pi)}{\theta^2\begin{bmatrix}1 & 0 & 0 & \cdots & 0\\ 1 & 1 & 0 & \cdots & 0\end{bmatrix}\left(\frac{1}{2}\pi^{(1)},\Pi\right)\theta^2\begin{bmatrix}1 & 0 & 0 & \cdots & 0\\ 0 & 1 & 0 & \cdots & 0\end{bmatrix}(\varphi(P),\Pi)}$$

$$= \frac{\theta^2\begin{bmatrix}0 & 0 & 0 & \cdots & 0\\ 0 & 1 & 0 & \cdots & 0\end{bmatrix}\theta^2\begin{bmatrix}1 & 0 & 0 & \cdots & 0\\ 1 & 1 & 0 & \cdots & 0\end{bmatrix}(\varphi(P),\Pi)}{\theta^2\begin{bmatrix}0 & 0 & 0 & \cdots & 0\\ 1 & 1 & 0 & \cdots & 0\end{bmatrix}\theta^2\begin{bmatrix}1 & 0 & 0 & \cdots & 0\\ 0 & 1 & 0 & \cdots & 0\end{bmatrix}(\varphi(P),\Pi)} \tag{4.2.2}$$

(the last equality by VI(1.4.5)).

Setting $P = P_{j+2}$ we get from (4.2.2) formulae for λ_j. These formulae are useful only for

$$j = 1, 2, 4, 6, \ldots, 2g - 2. \tag{4.2.3}$$

For other values of j, both the numerator and denominator vanish. While the limit can be calculated to obtain λ_j, the calculation involves the normalized abelian differentials of the first kind. For the values of j given in (4.2.3), nice formulae for λ_j can be obtained in terms of θ-constants only. For example,

$$\lambda_1 = \frac{\theta^2\begin{bmatrix} 0 & 0 & 0 & \cdots & 0 \\ 0 & 1 & 0 & \cdots & 0 \end{bmatrix}\theta^2\begin{bmatrix} 0 & 0 & 0 & \cdots & 0 \\ 0 & 0 & 0 & \cdots & 0 \end{bmatrix}}{\theta^2\begin{bmatrix} 0 & 0 & 0 & \cdots & 0 \\ 1 & 1 & 0 & \cdots & 0 \end{bmatrix}\theta^2\begin{bmatrix} 0 & 0 & 0 & \cdots & 0 \\ 1 & 0 & 0 & \cdots & 0 \end{bmatrix}}.$$

By replacing the point of order 2 that started this whole procedure (for example, use

$$\varphi(P_1 P_3 P_7 \cdots P_{2g+1}) + \mathscr{K}$$

instead of

$$\varphi(P_1 P_5 P_7 \cdots P_{2g+1}) + \mathscr{K}),$$

we can get similar formulae for λ_j for the other values of j. We have hence established the following

Theorem. *The branch points of the two sheeted representation of a hyperelliptic Riemann surface are holomorphic functions of the period matrix. Furthermore, the hyperelliptic surface is completely determined by its period matrix.*

VII.4.3. The fact that there are many ways to express the function z in terms of θ-functions leads to useful and interesting relations among θ-constants. We will, however, not pursue this fascinating subject.

VII.5. Examples of Prym Differentials

On the hyperelliptic surface M

$$w^2 = z(z-1)\prod_{k=1}^{2g-1}(z - \lambda_k), \qquad g \geq 2,$$

the differentials

$$\frac{dz}{w}, \; z\frac{dz}{w}, \; \ldots, \; z^{g-1}\frac{dz}{w}$$

form a basis for abelian differentials of the first kind.

On M we can construct (locally) the function y given by

$$y^2 = z(z-1)\prod_{k=1}^{2g-3}(z-\lambda_k),$$

and (locally) the differentials

$$\frac{dz}{y},\ z\frac{dz}{y},\ \ldots,\ z^{g-2}\frac{dz}{y}.$$

Continuation of these differentials along the curves $a_1, \ldots, a_g, b_1, \ldots, b_{g-1}$ of Figure VII.1 (interpreted correctly) leaves them invariant. However, continuation along b_g leads to a change of sign. We have hence constructed a basis for the Prym differentials with characteristic $\left[\begin{smallmatrix}0 & \cdots & 0 & 0\\ 0 & \cdots & 0 & 1\end{smallmatrix}\right]$ as defined in III.9.

The fascinating relation between the lifts of these differentials to a smooth two-sheeted cover and the theory of moduli will have to be pursued elsewhere.

VII.6. The Trisecant Formula

Although a purist might object that we are no longer considering examples, we will proceed nevertheless. The formula in the title of this section has turned out to have remarkable applications.

VII.6.1. The notation is the same as in Section VII.2. Let α be a nonsingular element of the theta divisor, Θ (thus $\theta(\alpha) = 0$ and $(\partial\theta/\partial z_j)(\alpha) \neq 0$ for at least one integer j, $1 \leq j \leq g$). For any such α, and any four points P_1, $\ldots$, P_4 on the Riemann surface consider the following expression:

$$\frac{\theta(\alpha + \int_{P_2}^{P_1})\theta(\alpha + \int_{P_4}^{P_3})}{\theta(\alpha + \int_{P_4}^{P_1})\theta(\alpha + \int_{P_2}^{P_3})} = \lambda(P_1,P_2,P_3,P_4).$$

In the above formula $\int_P^Q$ is used as an abbreviation for $\varphi(P) - \varphi(Q)$ to emphasize the independence of this difference on the base point for the map φ. In the above expression for λ, we may think of the four points P_i as being fixed. It is more useful, however, to think of the last three points as fixed and the point P_1 as a variable point on the surface M. We hence view λ as a multivalued meromorphic function of P_1 on the surface M. When the point P_1 is continued over cycles the expression picks up multiples by an exponential. We leave for the reader the task of computing the multipliers. The multivalued function λ has a zero at $P_1 = P_2$ and a pole at $P_1 = P_4$. If we restrict P_1 (and choose P_2, P_3, and P_4 also) to lie in a fundamental polygon for the surface (or in a simply connected region whose closure is the surface), then the value of the function (now single-valued) at $P_1 = P_3$ is one.

The above properties suggest calling $\lambda(P_1,P_2,P_3,P_4)$ *the generalized cross ratio of the four points* (taken in the specified order). We (should) quickly remark that it is possible for the expression not to be well defined (the cases resulting in either the numerator or denominator of the formula defining λ vanishing) for a given α and a given set of four points (without the use of some limiting procedure). However, if we are given α and a set of three generic (we will not define this term here) points P_2, P_3, P_4, on the surface M, there are at most $g-1$ points P_1 on the surface where λ is not well defined. It is therefore not too hard to show that the generalized cross ratio of the four distinct points can always be defined. This expression would not be of much use if it were to depend on the point α. Our next result shows that this is not the case.

Theorem. *The multivalued function $\lambda(P_1,P_2,P_3,P_4)$ is independent of the point α used to define it.*

PROOF. Let α and β be two distinct elements in the nonsingular set of the theta divisor. Consider the quotient of the two expressions, one formed with α and the other with β. This quotient is easily seen to be a (single-valued) meromorphic function of P_1 on the surface. It however has no zeros or poles and is therefore a constant. The constant is one since at $P_1 = P_3$ each of the two expressions has value one.

Consider now the meromorphic function G on the Jacobian variety $J(M)$ of the surface M given by (we are now fixing the four points P_1, P_2, P_3, P_4)

$$G(z) = \frac{\theta(z + \int_{P_2}^{P_1})\theta(z + \int_{P_4}^{P_3})}{\theta(z + \int_{P_4}^{P_1})\theta(z + \int_{P_2}^{P_3})}, \qquad z \in \mathbb{C}^g.$$

The reader will easily confirm that $G(z)$ is actually a meromorphic function on the torus $J(M)$ and that therefore the function

$$h(z) = G(z) - \lambda(P_1,P_2,P_3,P_4), \qquad z \in \mathbb{C}^g$$

is also such a function which however vanishes on the theta divisor (by the previous theorem).

If we now multiply h by the denominator of G we find

$$\theta\left(z + \int_{P_4}^{P_1}\right)\theta\left(z + \int_{P_2}^{P_3}\right)h(z)$$
$$= \theta\left(z + \int_{P_2}^{P_1}\right)\theta\left(z + \int_{P_4}^{P_3}\right) - \lambda(P_1,P_2,P_3,P_4)\theta\left(z + \int_{P_4}^{P_1}\right)\theta\left(z + \int_{P_2}^{P_3}\right).$$

In particular, we find that the right-hand side vanishes for $z \in \Theta$. It thus follows that

$$\frac{\theta(z + \int_{P_2}^{P_1})\theta(z + \int_{P_4}^{P_3}) - \lambda(P_1,P_2,P_3,P_4)\theta(z + \int_{P_4}^{P_1})\theta(z + \int_{P_2}^{P_3})}{\theta(z)}$$

is a holomorphic function of z on $\mathbb{C}^g$ but not on the torus. If we check, its behavior on the torus though we find that it has the same multiplicative character as the function $\theta(z + \int_{P_2}^{P_1} + \int_{P_4}^{P_3})$ and therefore must be a constant multiple of this function. This last point follows from the fact (which we have not proved here) that the dimension of the space of holomorphic functions on $\mathbb{C}^g$ with the given multiplicative behavior is one. It thus follows that

$$\theta\left(z + \int_{P_2}^{P_1}\right)\theta\left(z + \int_{P_4}^{P_3}\right) - \lambda(P_1,P_2,P_3,P_4)\theta\left(z + \int_{P_4}^{P_1}\right)\theta\left(z + \int_{P_2}^{P_3}\right)$$
$$= c\theta(z)\theta\left(z + \int_{P_2}^{P_1} + \int_{P_4}^{P_3}\right),$$

for some constant c. The preceding is an identity for all values of z. In particular, if we set $z = \alpha - \int_{P_2}^{P_1}$ we find that

$$-\lambda(P_1,P_2,P_3,P_4) = c\frac{\theta(\alpha - \int_{P_2}^{P_1})\theta(\alpha + \int_{P_4}^{P_3})}{\theta(\alpha - \int_{P_2}^{P_1} + \int_{P_4}^{P_1})\theta(\alpha - \int_{P_2}^{P_1} + \int_{P_2}^{P_3})}.$$

The evenness of the theta function and the fact that the expression defining λ is independent of the point α in the theta divisor gives us

$$-\lambda(P_1,P_2,P_3,P_4) = c\lambda(P_1,P_2,P_4,P_3).$$

We have therefore proved the following

Theorem. *Let P_i, $i = 1, \ldots, 4$ be four distinct points on a compact Riemann surface of positive genus. Then the following identity holds*:

$$\theta\left(z + \int_{P_2}^{P_1}\right)\theta\left(z + \int_{P_4}^{P_3}\right) - \lambda(P_1,P_2,P_3,P_4)\theta\left(z + \int_{P_4}^{P_1}\right)\theta\left(z + \int_{P_2}^{P_3}\right)$$
$$= -\frac{\lambda(P_1,P_2,P_3,P_4)}{\lambda(P_1,P_2,P_4,P_3)}\theta(z)\theta\left(z + \int_{P_2}^{P_1} + \int_{P_4}^{P_3}\right).$$

In order to make use of this result we need to undertake a careful study of the possible cross ratios of four points; in other words, how the order of the points affects the cross ratio. If we denote the expression $\lambda(P_i,P_j,P_k,P_l)$ by λ_{ijkl}, it follows immediately from the definitions that

$$\lambda_{ijkl} = \lambda_{jilk} = \lambda_{klij} = \lambda_{lkji}.$$

In addition, we also see that if we denote λ_{1234} by λ and λ_{1342} by μ and finally λ_{1423} by ν, then $\lambda_{1432} = 1/\lambda$, $\lambda_{1243} = 1/\mu$, and $\lambda_{1324} = 1/\nu$.

The reader can now easily check that

$$\lambda\mu\nu = \frac{\theta(\alpha + \int_{P_4}^{P_3})\theta(\alpha + \int_{P_2}^{P_4})\theta(\alpha + \int_{P_3}^{P_2})}{\theta(\alpha + \int_{P_3}^{P_4})\theta(\alpha + \int_{P_4}^{P_2})\theta(\alpha + \int_{P_2}^{P_3})}$$

and this is clearly equal to

$$\frac{\theta(\alpha + \int_{P_4}^{P_3})\theta(\alpha + \int_{P_2}^{P_4})\theta(\alpha + \int_{P_3}^{P_2})}{\theta(\alpha - \int_{P_4}^{P_3})\theta(\alpha - \int_{P_2}^{P_4})\theta(\alpha - \int_{P_3}^{P_2})}.$$

Let us now consider the case of α, a nonsingular point of order two, in the theta divisor and write $\alpha = I(\varepsilon'/2) + \tau(\varepsilon/2)$. By the exercise preceding Section VI.2.6 and the fact that $\int_{P_4}^{P_3} + \int_{P_2}^{P_4} + \int_{P_3}^{P_2} = 0$ (when all integrations take place along paths lying inside the fundamental polygon) we get that the above expression is equal to -1. The same argument, in fact, shows that for α as above,

$$\lambda_{ijkl} = \frac{\theta\begin{bmatrix}\mu\\ \mu'\end{bmatrix}\left(\int_{P_j}^{P_i}\right)\theta\begin{bmatrix}\mu\\ \mu'\end{bmatrix}\left(\int_{P_l}^{P_k}\right)}{\theta\begin{bmatrix}\mu\\ \mu'\end{bmatrix}\left(\int_{P_l}^{P_i}\right)\theta\begin{bmatrix}\mu\\ \mu'\end{bmatrix}\left(\int_{P_j}^{P_k}\right)}.$$

We have therefore just proved the following:

Lemma. *With λ, μ, and ν defined as above, we have $\lambda\mu\nu = -1$.*

Let us now return to our theorem which gave us the fundamental identity. In view of the lemma we have just proved we can rewrite the theorem and replace the quotient $-\lambda_{1234}/\lambda_{1243}$ which appears there by $-\lambda\mu$. This by the lemma is simply $1/\nu$, which is λ_{1324}.

For the convenience of the reader we rewrite the theorem in this new notation as

$$\theta\left(z + \int_{P_2}^{P_1}\right)\theta\left(z + \int_{P_4}^{P_3}\right) = \lambda_{1234}\theta\left(z + \int_{P_4}^{P_1}\right)\theta\left(z + \int_{P_2}^{P_3}\right) + \lambda_{1324}\theta(z)\theta\left(z + \int_{P_2}^{P_1} + \int_{P_4}^{P_3}\right).$$

The above formula is known as the trisecant formula. It has been shown to be a very useful formula. Here we content ourselves with an application to the case of genus 1 with period τ in the upper half-plane.

We can clearly choose a torus and four points on it so that $\int_{P_1}^{P_2} = \frac{1}{2}$, $\int_{P_2}^{P_3} = \tau/2$, and such that $\int_{P_1}^{P_4} = (1 + \tau)/2$. This is the usual choice of the four points of order two in the period parallelogram. If we now substitute this into the above formula and use the identity in the exercise preceding Section VI.2.6 we find that it reduces to the well-known elliptic theta identity

$$\theta^2\begin{bmatrix}0\\0\end{bmatrix}(0)\theta^2\begin{bmatrix}0\\0\end{bmatrix}(z) = \theta^2\begin{bmatrix}1\\0\end{bmatrix}(0)\theta^2\begin{bmatrix}1\\0\end{bmatrix}(z) + \theta^2\begin{bmatrix}0\\1\end{bmatrix}(0)\theta^2\begin{bmatrix}0\\1\end{bmatrix}(z).$$

By setting $z = 0$ this now becomes a theta constant identity. We are now at the beginning of another story; material for another book.

Bibliography

1. Accola, R. D. M.: Riemann surfaces, theta functions, and abelian automorphisms groups. Lecture Notes in Mathematics vol. 483. Springer: Berlin, Heidelberg, New York 1975
2. Ahlfors, L. V.: Lectures on quasiconformal mappings. Van Nostrand: New York 1966. Reprinted Brooks-Cole, 1987.
3. Ahlfors, L. V.: Conformal invariants: topics in geometric function theory. McGraw Hill: New York 1973
4. Ahlfors, L. V., Sario, L.: Riemann surfaces. Princeton Univ. Press: Princeton, New Jersey 1960
5. Alling, N. L., Greenleaf, N.: Foundations of the theory of Klein surfaces. Lecture Notes in Mathematics vol. 219. Springer: Berlin, Heidelberg, New York 1971
6. Appel, P., Goursat, E.: Theorie des fonctions algebriques et de leurs integrales: I. Etude des fonctions analytiques sur une surface de Riemann. Gauthier–Villars: Paris 1929
7. Baker, H.: Abel's theorem and the allied theory including the theory of theta functions. Cambridge Univ. Press: Cambridge 1897
8. Behnke, H., Sommer, F.: Theorie der analytischen Funktionen einer komplexen Veränderlichen (second edition). Springer: Berlin, Göttingen, Heidelberg 1962
9. Bers, L.: Riemann surfaces. Lecture Notes, New York University, Institute of Mathematical Science Lecture Notes, 1957–58
10. Chevalley, C.: Introduction to the theory of algebraic functions of one variable. American Mathematical Society, Providence, Rhode Island 1951.
11. Conforto, F.: Abelsche Funktionen und algebraische Geometrie. Springer: Berlin, Göttingen, Heidelberg 1956
12. Fay, J. D.: Theta functions on Riemann surfaces. Lecture Notes in Mathematics vol. 352. Springer: Berlin, Heidelberg, New York 1973
13. Ford, L.: Automorphic functions (second edition). Chelsea: New York 1951
14. Fricke, R., Klein, F.: Vorlesunger über die Theorie der automorphen Funktionen: I. Die gruppentheoretischen Grundlagen, Teubner: Leipzig 1897. II. Die Funktionen theoretischen Ausführungen und die Anwendungen;
    a) Engere Theorie der automorphen Funktionen, 1901.
    b) Kontinuitätsbetrachtungen im Gebiete der Hauptkreisgruppen, 1911.

15. Fuchs, W. H. J.: Topics in the theory of functions of one complex variable. Van Nostrand: Princeton, New Jersey 1967
16. Griffiths J. P., Harris, J.: Principles of algebraic geometry. Wiley: New York 1978
17. Gunning, R. C.: Lectures on Riemann surfaces. Mathematical Notes. Princeton Univ. Press: Princeton. New Jersey 1967
18. Gunning, R. C.: Lectures on vector bundles over Riemann surfaces. Mathematical Notes. Princeton Univ. Press: Princeton, New Jersey 1967
19. Gunning, R. C.: Lectures on Riemann surfaces: Jacobi varieties. Mathematical Notes.Princeton Univ. Press: Princeton, New Jersey 1972
20. Gunning, R. C.: Riemann surfaces and generalized theta functions. Springer: New York, Heidelberg, Berlin 1976
21. Hensel, K., Landsberg, G.: Theorie der algebraischen Funktionen einer Variabeln. Teubner: Leipzig 1902. Reprinted Chelsea, 1965.
22. Hurwitz, A., Courant, R.: Vorlesungen über allgemeine Funktionen theorie und elliptische Funktionen. Springer: Berlin 1929
23. Igusa, J.-I.: Theta functions. Springer: New York, Heidelberg, Berlin 1972
24. Klein, F.: Über Riemann's Theorie der algebraischen Funktionen und ihrer Integrale. Teubner: Leipzig 1882
25. Kra, I.: Automorphic forms and Kleinian groups. Benjamin: Reading, Massachusetts 1972
26. Krazer, A.: Lehrbuch der Thetafunktionen. Teubner: Leipzig 1903. Reprinted Chelsea, 1970.
27. Krushkal, S. L.: Quasiconformal mappings and Riemann surfaces. Winston & Sons: Washington, D.C. 1979
28. Kunzi, H. P.: Quasikonforme Abbildungen: Springer: Berlin, Göttingen, Heidelberg 1960
29. Lang, S.: Introduction to algebraic and abelian functions. Addison–Wesley: Reading, Massachusetts 1972
30. Lang, S.: Elliptic functions. Addison–Wesley: Reading, Massachusetts 1973
31. Lehner, J.: Discontinuous groups and automorphic functions. Mathematical Surveys, Number VIII. American Mathematical Society: Providence, Rhode Island 1964
32. Lehner, J.: A short course in automorphic functions. Holt: New York 1966
33. Lehto, O., Virtanen, K. I.: Quasiconformal mappings in the plane. Springer: New York, Heidelberg, Berlin 1973
34. Magnus, J. W.: Noneuclidean tesselations and their groups. Academic Press: New York 1974
35. Mumford, D.: Curves and their Jacobians. The Univ. of Michigan Press: Ann Arbor 1975
36. Nevanlinna, R.: Uniformisierung. Springer: Berlin, Göttingen, Heidelberg 1953
37. Pfluger, A.: Theorie der Riemannschen Flächen. Springer: Berlin, Göttingen, Heidelberg 1957
38. Rauch, H. E., Farkas, H. M.: Theta functions with applications to Riemann surfaces. Williams & Wilkins: Baltimore, Maryland 1974
39. Rauch, H. E., Lebowitz, A.: Elliptic functions, theta functions, and Riemann surfaces. Williams & Wilkins: Baltimore, Maryland 1973
40. Riemann, B.: Gesammelte Mathematische Werke. Dover: New York 1953
41. Schiffer, M., Spencer, D. C.: Functionals of finite Riemann surfaces. Princeton Univ. Press: Princeton, New Jersey 1954. Second edition, Chelsea.
42. Springer, G.: Introduction to Riemann surfaces. Addison–Wesley: Reading, Massachusetts 1957
43. Siegel, C. L.: Topics in complex function theory. Wiley–Interscience, New York. Vol. I, 1969, Elliptic functions and uniformization theory. Vol. II, 1971, Automorphic functions and abelian integrals. Vol. III, 1973, Abelian functions and modular functions of several variables

44. Swinnerton-Dyer, H. P. F.: Analytic theory of abelian varieties. Cambridge Univ. Press: Cambridge 1974
45. Walker, R.: Algebraic curves. Springer: New York, Heidelberg, Berlin 1978
46. Weyl, H.: Die Idee der Riemannschen Fläche. Teubner: Berlin 1923. (Reprint, Chelsea, 1947). English Translation: The concept of a Riemann surface. Addison-Wesley: Reading, Massachusetts 1955
47. Wirtinger, W.: Untersuchungen über Thetafunktionen. Teubner: Leipzig 1895
48. Zieschang, H., Vogt, E., Coldewey, H. D.: Flächen und ebene diskontinuierliche Gruppen. Lecture Notes in Mathematics vol. 122. Springer: Berlin, Heidelberg, New York 1970

Index

Graduate Texts in Mathematics

continued from page ii

119 ROTMAN. An Introduction to Algebraic Topology.
120 ZIEMER. Weakly Differentiable Functions: Sobolev Spaces and Functions of Bounded Variation.
121 LANG. Cyclotomic Fields I and II. Combined 2nd ed.
122 REMMERT. Theory of Complex Functions. *Readings in Mathematics*
123 EBBINGHAUS/HERMES et al. Numbers. *Readings in Mathematics*
124 DUBROVIN/FOMENKO/NOVIKOV. Modern Geometry—Methods and Applications. Part III.
125 BERENSTEIN/GAY. Complex Variables: An Introduction.
126 BOREL. Linear Algebraic Groups. 2nd ed.
127 MASSEY. A Basic Course in Algebraic Topology.
128 RAUCH. Partial Differential Equations.
129 FULTON/HARRIS. Representation Theory: A First Course. *Readings in Mathematics*
130 DODSON/POSTON. Tensor Geometry.
131 LAM. A First Course in Noncommutative Rings.
132 BEARDON. Iteration of Rational Functions.
133 HARRIS. Algebraic Geometry: A First Course.
134 ROMAN. Coding and Information Theory.
135 ROMAN. Advanced Linear Algebra.
136 ADKINS/WEINTRAUB. Algebra: An Approach via Module Theory.
137 AXLER/BOURDON/RAMEY. Harmonic Function Theory.
138 COHEN. A Course in Computational Algebraic Number Theory.
139 BREDON. Topology and Geometry.
140 AUBIN. Optima and Equilibria. An Introduction to Nonlinear Analysis.
141 BECKER/WEISPFENNING/KREDEL. Gröbner Bases. A Computational Approach to Commutative Algebra.
142 LANG. Real and Functional Analysis. 3rd ed.
143 DOOB. Measure Theory.
144 DENNIS/FARB. Noncommutative Algebra.
145 VICK. Homology Theory. An Introduction to Algebraic Topology. 2nd ed.
146 BRIDGES. Computability: A Mathematical Sketchbook.
147 ROSENBERG. Algebraic *K*-Theory and Its Applications.
148 ROTMAN. An Introduction to the Theory of Groups. 4th ed.
149 RATCLIFFE. Foundations of Hyperbolic Manifolds.
150 EISENBUD. Commutative Algebra with a View Toward Algebraic Geometry.
151 SILVERMAN. Advanced Topics in the Arithmetic of Elliptic Curves.
152 ZIEGLER. Lectures on Polytopes.
153 FULTON. Algebraic Topology: A First Course.
154 BROWN/PEARCY. An Introduction to Analysis.
155 KASSEL. Quantum Groups.
156 KECHRIS. Classical Descriptive Set Theory.
157 MALLIAVIN. Integration and Probability.
158 ROMAN. Field Theory.
159 CONWAY. Functions of One Complex Variable II.
160 LANG. Differential and Riemannian Manifolds.
161 BORWEIN/ERDÉLYI. Polynomials and Polynomial Inequalities.
162 ALPERIN/BELL. Groups and Representations.
163 DIXON/MORTIMER. Permutation Groups.
164 NATHANSON. Additive Number Theory: The Classical Bases.
165 NATHANSON. Additive Number Theory: Inverse Problems and the Geometry of Sumsets.
166 SHARPE. Differential Geometry: Cartan's Generalization of Klein's Erlangen Program.
167 MORANDI. Field and Galois Theory.
168 EWALD. Combinatorial Convexity and Algebraic Geometry.
169 BHATIA. Matrix Analysis.
170 BREDON. Sheaf Theory. 2nd ed.
171 PETERSEN. Riemannian Geometry.
172 REMMERT. Classical Topics in Complex Function Theory.
173 DIESTEL. Graph Theory.
174 BRIDGES. Foundations of Real and Abstract Analysis.
175 LICKORISH. An Introduction to Knot Theory.
176 LEE. Riemannian Manifolds.
177 NEWMAN. Analytic Number Theory.
178 CLARKE/LEDYAEV/STERN/WOLENSKI. Nonsmooth Analysis and Control Theory.

Printed in the United States
By Bookmasters